Second Edition

A Visual Analogy Guide to Human Physiology

Paul A. Krieger
Grand Rapids Community College

MORTON
PUBLISHING
925 W. Kenyon Avenue, Unit 12
Englewood, CO 80110

www.morton-pub.com

Book Team

Publisher	Douglas N. Morton
President	David M. Ferguson
Illustration	Paul A. Krieger
Production Manager	Joanne Saliger
Composition	Will Kelley
Project Manager	Melanie Stafford
Associate Project Manager	Rayna Bailey

Printed in the United States of America

10 9 8 7 6 5 4 3 2 1

ISBN-13: 978-1-61731-240-3

Library of Congress Control Number: 2013956073

To those teachers, students,
and friends who have inspired me

Acknowledgments

Many people contributed to this book. First, I acknowledge David Ferguson, president of Morton Publishing, for signing me to my first book contract while working as an editor. David graciously gave me the extra time and autonomy I needed to complete the book the way I originally envisioned it. Doug Morton, founder of Morton Publishing, for agreeing to take a chance on my new book concept and publish it. Working as an author for Morton Publishing has always been a pleasure because everyone treats you with respect and works as a team to create the best book possible. It certainly makes me proud that they remain true to their mission of developing very high-quality learning materials at sensible prices. Marta Martins, acquisitions editor, did a wonderful job compiling all the reviewer comments in a spreadsheet and promptly responded to all my inquiries. To Dan Matusiak, for his diligence in faithfully creating the glossary. I would also like to offer a special thank you to the reviewers listed on the next page for their valuable feedback during the writing process. The production design team at Morton—Joanne Saliger and Will Kelley—and project manager Melanie Stafford deserve special recognition for the terrific work they have consistently done in producing all of the *Visual Analogy Guides*. Will Kelley designed the dazzling new book cover and added many new features to this edition such as the illustrations on the section dividers. Rayna Bailey, editorial assistant, thoroughly proofed the manuscript. To all the dedicated sales representatives at Morton for introducing my books to students all over the country. Without them, there would not be a new edition. To indexer Doug Easton for carefully creating the helpful index. My wonderful wife, Lily, kindly accepted all my long working hours and offered valuable feedback. Without her help and encouragement this book would not have been written. My son, Ethan, inspired me with his passion for life and his good humor. Thanks to veteran author Mike Timmons for helping me crystallize my initial ideas about the visual analogy guides to turn them into a tangible reality. To my fellow author and friend Kevin Patton for graciously responding to all my annoying authoring questions along the way. Thanks to Jim Gibson for teaching me how to improve the graphic layout and page design. To my father, Eugene Krieger, and my mother, Norma Krieger, for emphasizing the need for a good education and for giving me the freedom to pursue my own path. Finally, for my students, colleagues, friends in the Human Anatomy & Physiology Society (HAPS) and the Michigan Community College Biologists (MCCB), and everyone else who offered suggestions, support, and encouragement—thanks to all of you. I am truly grateful.

Reviewers

Steve Bassett, Southeast Community College (first edition)

Neelima Bhavsar, Ph.D., Saint Louis Community College–Florissant Valley

David Canoy, Chemeketa Community College (first edition)

John E. Fishback, Ozarks Technical Community College

Cynthia Herbrandson, Kellogg Community College (first edition)

Dr. Tony Lamanna, Paradise Valley Community College

Darren Mattone, Muskegon Community College

Suzanne Pundt, University of Texas–Tyler

Dr. Andrea Scollard, Baton Rouge Community College

Marilyn Shopper, Ed.D., Johnson Community College

Alan J. Spindler, M.D., D.C., A.R.N.P., Brevard Community College–Melbourne

Chris Sullivan, California State University–Sacramento (first edition)

Faith Vruggink, Kellogg Community College (first edition)

Contents

How to Use This Book

Purpose

This book was written primarily for students of human physiology; however, it will be useful for teachers or anyone else with an interest in this topic. It was designed to be used in conjunction with any of the major physiology or anatomy and physiology textbooks. What makes it unique, creative, and fun is the **visual analogy** learning system. This will be explained later. The modular format allows you to focus on one key concept at a time. Each module has a text page on the left with corresponding illustrations on the facing page. Most illustrations are unlabeled so that you can quiz yourself on the structures. A handy key to the illustration is often provided on the text page. Although this book covers most of the major organ systems, the topics are weighted more toward areas that typically give students difficulty. It uses a variety of learning activities such as labeling, coloring, and mnemonics to help instruct. In addition, it offers special study tips for mastering difficult topics.

What Are Visual Analogies?

A visual analogy is a helpful way to learn new material based on what you already know from everyday life. It compares a physiology concept to something familiar. For example, the concept of a membrane potential is compared with a battery. Making this comparison allows you to correlate the membrane potential (unknown) with the battery (known). Doing this accomplishes several things.

1. It reduces your anxiety about learning the material and helps you focus on the task at hand.
2. It forces you to use a contextualized approach in learning physiology concepts.
3. It makes the learning more fun, relevant, and meaningful so you can better retain the information.

Whenever a visual analogy is used in this book, a small picture of it appears in the upper right-hand corner of the illustration page for easy reference. This allows you to quickly reference a page visually by simply flipping through the pages.

Icons Used

The following icons are used throughout this book:

Microscope icon—indicates any illustration that is microscopic.

Crayon icon—indicates illustrations that were specially made for coloring. Even though students may color any of the illustrations to enhance their learning, they may benefit more by referring to this icon. In some cases, written instructions appear next to this icon with directions about exactly what to color or how to color it.

Scissors icon—indicates that something is either cut or broken. For example, it may be used to show that a chemical bond between two molecules is broken.

Abbreviations and Symbols

The following abbreviations are commonly used throughout this book:

l.	= left	n.	= nerve
r.	= right	ex.	= example
a.	= artery	sing.	= singular
v.	= vein	pl.	= plural
m.	= muscle		

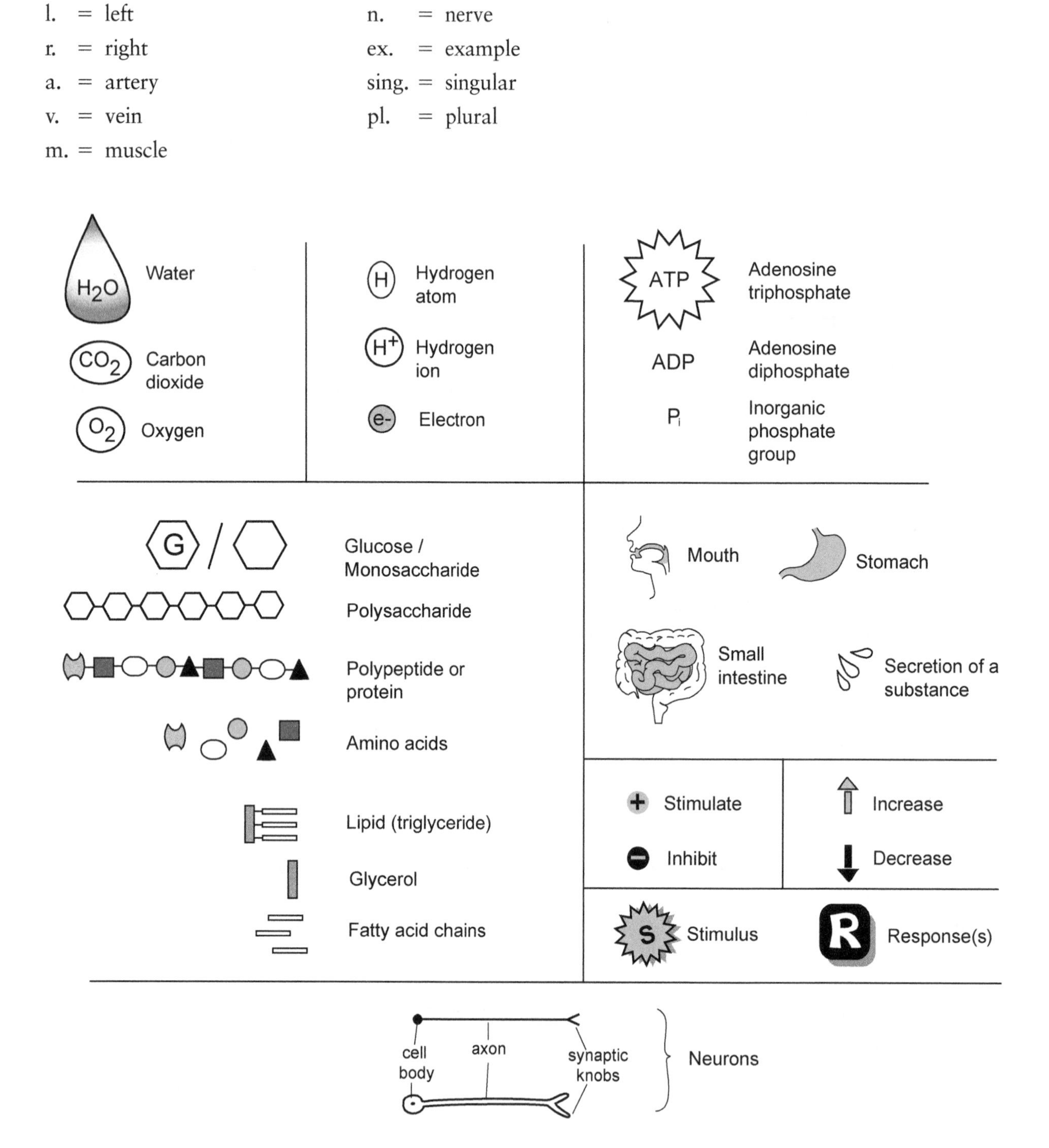

I sincerely hope that this book will be a fun, helpful tool for anyone interested in learning human physiology. Many of the visual analogies used in the book have been tested with students in both lab and lecture settings to ensure that they are useful.

Enjoy learning with visual analogies!

Visual Analogy Index

Visual Analogy Index

Visual Analogy Index

TOPIC	ANALOGY	ICON(S)	PAGE NO.
21. Carbohydrates: Functions of Carbohydrates	Like **gasoline for your car**, carbohydrates are used as energy for body cells		40–41
22. Carbohydrates: Functions of Carbohydrates	Some carbohydrates serve as **markers** on cell surfaces		40–41
23. Carbohydrates: Functions of Carbohydrates	Stored carbohydrates are held in reserve like **items in a storage box**	CARBS	40–41
24. Lipids: Functions of Lipids	Stored lipids are held in reserve like **items in a storage box**	LIPIDS	48
25. Lipids: Functions of Lipids	Like a **cork floating in water**, lipids provide buoyancy to an organism		48
26. Lipids: Functions of Lipids	Lipids act as a **vehicle to transport their lipid cargo**		48
27. Lipids: Functions of Lipids	Like a toy **sailboat can become a different boat by removing the sail**, a lipid precursor can be converted into another lipid		49
28. Lipids: Functions of Lipids	Like **packing peanuts** protect items during shipment, fat pads cushion organs		49
29. Lipids: Functions of Lipids	Like **attic insulation** helps a home retain heat, lipids help your body retain heat		49
30. Cells: Epithelial Cells	Tight junctions are like **rivets**		55

Visual Analogy Index

TOPIC	ANALOGY	ICON(S)	PAGE NO.
31. Cells: Plasma Membrane Structure	A plasma membrane is like a **water bed**	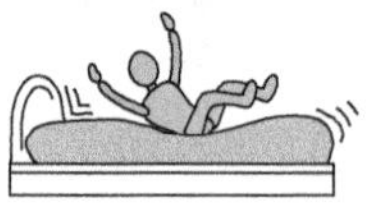	57
32. Cellular Physiology: DNA Replication	DNA opens up like a **zipper unzipping**	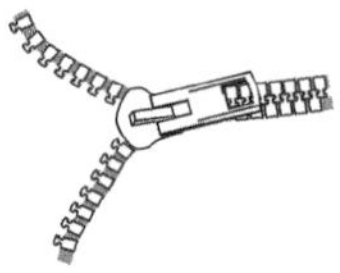	61
33. Cellular Physiology: Protein Synthesis	DNA is like the **master blueprint**	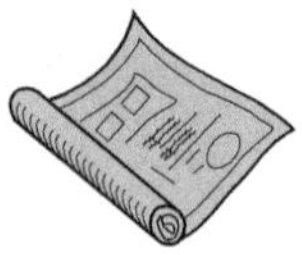	62–63
34. Cellular Physiology: Role of ATP	ATP hydrolysis is like a **good investment**	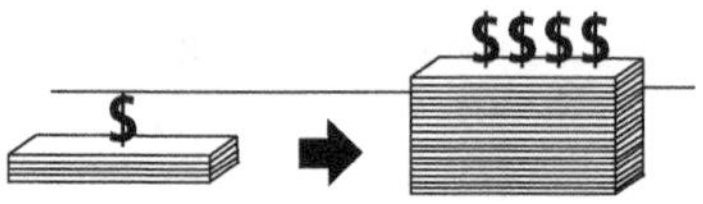	65
35. Cellular Physiology: Simple Diffusion	Simple diffusion is like **people moving out of a crowded room**	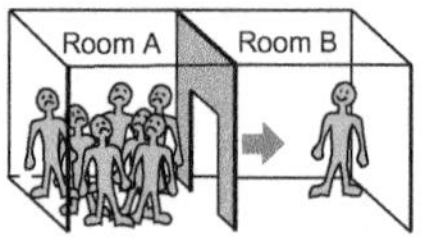	69
36. Cellular Physiology: Filtration	Filtration is like **making coffee in a coffeemaker**	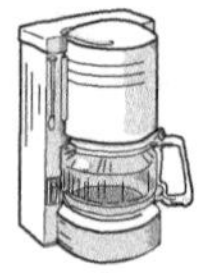	73
37. Cellular Physiology: Active Transport	Active transport is like a **sump pump**	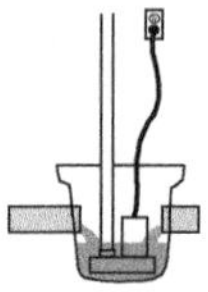	75
38. Cellular Physiology: Membrane Potentials	A membrane potential is like a **battery**	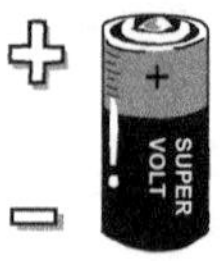	79
39. Cellular Physiology: Sodium-Potassium Pump	Sodium-potassium pump is like a **revolving door**	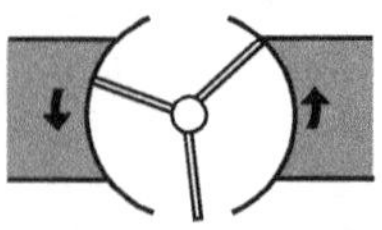	80–81
40. Nerve and Muscle: Multipolar Neuron	Axon of a neuron is like an **electrical cord**	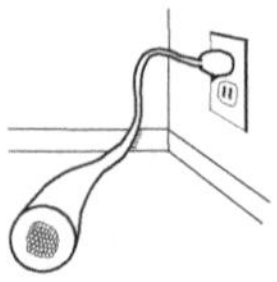	89

Visual Analogy Index

Visual Analogy Index

Visual Analogy Index

Visual Analogy Index

TOPIC	ANALOGY	ICON(S)	PAGE NO.
71. Nervous System: Reflexes	A reflex arc involves input, processing, and output, like **typing a letter on your laptop**	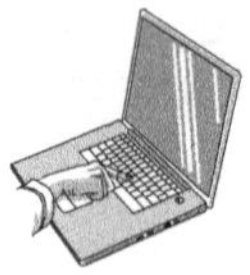	255
72. Nervous System: Sympathetic Division of ANS	Sympathetic division is like **stepping on the gas**	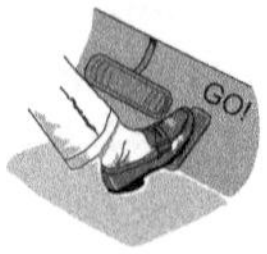	273
73. Nervous System: Parasympathetic Division of ANS	Parasympathetic division is like **stepping on the brake**	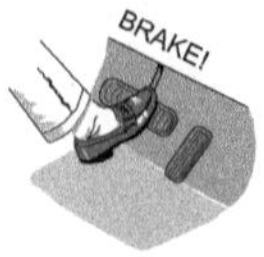	275
74. Female Reproductive System: Ovary	Ovulation is like a **water balloon bursting**	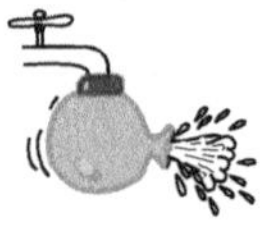	317

PHYSIOLOGY OVERVIEW

Description

The human body has different levels of structural organization. **Atoms** (*ex: oxygen*) combine to form **molecules** (*ex: water*). Very large molecules called **macromolecules** (*ex: carbohydrates, lipids, proteins, nucleic acids*) are the building blocks used to form any **cell** (*ex: cardiac muscle cell*). A group of similar cells working together is called a **tissue** (*ex: cardiac muscle tissue*). A variety of different tissues working together forms an **organ** (*ex: the heart*), and a group of organs working together forms a complex **organ system** (*ex: cardiovascular system = heart and the blood vessels*). These complex organ systems all working together form an **organism** (*ex: human organism*). For convenience, teachers and textbook authors sometimes discuss each organ system independent of the others. But it should be emphasized that no organ system exists as an island unto itself, because they are all interdependent. Substances such as water, nutrients, and waste products have to move between organ systems to maintain **homeostasis** (see pp. 4–5) for the entire organism. Keep in mind that when studying physiology it is common to jump back and forth between these different levels of structure when learning about physiological mechanisms.

The following 10 organ systems are shown in the illustration on the facing page: **integumentary system, musculoskeletal system, nervous** and **endocrine systems, cardiovascular system, lymphatic/immune system, respiratory system, digestive system, urinary system,** and **reproductive systems.** This schematic illustration attempts to show the interrelationships between organ systems. The white rectangular band around the periphery represents the **integumentary system.** This includes the skin and all its components, such as hair follicles, oil glands, and a host of other structures. Our skin functions as a thin mantle to protect us from our external environment. The **musculoskeletal system** provides an internal framework for support and allows for body movements. Consider that movement would be impossible without muscles acting on bones as a kind of lever system.

Two organ systems work together to act as the regulators of body function—the **nervous** and **endocrine** systems. Because they both serve a similar purpose, they were placed together in the same box. Without the nervous system we would have no way to detect stimuli from either our external or internal environments, no way to process this information, and no way to respond to it. The endocrine system is composed of glands (*ex: pancreas*) that release chemical messengers called **hormones** (*ex. insulin*) that travel through the bloodstream to target a particular structure such as an organ to induce a response. Working together, these two systems help maintain the vital activities that keep us alive.

The **cardiovascular system** consists of the heart and all the blood vessels (*ex. arteries, veins, arterioles, venules,* and *capillaries*). The heart functions as a pump to constantly circulate blood throughout the body. The blood vessels must connect to all the other organ systems to deliver nutrients (*ex. oxygen*) to all the body cells via the bloodstream.

The following four organ systems contain organs/structures that are hollow: **respiratory system, digestive system, urinary system,** and **reproductive systems.** Structurally speaking, all of these systems have an internal chamber called a **lumen** that extends to the external environment. Any substance within this lumen is still considered part of the external environment until it crosses the wall of that structure and enters body tissues. The major function of the **respiratory system** is to exchange the respiratory gases, oxygen and carbon dioxide, between the body and the external environment. Oxygen is brought in from the external environment, transferred to the blood, and delivered to body cells. Carbon dioxide is a normal waste product made by body cells. It is transferred to the blood, then sent back to the respiratory system, where some of it is released to the external environment.

The **digestive system** functions to take in nutrients, break them down to their simplest components, and absorb them into the blood. Then they are delivered to cells. In addition, the digestive system eliminates waste products from the body. The **urinary system** constantly filters and processes the blood to eventually form urine, which contains nitrogen wastes.

The **lymphatic** and **immune systems** typically are grouped together (*even though they are not illustrated that way*). One major function of the **lymphatic system** is to take the **interstitial fluid** found between cells and to cleanse it of debris and possible pathogens, then to deliver it back to the bloodstream. The **immune system** functions to protect the body from foreign pathogens such as bacteria and viruses. For example, when a break in the skin occurs, pathogens may enter the body and the immune system must launch a response akin to soldiers going to war. It accomplishes this by making antibodies and other substances that help fight off these invaders.

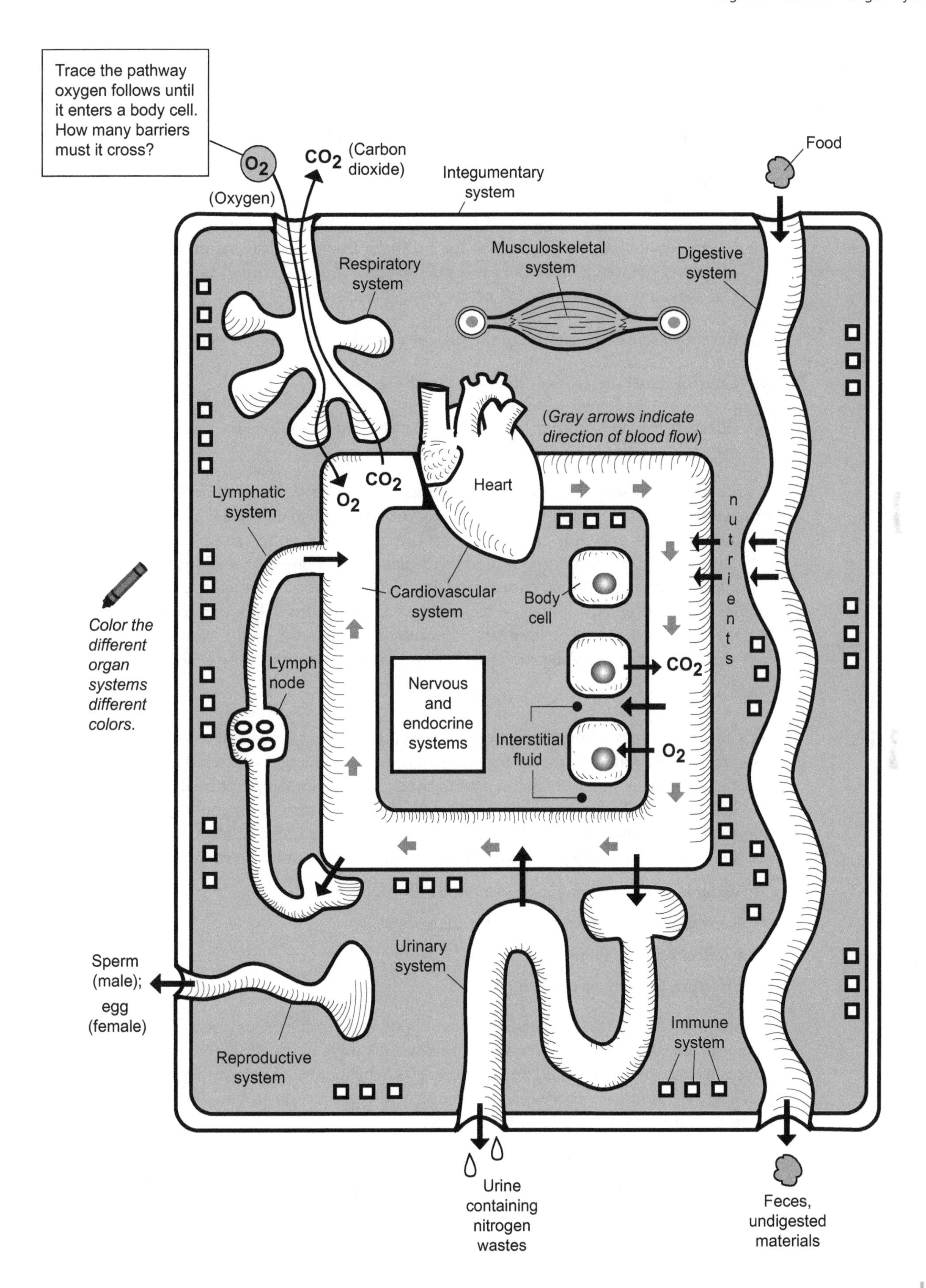
Trace the pathway oxygen follows until it enters a body cell. How many barriers must it cross?
O_2
(Oxygen)
CO_2 (Carbon dioxide)
Integumentary system
Food
Respiratory system
Musculoskeletal system
Digestive system
(Gray arrows indicate direction of blood flow)
CO_2
O_2
Heart
Lymphatic system
nutrients
Cardiovascular system
Body cell
CO_2
Lymph node
Nervous and endocrine systems
Interstitial fluid
O_2
Color the different organ systems different colors.
Urinary system
Sperm (male); egg (female)
Reproductive system
Immune system
Urine containing nitrogen wastes
Feces, undigested materials

Description

Homeostasis is defined as maintaining a stable, constant internal environment within an organism. As changes constantly occur in the external environment, organisms must have control systems so they can detect these changes and respond to them. Anything that must be maintained in the body within a normal range must have a control system. For example, body temperature, blood pressure, and blood glucose levels all have to be regulated.

A **control system** consists of the following four parts:

1. **Variable:** the item to be regulated. Consider the variable like a teeter-totter. If it is perfectly balanced in the horizontal position, it is at the normal value called the **set point.** Just as a single person resting on the teeter-totter can cause an imbalance, various stimuli can cause the variable to either rise above normal levels or fall below normal levels.

2. **Receptor:** senses the change in the variable; provides input to the control center.

3. **Control center:** determines the value for the set point.

4. **Effector:** provides output to influence the stimulus; uses feedback to return the system to the original set point.

Feedback mechanisms may be either positive or negative. **Negative feedback** mechanisms are the more common in the human body. As a city works, they regulate the day-to-day activities of the body. The output in this system is used to turn off the stimulus. A good example of this is explained below, with the cooling system in a room used to regulate room temperature. **Positive feedback** mechanisms, in contrast, are like the emergency responders. They are not as common, and they use the responses to intensify the original stimulus. A biological example of this is labor contractions during childbirth. A hormone called oxytocin stimulates the muscle in the wall of the woman's uterus to strengthen contractions so they become progressively more forceful until the baby is born.

Analogy

Let's use the analogy of regulating room temperature as an example of a control system. If we set the room temperature at 68°F, then it falls below normal, the change is detected and triggers the furnace to heat the room back to the set point. Alternatively, if the room temperature increases above this level, it also is detected and triggers the central air conditioning to turn on to cool the room temperature back to normal. The illustration on the facing page shows the example of a room getting warmer on a hot summer day. Let's identify each of the four elements in this control system.

- **Variable:** Room temperature. Set point? 68°F.
- **Receptor:** Thermometer within the thermostat.
- **Control center:** Thermostat.
- **Effector:** Central air conditioning.

Similarly, the body must regulate its own body temperature. Our normal set point is 98.6°F. When we are overheating our body responds in numerous ways in an effort to cool itself, such as sweating, increasing our breathing rate, and increasing our heart rate. Alternatively, if our body temperature is falling, we stimulate processes that conserve or generate heat, such as shivering, decreasing our breathing rate, and decreasing our heart rate.

- **Variable:** Body temperature. Set point? 98.6°F.
- **Receptor:** Hot and cold temperature sensors in the skin.
- **Control center:** Hypothalamus in the brain.
- **Effector:** Skeletal muscles—shivering.

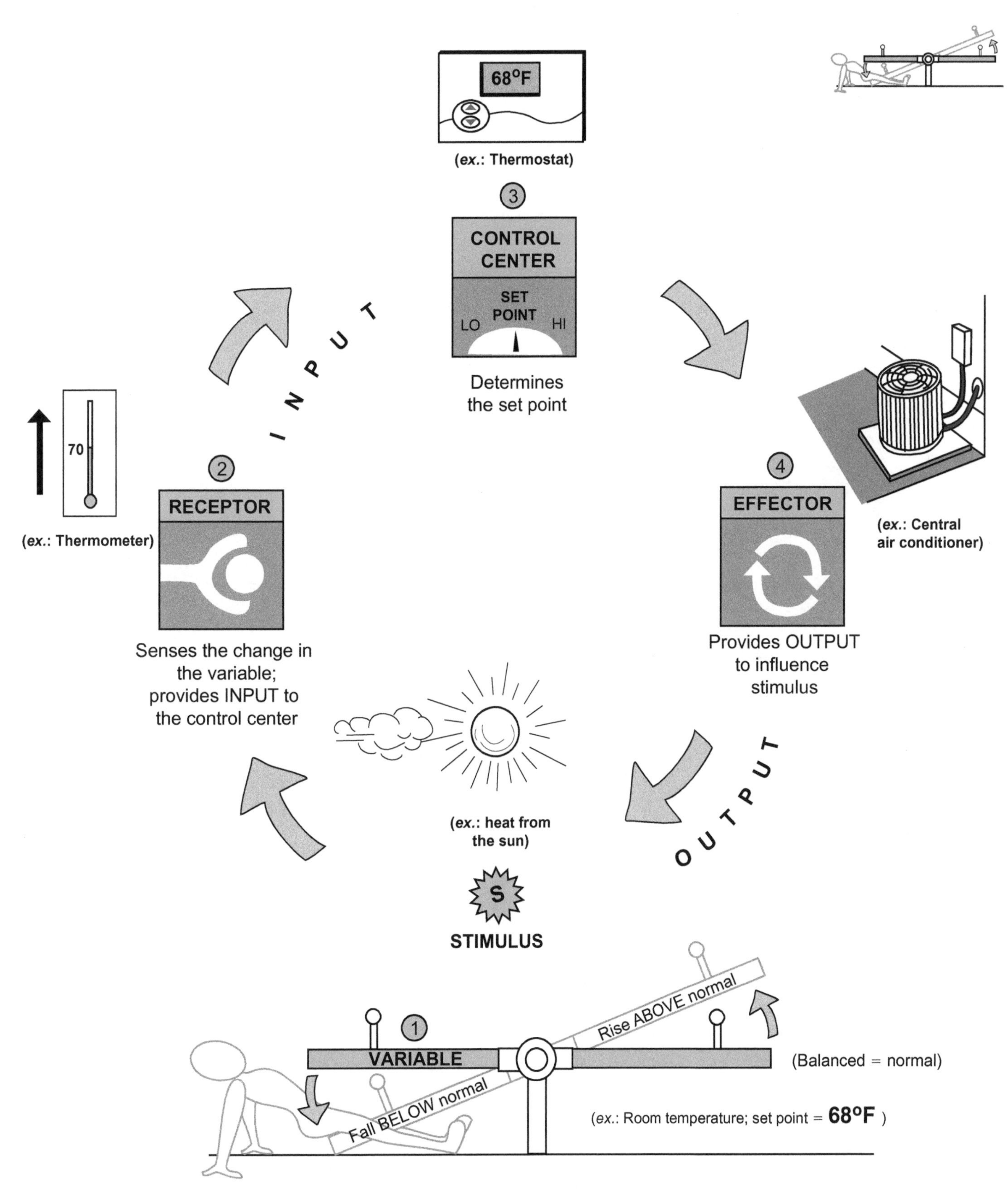

Control system: This illustration shows all the features of a control system. As an analogy, regulating room temperature on a hot day is shown.

Notes

CHEMISTRY

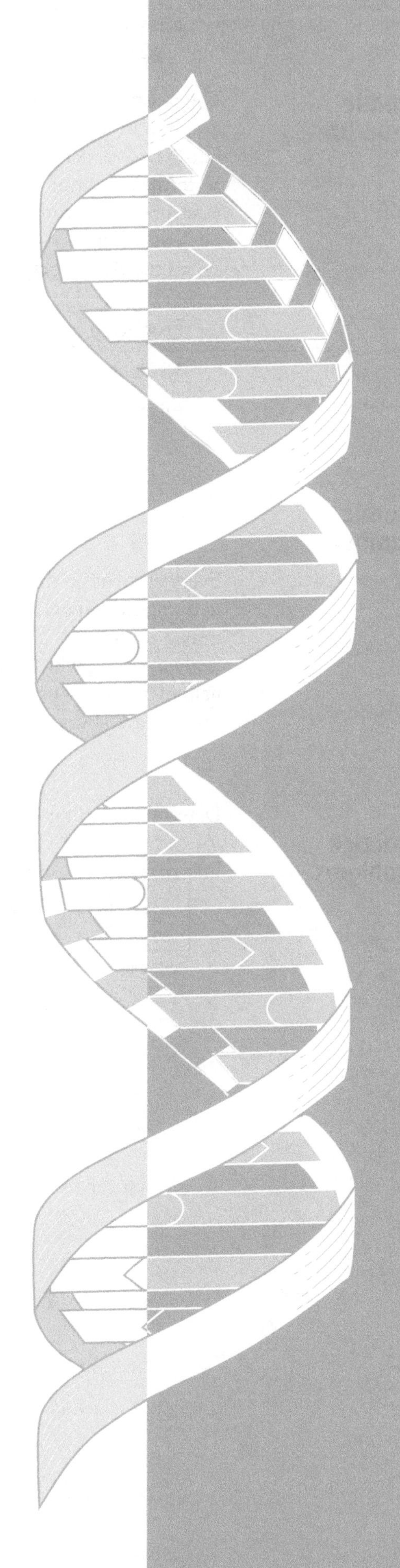

Atomic Structure

The illustration on the facing page shows a cylinder of **helium** gas to represent the element helium and graphite in a pencil to represent the element **carbon.** Helium gas is used to fill balloons, and we all know how pencils are used. Each of these elements, like all others, is made of many identical **atoms.** With the exception of hydrogen, each atom contains three subatomic particles: **protons, neutrons,** and **electrons.** The protons and neutrons are packed tightly together at the center of the atom called the **nucleus.** Electrons are particles found moving at high speed in a spherical region around the nucleus called the **electron cloud.** Here is a summary of the properties of each subatomic particle:

- **Proton**—a particle that has a **positive** (+) charge (remember the phrase "**P**rotons are **P**ositive")
- **Neutron**—a particle that is **neutral** (no charge) (remember the phrase "**N**eutrons are **N**eutral")
- **Electron**—a small, almost massless particle with a **negative** charge

Atomic Number

Atomic number = number of protons in an atom

Every element on the periodic table has a different atomic number that indicates the number of protons in a single atom of that element. The illustrations show helium as atomic number 2, telling us that it has only 2 protons (+2); carbon, with atomic number 6, has 6 protons (+6). The net charge on any atom is always zero, which means that the number of protons and electrons is always equal. Notice that the helium atom has 1 proton and 1 electron, and the carbon atom has 6 protons and 6 electrons. Because protons are always positive, and electrons are always negative, the net charge of both atoms is zero.

Practice Problems

Determine the number of protons in an atom of the following elements:

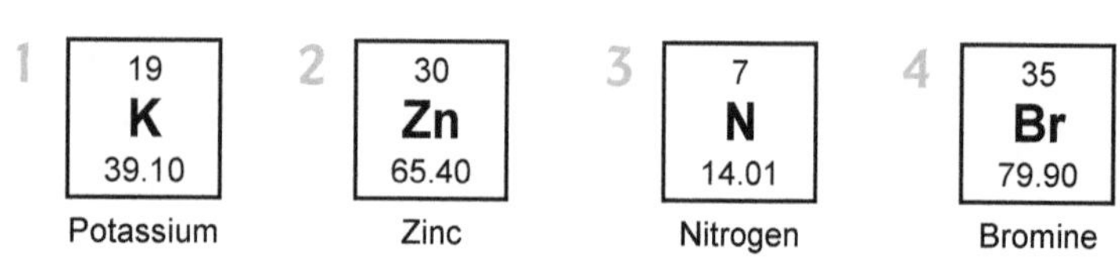

5 Are the total number of protons and electrons in an atom the same or different? Explain.

6 Which subatomic particle has no charge?

Answers

1 19 2 30 3 7 4 35 5 They are the same. The net charge on any atom is always zero, which requires the number of protons and electrons to be equal. 6 neutron

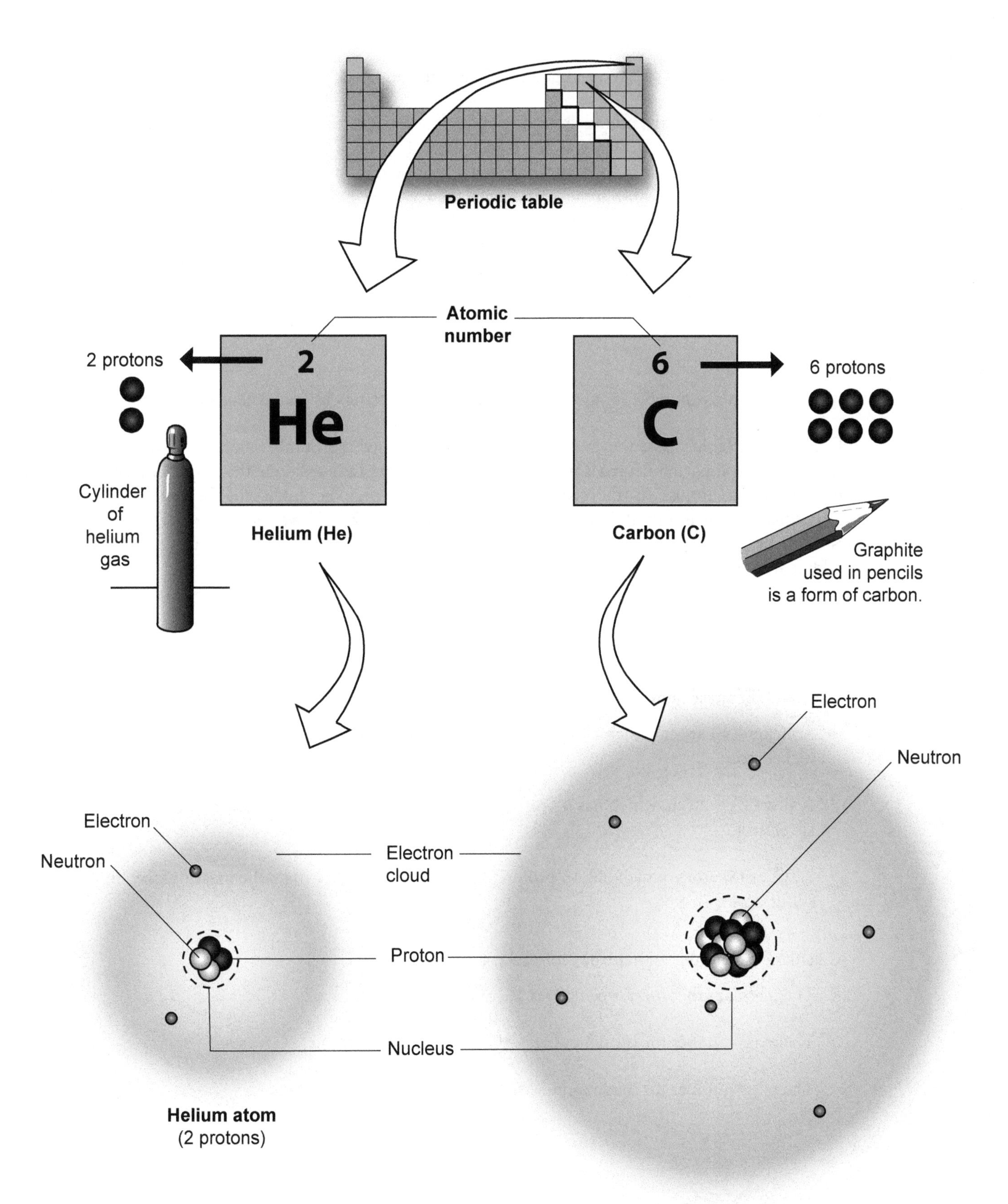

Helium atom
(2 protons)

Carbon atom
(6 protons)

Atomic Mass

Atomic mass = average mass of all the naturally occurring isotopes of that element

On the periodic table, the atomic mass number is found below the symbol of the element. For example, on the facing page, helium (He) has an atomic mass of **4.003.** The units used for measuring very small particles are called atomic mass units (amu). These represent the atomic mass expressed in grams (g). Why isn't the atomic mass a whole number? It is calculated as an *average* of all the naturally occurring **isotopes** of that element. By comparison, the atomic mass for carbon is given as **12.011.** Notice that it also is not a whole number for the reason already mentioned. The atomic mass tells us the mass of the nucleus of an atom, which contains both protons and neutrons. Why not consider the electrons? They have a negligible mass, so they contribute almost nothing to the overall mass of the atom.

Mass Number

Mass number = number of **protons and neutrons** in the nucleus of an atom

Unlike the atomic mass, the mass number is always a whole number because it tells us the total number of protons and neutrons in the nucleus of an atom. Let's examine the mass number for each of the illustrated atoms. As the second element on the periodic table, the helium atom has only 2 protons and 2 neutrons. By adding these numbers together (2 + 2) we determine that the mass number is 4. In comparison, the carbon atom has 6 protons and 6 neutrons, giving it a mass number of 12 (by adding 6 + 6).

Practice Problems

Calculate the **mass number** for each of the following:

1 A nickel (Ni) atom, which has 28 protons and 31 neutrons. Calculate the **mass number** for this nickel atom.

2 A silver (Ag) atom, which has 47 protons and 61 neutrons. Calculate the **mass number** for this silver atom.

3 A sulfur (S) atom, which has 16 protons and 16 neutrons. Calculate the **mass number** for this sulfur atom.

Calculate the number of **neutrons** for each of the following:

4 The mass number for oxygen (O) is 16, and it has 8 protons.

5 The mass number for bromine (Br) is 80, and it has 35 protons.

6 The mass number for magnesium (Mg) is 24, and it has 12 protons.

Answers

1 59 2 108 3 32 4 8 5 45 6 12

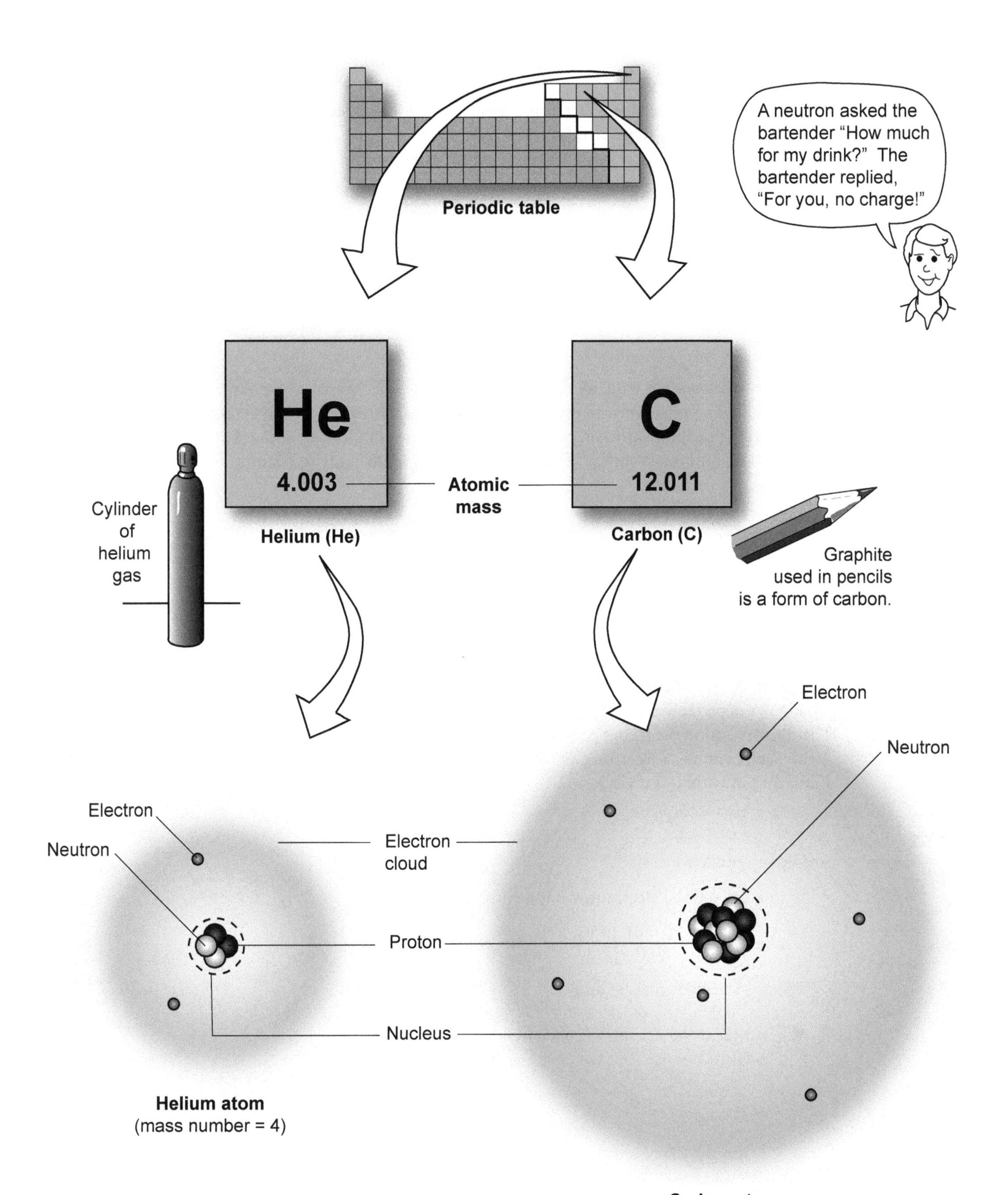
Periodic table
A neutron asked the bartender "How much for my drink?" The bartender replied, "For you, no charge!"
He
4.003
Atomic mass
C
12.011
Cylinder of helium gas
Helium (He)
Carbon (C)
Graphite used in pencils is a form of carbon.
Electron
Neutron
Electron
Neutron
Electron cloud
Proton
Nucleus
Helium atom
(mass number = 4)
Carbon atom
(mass number = 12)

Ions

Ion = atom that has either *gained* or *lost* one or more electrons

When an atom gains or loses electrons, it is called an **ion** instead of an atom. Recall that atoms are electrically neutral, so they have no charge. Atoms that lose one or more electrons become positively charged and are called **cations**. When an atom gains one or more electrons, it becomes negatively charged and is called an **anion**.

Positive (+) Ions (Cations)

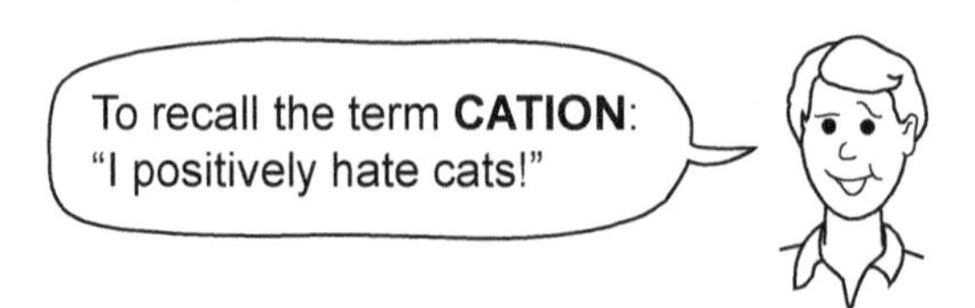

If an atom loses one electron, its charge is +1, if it loses 2 electrons, its charge is +2, and so on. Electrons are always negatively charged, so the *loss* of this negative charge results in a positively (+) charged ion. The illustration on the facing page shows an atom of sodium (Na). Sodium has only 1 electron in its outermost shell, so it tends to give it up quite easily to become an ion with a +1 charge.

Negative (-) Ions (Anions)

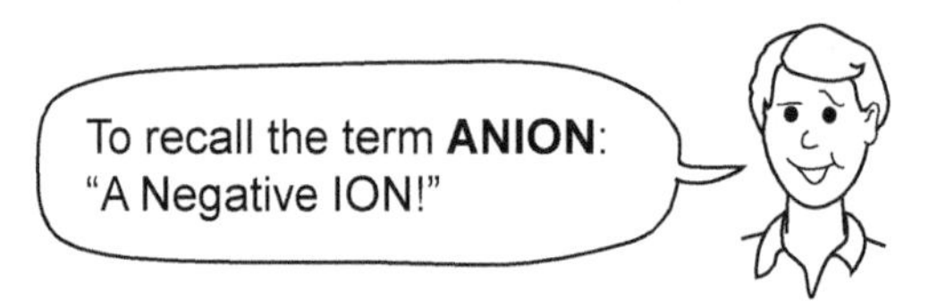

If an atom gains one electron, its charge is –1, if it gains 2 electrons, its charge is –2, and so on. Electrons are always negatively charged, so the *gain* of a negative charge results in a negatively (−) charged ion. The illustration on the facing page shows an atom of chlorine. It has 7 electrons in its outermost shell. A full shell contains 8 electrons, so chlorine tends to readily accept an electron to become full. The result is an ion with a –1 charge.

Practice Problems

Answer the following:

1. If an atom *gains* 3 electrons, what is the charge?
2. If an atom *loses* 2 electrons, what is the charge?
3. What is the charge on an atom?
4. What is the difference between a **cation** and an **anion**?

Answers

1 –3 2 +2 3 no charge 4 cations are positively (+) charged and anions are negatively (−) charged.

Positive (+) Ions
(cations)

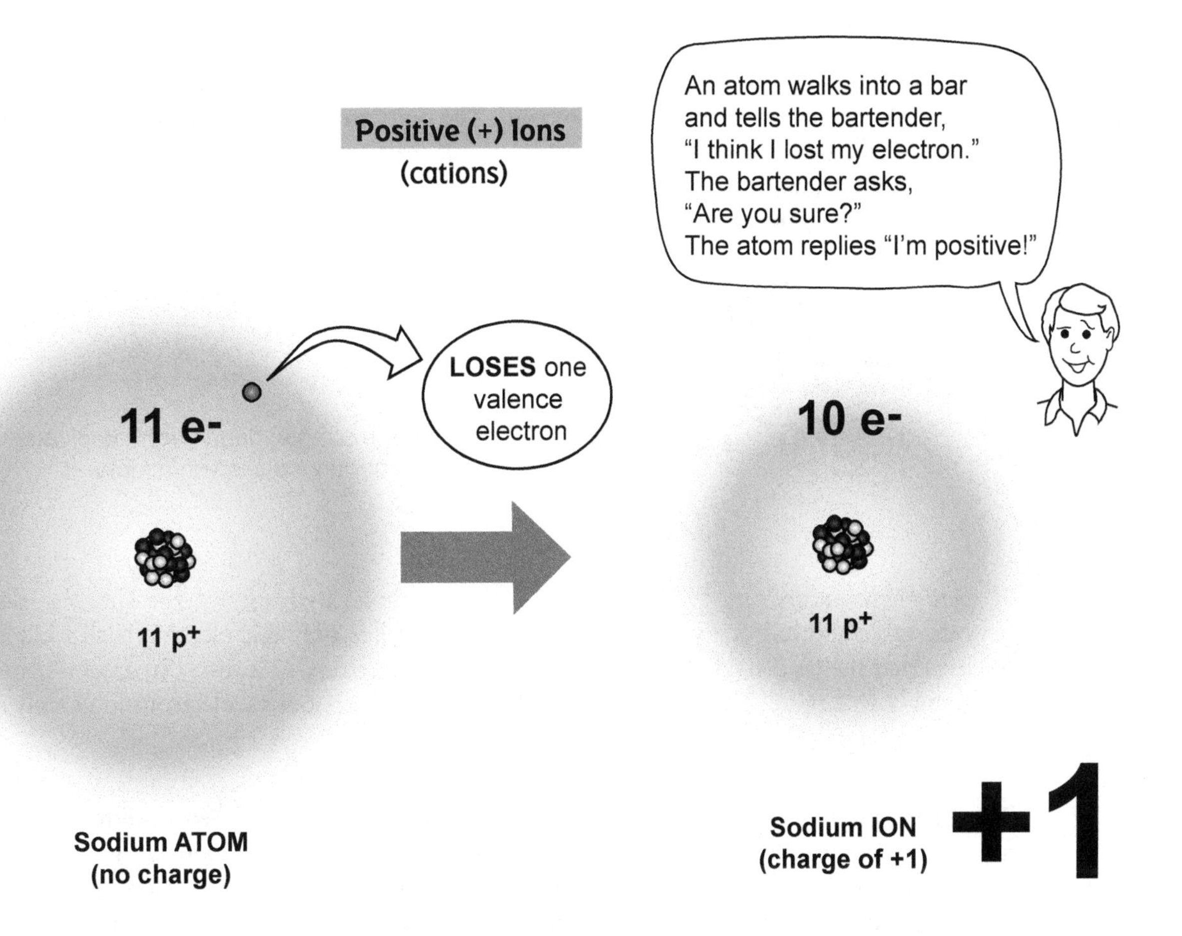

Negative (-) Ions
(anions)

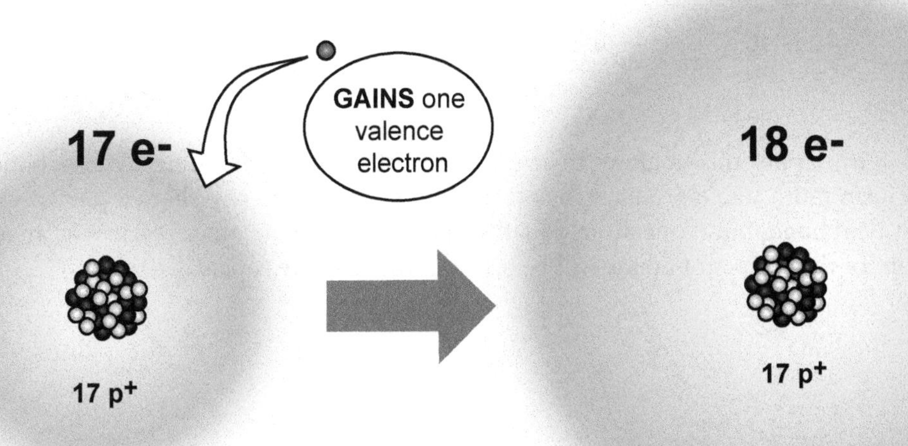

Chlorine ATOM
(no charge)

Chloride ION
(charge of –1)

-1

Octet Rule and Exceptions

Matter always wants to be in its most stable state. But what determines this stability? One measure is an atom's total number of valence electrons. The **octet rule** describes the *general* rule for stability for an atom: 8 electrons in its outermost shell (or 8 valence electrons). Why 8? This is the typical maximum number of electrons that makes most electron shells "full." To remember the octet rule, think of the eight ball used in playing pool.

EIGHT BALL = octet rule

For example, neon (Ne) has 8 valence electrons, so it is very stable.

Are there exceptions to this rule? Yes. For example, hydrogen is the simplest element on the periodic table and has one valence electron. Because it has only one electron shell located close to the nucleus, that shell is considered full when it contains 2 electrons. As a result, hydrogen's "magic number" for stability is 2, not 8. The same is true for helium, which contains 2 electrons in its only electron shell.

Analogy

This octet rule is a driving force for chemical bonding. Unstable atoms combine with other unstable atoms to form a more stable product. As an analogy, let's classify a lonely, single guy looking for a spouse as "unstable." Ideally speaking, if he bonds with a suitable companion, and they work hard at their relationship, they have formed a more "stable" couple. Obviously, there are LOTS of exceptions to this rule, but you get the point, right?

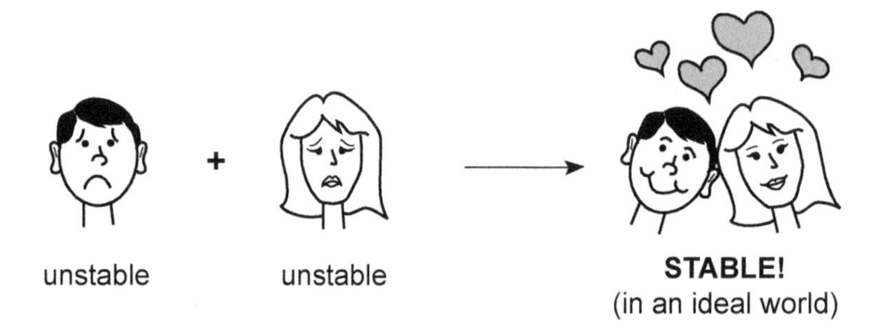

Let's apply this analogy to the concept of chemical bonding. Atoms bond with other atoms to form more stable products in several different ways. If they share a pair of electrons, they form a **covalent bond**, but if one atom donates an electron and the other receives it, an **ionic bond** is formed. Both of these types of bonds will be discussed as separate topics.

Example of Octet Rule

Let's use the example of an ionic bond in the formation of table salt to understand this further.

Word description

A sodium atom (Na) plus a chlorine atom (Cl) form sodium chloride salt (NaCl)

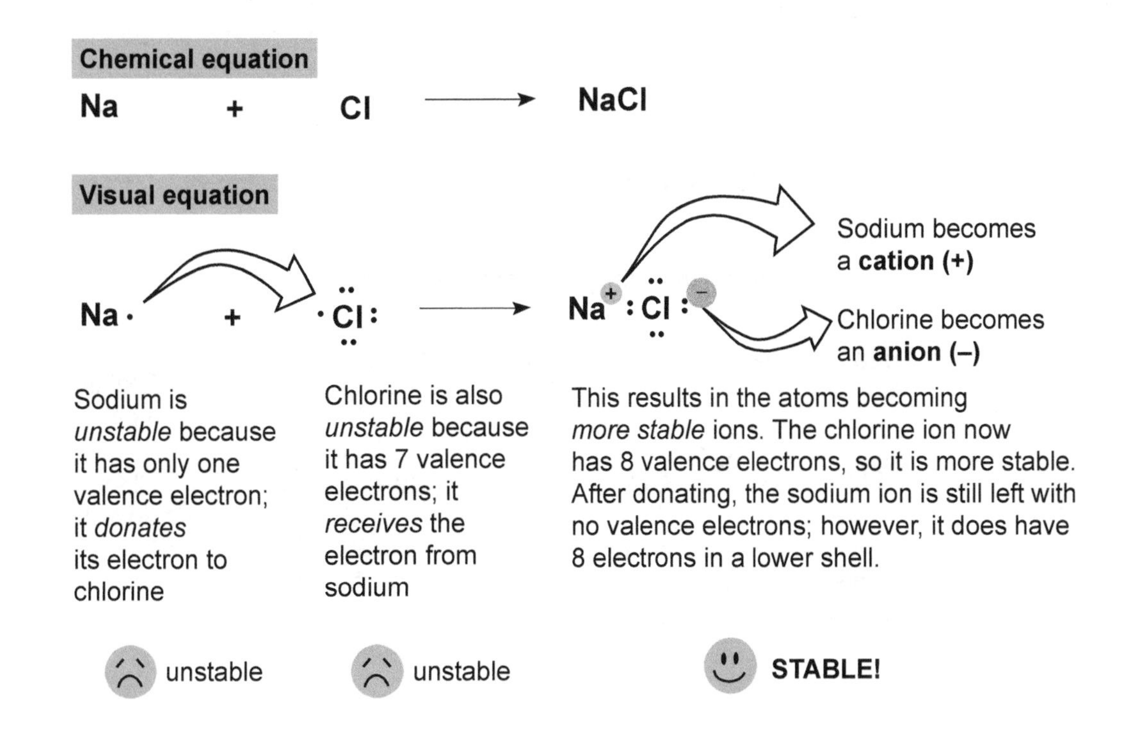

Practice Problems

On a separate sheet of paper, answer the following:

1. Alkali metals (1A) such as lithium and sodium each have 1 valence electron. Based on what you know about chemical bonding, predict if they are more chemically reactive or less chemically reactive. Explain.
2. The noble gases (8A) such as neon and argon all have 8 valence electrons. Based on what you know about chemical bonding, predict if they are more reactive or less reactive. Explain.
3. Calcium has 2 valence electrons. If it formed an ionic bond, would it more likely be an electron donor or a recipient? Explain.
4. Oxygen has 6 valence electrons. How many more does it need to have a full outermost shell?

Answers

1 Alkali metals need 8 valence electrons to be stable, but have only 1, so they are unstable. This makes them more chemically reactive because they have to lose an electron to become more stable. 2 Noble gases have a full outermost shell with 8 valence electrons, so they are very stable and less chemically reactive. They do not have to bond to become stable. 3 Calcium has only 2 valence electrons but needs 8, so it would be easier for it to donate 2 than receive 6. Thus, it would be an electron donor. 4 Oxygen needs 8 valence electrons to have a full outermost shell. It already has 6, so it would need 2 more. Thus, it would be an electron acceptor.

Description

Atoms bond to be in a more stable state. We're going to examine two common types of bonds: **ionic bonds** and **covalent bonds.**

Ionic bonds = one atom *donates* its electron; the other *receives* that same electron

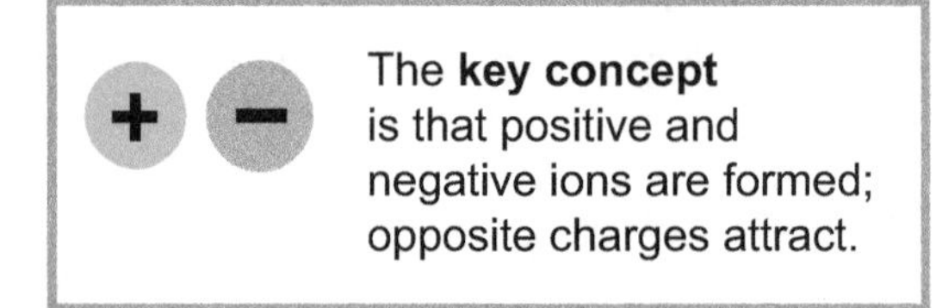

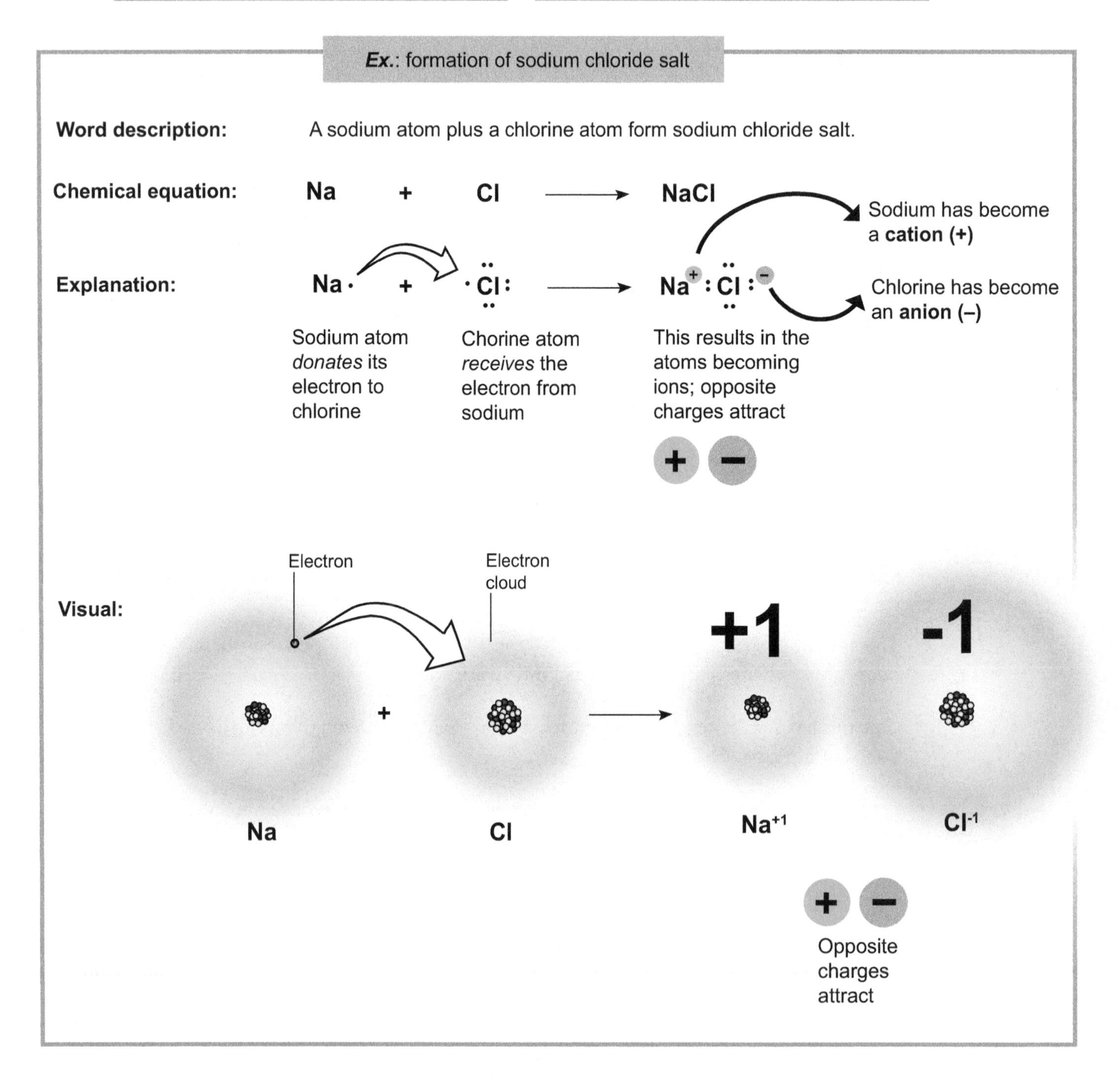

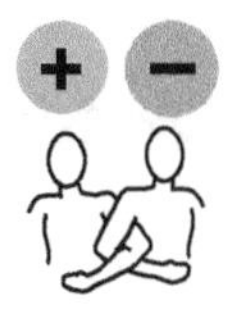

Covalent bonds = **shared pair** of electrons (like a couple in love, the word *love* is scrambled in the word *covalent*)

The **key concept** is that a covalent bond links two atoms like two people locking arms.

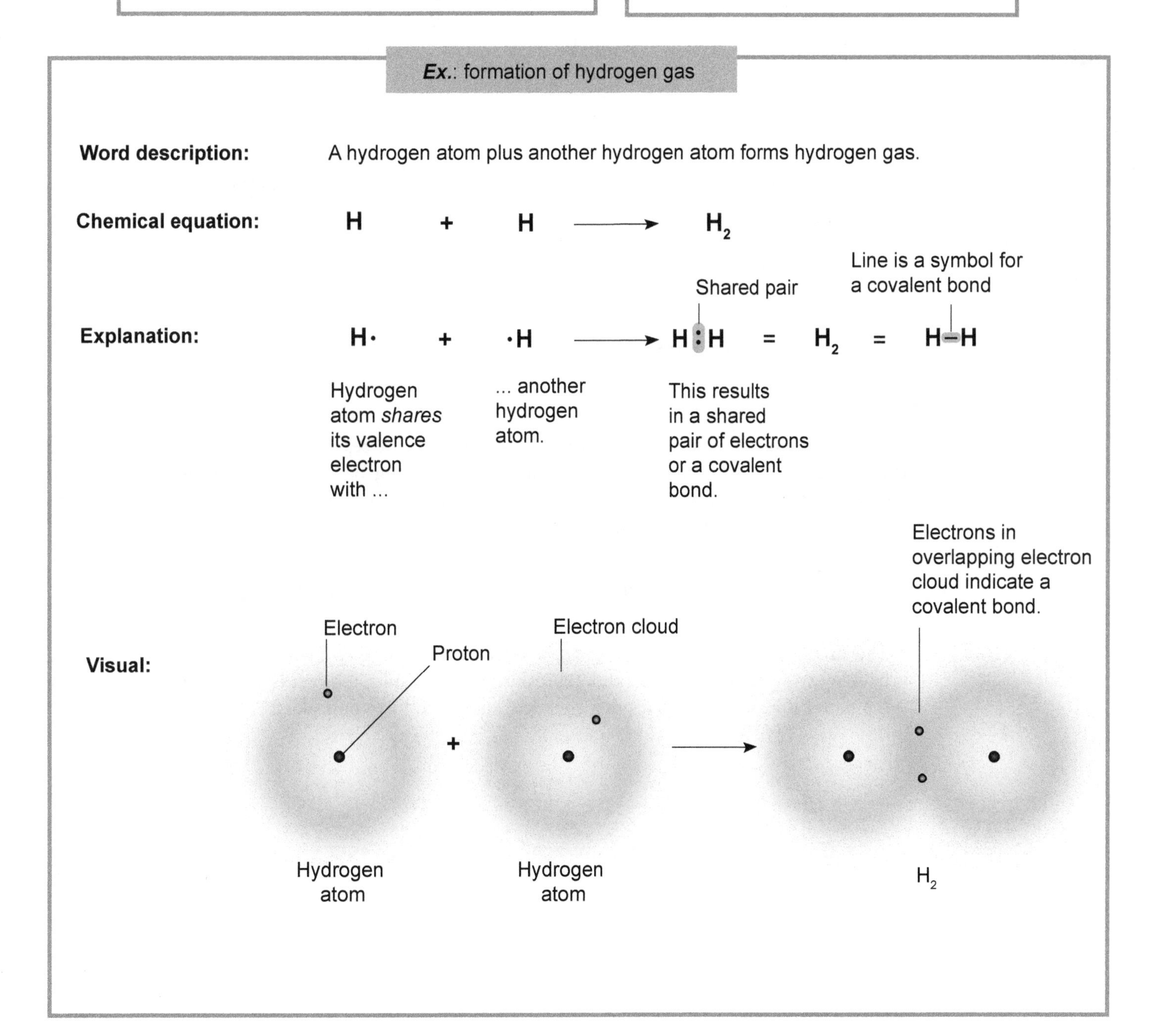

Description

Knowing the terms and symbols used in chemical equations is essential to understanding chemistry. Let's use making pancakes as an analogy. Mixing ingredients together in a recipe is similar to what occurs in some chemical reactions. By combining two substances, then heating the batter, you form something new.*

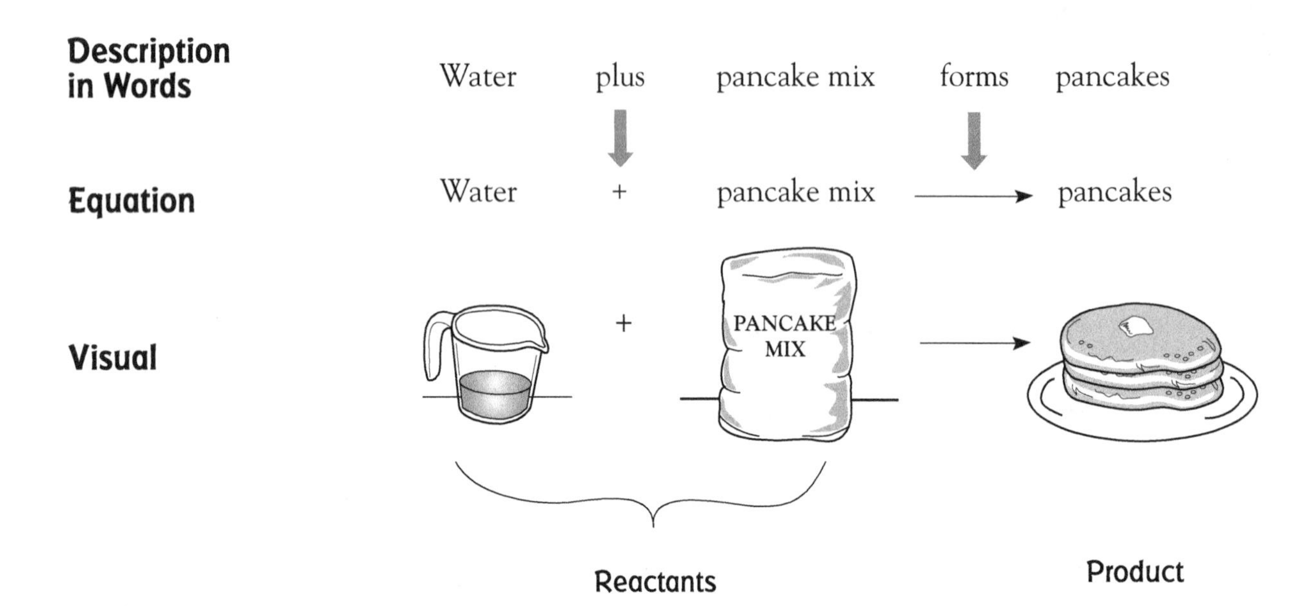

In the above example, the *water* and the *pancake mix* are the **reactants.** They always are written to the left of the arrow. After the batter is heated in a frying pan, the final **product**, pancakes, is formed. The product always is written to the right of the arrow. Knowing this, let's write the general formula for any chemical equation:

Reactant #1 + Reactant #2 ⟶ Product(s)

* This analogy works only in the broadest sense of combining two substances to make something new. When water is combined with pancake mix, the resulting batter is actually a mixture. When the batter is heated, it undergoes a chemical change to form pancakes.

Practice Problems

Identify the **REACTANTS** in the following chemical equations:

1 $H_2CO_3 \longrightarrow H_2O + CO_2$

2 $BaCl_2 + Na_2SO_4 \longrightarrow 2NaCl + BaSO_4$

3 $C + O_2 \longrightarrow CO_2$

Identify the **PRODUCTS** in the following chemical equations:

4 $CaCO_3 + 2HCl \longrightarrow CaCl_2 + CO_2 + H_2O$

5 $2KClO_3 \longrightarrow 2KCl + 3O_2$

6 $CH_4 + 2O_2 \longrightarrow CO_2 + 2H_2O$

Change the following word descriptions into chemical equations:

7 Hydrochloric acid plus sodium hydroxide forms sodium chloride plus water.

8 Glucose plus oxygen forms carbon dioxide plus water.

9 Chlorine plus sodium bromide forms sodium chloride plus bromine.

Answers

1 H_2CO_3 = carbonic acid

2 $BaCl_2, Na_2SO_4$ = barium chloride, sodium sulfate

3 C, O_2 = carbon, oxygen

4 $CaCl_2, CO_2, H_2O$ = calcium chloride, carbon dioxide, water

5 KCl, O_2 = potassium chloride, oxygen

6 CO_2, H_2O = carbon dioxide, water

7 $HCl_{(aq)} + NaOH \longrightarrow NaCl + H_2O$

8 $C_6H_{12}O_6 + 6O_2 \longrightarrow 6CO_2 + 6H_2O$

9 $Cl_2 + 2NaBr \longrightarrow 2NaCl + Br_2$

Description

Chemical reactions have predictable patterns depending on their type. Let's examine three basic types of chemical reactions: *combination reactions*, *decomposition reactions*, and *replacement reactions*.

Combination Reactions

Two or more reactants combine to form a **single** product—the opposite of a decomposition reaction.

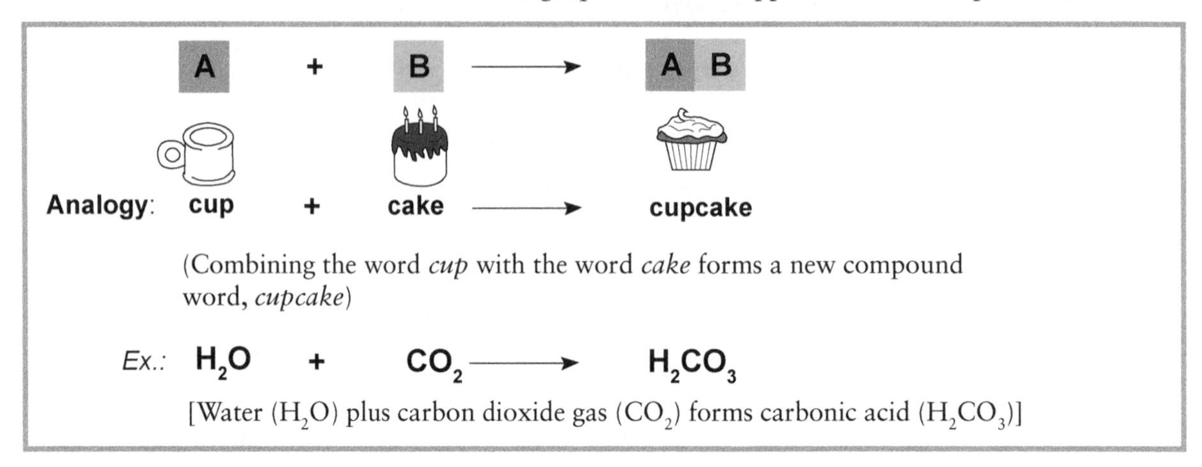

Decomposition Reactions

A compound breaks down into its component parts—the opposite of a combination reaction.

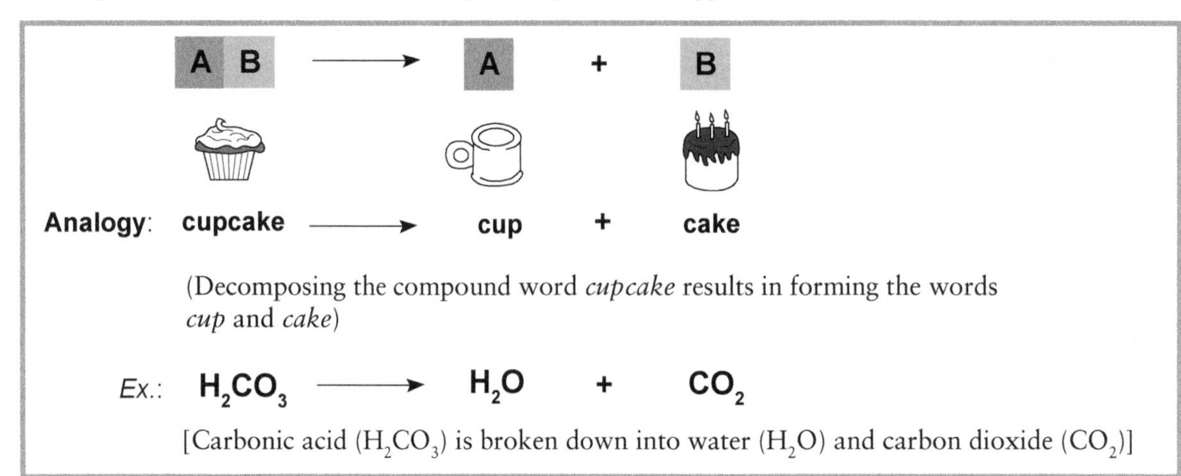

Replacement Reactions

Like changing dance partners, components recombine to form new products. There are two types: **single replacement** and **double replacement.**

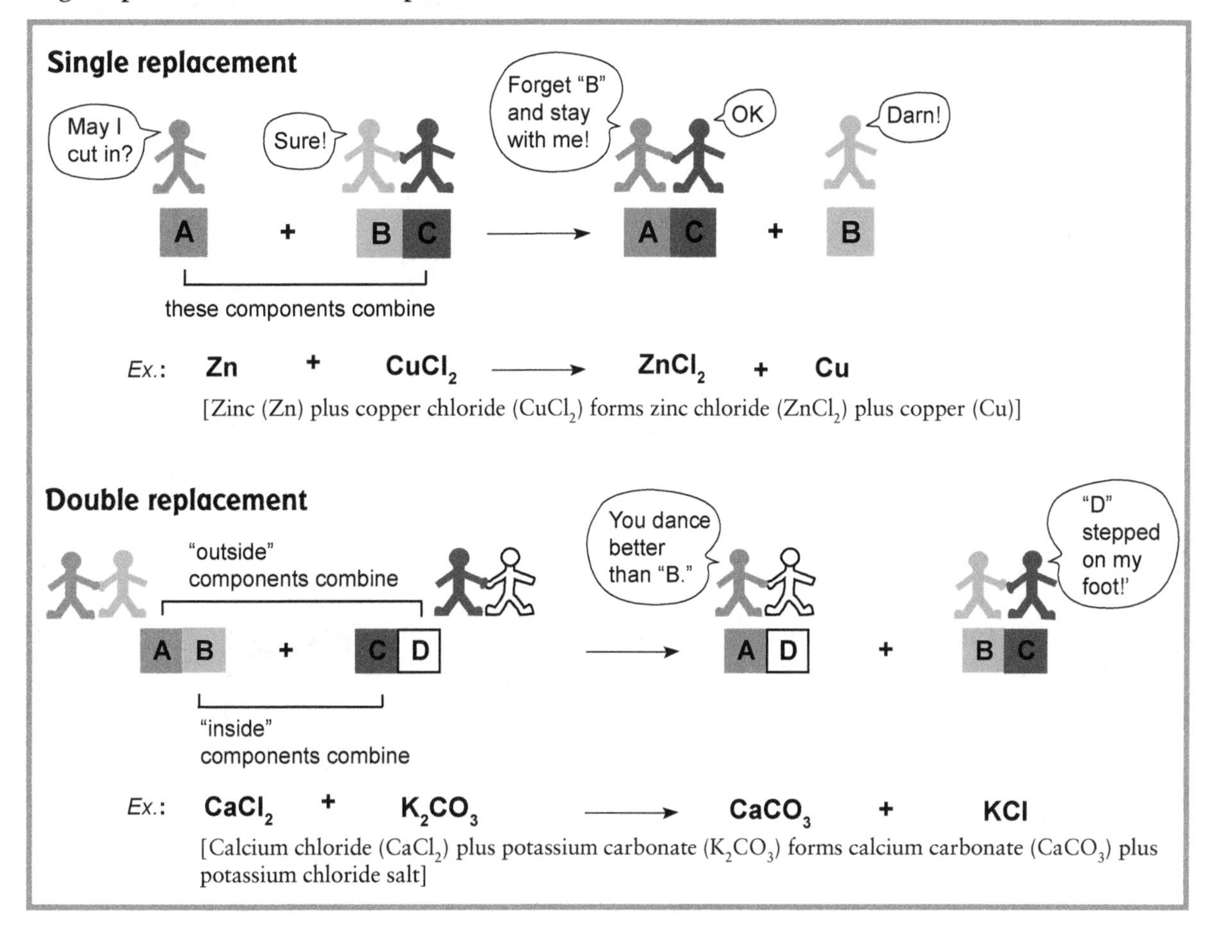

Classify each of the following reactions as *combination*, *decomposition*, *single replacement*, or *double replacement*:

1. $2Mg + O_2 \rightarrow 2MgO$
2. $NaCl + AgNO_3 \rightarrow NaNO_3 + AgCl$
3. $Cl_2 + 2NaBr \rightarrow 2NaCl + Br_2$
4. $SO_2 + H_2O \rightarrow H_2SO_3$

Answers

1 Combination 2 Double replacement 3 Single replacement 4 Combination

Description

A **solution** is a homogeneous mixture consisting of a substance (solute) dissolved and uniformly distributed in another (solvent). A common example is a **saline** (salt) **solution.** Water is the **solvent** that dissolves sodium chloride (NaCl), the **solute.**

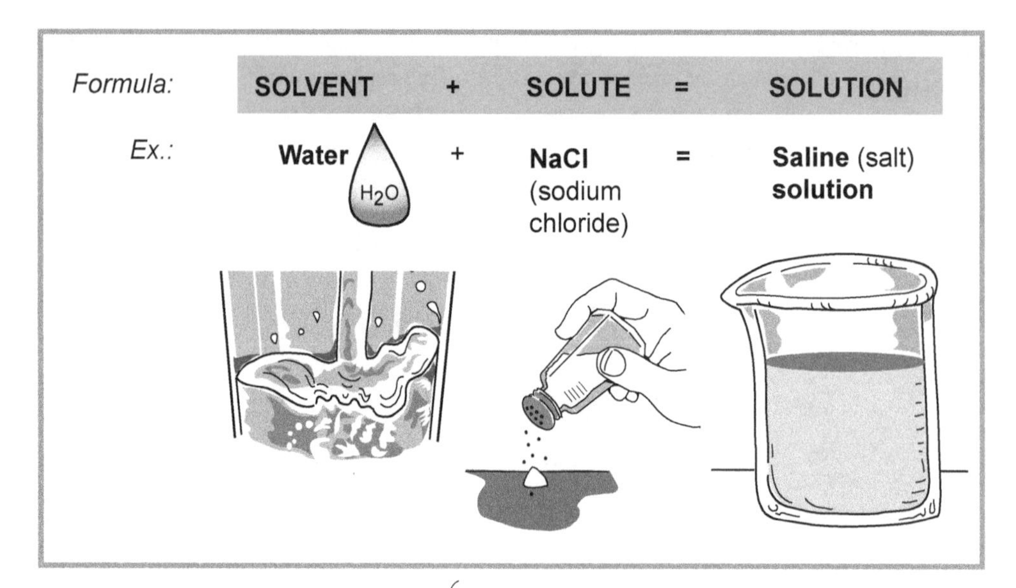

Liquid Solutions

Water serves as the solvent for most **liquid solutions.** These solutions are called aqueous. Solutes can be gases (*ex.:* CO_2), solids (*ex.:* NaCl), and/or liquids (*ex.:* acetic acid) dissolved in water. Wine, vinegar, and seawater are all examples of liquid solutions.

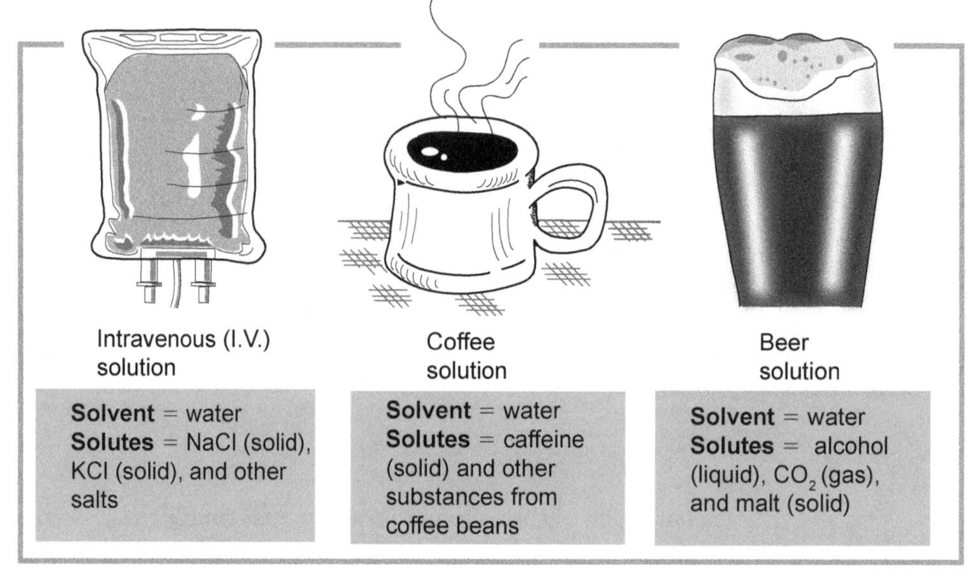

Intravenous (I.V.) solution

Solvent = water
Solutes = NaCl (solid), KCl (solid), and other salts

Coffee solution

Solvent = water
Solutes = caffeine (solid) and other substances from coffee beans

Beer solution

Solvent = water
Solutes = alcohol (liquid), CO_2 (gas), and malt (solid)

Gas and Solid Solutions

In additon to liquid solutions, there are also **gas solutions** (the solvent and the solute are both gases) and **solid solutions** (the solvent and the solute are both solids). The solvent is the substance present in a greater amount and surrounds the solute.

Gas solution

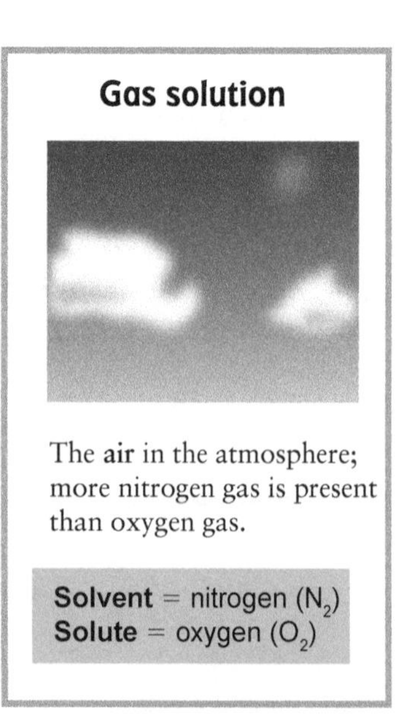

The **air** in the atmosphere; more nitrogen gas is present than oxygen gas.

Solvent = nitrogen (N_2)
Solute = oxygen (O_2)

Solid solution

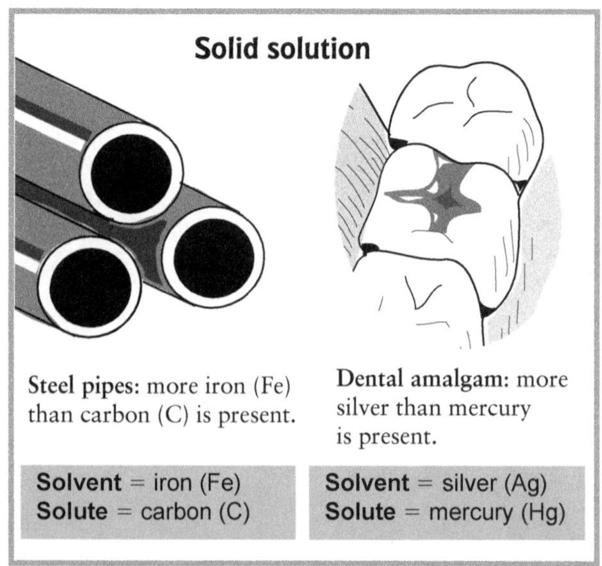

Steel pipes: more iron (Fe) than carbon (C) is present.

Solvent = iron (Fe)
Solute = carbon (C)

Dental amalgam: more silver than mercury is present.

Solvent = silver (Ag)
Solute = mercury (Hg)

Water as a Solvent

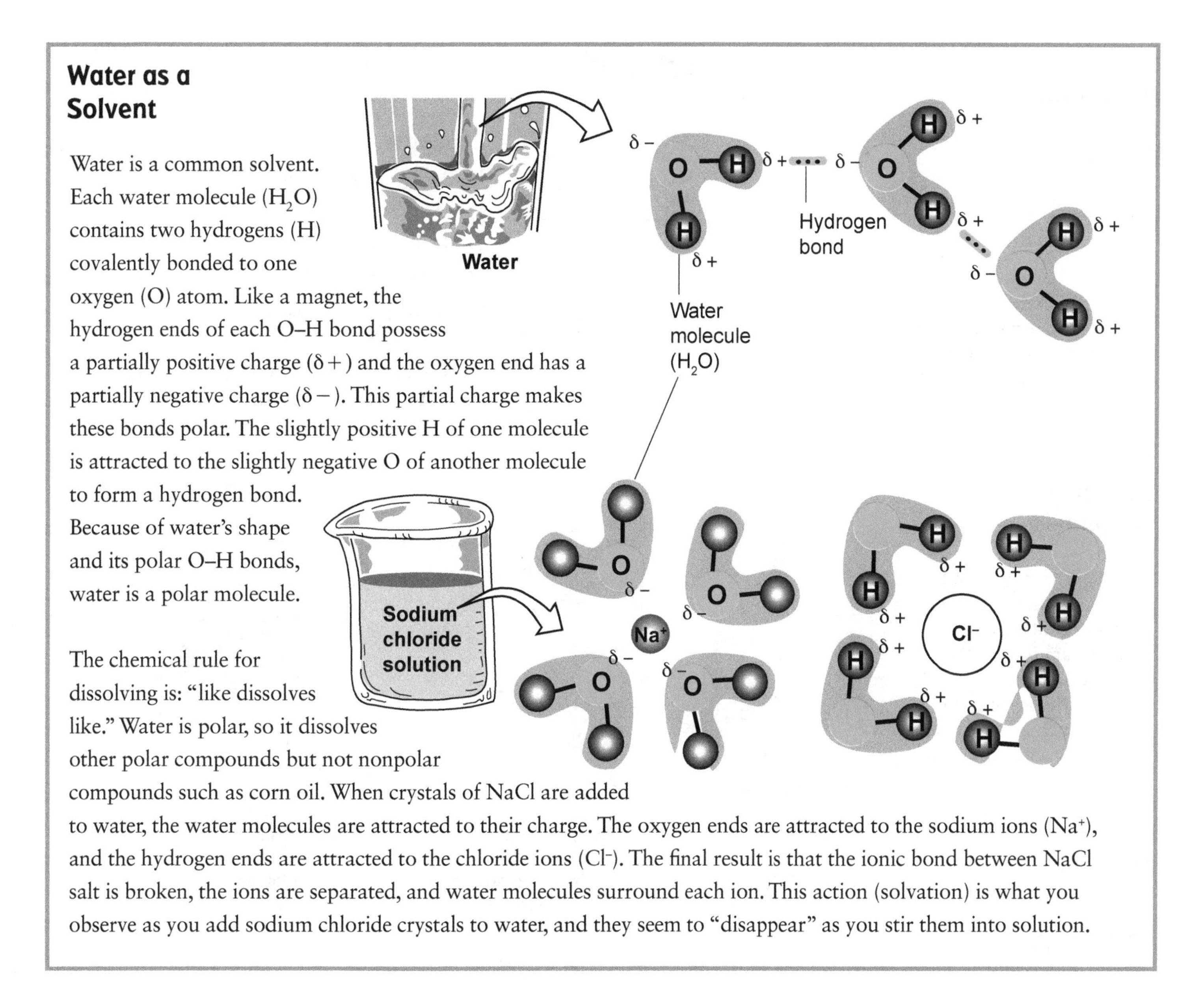

Water is a common solvent. Each water molecule (H_2O) contains two hydrogens (H) covalently bonded to one oxygen (O) atom. Like a magnet, the hydrogen ends of each O–H bond possess a partially positive charge (δ +) and the oxygen end has a partially negative charge (δ –). This partial charge makes these bonds polar. The slightly positive H of one molecule is attracted to the slightly negative O of another molecule to form a hydrogen bond. Because of water's shape and its polar O–H bonds, water is a polar molecule.

The chemical rule for dissolving is: "like dissolves like." Water is polar, so it dissolves other polar compounds but not nonpolar compounds such as corn oil. When crystals of NaCl are added to water, the water molecules are attracted to their charge. The oxygen ends are attracted to the sodium ions (Na^+), and the hydrogen ends are attracted to the chloride ions (Cl^-). The final result is that the ionic bond between NaCl salt is broken, the ions are separated, and water molecules surround each ion. This action (solvation) is what you observe as you add sodium chloride crystals to water, and they seem to "disappear" as you stir them into solution.

Practice Problems

You make a sugar solution to put in your hummingbird feeder. Answer the following:

1 What is the **solvent** in your sugar solution?

2 What is the **solute** in your sugar solution?

3 What type of solution did you make? Explain.

Answers

1 water 2 sugar 3 It's an aqueous solution, a type of liquid solution, because the solid solute (sugar) was dissolved in a liquid solvent (water).

COLLOIDS

Definition and Properties

Definition:

A **colloid** is a heterogeneous mixture containing either large solute particles or solid, liquid, or gas aggregates scattered within a dispersing agent such as a gas, a liquid, or a solid.

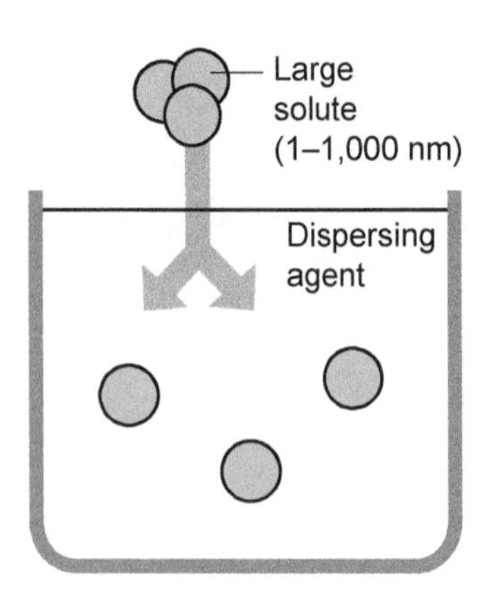

Properties of a colloid:

- **Particle size**—solute particle is large, roughly 1 to 1,000 nm
- **Turbidity**—scatters light so it appears turbid or cloudy
- **Settling**—solute particles do **NOT** settle
- **Separation**—solute particles **CANNOT** be separated by filters

General Examples of Colloids

Colloids can be found in the kitchen and other common places. It's important to note that they can involve substances in different states such as a gas in a liquid or a gas in a solid.

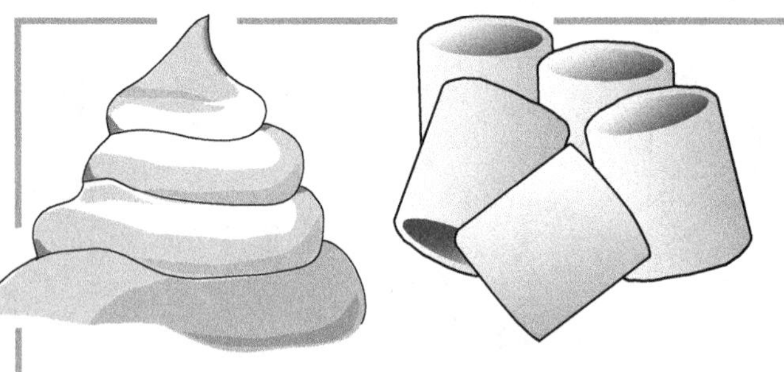

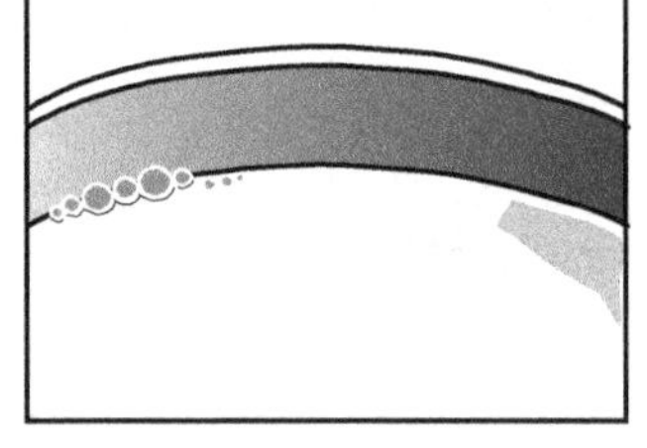

Whipped cream	Marshmallows	Milk
Gas in a liquid	**Gas in a solid**	**Liquid in a liquid**
(Nitrous oxide [N_2O] gas is dispersed in liquid cream.)	(Air is dispersed in a solid, gelatinous sugar mixture.)	(Lipid globules are dispersed in water.)

Biological Examples

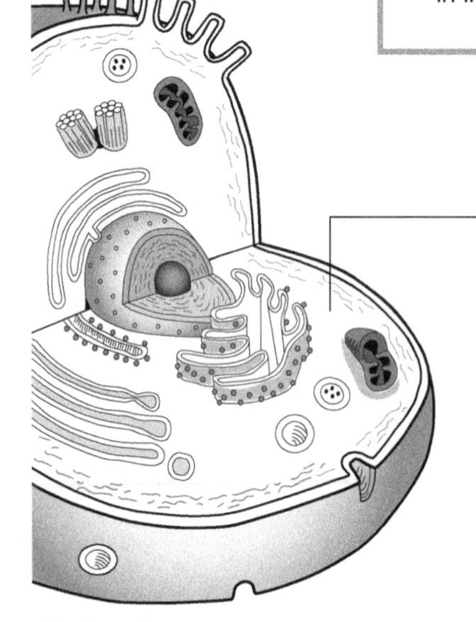

Body cell

The **cytosol** is the liquid portion of the cell (excluding the organelles), and it is both a solution and a colloid. It's a solution because it contains many different ions dissolved in water. But it also contains large organic substances like proteins that are dispersed in the water, which makes it a colloid.

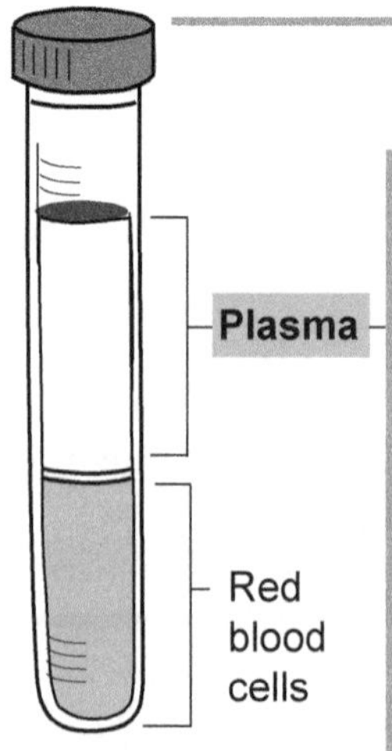

Tube of centrifuged blood

The **plasma** is an example of a colloid. When whole blood is spun in a centrifuge it separates into its liquid portion—the plasma—and its solid portion—mostly red blood cells. Various proteins are scattered in water, which gives the plasma a pale yellow color.

SUSPENSIONS

Definition and Properties

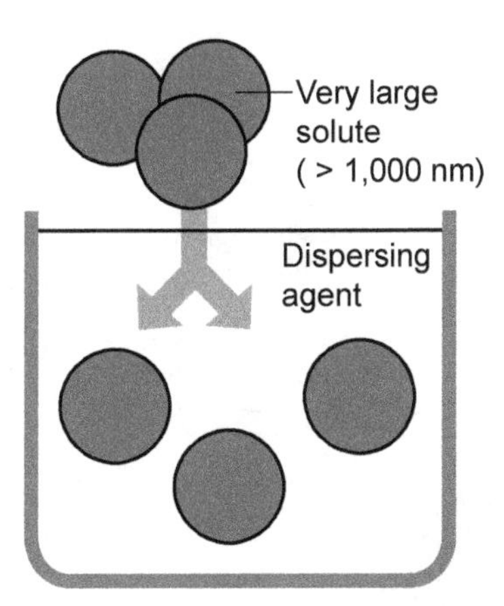

Definition:

A **suspension** is a mixture containing very large solute particles evenly distributed in a dispersing agent by mechanical means that will eventually settle out.

Properties of a suspension:

- **Particle size**—solute particle is very large, > 1,000 nm
- **Turbidity**—scatters light so it appears very turbid, or cloudy
- **Settling**—solute particles settle quickly
- **Separation**—solute particles can be separated by filters

General Examples of Suspensions

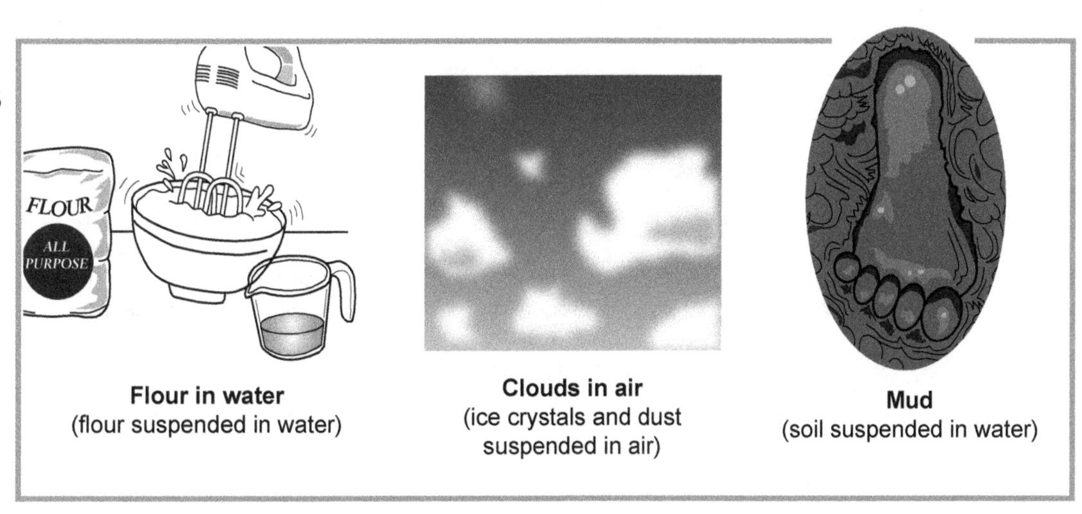

Flour in water
(flour suspended in water)

Clouds in air
(ice crystals and dust suspended in air)

Mud
(soil suspended in water)

Biological Examples

Cytoplasm is composed of the liquid portion inside the cell called the cytosol and all the very large organelles, such as the mitochondrion. These organelles are suspended in the cytosol, making the cytoplasm a suspension.

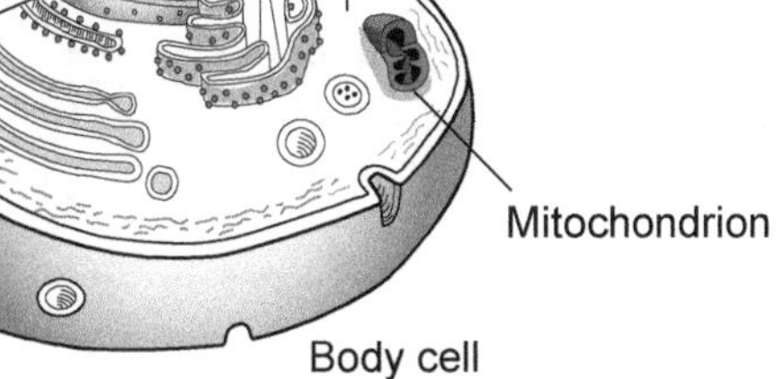

Whole blood is a suspension because red and white blood cells are suspended in the liquid portion of the blood called the plasma. Cell fragments called platelets are another large solute in this suspension.

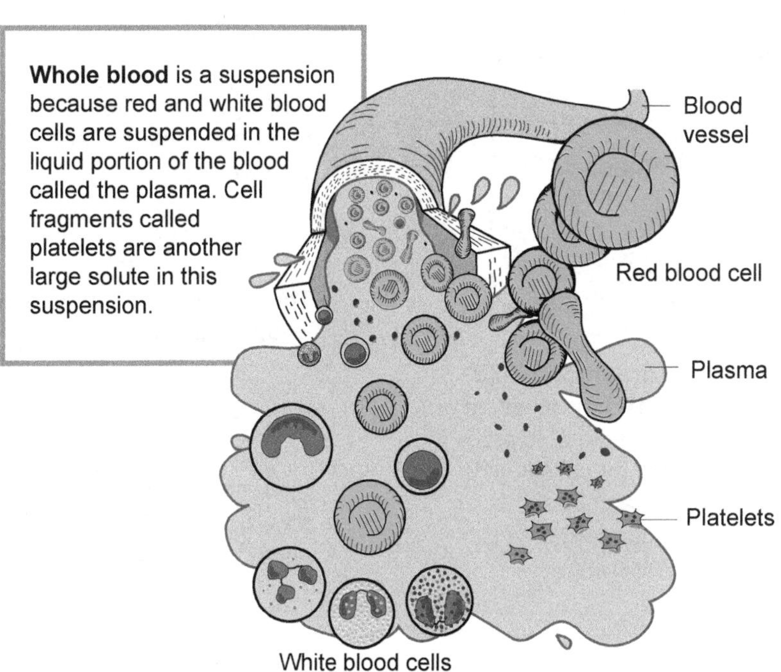

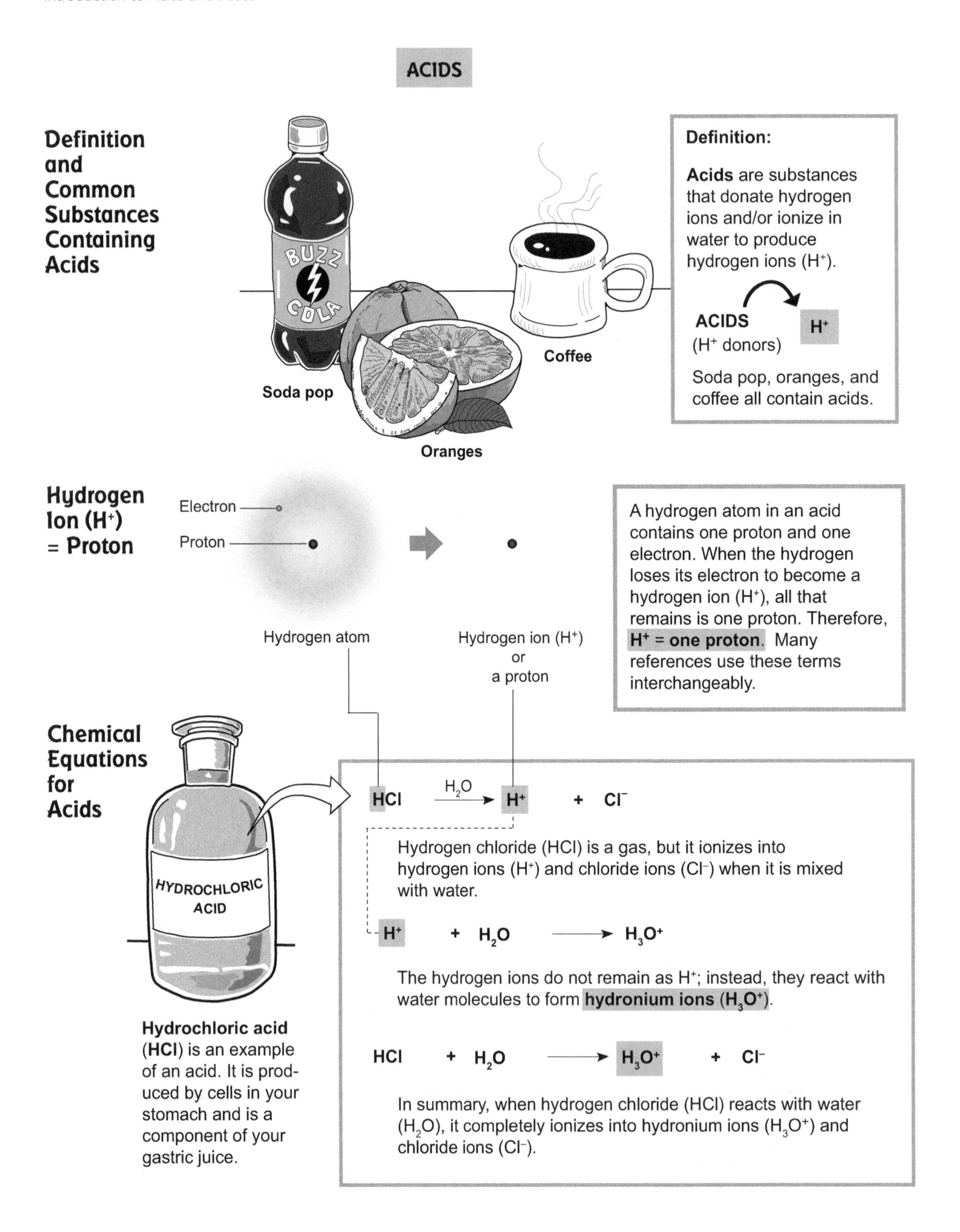
ACIDS
Definition and Common Substances Containing Acids
BUZZ COLA
Soda pop
Oranges
Coffee
Definition:
Acids are substances that donate hydrogen ions and/or ionize in water to produce hydrogen ions (H^+).
ACIDS
(H^+ donors)
H^+
Soda pop, oranges, and coffee all contain acids.
Hydrogen Ion (H^+) = Proton
Electron
Proton
Hydrogen atom
Hydrogen ion (H^+) or a proton
A hydrogen atom in an acid contains one proton and one electron. When the hydrogen loses its electron to become a hydrogen ion (H^+), all that remains is one proton. Therefore, H^+ = one proton. Many references use these terms interchangeably.
Chemical Equations for Acids
HYDROCHLORIC ACID
Hydrochloric acid (HCl) is an example of an acid. It is produced by cells in your stomach and is a component of your gastric juice.
$HCl \xrightarrow{H_2O} H^+ + Cl^-$
Hydrogen chloride (HCl) is a gas, but it ionizes into hydrogen ions (H^+) and chloride ions (Cl^-) when it is mixed with water.
$H^+ + H_2O \longrightarrow H_3O^+$
The hydrogen ions do not remain as H^+; instead, they react with water molecules to form hydronium ions (H_3O^+).
$HCl + H_2O \longrightarrow H_3O^+ + Cl^-$
In summary, when hydrogen chloride (HCl) reacts with water (H_2O), it completely ionizes into hydronium ions (H_3O^+) and chloride ions (Cl^-).

BASES

Definition and Common Substances Containing Bases

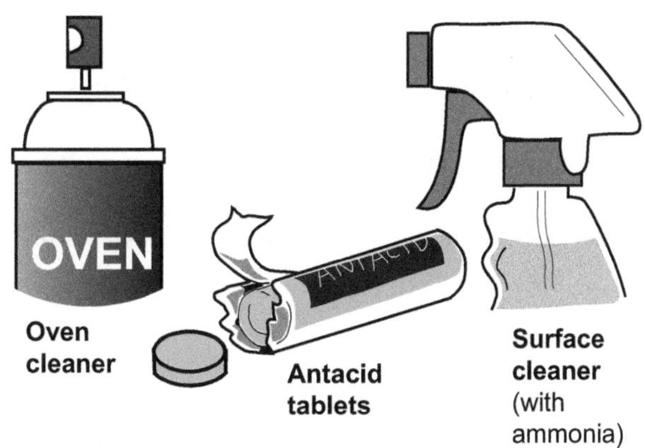

Definition:

Bases are substances that accept hydrogen ions and/or ionize in water to produce hydroxide ions (OH^-).

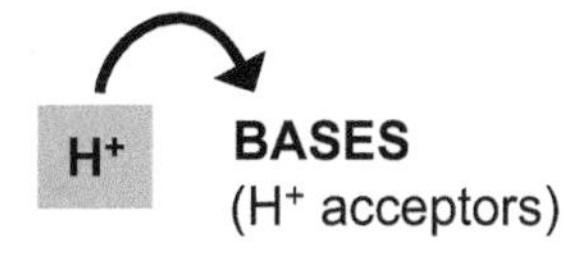

Oven cleaner, antacid tablets, and any surface cleaner containing ammonia all contain bases.

Chemical Equations for Bases

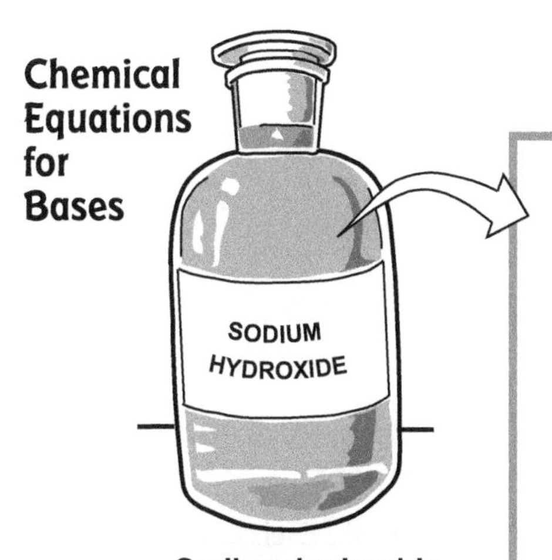

Sodium hydroxide (**NaOH**) is an example of a base that is a metal hydroxide.

$$NaOH \xrightarrow{H_2O} Na^+ + OH^-$$

Sodium hydroxide (**NaOH**) is a solid, but it ionizes to form sodium ions and hydroxide ions (OH^-) when it is mixed with water. The H_2O above the arrow indicates that water is involved in the ionization process but does not serve as a chemical reactant.

Many bases are metal hydroxides, such as NaOH, KOH, and $Mg(OH)_2$, but some are not.

Let's look at an example of another base that is not a metal hydroxide—namely, **ammonia** (**NH_3**).

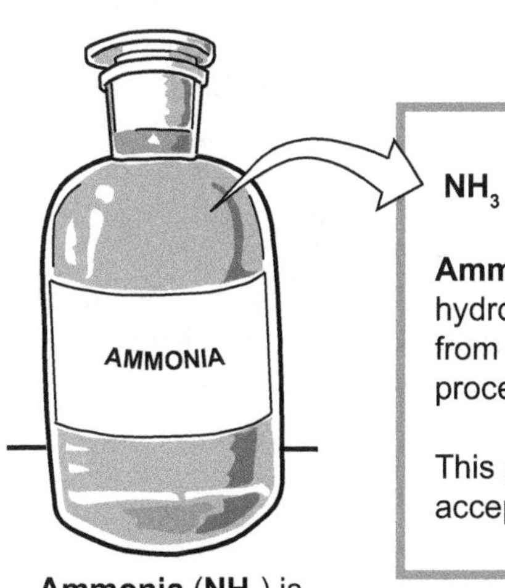

Ammonia (**NH_3**) is an example of a base that is *not* a metal hydroxide.

$$NH_3 + H_2O \rightleftharpoons NH_4^+ + OH^-$$

Ammonia (**NH_3**) is a gas that forms ammonium ions (NH_4^+) and hydroxide ions (OH^-) when it reacts with liquid water. NH_3 accepts H^+ from a water molecule to form NH_4^+. This leaves behind an OH^- in the process.

This perfectly fits the definition of a base as being both a substance that accepts hydrogen ions (H^+) and produces hydroxide ions (OH^-).

Description

The pH scale stands for "potential hydrogen" scale because it measures the molar concentration of hydrogen ions (H^+) in solution. Brackets [ ] are used to indicate molar concentration, so the phrase "molar concentration of hydrogen ions in solution" can be abbreviated as [H^+]. But hydrogen ions immediately bind to water molecules to form hydronium ions (H_3O^+), so pH is really measuring [H_3O^+] in solution. To avoid confusion, because a hydrogen ion is a proton, it's important to note that all three of these are equivalent when discussing hydrogen ions in solution:

hydrogen ion (H^+) = proton = hydronium ion (H_3O^+)

The pH values identify solutions as being acidic, basic (alkaline), or neutral.

pH Scale Facts

Range:	0–14
Acid:	< 7.0
Base:	> 7.0
Neutral point:	7.0; this is where [H_3O^+] = [OH^-].
Increments:	Each number represents a 10× (tenfold) change in [H_3O^+]
Units:	[H_3O^+] per liter of solution

1. pH is mathematically calculated as: **$pH = -\log [H_3O^+]$.**
2. CpH is a logarithmic scale like the Richter scale used to measure the intensity of earthquakes. An earthquake at 6 is 1,000 times more intense than an earthquake measured at 3.
3. Changes in pH reflect exponential changes in the [H_3O^+] of the solution. Each unit on the scale represents a 10× (tenfold) change in [H_3O^+], so changes occur as 10^X, where x = the number of units changed on the scale.

For example, a solution with a pH of 5 is moderately acidic, like coffee. If we add acid and the pH decreases from 5 to 3, how much of a change is it? Well, $5-3 = 2$, so the solution changes 2 pH units. Therefore, $10^2 = 10 \times 10 = 100$, so there is a 100× change in [H_3O^+]. But is it an increase or a decrease in [H_3O^+]? If the pH value decreases, the solution is more acidic. Likewise, if the pH value of a solution increases, it is becoming more basic and therefore less acidic.

Now, if you add a base to your pH = 3 solution, and the pH increases to 7, you are decreasing the [H_3O^+] by 4 pH units. Similarly, $3-7 = -4$, so $10^{-4} = 10^{-1} \times 10^{-1} \times 10^{-1} \times 10^{-1} = 1/10{,}000$. At a pH of 7, the solution has 1/10,000 the [H_3O^+] as it did at pH = 3.

So if you are hired as a lab technician and are asked to make a solution with a pH of 5, but you are a little sloppy and make a solution with a pH of 6 instead, you can see that you have made a big error. Your solution has a [H_3O^+] 10 times lower than it is supposed to be. Let's hope you don't get fired!

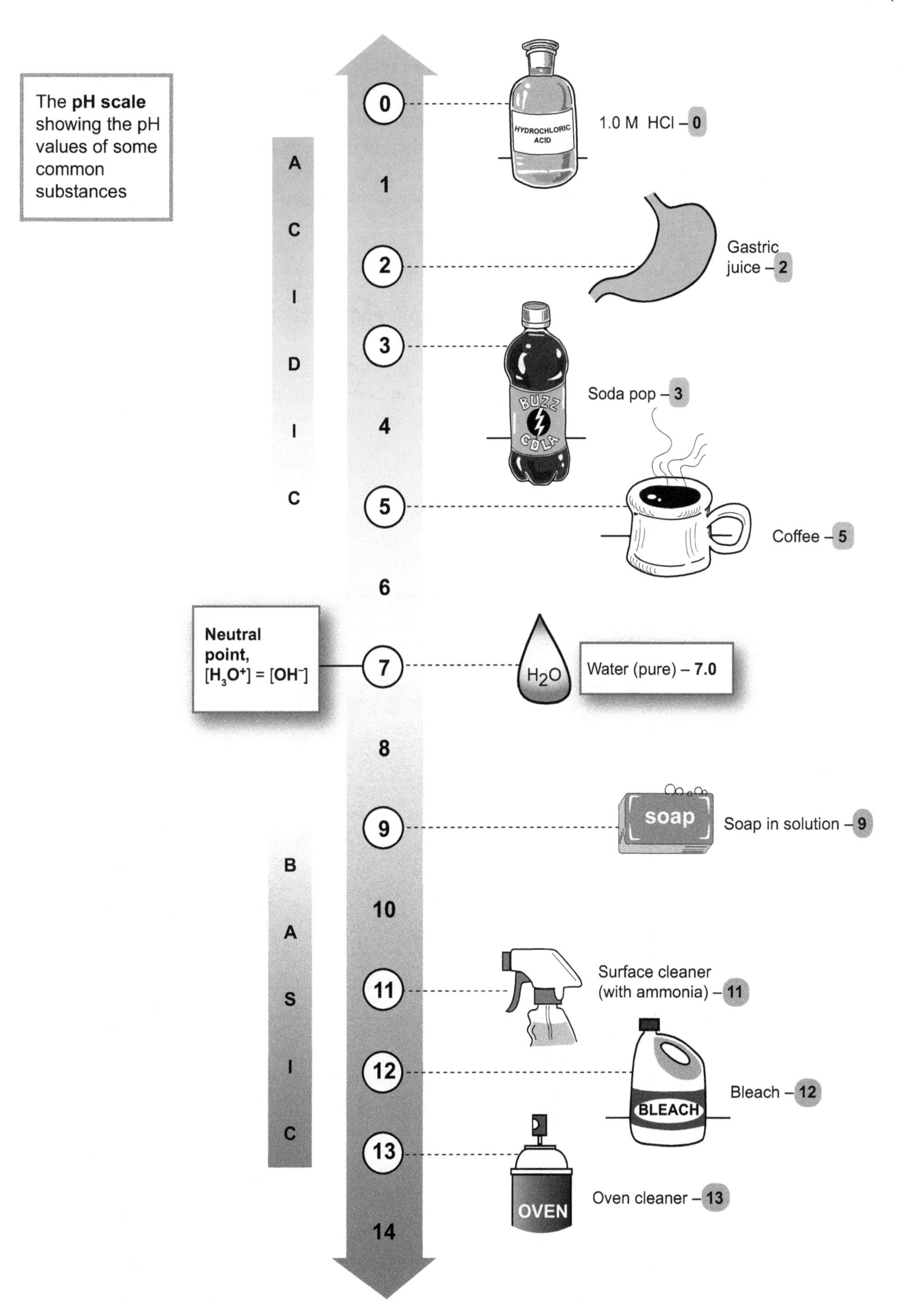
The **pH scale** showing the pH values of some common substances
ACIDIC
BASIC
0
1
2
3
4
5
6
7
8
9
10
11
12
13
14
HYDROCHLORIC ACID
1.0 M HCl – 0
Gastric juice – 2
BUZZ COLA
Soda pop – 3
Coffee – 5
Neutral point,
$[H_3O^+] = [OH^-]$
H_2O
Water (pure) – **7.0**
soap
Soap in solution – 9
Surface cleaner (with ammonia) – 11
BLEACH
Bleach – 12
OVEN
Oven cleaner – 13

Description

Organic compounds contain the element carbon (C). They typically also contain hydrogen (H) and also may contain other elements such as oxygen (O) and nitrogen (N). Within all organic compounds are groups of atoms called **functional groups** that determine many of the chemical and physical properties of that compound. Functional groups allow us to predict chemical behaviors and thus serve as a basis for classification of the different types of organic compounds. We will examine the following four common types of functional groups: hydroxyl, carbonyl, carboxyl, and amino groups.

Four Common Functional Groups

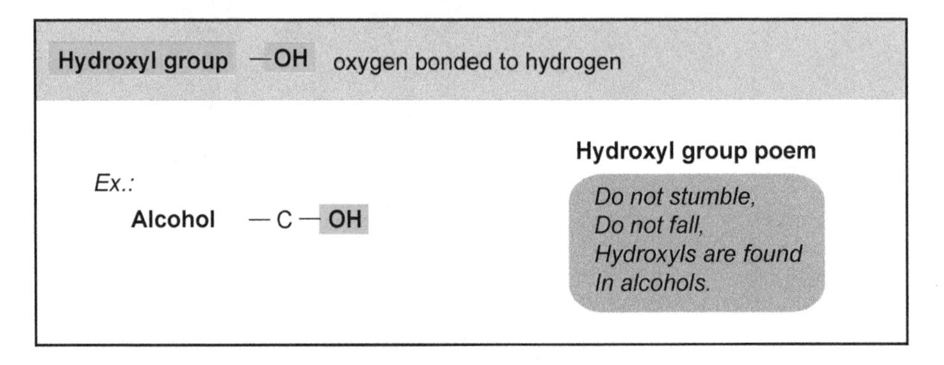

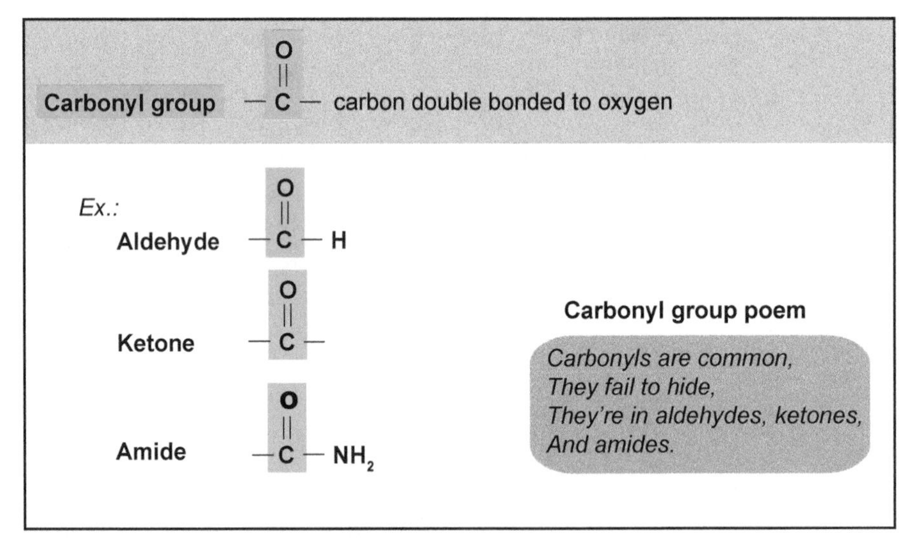

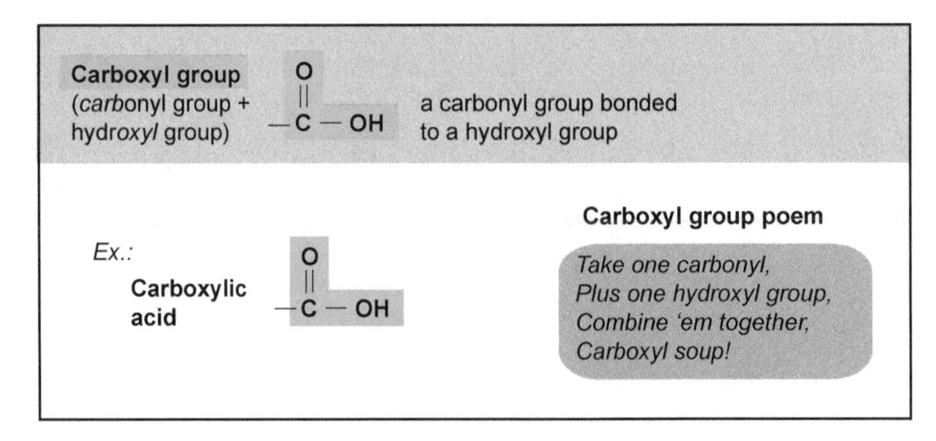

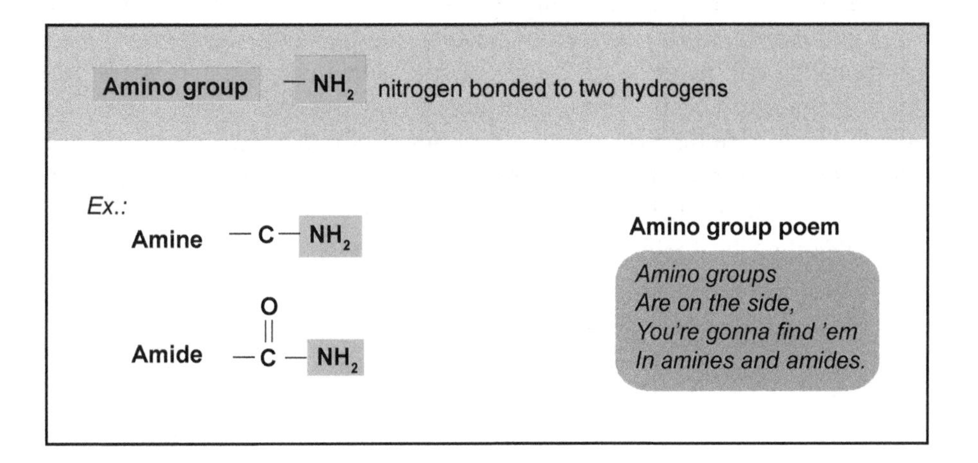

Practice Problems

1. What type of functional group is found in a **ketone**?
2. What type of functional group is found in an **alcohol**?
3. Identify the two different types of functional groups found in an **amide**?
4. What type of functional group is found in a **carboxylic acid**?

Answers

1 carbonyl group 2 hydroxyl group 3 carbonyl group and amino group 4 carboxyl group

Description

Polymers are large, long-chain molecules formed by linking building blocks called **monomers.** Think of a children's necklace formed by pushing plastic beads together. In this analogy, the beads are the monomers, and the necklace is the polymer. Some polymers are made from the same monomer repeated again and again. This is indicated here with beads having the same color. Other polymers are made from more than one type of monomer, as indicated with beads of different colors. Synthetic polymers, derived from alkenes, are the foundation of the plastics industry. In contrast, our bodies have the ability to make natural polymers, and they also are found in many of the foods we ingest. Let's examine both synthetic and natural polymers.

Analogy

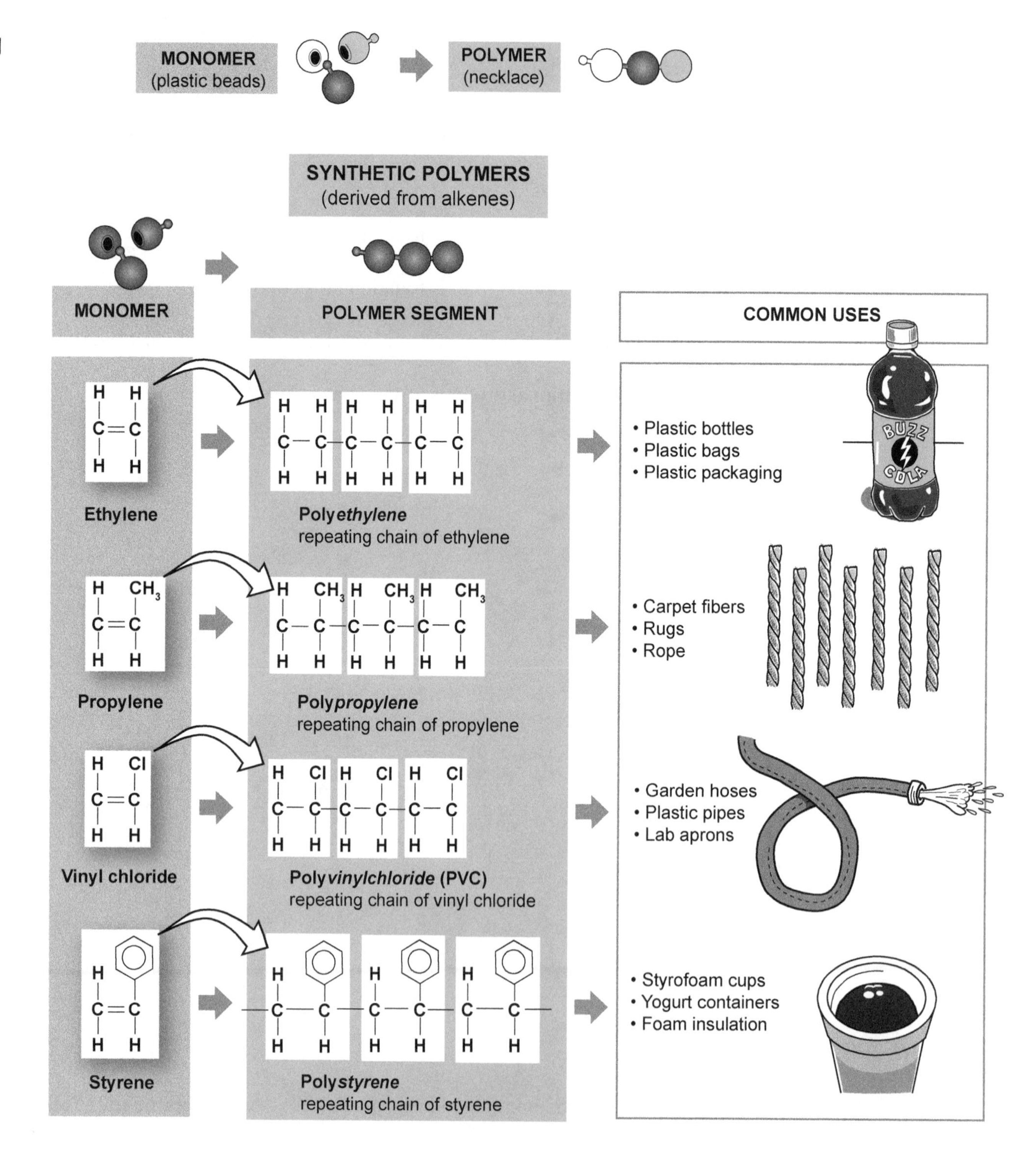

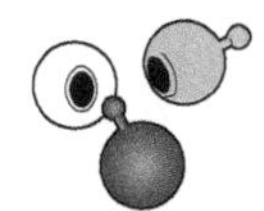
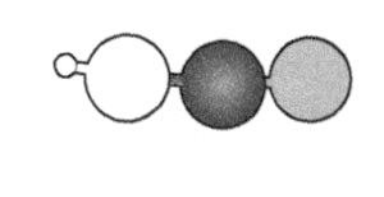

Natural Polymers

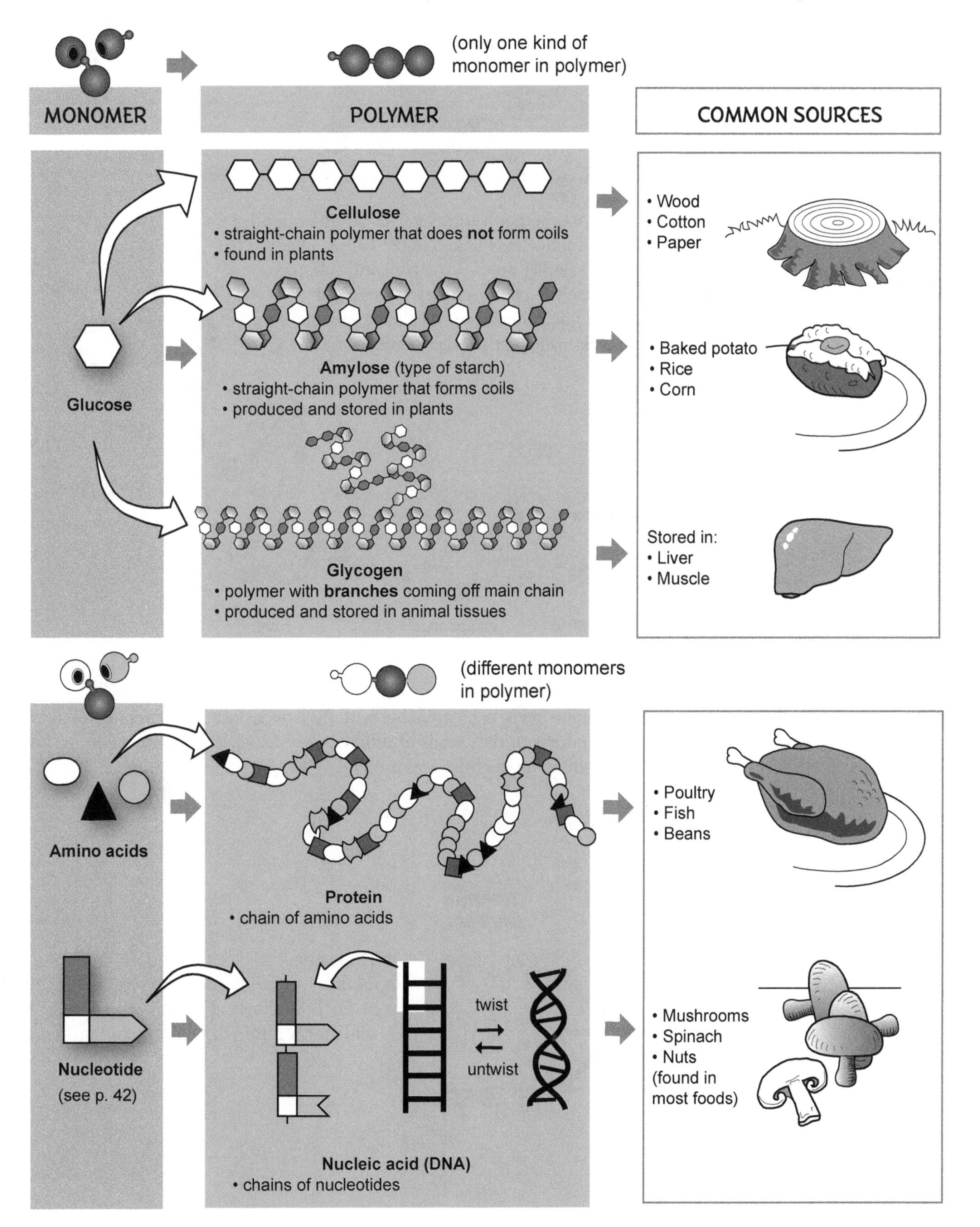

Description

Recall that **polymers** are large, long-chain molecules formed by linking building blocks called **monomers. Polypeptides** are natural polymers formed from **amino acid** monomers. It's like how a children's necklace is formed by pushing plastic beads together. In this analogy, the beads are the amino acid monomers, and the necklace is the polypeptide polymer.

Analogy 1

Overview: Building the Chain

When one amino acid links to another it forms a C-N covalent bond called a **peptide bond.** Two amino acids linked together are called a **dipeptide**, and three linked together are called a **tripeptide.** As the chain grows in length, it becomes a **polypeptide.** A chain with 50 or more amino acids *and* biological function is called a **protein.** (*Note:* There is no clear consensus on the exact number of amino acids to define a protein).

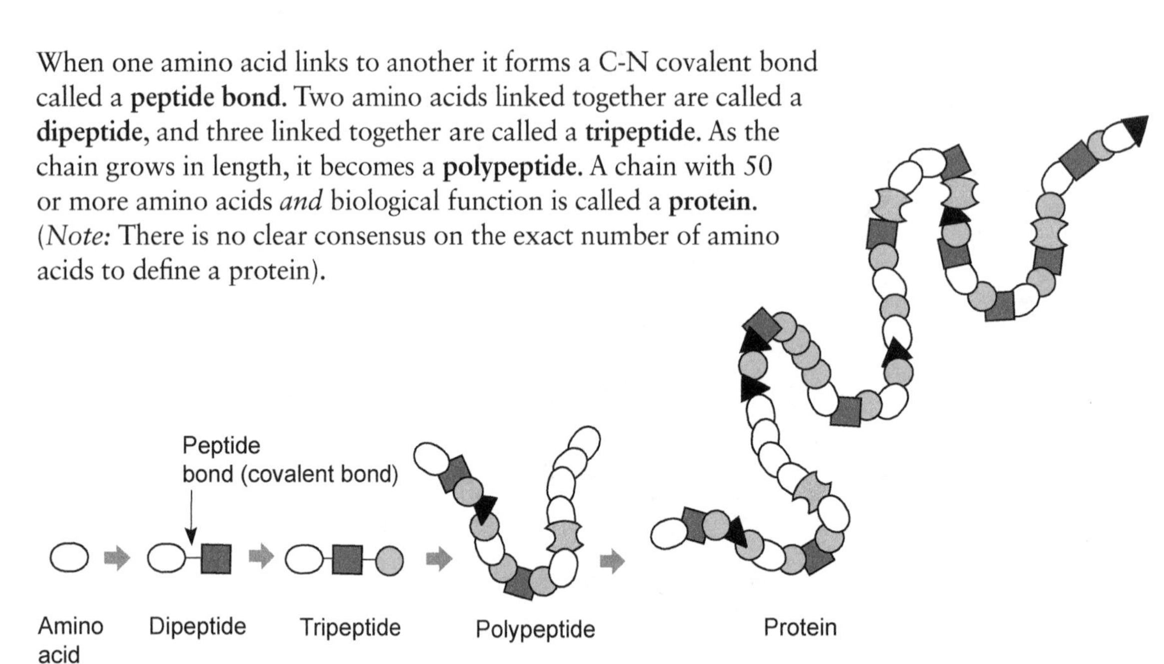

Analogy 2

Think of the 20 different amino acids as an alphabet with 20 letters. Just as different combinations of letters are used to make hundreds of thousands of different words, various amino acids are linked in different sequences to form different peptides and proteins.

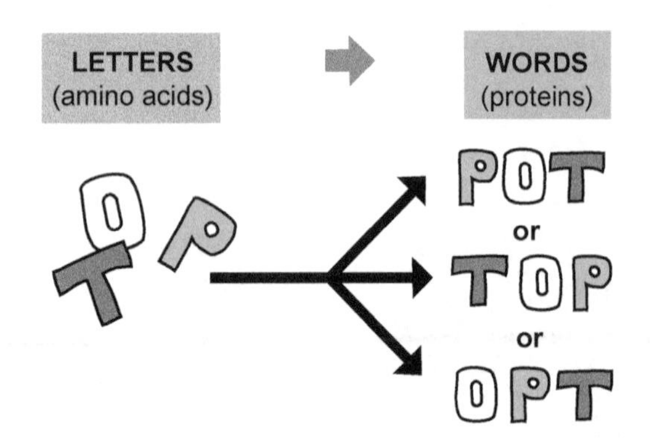

The same three letters—O, T, and P—can be used to form three different words: POT or TOP or OPT. Similarly, different sequences of amino acids form different peptides and proteins.

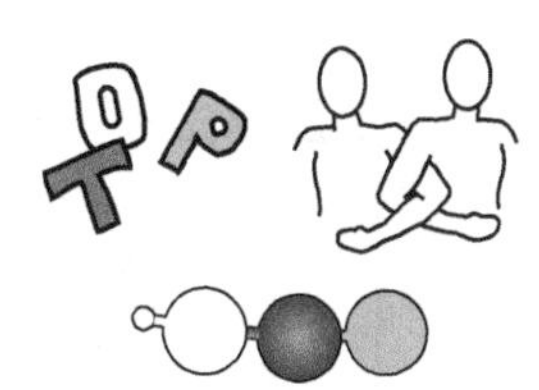

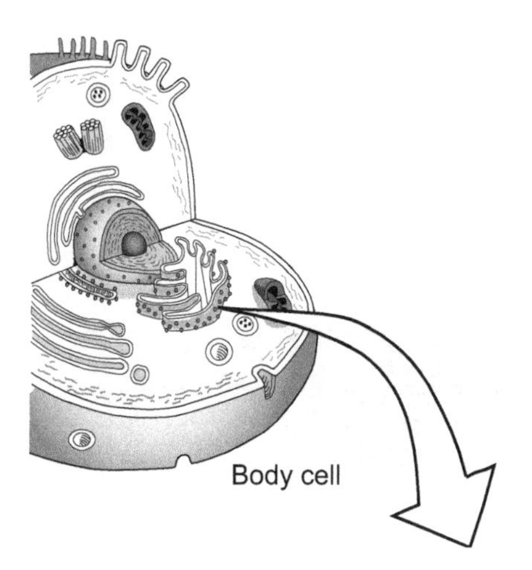

Forming the Peptide Bond

The process of creating proteins from amino acids occurs in body cells and is called **protein synthesis.** As a preview, let's look at the formation of a peptide bond between two amino acids. Using the same icons on the facing page, imagine we are forming a dipeptide. When one amino acid links to another, it forms a dipeptide, and one water molecule is produced as a product:

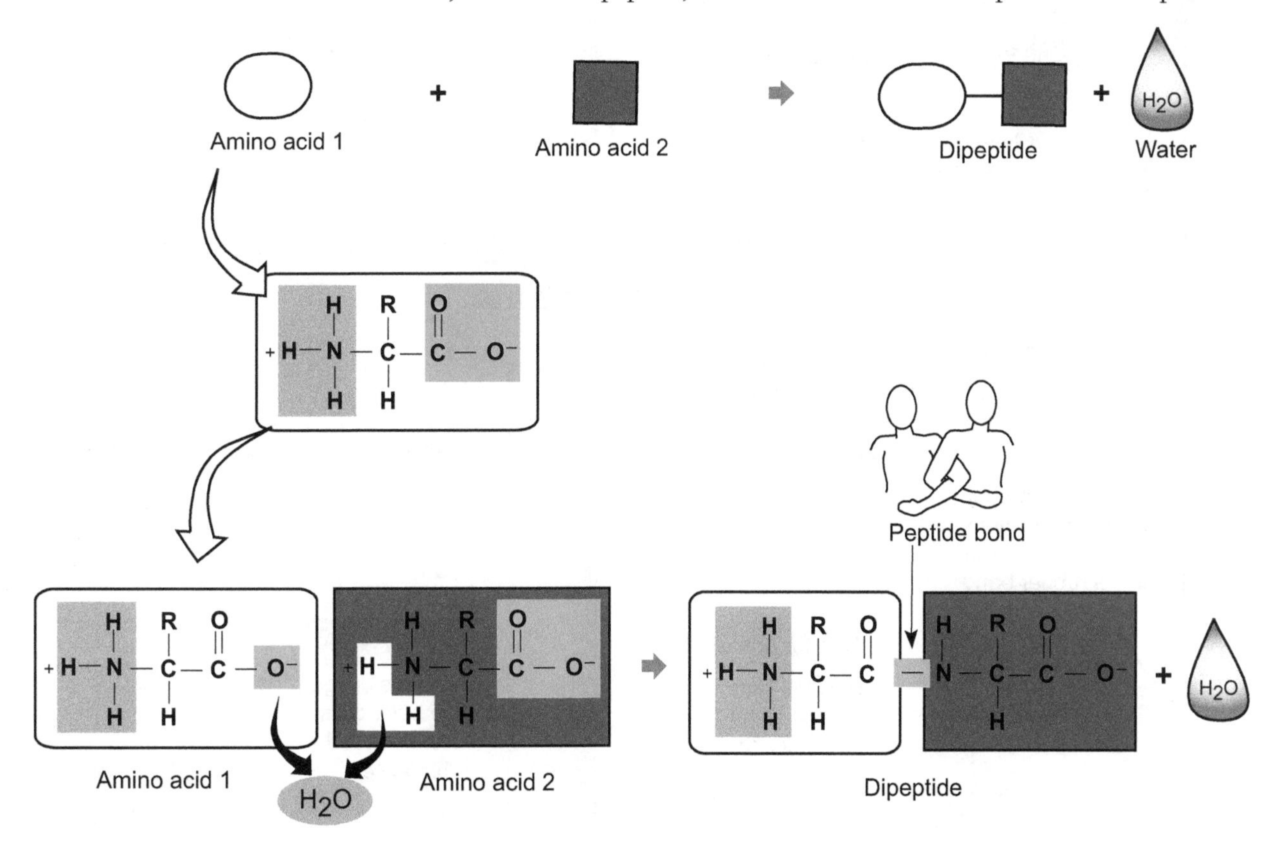

The oxygen is removed from amino acid 1, which combines with two hydrogens from amino acid 2 to form H_2O. The carbon and nitrogen form a peptide bond.

Notice that the peptide bond is a covalent bond between the carbon (C) of amino acid 1 and the nitrogen (N) of amino acid 2. Because all covalent bonds are a shared pair of electrons, it is like two people locked arm in arm.

To form a tripeptide and then a polypeptide, the process simply repeats itself.

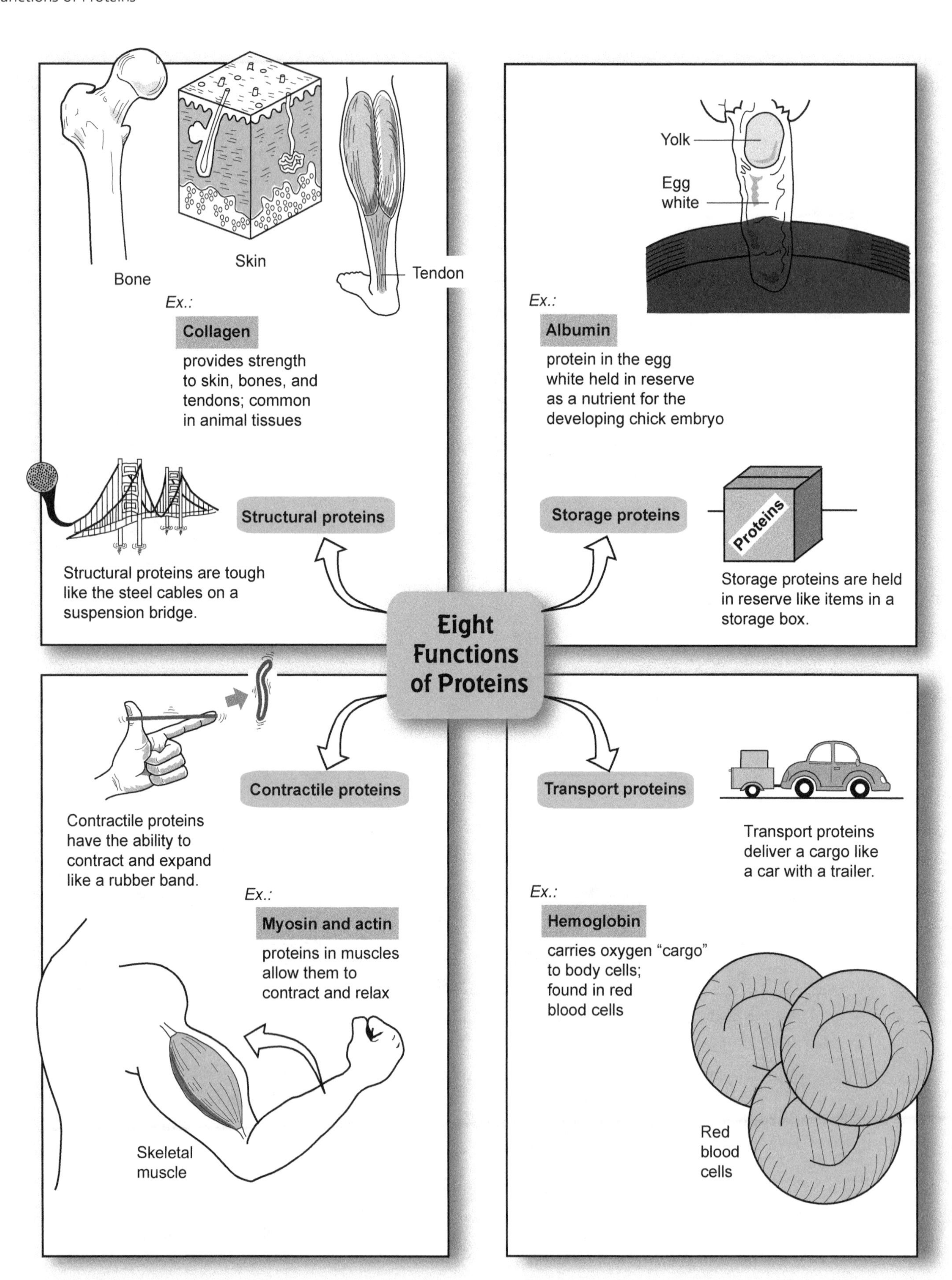
Bone
Skin
Tendon
Ex.:
Collagen
provides strength to skin, bones, and tendons; common in animal tissues
Structural proteins
Structural proteins are tough like the steel cables on a suspension bridge.
Yolk
Egg white
Ex.:
Albumin
protein in the egg white held in reserve as a nutrient for the developing chick embryo
Storage proteins
Proteins
Storage proteins are held in reserve like items in a storage box.
Eight Functions of Proteins
Contractile proteins
Contractile proteins have the ability to contract and expand like a rubber band.
Ex.:
Myosin and actin
proteins in muscles allow them to contract and relax
Skeletal muscle
Transport proteins
Transport proteins deliver a cargo like a car with a trailer.
Ex.:
Hemoglobin
carries oxygen "cargo" to body cells; found in red blood cells
Red blood cells

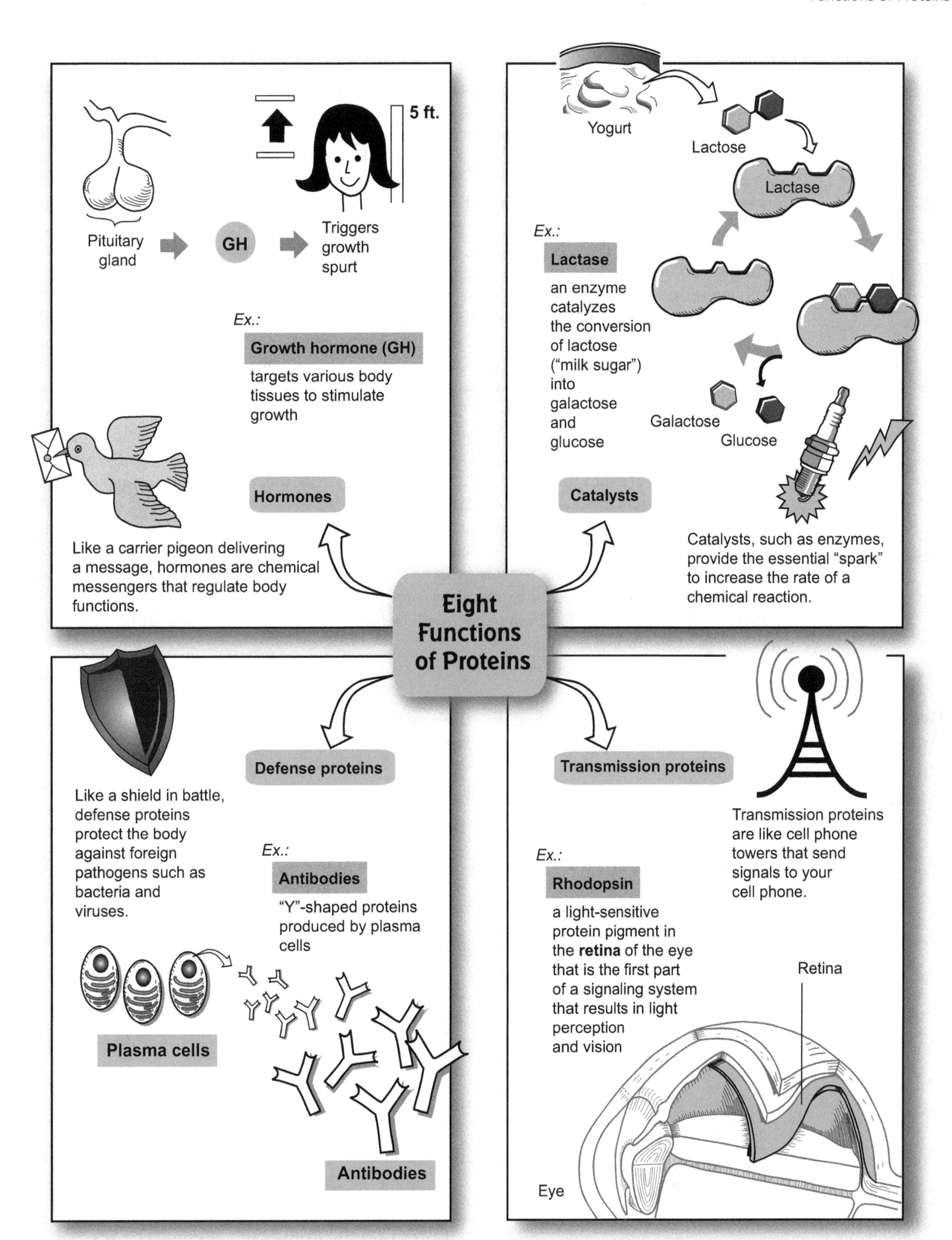
5 ft.
Pituitary gland
GH
Triggers growth spurt
Ex.:
Growth hormone (GH)
targets various body tissues to stimulate growth
Hormones
Like a carrier pigeon delivering a message, hormones are chemical messengers that regulate body functions.
Yogurt
Lactose
Lactase
Ex.:
Lactase
an enzyme catalyzes the conversion of lactose ("milk sugar") into galactose and glucose
Galactose
Glucose
Catalysts
Catalysts, such as enzymes, provide the essential "spark" to increase the rate of a chemical reaction.
Eight Functions of Proteins
Defense proteins
Like a shield in battle, defense proteins protect the body against foreign pathogens such as bacteria and viruses.
Ex.:
Antibodies
"Y"-shaped proteins produced by plasma cells
Plasma cells
Antibodies
Transmission proteins
Transmission proteins are like cell phone towers that send signals to your cell phone.
Ex.:
Rhodopsin
a light-sensitive protein pigment in the retina of the eye that is the first part of a signaling system that results in light perception and vision
Retina
Eye

Description

Enzymes are proteins that act as **catalysts** to increase the rate of chemical reactions. Astonishingly, some raise the reaction rate by as much as 1 million times! Without enzymes, the chemical reactions needed to sustain life would not be possible. Consequently, they are found throughout your body in such places as the plasma, the digestive tract, and inside your body cells.

The three-dimensional shape of an enzyme is vital to its function. Every enzyme has a cleft or depression called the **active site**. This is where it binds the **substrate**, the substance upon which the enzyme acts. Every enzyme has a shape that specifically fits with only one kind of substrate. Digestive enzymes are common, so they are a good example to use. Every time you eat food, enzyme-producing cells either lining your digestive tract or within glands secrete many different digestive enzymes to assist in breaking down food into smaller substances that can be used by body cells.

Lactose is a disaccharide, or "double sugar," that breaks down into its two component sugars, galactose and glucose. Imagine that you just ate some yogurt for breakfast. It contains many nutrients, but let's focus only on the lactose. After the yogurt is swallowed, it moves down your esophagus, into your stomach, then into your small intestine. Enzyme-producing cells lining the small intestine secrete the enzyme **lactase**. As lactase mixes with the food inside the intestine, the lactase comes in contact with the lactose, and the two bind together. This is called the **enzyme-substrate complex**. Next, a type of chemical reaction called a **hydrolysis** reaction occurs in which a water molecule helps to break the covalent bond between the two sugars. The two sugars, glucose and galactose, are then released from the enzyme, absorbed into the blood, and delivered to body cells. Once inside body cells, they can either be used as a fuel immediately or stored for later use.

Have you heard of lactose intolerance? This is the condition in which a person is unable to digest lactose. As you might suspect, the cause of this disorder is a deficiency in the production of the enzyme lactase. To compensate for this problem, lactase can be purchased from stores and added to dairy products prior to ingestion. Some dairy products such as milk claim to be "lactose-free." This indicates that milk already has been treated with lactase and that the lactose already has been digested to galactose and glucose.

Two Models for Enzyme-Substrate Interaction

Two models commonly are used to describe how the substrate binds to the enzyme: (1) **lock and key model**, and (2) **induced fit model**.

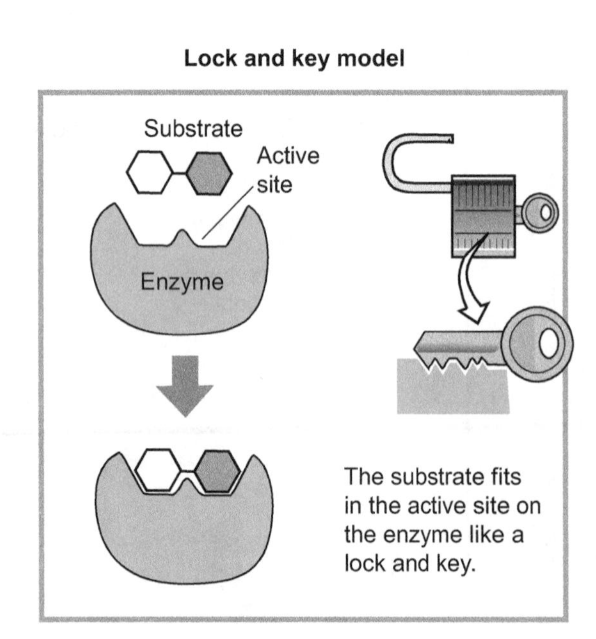

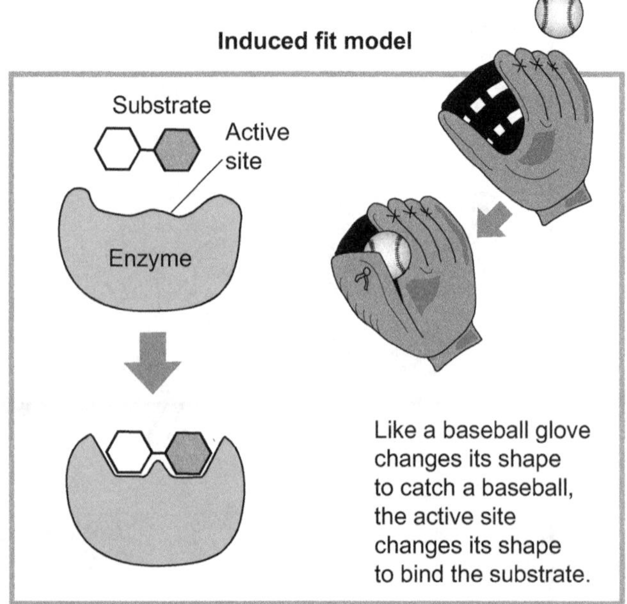

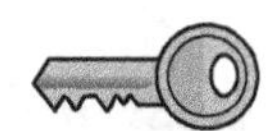

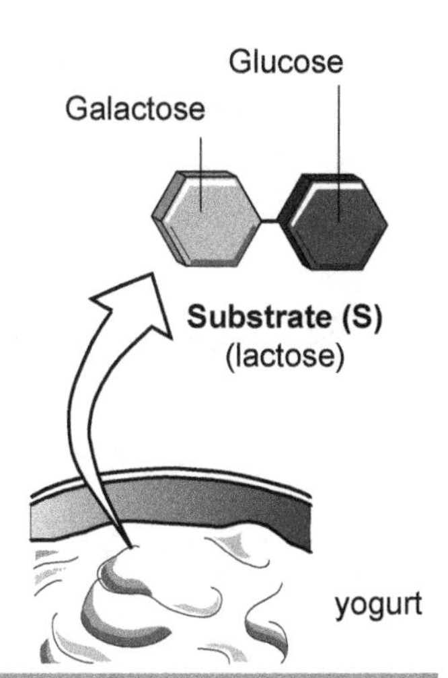

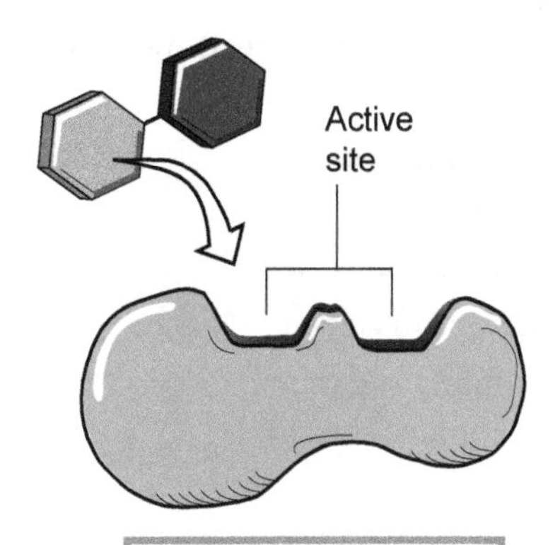

① **Lactose** is the sugar found in dairy products such as milk and yogurt. We break it down in our small intestine into its two component sugars—galactose and glucose.

② **Enzyme (E)**
(lactase)

The enzyme lactase is made by cells in your small intestine and digests *only* the substrate lactose.

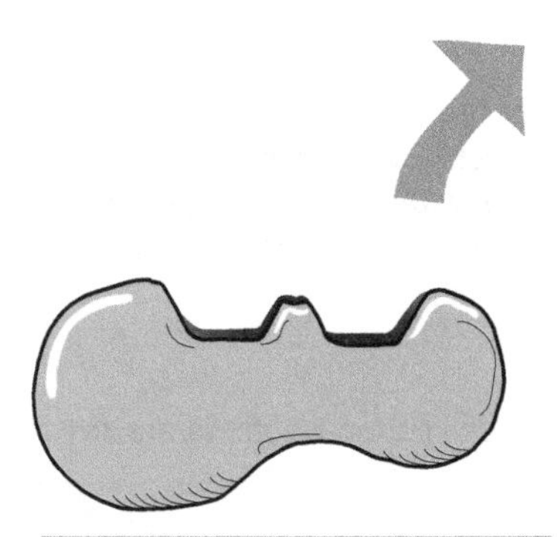

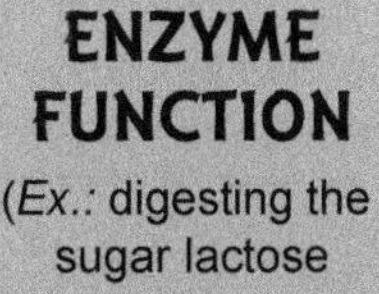

ENZYME FUNCTION
(*Ex.:* digesting the sugar lactose with the enzyme lactase)

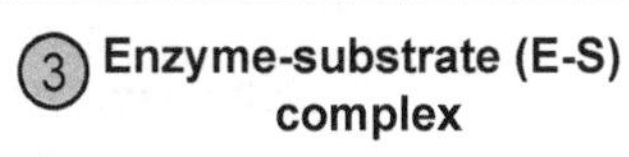

⑤ **Free enzyme**

The lactase enzyme is now free to bind another lactose molecule and repeat the cycle.

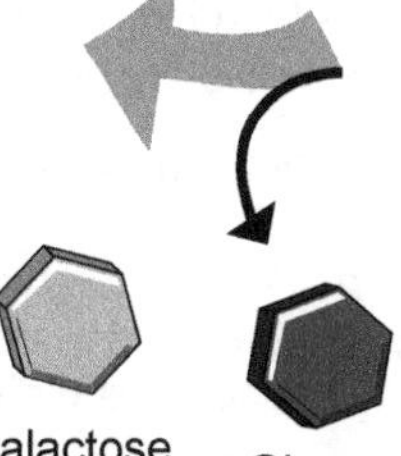

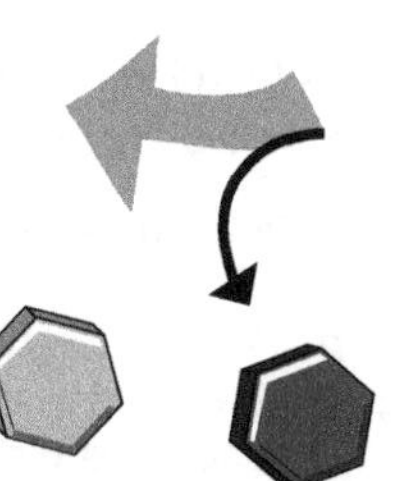

③ **Enzyme-substrate (E-S) complex**

Inside the small intestine, lactose binds to the active site on lactase. Then, with the help of a water molecule, the covalent bond shown is broken, forming the products glucose and galactose.

④ **Products formed**

The products galactose and glucose are released from the enzyme. These sugars are small enough to be absorbed into the blood and delivered to body cells where they can be used as fuel or stored.

Description

When runners talk about "carb loading" or advertisers talk about a "low-carb diet," they are referring to carbohydrates. In simplest terms, carbohydrates are "sugars," or polymers made from sugars, that often have names ending in the suffix *–ose*, such as glucose. They are composed of the elements carbon, hydrogen, and oxygen. From simplest to most complex, carbohydrates are divided into the following three major types:

- **Monosaccharides** (single sugars)
  Ex.: glucose (blood sugar)
- **Disaccharides** (double sugars)
  Ex.: maltose (grain sugar)
- **Polysaccharides** (many sugars)
  Ex.: amylose (plant starch)

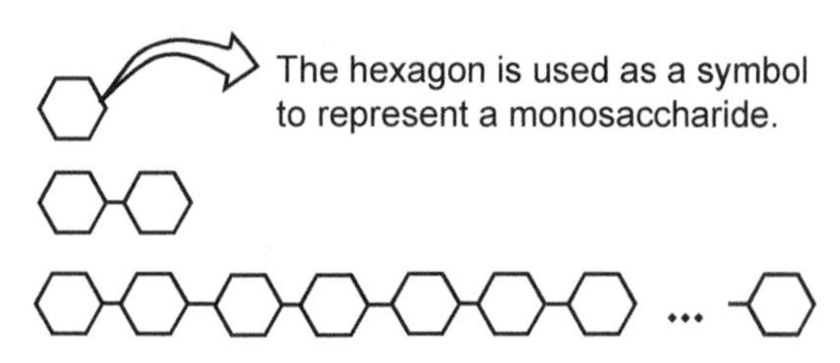

Carbohydrates have several functions vital to all living organisms. As an overview, let's put them into three functional categories: energy, stored energy, and structural components.

Major Functional Types

These three functional categories are presented visually on the facing page.

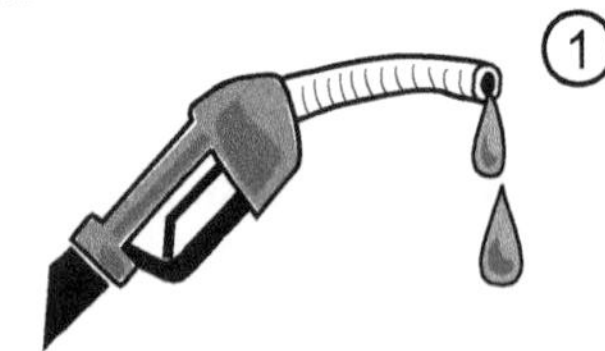

① **Energy:** Like gasoline for your car, monosaccharides are used as energy for body cells.

Ex.:

- **Glucose** ("blood sugar"): The oxidation of the monosaccharide glucose within body cells produces carbon dioxide and water. During this process, free energy is released that can be used to power cellular work such as muscle contraction.

② **Structural components:** Some serve as markers on cell surfaces; others are key parts of molecular structures.

Ex.:

- **Cellulose:** This polysaccharide is found in plants and is composed of long, straight chains of glucose molecules. It makes the cell walls around plant cells rigid.
- **DNA:** The backbone of the DNA double helix is made of deoxyribose sugars and phosphate groups.
- **Antigens for ABO blood groups:** Your blood type (A, B, AB, or O) is determined by proteins on the surface of red blood cells. Attached to these proteins are short-chain carbohydrates.

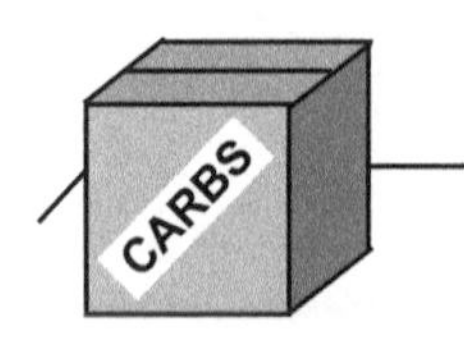

③ **Stored energy:** Carbohydrates are held in reserve like items in a storage box.

Ex.:

- **Glycogen:** This polysaccharide is found in animal tissues (such as liver and muscle cells) and is composed of long chains of glucose molecules. When we ingest excess glucose not immediately used for energy, glycogen is produced.
- **Amylose** (type of starch): This polysaccharide is found in plants and is composed of a long chain of glucose molecules that functions as an important energy storage molecule.

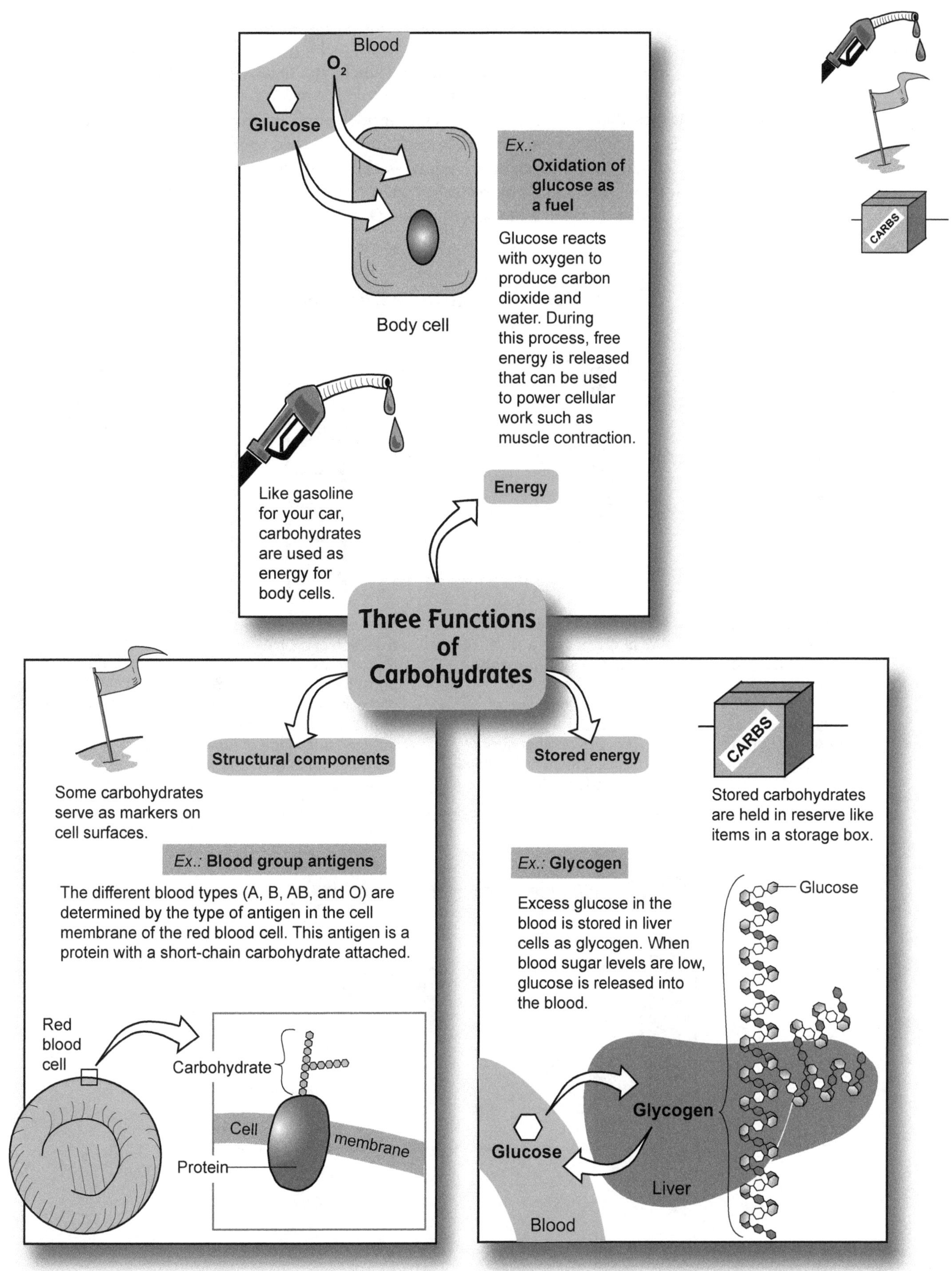
Blood
O_2
Glucose
Ex.:
Oxidation of glucose as a fuel
Glucose reacts with oxygen to produce carbon dioxide and water. During this process, free energy is released that can be used to power cellular work such as muscle contraction.
Body cell
Like gasoline for your car, carbohydrates are used as energy for body cells.
Energy
Three Functions of Carbohydrates
Structural components
Some carbohydrates serve as markers on cell surfaces.
Ex.: Blood group antigens
The different blood types (A, B, AB, and O) are determined by the type of antigen in the cell membrane of the red blood cell. This antigen is a protein with a short-chain carbohydrate attached.
Red blood cell
Carbohydrate
Cell
membrane
Protein
Stored energy
CARBS
Stored carbohydrates are held in reserve like items in a storage box.
Ex.: Glycogen
Glucose
Excess glucose in the blood is stored in liver cells as glycogen. When blood sugar levels are low, glucose is released into the blood.
Glycogen
Glucose
Liver
Blood

Description

Nucleotides are the monomers, or building blocks, for making long-chained nucleic acid polymers such as **DNA** and **RNA.** If you ate a spinach salad for lunch, you ingested DNA from inside the nucleus of the plant cells. In your digestive system, DNA is broken down into nucleotides, which then are transported through the blood to all your cells. Once inside your cells, the free nucleotides float around like raw materials at a construction site. These nucleotides are used to synthesize new DNA in your cells. A DNA nucleotide from a plant or an animal is chemically the same, so the original source doesn't matter. It just reminds you that you really are what you eat! Let's examine in more detail how a DNA nucleotide is used as a building block for making DNA.

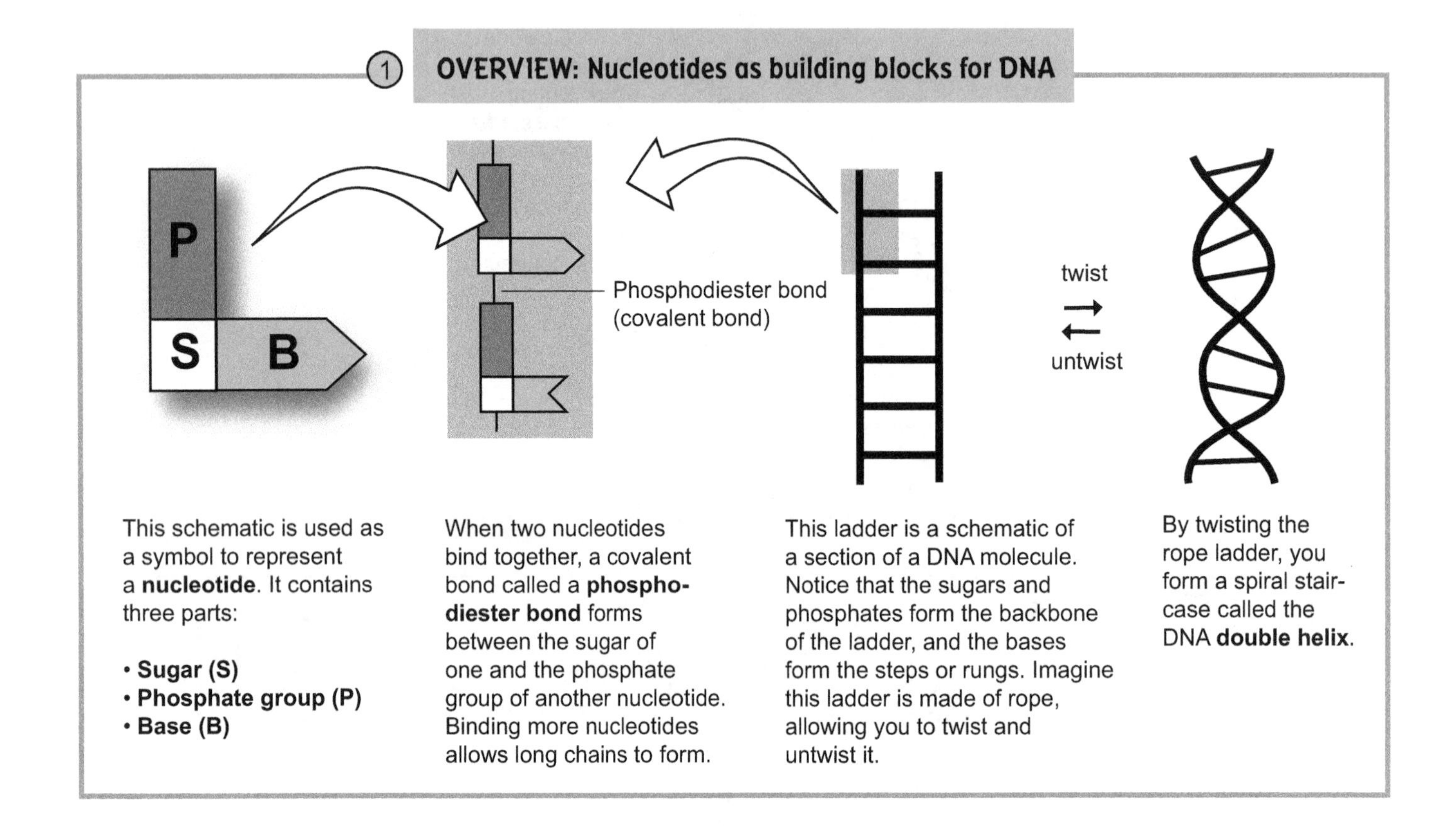

This schematic is used as a symbol to represent a **nucleotide**. It contains three parts:

- **Sugar (S)**
- **Phosphate group (P)**
- **Base (B)**

When two nucleotides bind together, a covalent bond called a **phosphodiester bond** forms between the sugar of one and the phosphate group of another nucleotide. Binding more nucleotides allows long chains to form.

This ladder is a schematic of a section of a DNA molecule. Notice that the sugars and phosphates form the backbone of the ladder, and the bases form the steps or rungs. Imagine this ladder is made of rope, allowing you to twist and untwist it.

By twisting the rope ladder, you form a spiral staircase called the DNA **double helix**.

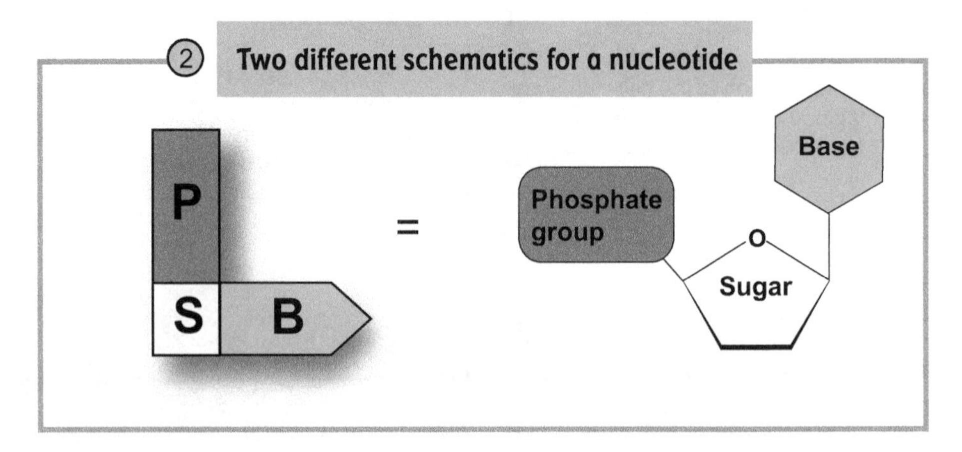

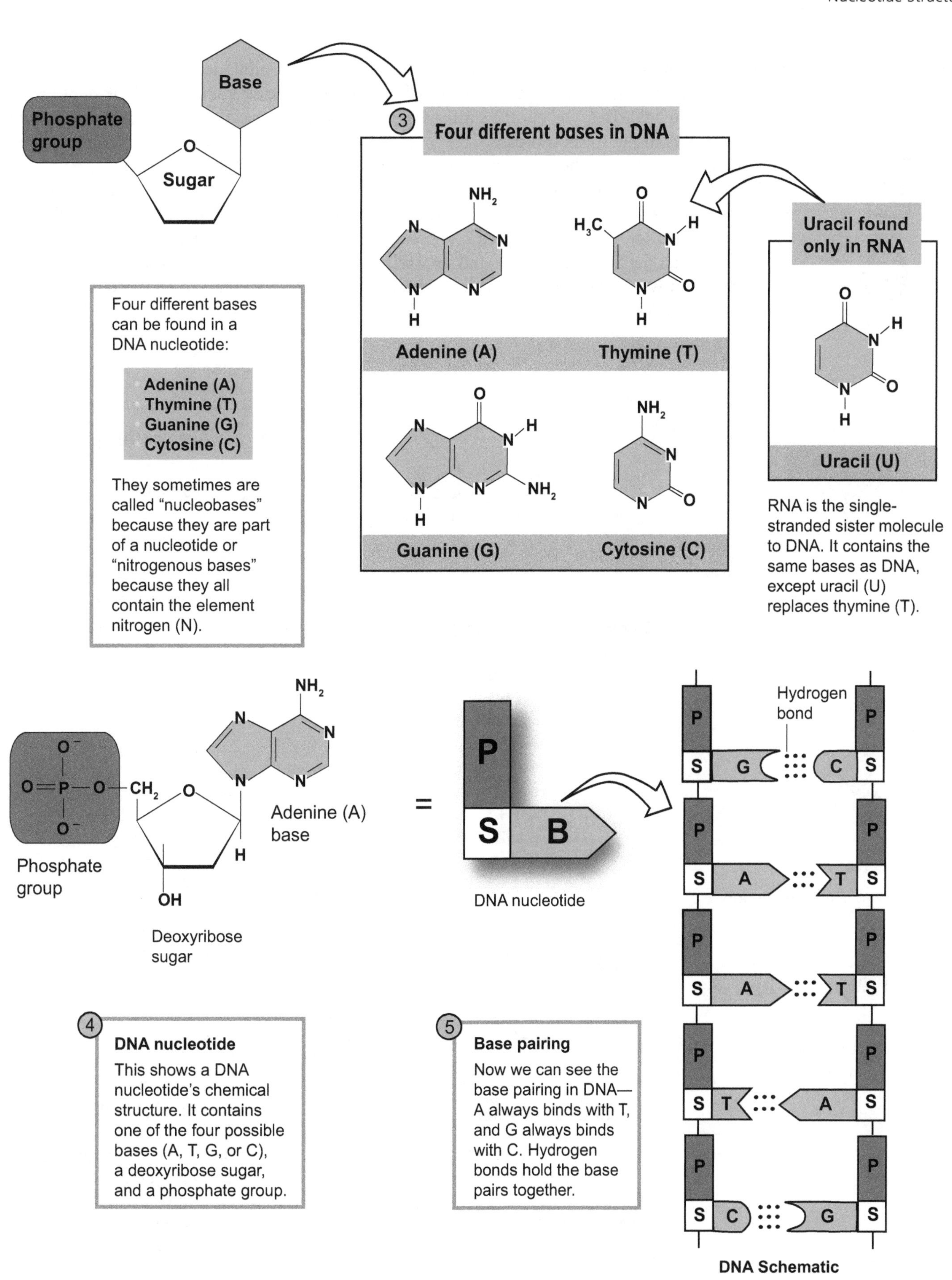
Base
Phosphate group
O
Sugar
3
Four different bases in DNA
NH2
N
N
N
N
H
Adenine (A)
O
H3C
N
H
N
O
H
Thymine (T)
O
N
N
H
N
N
NH2
H
Guanine (G)
NH2
N
N
O
Cytosine (C)
Uracil found only in RNA
O
N
H
N
O
H
Uracil (U)
RNA is the single-stranded sister molecule to DNA. It contains the same bases as DNA, except uracil (U) replaces thymine (T).
Four different bases can be found in a DNA nucleotide:
Adenine (A)
Thymine (T)
Guanine (G)
Cytosine (C)
They sometimes are called "nucleobases" because they are part of a nucleotide or "nitrogenous bases" because they all contain the element nitrogen (N).
O−
O=P—O—CH2
O−
O
H
OH
Phosphate group
Deoxyribose sugar
NH2
Adenine (A) base
=
P
S
B
DNA nucleotide
Hydrogen bond
P
S
G
C
A
T
DNA Schematic
4
DNA nucleotide
This shows a DNA nucleotide's chemical structure. It contains one of the four possible bases (A, T, G, or C), a deoxyribose sugar, and a phosphate group.
5
Base pairing
Now we can see the base pairing in DNA—A always binds with T, and G always binds with C. Hydrogen bonds hold the base pairs together.

Description

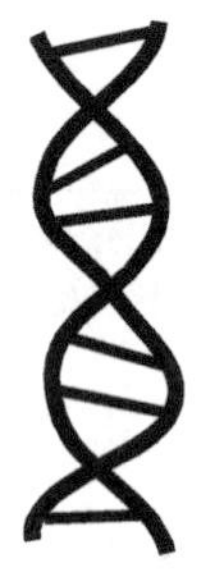

The most common, stable form of DNA—the double helix—resembles a spiral staircase. In terms of its structure, the railing of the staircase is made of a sugar-phosphate backbone, and the steps are made of base pairs. Imagine that you cut the staircase straight down the middle through the steps. Now you have two halves of the spiral staircase. This more accurately reflects the structure of the double helix because the two halves are held together by hydrogen bonds between the base pairs.

DNA is a nucleic acid—a natural polymer made from long chains of nucleotides. It is found in the nucleus of your cells and functions as the master blueprint for making all the different proteins found throughout the body. It is tightly coiled and condensed into structures called chromosomes. Imagine that you took all the DNA in one cell and unraveled it. If you were microscopically small, guess how many steps you would have to take to climb the spiral staircase from end to end? Roughly 3 billion! I'm tired just thinking about it.

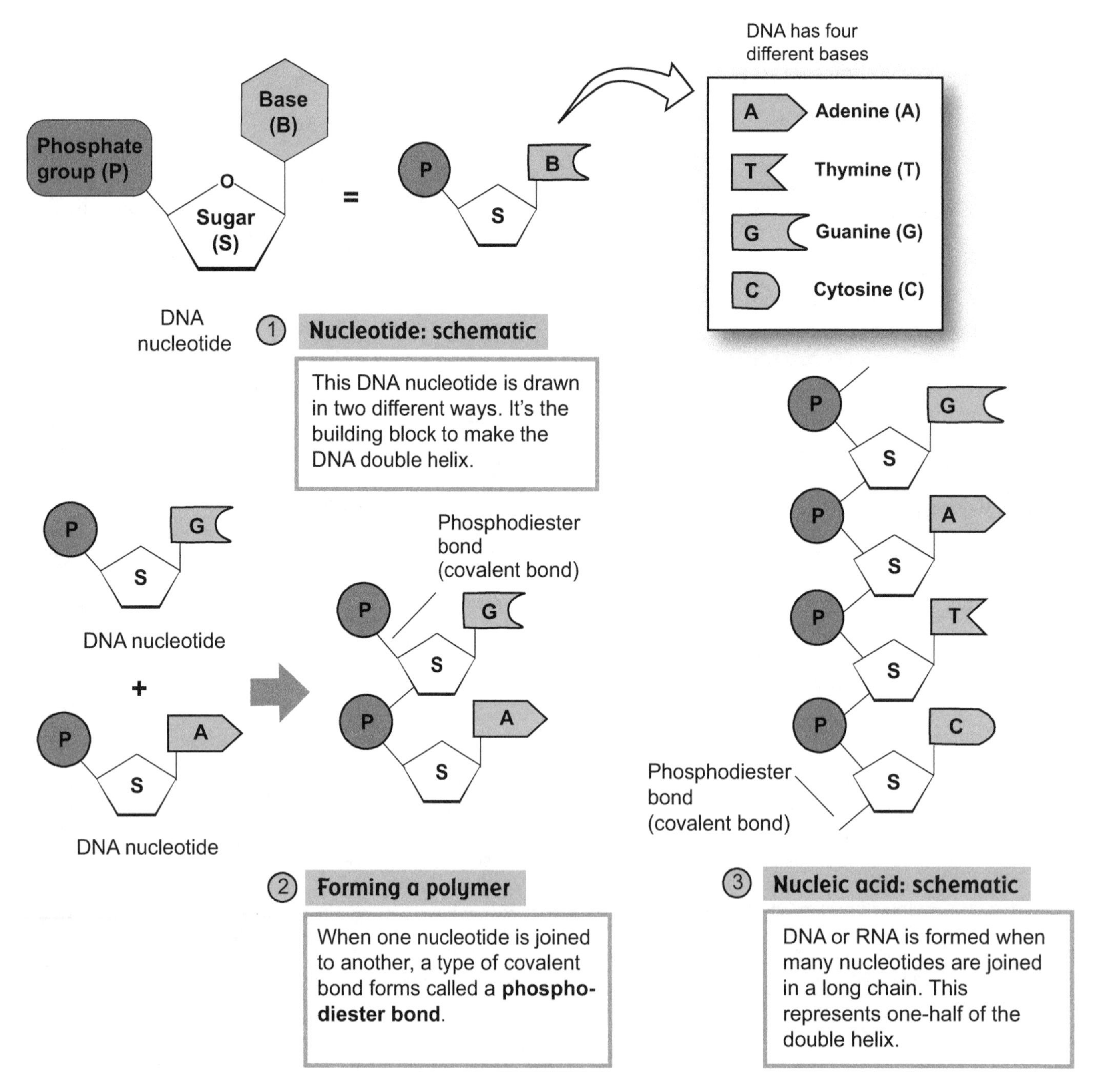

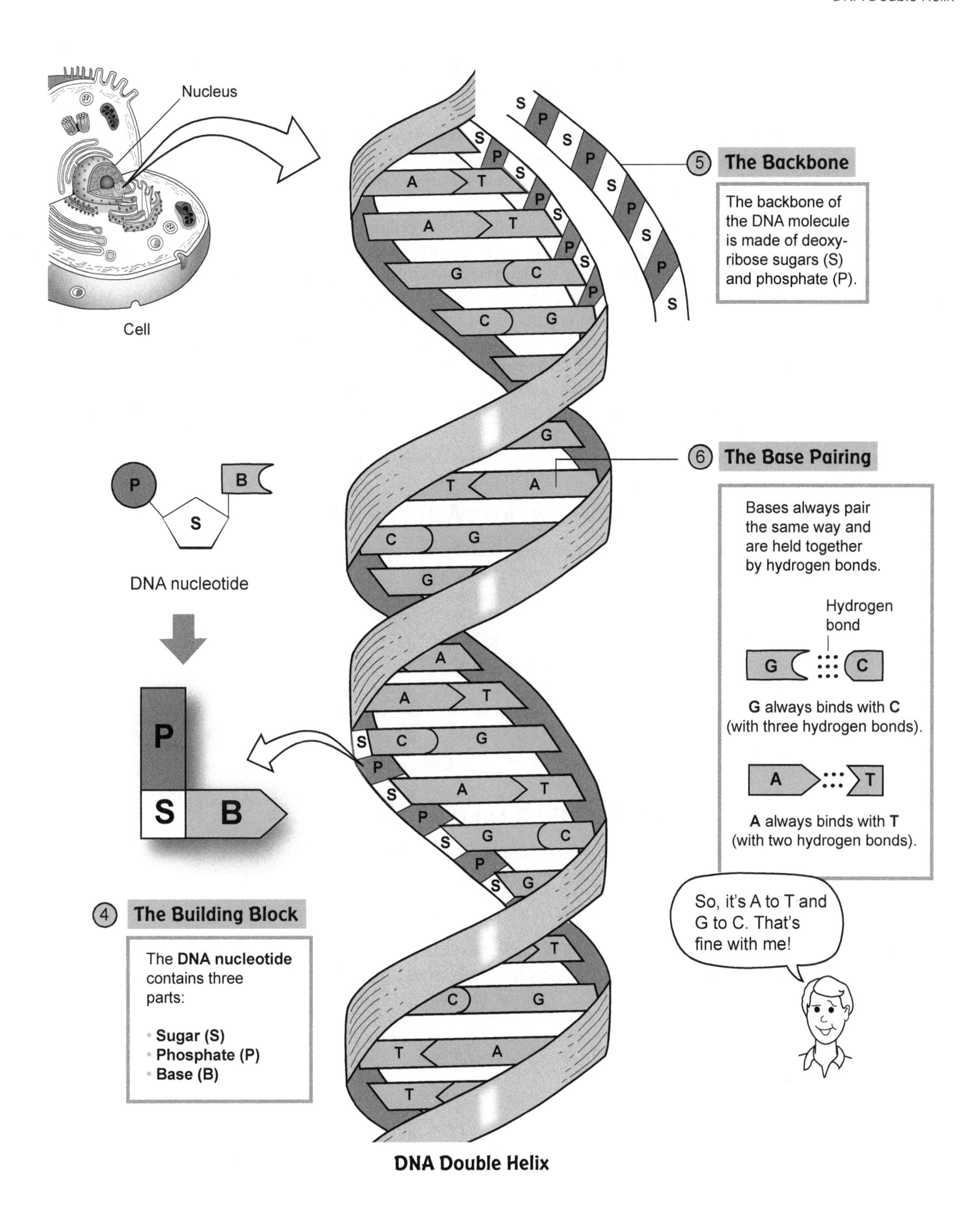

DNA Double Helix

Description

Lipids (*lipos* = fat) consist of a broad category of organic compounds that include waxes, fats and oils, phospholipids, and steroids. They are made of carbon, hydrogen, and oxygen and are not soluble in water. Most lipids, except steroids, contain fatty acid chains. The fatty acid chains are composed of carbon and hydrogen ranging from 10–20 carbons in length. The total number of fatty acid chains in a lipid molecule ranges from one to three, depending on the type of lipid. For example, phospholipids contain two fatty acid chains per molecule, and triglycerides (in fats and oils) contain three fatty acid chains per molecule.

Let's consider some practical examples. Shoe polish is a type of wax. The same is true of waxes for your car and your wood furniture. Waxes also are found in candles, soap, and some cosmetics. Fats and oils contain lipids called triglycerides (triacylglycerols). Most fats we consume are derived from animals, and oils are derived from plants. For example, when cooking, you could choose either lard (fat) or corn oil (oil) to "grease" the frying pan. The healthier choice would be corn oil because it contains unsaturated fatty acids, which are less likely to lead to coronary artery disease. Consuming large amounts of cholesterol is often reported to be bad for your health. But what isn't mentioned is that this steroid is made in small amounts by liver cells and is a vital component in your plasma membranes. To promote good health, your multivitamin contains other lipids—the fat-soluble vitamins (A, D, E, and K). Lipids are common in everyday life.

As an overview, let's compare the major classes of lipids with and without fatty acid chains.

LIPIDS WITH FATTY ACID CHAINS	
Waxes	Found in commercial products such as shoe polish, furniture polish, car wax, candles, soaps, and cosmetics
Fats and oils (contain triglycerides)	Found in commercial products such as lard, corn oil, olive oil, and canola oil
Phospholipids	Found in plasma membranes

LIPIDS WITHOUT FATTY ACID CHAINS	
Steroids	Examples include sex hormones such as progesterone, estrogen, and testosterone

As an introduction to lipids, the facing page illustrates and compares the structure and function of three classes of lipids: triglycerides (triacylglycerols), phospholipids, and steroids.

Comparison of three different classes of lipids: triglycerides, phospholipids, and steroids

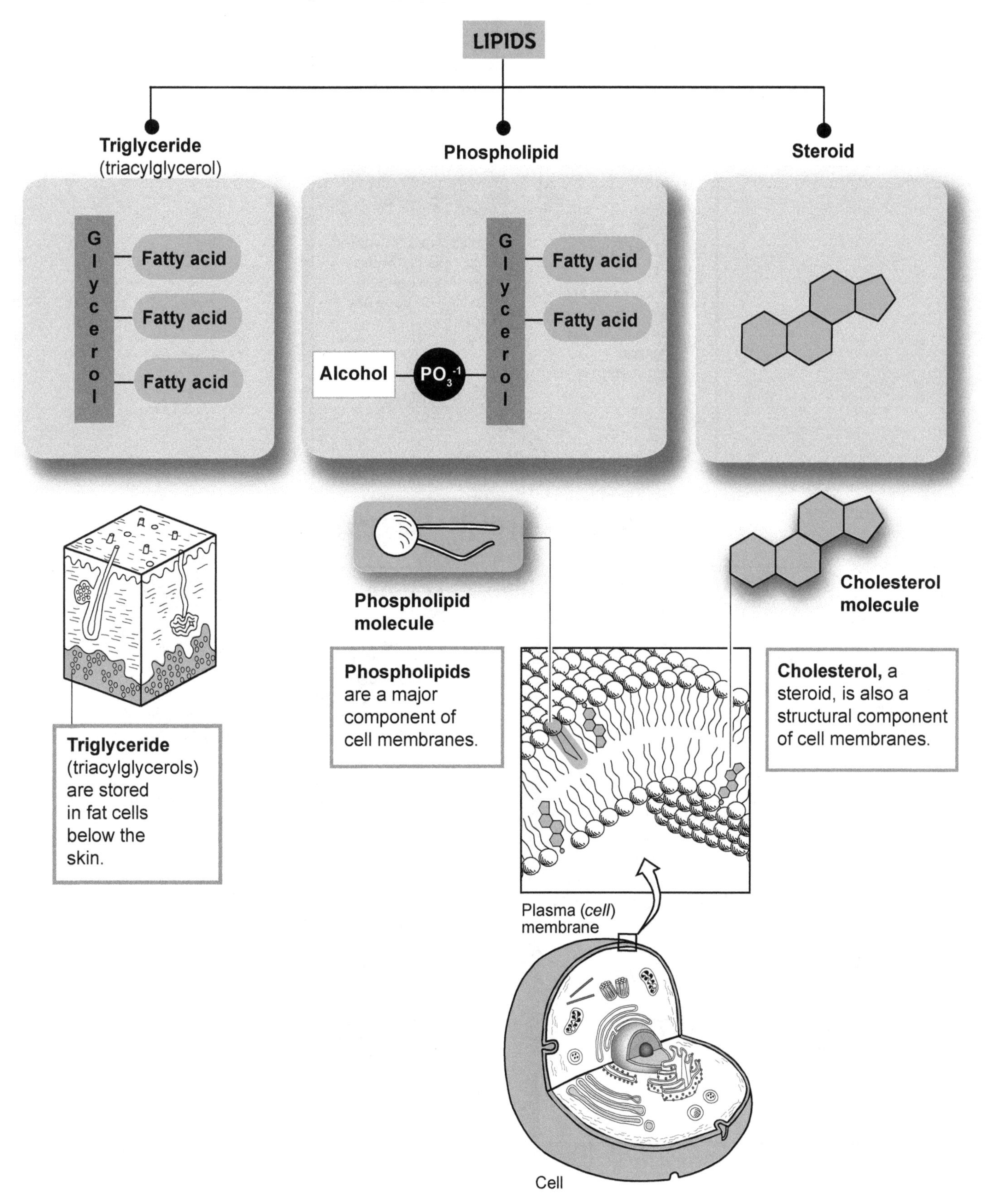

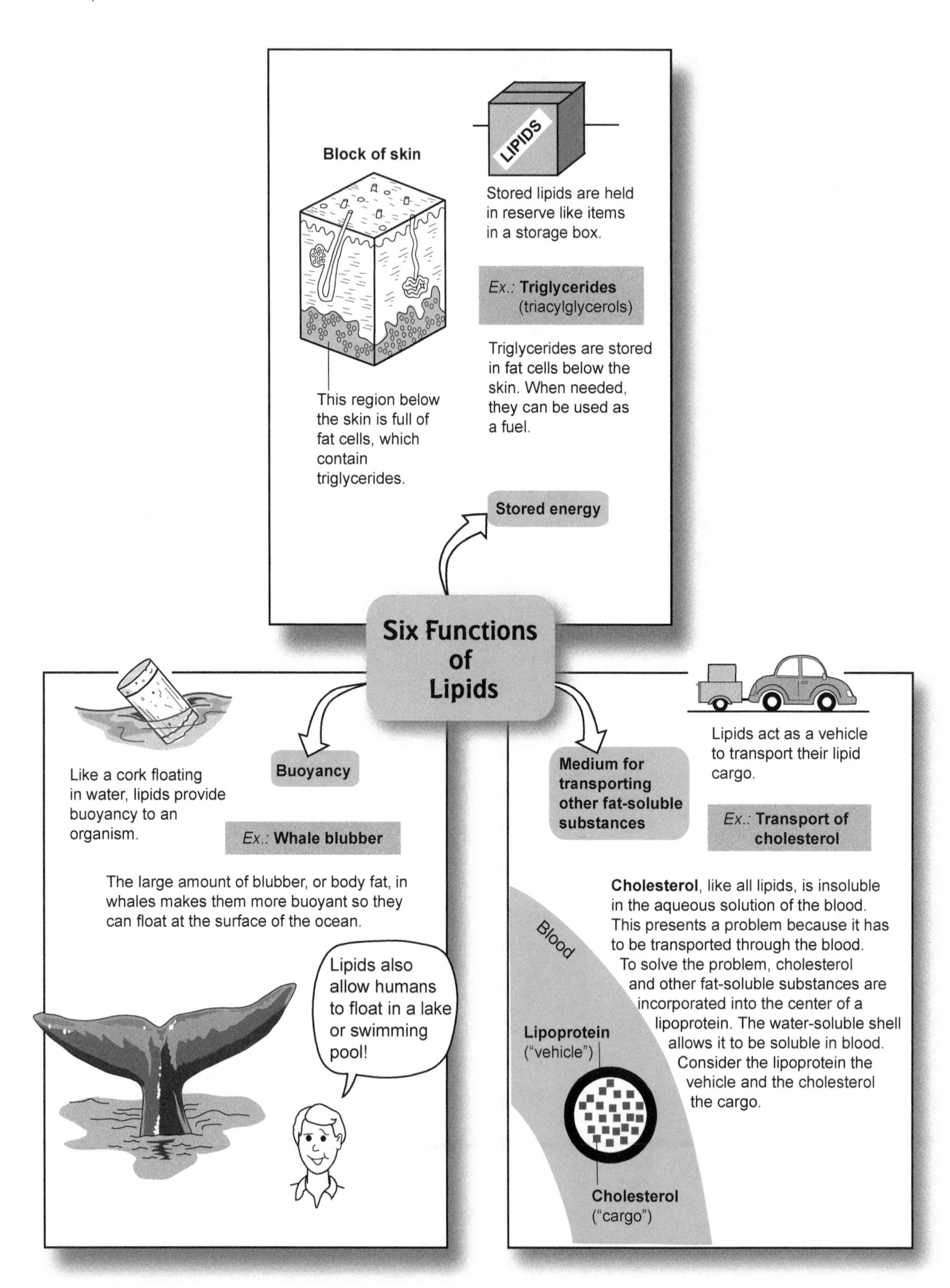
Six Functions of Lipids
Stored energy
Block of skin
LIPIDS
Stored lipids are held in reserve like items in a storage box.
Ex.: Triglycerides (triacylglycerols)
Triglycerides are stored in fat cells below the skin. When needed, they can be used as a fuel.
This region below the skin is full of fat cells, which contain triglycerides.
Buoyancy
Like a cork floating in water, lipids provide buoyancy to an organism.
Ex.: Whale blubber
The large amount of blubber, or body fat, in whales makes them more buoyant so they can float at the surface of the ocean.
Lipids also allow humans to float in a lake or swimming pool!
Medium for transporting other fat-soluble substances
Lipids act as a vehicle to transport their lipid cargo.
Ex.: Transport of cholesterol
Cholesterol, like all lipids, is insoluble in the aqueous solution of the blood. This presents a problem because it has to be transported through the blood. To solve the problem, cholesterol and other fat-soluble substances are incorporated into the center of a lipoprotein. The water-soluble shell allows it to be soluble in blood. Consider the lipoprotein the vehicle and the cholesterol the cargo.
Blood
Lipoprotein ("vehicle")
Cholesterol ("cargo")

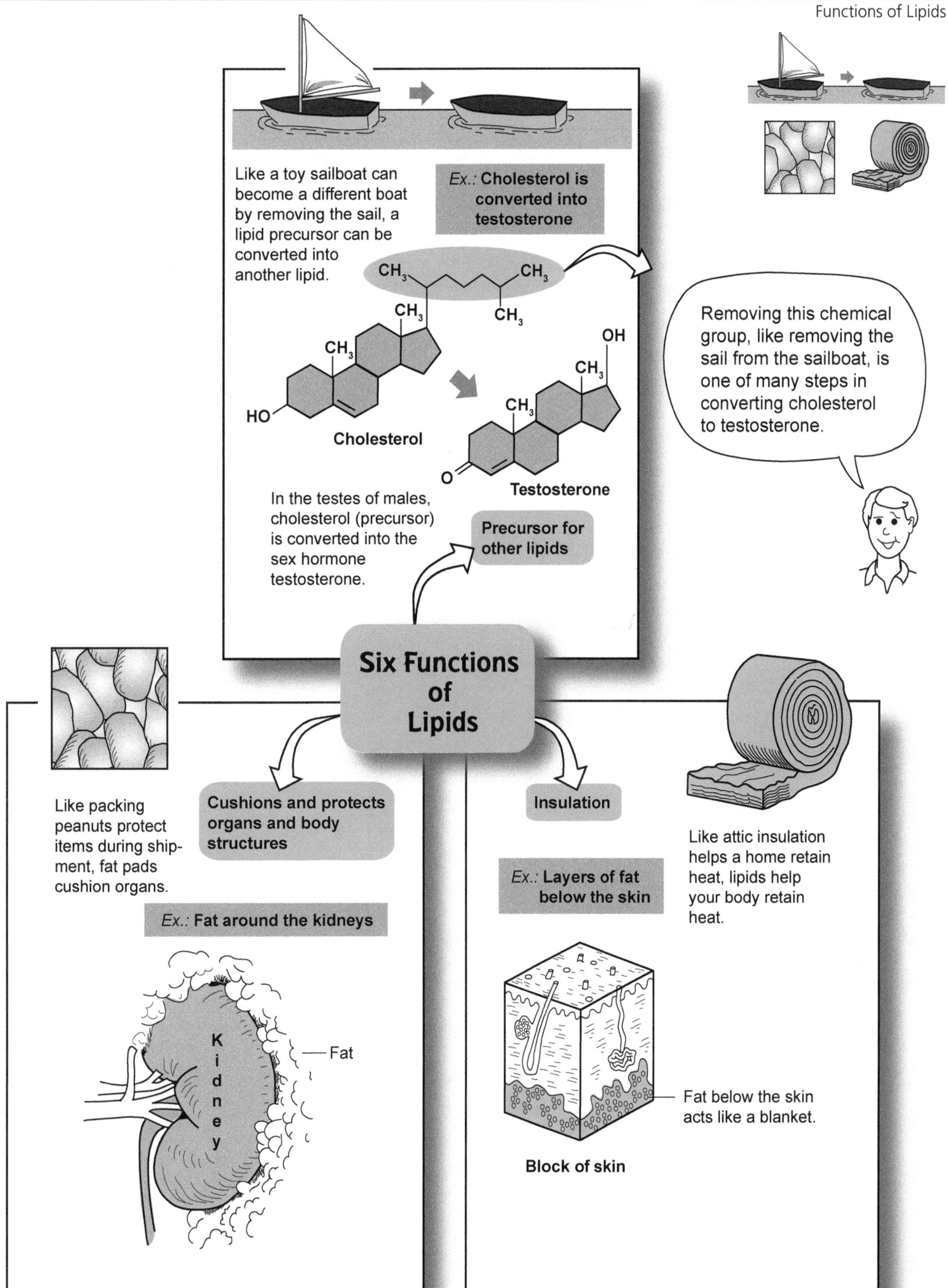
Like a toy sailboat can become a different boat by removing the sail, a lipid precursor can be converted into another lipid.
Ex.: Cholesterol is converted into testosterone
CH3
CH3
CH3
CH3
CH3
HO
Cholesterol
OH
CH3
CH3
O
Testosterone
In the testes of males, cholesterol (precursor) is converted into the sex hormone testosterone.
Precursor for other lipids
Removing this chemical group, like removing the sail from the sailboat, is one of many steps in converting cholesterol to testosterone.
Six Functions of Lipids
Like packing peanuts protect items during shipment, fat pads cushion organs.
Cushions and protects organs and body structures
Ex.: Fat around the kidneys
Kidney
Fat
Insulation
Like attic insulation helps a home retain heat, lipids help your body retain heat.
Ex.: Layers of fat below the skin
Fat below the skin acts like a blanket.
Block of skin

Notes

CELLS

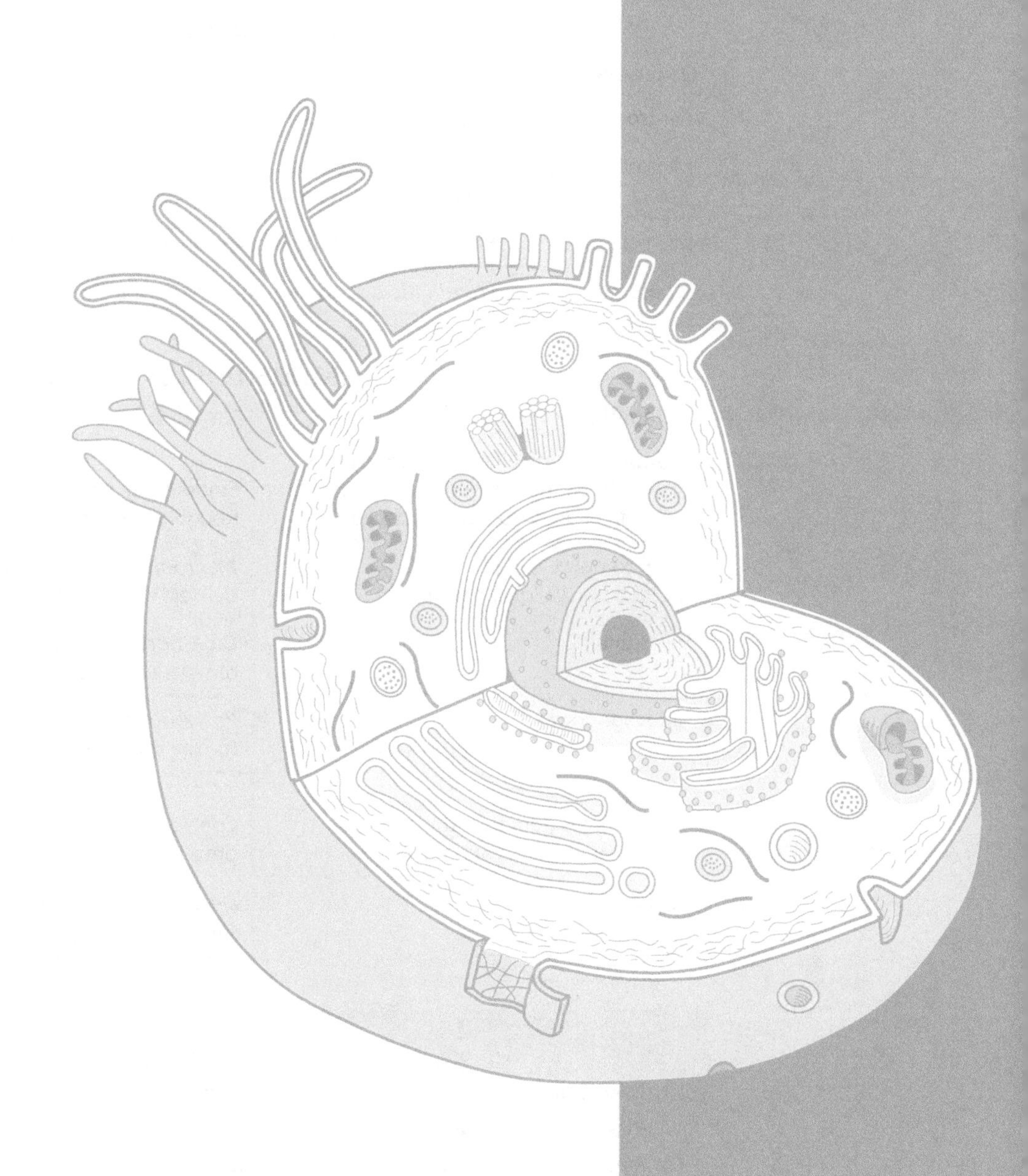

Description

The **cell** is the basic unit of life, and all cells have many common structures. Each cell is surrounded by a **plasma** (*cell*) **membrane**—the *gatekeeper* that only allows certain substances to cross it. In various epithelial cells, there are two types of possible extensions of the plasma membrane—either shorter **microvilli** used to absorb nutrients or much longer **cilia** used to move materials such as mucus over the cell surface. Beneath the plasma membrane is the **cytoskeleton**—a complex network of protein filaments that provide strength and shape to the cell. Many **organelles** (*small organs*) are scattered within the cell, with the largest one being the **nucleus. Membranous organelles** are covered by a membrane, and **nonmembranous organelles** are not. The region between the plasma membrane and the nucleus is called the **cytoplasm,** which is composed of the fluid portion called the **cytosol** and the **organelles.**

NONMEMBRANOUS ORGANELLES

Organelle	Function
Centrosome	Composed of two centrioles that each move to opposite ends of the cell during cell division
Microfilaments	Thin protein fibers commonly found at the periphery of the cell; anchor the cytoskeleton to the plasma membrane
Microtubules	Hollow tubes of protein that can act as tracks along which organelles move
Ribosome	*Protein factory*; composed of two subunits that function in protein synthesis

MEMBRANOUS ORGANELLES

Organelle	Function
Endoplasmic reticulum (ER)—(two types) Smooth ER (has no ribosomes attached) Rough ER (has ribosomes attached)	 Smooth ER: lipid and carbohydrate synthesis Rough ER: modification and packaging of newly made proteins
Golgi complex (Golgi body; Golgi apparatus)	*Mailroom*; receives proteins from the rough ER, then delivers them to the cytosol in a vesicle or secretes them
Lysosome	*Digestion center*; vesicles containing digestive enzymes to remove pathogens and broken organelles
Mitochondrion	*Powerhouse*; site of aerobic cellular respiration; produces ATP for the cell
Nucleus • nucleolus (region within nucleus containing DNA and RNA)	*Control center*; contains DNA—the genetic material for making proteins
Peroxisome	*Detox center*; vesicles containing enzymes to break down substances such as hydrogen peroxide, fatty acids, amino acids; detoxifies toxic substances

Key to Illustration

1. Microvilli
2. Plasma (*cell*) membrane
3. Smooth endoplasmic reticulum
4. Nuclear pores
5. Peroxisome
6. Nucleus
7. Nucleoplasm
8. Nucleolous
9. Ribosomes
10. Rough endoplasmic reticulum
11. Microfilament
12. Mitochondrion
13. Lysosome
14. Vacuole
15. Cytoskeleton
16. Golgi complex
17. Cytosol
18. Cilia
19. Microtubule
20. Centrioles

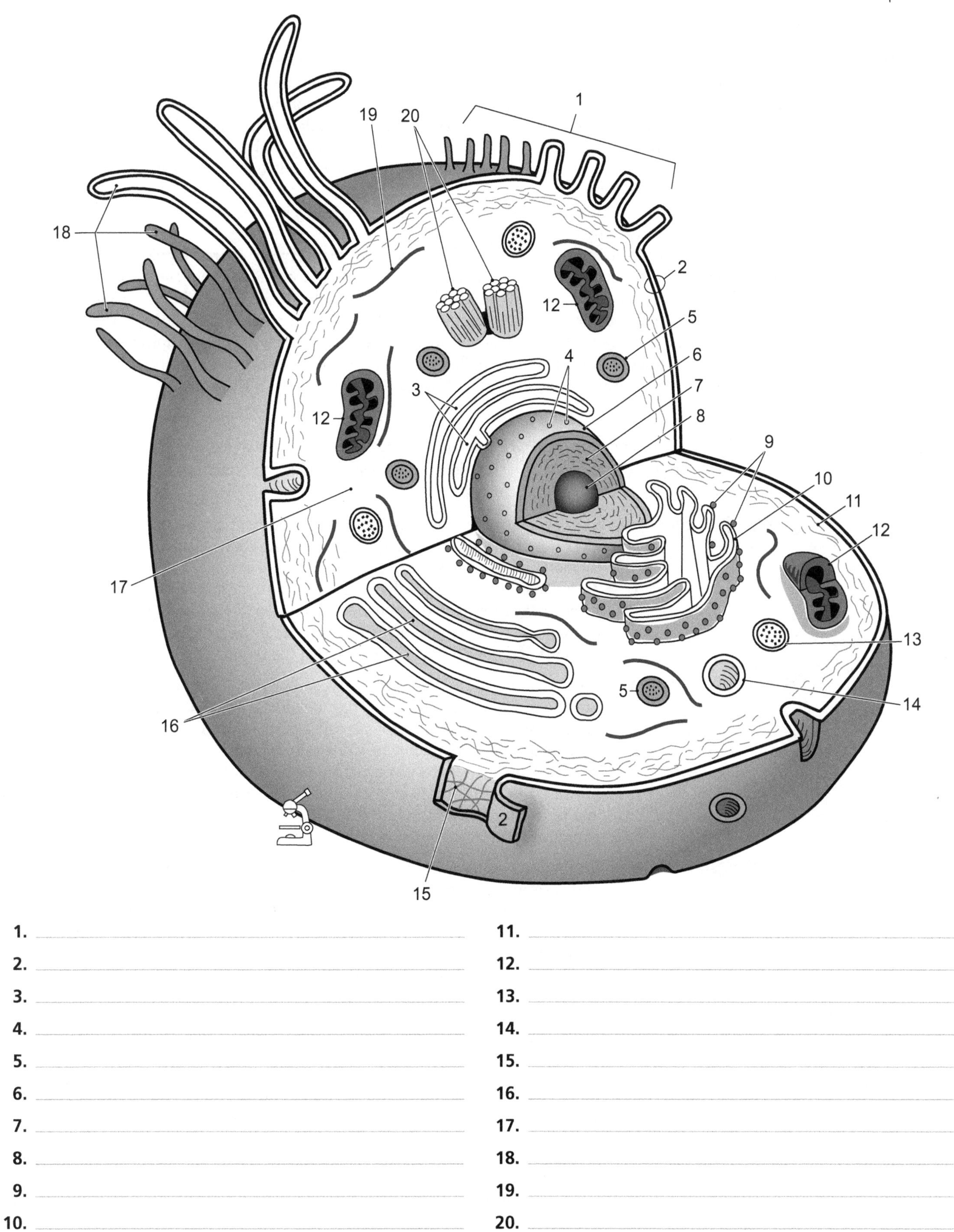

1. ____________________

2. ____________________

3. ____________________

4. ____________________

5. ____________________

6. ____________________

7. ____________________

8. ____________________

9. ____________________

10. ____________________

11. ____________________

12. ____________________

13. ____________________

14. ____________________

15. ____________________

16. ____________________

17. ____________________

18. ____________________

19. ____________________

20. ____________________

Description

Tissues are a group of similar cells working together for a common purpose. The four different types of tissues are: (1) **epithelial**, (2) **connective**, (3) **muscular**, and (4) **nervous**. Let's consider only the first tissue type—**epithelial tissues**. There are many different kinds of epithelial tissues, and they are often named after their cell shape. They are found lining internal body cavities and passageways and covering body surfaces. Epithelial tissues are composed mostly of cells that rest on a thin **basement membrane**. In physiology, epithelial cells are an important theme because substances have to pass through them before they can enter the blood. Every organ system contains epithelial tissues. For example, they line (*are inside of*) the following structures: blood vessels, digestive organs, the urinary bladder, and the microscopic air sacs in the lungs. Moreover, they cover the surface of the skin. In short, they are everywhere in the body.

Each cell in an epithelial tissue has three different surfaces that serve as reference points on the cell. Each surface represents a different side of the cell and is composed of the **plasma membrane**. The top of the cell, which faces the lumen or body surface, is called the **apical surface**. The sides of the cell are called the **lateral surfaces**. The bottom of the cell, which rests on the basement membrane and is closer to the blood, is called the **basal surface**. The typical pathway a substance follows through an epithelial cell is as follows: **apical surface** to **cytosol** to **basal surface** to **basement membrane** to **capillary**. (**Note:** A good way to distinguish the apical surface from the basal surface is to remember the alliteration "B*asal*, B*asement*, B*lood*." These three structures are in close proximity to each other.)

Intercellular Junctions

The lateral surfaces of adjacent epithelial cells are connected one to another by different kinds of junctions. The four major types of junctions are: (1) **tight junctions**, (2) **adhering junctions**, (3) **desmosomes**, and (4) **gap junctions**. Each is summarized below.

① Tight junctions

loop around the whole cell and are located near the apical surface. They look like lines of rivets that form a seal between adjacent cells by stapling their plasma membranes together. Each rivet-like structure is actually a **transmembrane protein**. Tight junctions prevent substances from passing between epithelial cells. This forces substances to move through cells instead of squeezing between them. For example, tight junctions in the digestive tract keep digestive enzymes inside the intestine and prevent them from entering the blood.

② Adhering junctions

function like seam welds. They are typically located below tight junctions, so they are also near the apical surface. They look like a belt that wraps around the whole cell and contain band-like proteins called **plaques**. Running along the length of the plaques are thin, contractile proteins called **microfilaments**. Together, the microfilaments and plaques form a structure called an **adhesion belt**. Adhering junctions prevent lateral surfaces from separating while providing a space for substances to enter. For example, a substance that already passed through the apical surface can enter this space as it continues on to the basal surface.

③ Desmosomes

are like spot welds. They consist of disklike proteins called **plaques** held together and spaced apart by **linker proteins**. Instead of surrounding the entire cell, desmosomes are found at discrete points on the cell. They function as structural reinforcements at specific stress points along the cell. **Intermediate filaments** are long, strong cable-like proteins that extend into the cytosol and connect the plaque of one desmosome to the plaque of another desmosome on the opposite side of the cell. This helps maintain the structural integrity of the cell and the tissue. Desmosomes are common in the cells within the epidermis of the skin.

Half-desmosomes or **hemidesmosomes** are located on the basal surface, where they serve to anchor this surface to the basement membrane.

④ Gap junctions

allow for small substances to be transported between adjacent cells. Each gap junction consists of a small group of tubular structures. Each tubular structure is composed of a cluster of 6 proteins called **connexins** and has a fluid-filled **channel** running down its center. This channel allows small substances such as ions and glucose to travel between adjacent cells. In ciliated epithelial cells, this may help to coordinate the movement of the cilia. Gap junctions are also found in other tissues such as cardiac and smooth muscle.

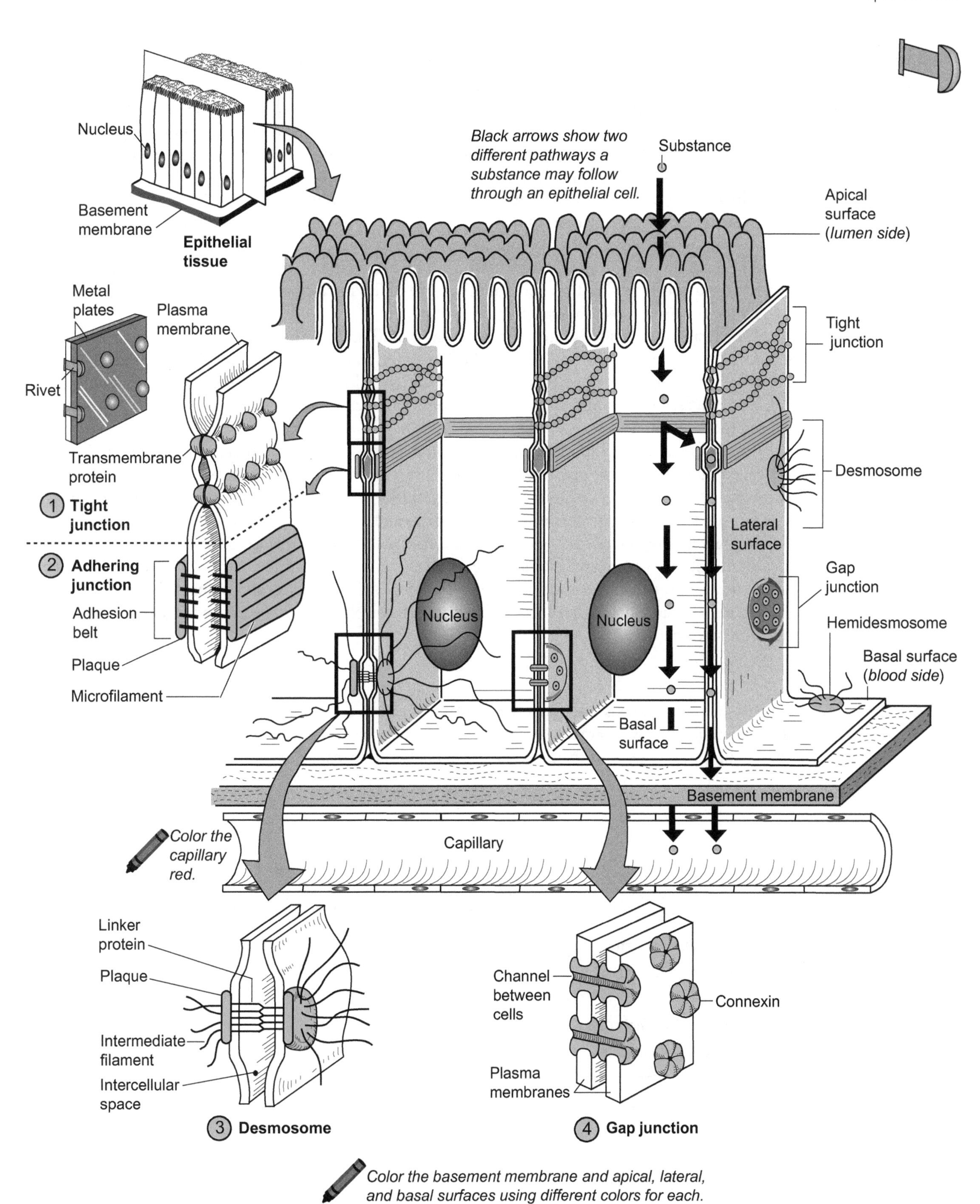
Nucleus
Basement membrane
Epithelial tissue
Black arrows show two different pathways a substance may follow through an epithelial cell.
Substance
Apical surface (lumen side)
Metal plates
Plasma membrane
Rivet
Tight junction
Transmembrane protein
1 Tight junction
2 Adhering junction
Adhesion belt
Plaque
Microfilament
Desmosome
Lateral surface
Gap junction
Hemidesmosome
Basal surface (blood side)
Nucleus
Nucleus
Basal surface
Basement membrane
Capillary
Color the capillary red.
Linker protein
Plaque
Intermediate filament
Intercellular space
3 Desmosome
Channel between cells
Connexin
Plasma membranes
4 Gap junction
Color the basement membrane and apical, lateral, and basal surfaces using different colors for each.

Description

Every cell in your body is enclosed by an envelope called a **plasma** (*cell*) **membrane.** It is the gateway through which substances enter or exit any cell. A plasma membrane is a double layer of lipids referred to as the **phospholipid bilayer** or, more simply, the **lipid bilayer.** The **phospholipid molecule** is the repeating unit within the cell membrane. Each molecule has a **polar head** and a **nonpolar tail** made of two fatty acid chains—typically one chain is saturated and straight, and the other is unsaturated and bent. The polar head is hydrophilic (*water loving*), so it is attracted to water molecules, and the nonpolar tail is hydrophobic (*water fearing*), so it is not attracted to water molecules. As a result, the nonpolar tails position themselves away from water in the middle of the membrane, and the polar heads position themselves toward polar water molecules on the outer and inner surfaces of the plasma membrane. Some phospholipids have a carbohydrate chain attached to them and are called **glycolipids.** Embedded and scattered within the lipid bilayer is another type of lipid—**cholesterol molecules.** They provide some rigidity for the plasma membrane.

The plasma membrane is often referred to as a fluid mosaic model: *fluid* because the phospholipid molecules move from side to side due to their bent, unsaturated fatty acid chains, and *mosaic* because the membrane is made of more than one type of macromolecule (lipids, proteins, and carbohydrates). If you were microscopically small and could fall on its surface, it would feel more like a water bed than a hard wooden floor.

Proteins are another important part of the plasma membrane. Those that span the entire lipid bilayer are called **integral proteins** and serve a variety of functions. Some act as *channels* through which only particular types of ions pass. Others act as *receptors* for hormones or neurotransmitters. Some integral proteins have carbohydrate chains attached to them and are called **glycoproteins.**

Peripheral proteins do not span the bilayer. Instead, they only connect to one surface of the membrane. Some act as enzymes, and others may serve as a structural component of the cytoskeleton. The **cytoskeleton** is composed of different types of protein filaments and is located beneath the lipid bilayer. It serves as a kind of scaffolding that supports the plasma membrane and gives the cell a defined structure and shape.

Function

The plasma membrane is a selectively permeable membrane, meaning that not all substances can pass through it. It controls what can enter and exit the cell based on factors such as size (molecular weight), charge, lipid solubility, and the presence of channels and transporters.

Key to Illustration

Plasma (*Cell*) Membrane

1. Polysaccharide chain
2. Glycoprotein
3. Phospholipid monolayer
4. Cholesterol molecule
5. Filaments of cytoskeleton
6. Peripheral protein
7. Integral protein
8. Glycolipid
9. Phospholipid bilayer

Phospholipid Molecule

10. Headgroup of phospholipid
11. Tailgroup of phospholipid

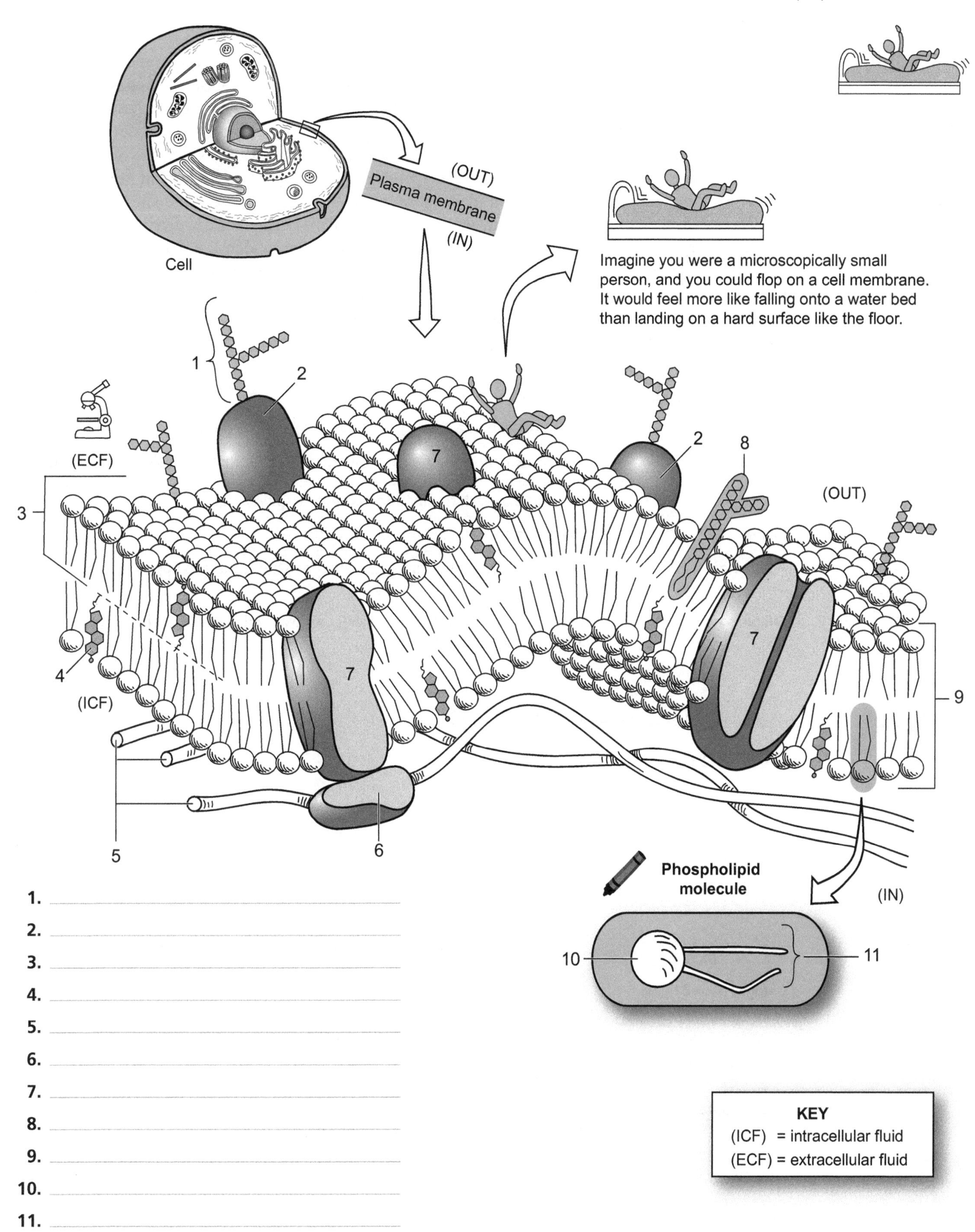

1. ______________________

2. ______________________

3. ______________________

4. ______________________

5. ______________________

6. ______________________

7. ______________________

8. ______________________

9. ______________________

10. ______________________

11. ______________________

Description

The life cycle of a cell, called the **cell cycle**, is divided into two major parts: interphase and mitosis. Interphase is subdivided into G_1, S, and G_2, and mitosis is subdivided into prophase, metaphase, anaphase, and telophase. The duration of the cell cycle varies widely from one cell type to another. Assuming a total cell cycle time of 24 hours, interphase may last roughly 22 hours and mitosis about 2 hours.

Interphase involves lots of activity. First, during G_1 (Gap 1), the longest phase, the cell is growing and synthesizing proteins. Next, in the S phase (synthesis), DNA replication occurs in preparation for cell division, so that each new cell has an identical set of DNA in its chromosomes. The final phase, G_2 (Gap 2) is relatively short in duration. Enzymes and other proteins needed for cell division are produced.

After interphase cells prepare to undergo cell division to form two identical daughter cells. Mitosis specifically refers to the duplication, separation, and reorganization of chromosomes into two identical nuclei. During mitosis a related process called cytokinesis occurs in which the cytoplasm is divided. Cytokinesis can begin in late anaphase, continues through telophase, and ends after mitosis is completed.

A few key events to identify each stage of the cell cycle are listed in the table below:

Phase of Cell Cycle	Key Event
Interphase	Nucleus remains intact; cell growth; DNA replication; protein synthesis
Prophase	Nuclear membrane disappears; duplicated chromosomes called **sister chromatids** become visible; centrioles move to opposite ends of the cell; **spindle fibers** appear; **kinetochore** of each chromatid attaches to a spindle fiber
Metaphase	Sister chromatids align at the middle of the cell
Anaphase	Spindle fibers pull sister chromatids apart and move them to opposite poles of the cell; early cleavage furrow forms
Telophase	Cleavage furrow deepens; cell pinches into two and divides cytoplasm (**cytokinesis**); nuclear membranes re-form

Study Tips

To recall the phases of the cell cycle, use this mnemonic:

IPMAT = "I Passed My Anatomy Test"

Key to Illustration

Stages of Cell Cycle
1. Interphase
2. Prophase
3. Metaphase
4. Anaphase
5. Telophase (*and cytokinesis*)

Final Products
6. Daughter cells

Significant Structures
a. Plasma (*cell*) membrane
b. Nucleolus
c. Nucleus
d. Centriole
e. Sister chromatids
f. Kinetochore
g. Spindle fibers
h. Cleavage furrow
i. Nuclear membrane (*still forming*)

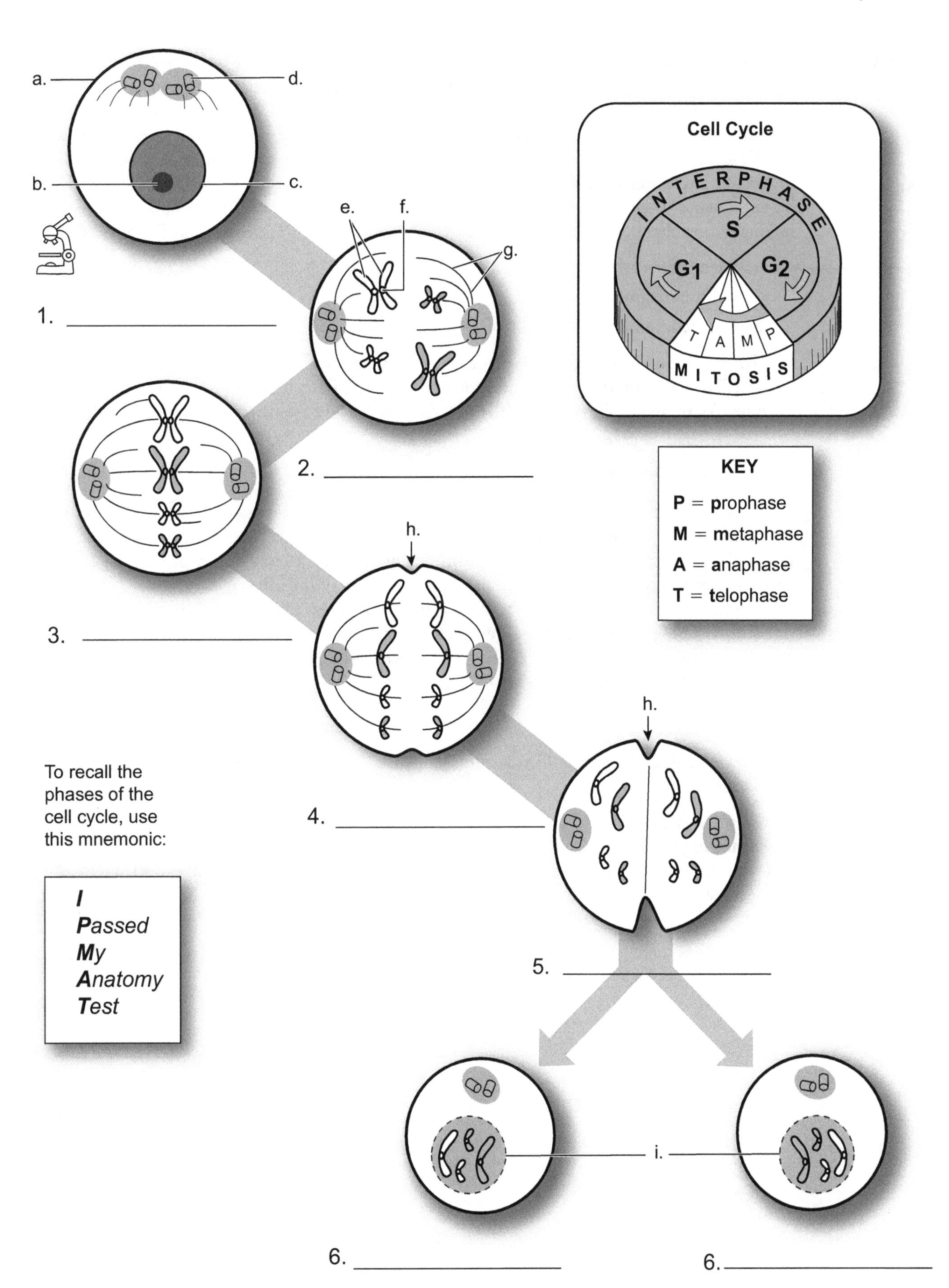
a.
d.
b.
c.
e.
f.
g.
1. ____________
2. ____________
h.
3. ____________
h.
4. ____________
5. ____________
i.
6. ____________
6. ____________
Cell Cycle
INTERPHASE
S
G1
G2
T
A
M
P
MITOSIS
KEY
P = prophase
M = metaphase
A = anaphase
T = telophase
To recall the phases of the cell cycle, use this mnemonic:
I
Passed
My
Anatomy
Test

Description

Because DNA (deoxyribonucleic acid) is THE genetic material, it has to be replicated in preparation for normal cell division so that a copy of it can be placed in each new daughter cell (see p. 58). This replication occurs during the S phase of the cell cycle. Structurally, DNA is a complex macromolecule located in the cell nucleus in a double helix form resembling an extremely tall spiral staircase.

To discuss its structure, we will simplify it. Imagine that the spiral staircase is flexible so it can be untwisted into a tall ladder. Then cut the ladder down the middle of the rungs so it is in two equal halves. Each half-ladder is bonded to the other by hydrogen bonds. Each half-ladder is created by linking a long chain of building blocks called **nucleotides.** Each nucleotide has three parts:

1. **Sugar** (five-carbon)
2. **Phosphate group**
3. **Base** (containing nitrogen).

The backbone of the ladder is made of alternating sugars and phosphate groups bonded to one another. The rungs on the ladder are made of base pairs. Each base can be one of four different types: adenine (A), thymine (T), guanine (G), or cytosine (C). The base pairing is as follows:

- **A** bonds to **T** (or **T** to **A**)
- **G** bonds to **C** (or **C** to **G**)

This base pairing is the basis for the genetic code. This message directs the creation of proteins by specifying the sequence of amino acids to be joined to make that protein. The type of proteins produced determines the type of organism, such as a goldfish, maple tree, or human. These proteins also account for individual differences between members of the same species.

Replication occurs with the help of specialized enzymes called **DNA polymerases.** They attach to one section of DNA, and it unzips. Free nucleotides within the cytoplasmic soup are paired with their proper bonding partners (A to T, G to C) to synthesize a new strand. Each new DNA molecule contains one strand from the parent DNA molecule.

Analogies

- The DNA **double helix** is like a spiral **staircase.** The **DNA molecule** (simplified version) is like a **ladder.**
- When DNA prepares to replicate itself, it is said to **UNZIP.** This is an accurate description. When you **unzip** your coat, each **half of the zipper** is like one-half of the DNA molecule, called the **DNA template,** used to synthesize a new strand. As the new strand is synthesized, DNA **ZIPS** back together again.

Location

DNA is found in the nucleus of all body cells.

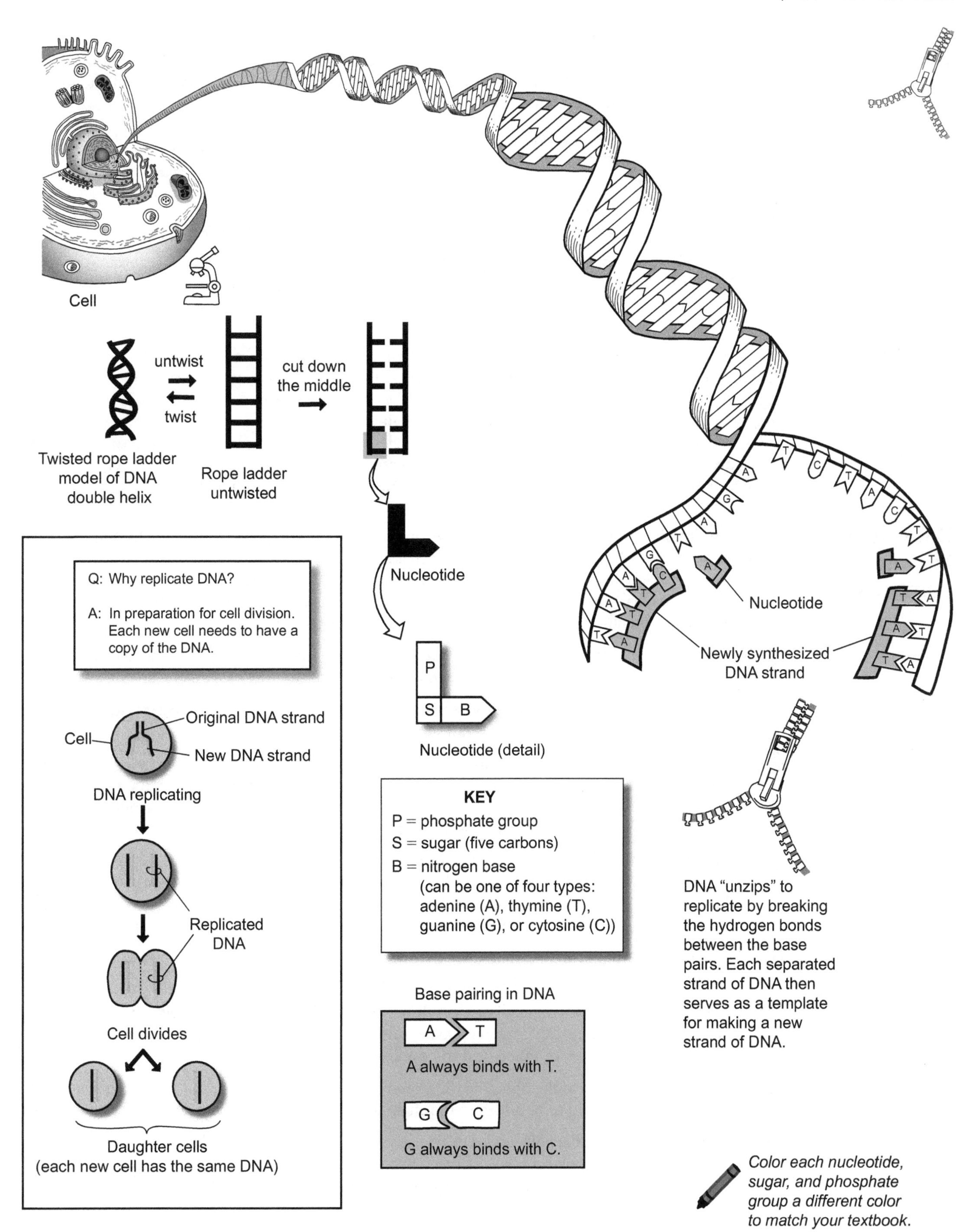

Color each nucleotide, sugar, and phosphate group a different color to match your textbook.

Description

Protein synthesis is a major function in most cells. It requires that DNA first be transcribed into a similar molecule called messenger RNA, or mRNA. Then the mRNA must be translated from the nucleotide "language" of RNA into the amino acid "language" of proteins.

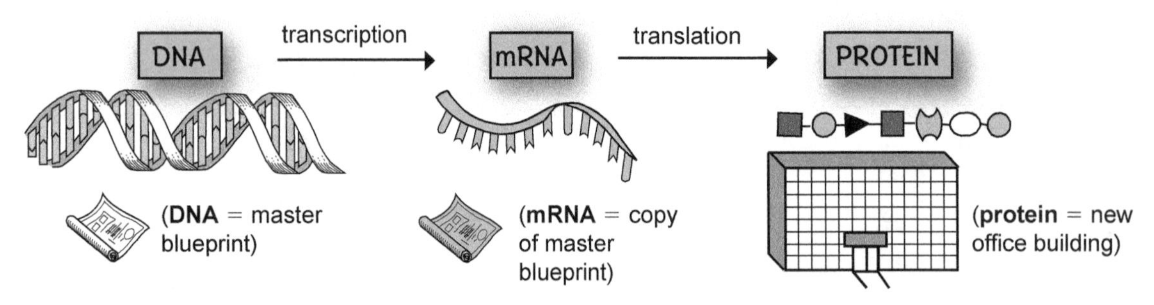

Analogy

The process of protein synthesis in a cell is like constructing a new office building. The cytoplasm is like the construction site, and the DNA is like the master blueprint. The master blueprint for the office building is stored in the trailer on the construction site, just as DNA is stored in the nucleus (because it is too large to leave). Just as copies can be made of the original blueprint, a copy of DNA is made in the form of mRNA. The mRNA is small enough to leave the nucleus, enter the cytoplasm, and bind to the ribosomes, which are like the construction workers. The raw materials for the office building are steel, concrete, and glass, to name a few. To build a protein, the needed raw materials are amino acids.

Basic Steps in Protein Synthesis

① **Transcription:** The enzyme RNA polymerase travels along an "unzipped" section of DNA for the purpose of making mRNA. The building blocks for making mRNA are called RNA nucleotides. The enzyme uses one side of the DNA molecule as a template to match the base of the RNA nucleotide to the corresponding base in DNA. As the RNA nucleotides are lined up next to each other, they are covalently bonded to eventually form a short, single-stranded mRNA molecule.

② **RNA Processing:** Like old film being spliced together during editing, the single-stranded mRNA molecule also is processed. Small segments called introns are removed from the molecule.

③ The single-stranded mRNA is small enough to leave the nucleus through the nuclear pore and move into the cytoplasm, where it binds to the small subunit of a ribosome.

④ A transfer RNA (tRNA) molecule has an amino acid binding site section at one end and a three-base section at the other end called the anticodon. Just as a taxi transports passengers, the tRNA transports the proper amino acid to the ribosome. With the help of an enzyme and free energy from ATP hydrolysis, one specific amino acid is bonded to one specific tRNA molecule based on its anticodon sequence.

⑤ The mRNA message is read in groups of three bases. Each triplet group is called a codon. The tRNA with the amino acid called methionine always binds to the beginning of the message called the start codon. Next, the large ribosomal subunit binds to the small ribosomal subunit to activate the ribosome. The large subunit contains two binding sites for tRNAs called the P site and the A site.

⑥ The initial tRNA binds to the P site. One by one, amino acids are positioned next to each other, then bonded together. This results in a growing polypeptide chain. Here is how it works:

⑦ A tRNA with the anticodon matching the mRNA codon arrives at the A site with the next amino acid to be added to the chain. The growing polypeptide chain at the P site is transferred to the tRNA at the A site as it binds to the next amino acid to be added to the chain.

⑧ As the ribosome moves to the right, the tRNA moves from the A site over to the P site. Then the tRNA with the next amino acid to be added to the chain arrives at the A site, and the process repeats itself.

⑨ The process of growing the polypeptide chain ends when a special codon called the stop codon is reached at the A site. The final polypeptide is released from the ribosome along with the last tRNA. Finally, the ribosome separates into its large and small subunits. This marks the end of protein synthesis.

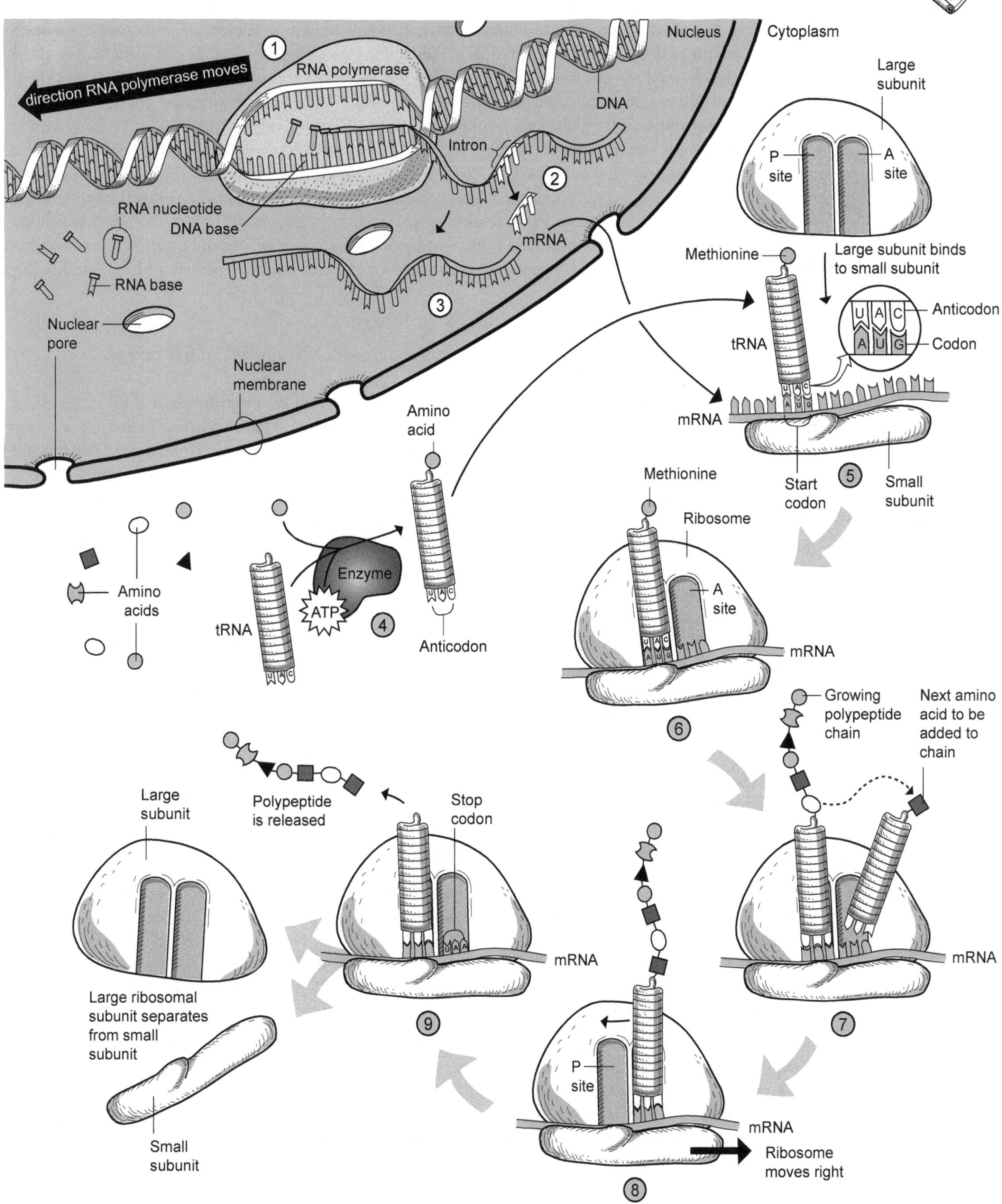
1
direction RNA polymerase moves
RNA polymerase
Nucleus
Cytoplasm
DNA
Intron
2
RNA nucleotide
DNA base
mRNA
RNA base
3
Nuclear pore
Nuclear membrane
Large subunit
P site
A site
Methionine
Large subunit binds to small subunit
Anticodon
Codon
tRNA
mRNA
Start codon
5
Small subunit
Amino acid
Amino acids
Enzyme
ATP
4
tRNA
Anticodon
Methionine
Ribosome
A site
mRNA
6
Growing polypeptide chain
Next amino acid to be added to chain
mRNA
7
Polypeptide is released
Stop codon
Large subunit
mRNA
9
Large ribosomal subunit separates from small subunit
Small subunit
P site
mRNA
Ribosome moves right
8

Description

Adenosine triphosphate (**ATP**) is the universal energy currency for all cells. Look at the illustration of ATP on the facing page to understand its structure. It is classified as a nucleotide and consists of three parts: (1) **adenine base**, (2) **ribose sugar**, and (3) **phosphate groups**—three total. The term "adenosine" refers to the adenine base and ribose sugar bonded together, and the term "triphosphate" comes from the three phosphate groups covalently bonded to each other. There is a net negative charge on each of the phosphate groups, so they repel each other and make ATP relatively unstable.

Cells are highly organized structures constantly doing work to maintain their physical structure and carry out their general functions. This cycle of regular work requires energy. In biological systems, it is common to couple a spontaneous reaction with a nonspontaneous reaction. A metal rusting when exposed to moist air is an example of a spontaneous reaction (it occurs all by itself). A muscle cell attempting to contract is an example of a nonspontaneous reaction (it occurs only with the input of additional energy). ATP hydrolysis occurs as a spontaneous reaction and is represented by the following chemical equation:

$$ATP + H_2O \xrightarrow{\text{ATPase}} ADP + P_i \text{ (inorganic phosphate group)} + \textbf{free energy}$$

This reaction requires the assistance of an enzyme called an **ATPase** that binds an ATP molecule to itself with a shape-specific fit like a lock and key. In any hydrolysis ("water-splitting") reaction, a water molecule is used to cleave a single covalent bond. In this case, it cleaves the covalent bond linking the terminal phosphate group to the second phosphate group. During this process, water is split into a **hydroxyl group** (OH^-) and a **hydrogen ion** (H^+). The hydroxyl group binds to the phosphorus atom (P) in the terminal phosphate group, and the hydrogen binds to the oxygen atom on the second phosphate group (see illustration). As a result, a more stable molecule called **adenosine diphosphate** (**ADP**) is formed as a result of the decreased repulsion between the phosphate groups. The free phosphate group is often transferred to another substrate or to an enzyme. In addition, some free energy is released in the process. The free energy is used to drive processes that must occur in cells, such as muscle contraction, active transport of substances across cell membranes, movement of a sperm cell's tail, manufacturing a hormone, or anything else a cell has to do.

ATP is manufactured inside your cells via the energy derived from foods ingested. During this process, called cellular respiration, ATP is formed by bonding ADP and P_i through a series of **oxidation-reduction** reactions.

Analogies

ATP hydrolysis is also like an **investment in a profitable stock or mutual fund.** Though you must make an initial investment of your own money, you may reap a greater reward (dividend) in the end. Similarly, it takes an initial investment of energy to break the covalent bond in ATP, but the result is a larger amount of free energy when the new bonds in the final products are formed.

Study Tips

Unfortunately, the function of ATP and the process of ATP hydrolysis often are explained incorrectly in many textbooks. Let's correct some common misconceptions:

Misconception 1: **Breaking chemical bonds releases energy.**
Actually, just the opposite is true. During a chemical reaction, the formation of new chemical bonds in a more stable product results in the release of energy.

Misconception 2: **ATP has a special "high-energy phosphate bond" between the second phosphate group and the terminal phosphate group.**
The bond here is actually a covalent bond. This statement gives the false impression that this bond is ready to fly apart like a "jack-in-the-box." In fact, bonds are a force that hold atoms together so an input of energy actually is required to break a chemical bond. Instead, the three phosphate groups in ATP are all linked by relatively strong bonds called covalent bonds that represent shared pairs of electrons. It takes energy to break these bonds.

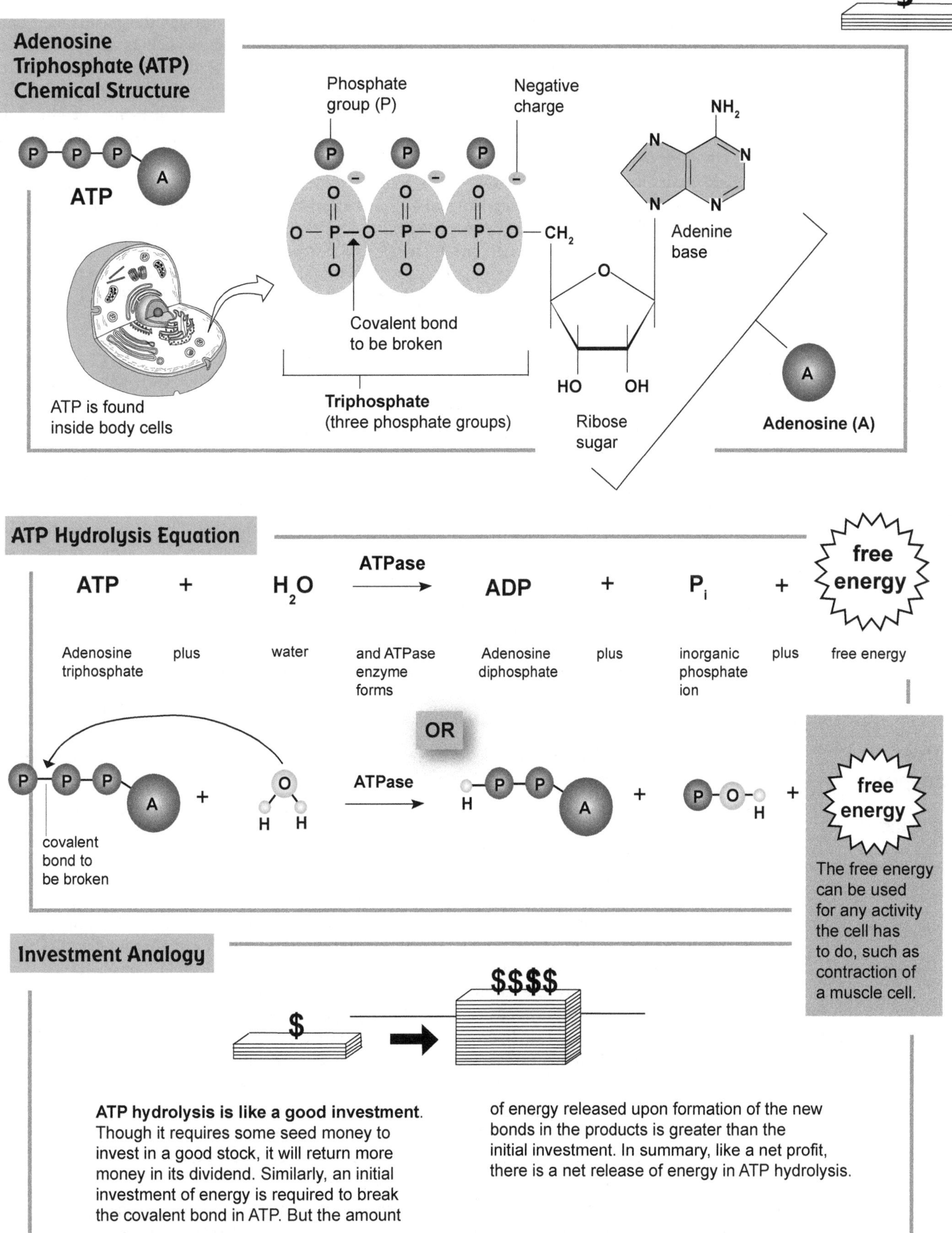

ATP hydrolysis is like a good investment. Though it requires some seed money to invest in a good stock, it will return more money in its dividend. Similarly, an initial investment of energy is required to break the covalent bond in ATP. But the amount of energy released upon formation of the new bonds in the products is greater than the initial investment. In summary, like a net profit, there is a net release of energy in ATP hydrolysis.

Description

Any substance able to either enter or leave the cell must do so by passing through the plasma membrane. In general, nutrients such as oxygen have to pass into cells, and waste products such as carbon dioxide have to pass out of them. Because the cell membrane is selectively permeable, it allows some substances to pass through but prevents others from doing so. Factors that determine whether a substance can pass through include: (1) size (molecular weight), (2) lipid solubility, (3) charge, and (4) presence of channels and transporters. Therefore, if a substance is smaller, more lipid-soluble, has no charge, and has the appropriate protein channels or transporters available, it is more likely to be able to cross the membrane. The smaller the size, the easier it is for a molecule to cross. The more lipid-soluble the better, because the cell membrane is made of phospholipids. As for charge, uncharged substances have an easier time crossing directly through the membrane (without using a membrane protein).

Transport of a substance through the membrane is categorized as either passive or active. **Passive transport** refers to processes that occur spontaneously without any energy investment by the cell via ATP hydrolysis. **Active transport** refers to any transport process that occurs only when the cell invests energy from ATP hydrolysis to force transport to occur.

Passive Transport

The illustration on the facing page shows three types of passive transport through the plasma membrane. All of them deal with the diffusion process, in which a solute particle spontaneously moves from an area of higher solute concentration to an area of lower solute concentration. Each is described below:

① In **simple diffusion** a small, nonpolar, uncharged particle diffuses directly through the phospholipid bilayer of the plasma membrane. Oxygen gas enters our cells by simple diffusion, and carbon dioxide gas leaves our cells by the same process. Other substances that diffuse across plasma membranes include fatty acids, steroids, and fat-soluble vitamins (A, D, E, and K).

② **Simple diffusion** also can be a channel-mediated-type **protein channel.** This is how small substances such as ions diffuse through a plasma membrane. For example, sodium ions diffuse through sodium channels to enter a cell. Similarly, potassium ions pass through potassium channels to leave the cell.

③ **Facilitated diffusion** differs from simple diffusion in the following ways: (1) involves the transport of larger solutes, (2) uses a protein carrier that has a shape-specific fit for a specific solute, (3) involves a shape change in the protein carrier. The solute must bind to the shape-specific binding site. This puts a limitation on how quickly diffusion can occur. The action of the binding of the solute to its binding site induces a shape change in the protein carrier. This allows the solute to be released on the opposite side of the plasma membrane. Glucose enters cells via facilitated diffusion.

Active Transport

The illustration on the facing page shows three types of active transport processes through the plasma membrane. Each is described below:

④ **Primary active transport** refers to the movement of a specific substance against its concentration gradient. In other words, a solute particle is moved from an area of low solute concentration to an area of high solute concentration. Like paddling against the current in a river, this requires energy. The protein used in this process is generally referred to as a pump, which appropriately implies the need for energy. Free energy from ATP hydrolysis is the fuel that keeps the pump working. This protein pump always contains a binding site for the substance to be transported. For example, the iodine pump actively transports iodine out of the blood and into the cells in our thyroid gland. The thyroid requires iodine as a raw material for manufacturing the hormones it normally produces.

Unlike primary active transport, **secondary active transport** does not directly use free energy from ATP hydrolysis. Instead, it gets its energy indirectly from the established concentration gradients from either sodium ions (Na^+) or hydrogen ions (H^+). These ion gradients were established by primary active transport. Movement of these ions down their concentration gradients is then coupled to the simultaneous transport of another substance via either symport or antiport.

⑤ **Cotransport** refers to the simultaneous transport of two substances in the same direction across the membrane with the aid of a specific type of membrane protein. One example is the sodium-glucose symporter located in the epithelial cells in the intestinal mucosa. This transport process helps glucose get absorbed into the blood from the digestive tract.

⑥ **Countertransport** refers to the simultaneous transport of two substances in the opposite direction across the membrane with the aid of a specific type of membrane protein. One example is the sodium-calcium antiporter, which keeps calcium levels low inside cells by pumping it out.

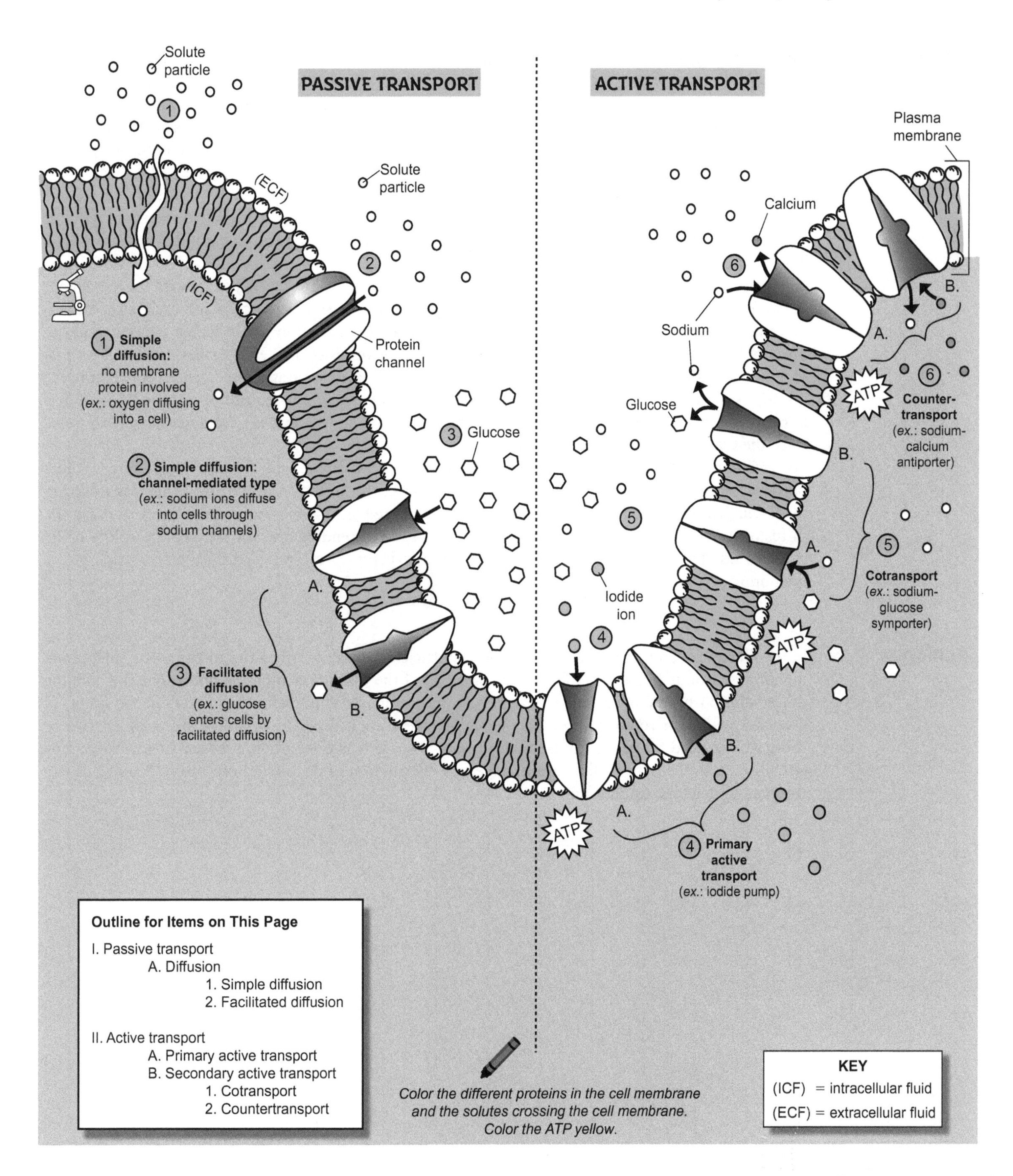

Outline for Items on This Page

I. Passive transport
- A. Diffusion
  - 1. Simple diffusion
  - 2. Facilitated diffusion

II. Active transport
- A. Primary active transport
- B. Secondary active transport
  - 1. Cotransport
  - 2. Countertransport

Color the different proteins in the cell membrane and the solutes crossing the cell membrane. Color the ATP yellow.

KEY

(ICF) = intracellular fluid

(ECF) = extracellular fluid

Description

The plasma membrane is selectively permeable because it controls which substances enter and exit the cell. Nutrients and waste products both must pass through this structure, but not all substances are able to cross the plasma membrane. Factors that determine whether a substance can pass through include: (1) size (molecular weight), (2) lipid solubility, (3) charge, and (4) presence of channels and transporters. Therefore, if a substance is smaller, more lipid-soluble, has no charge, and has the appropriate protein channels or transporters available, it is more likely to be able to cross the membrane.

Transport of a substance is classified as either **passive transport** or **active transport.** In passive transport, the cell does not have to expend energy for the process to occur. Active transport requires the use of cellular energy. Typically, this energy is in the form of the free energy liberated from ATP hydrolysis.

Simple diffusion falls under the category of **passive transport.** Typically, it deals with the movement of a solute particle. For example, this process allows oxygen to enter the cells and carbon dioxide to exit them. In diffusion, substances tend to move from areas where they are in higher concentration to areas of lower concentration. The driving force for simple diffusion is the movement toward a state of dynamic equilibrium, in which concentrations of the designated substance are in equal concentrations on both sides of the plasma membrane. This **driving force** is like a **room full of crowded people at a party who all need their personal space.** As the illustration shows, this *need* induces people to move from the crowded room A over to room B.

Facilitated diffusion is also a **passive transport** process, but it typically deals with larger solute particles than those in simple diffusion. It differs from simple diffusion primarily in that it is a **carrier-mediated** process. This means that a protein carrier in the plasma membrane is needed to assist in the diffusion process. The carrier protein has a shape-specific binding site for the specific solute to be transported. The solute binds to this site temporarily before being released on the other side of the membrane.

Analogy

In the illustration, **a person** represents **a solute particle.** The **wall** between the rooms is like the **plasma membrane. Room A** is like a **solution** on one side of the plasma membrane, and **room B** is like the **other solution** on the other side of the plasma membrane. In the "Before" illustration, the **group of crowded people in room A** represents a **high-concentration gradient** of solute particles relative to room A. The **need** for people to have their personal space is like the **driving force** to achieve dynamic equilibrium. In the "After" illustration, the **equal distribution of people** in both rooms A and B is like the **state of dynamic equilibrium.**

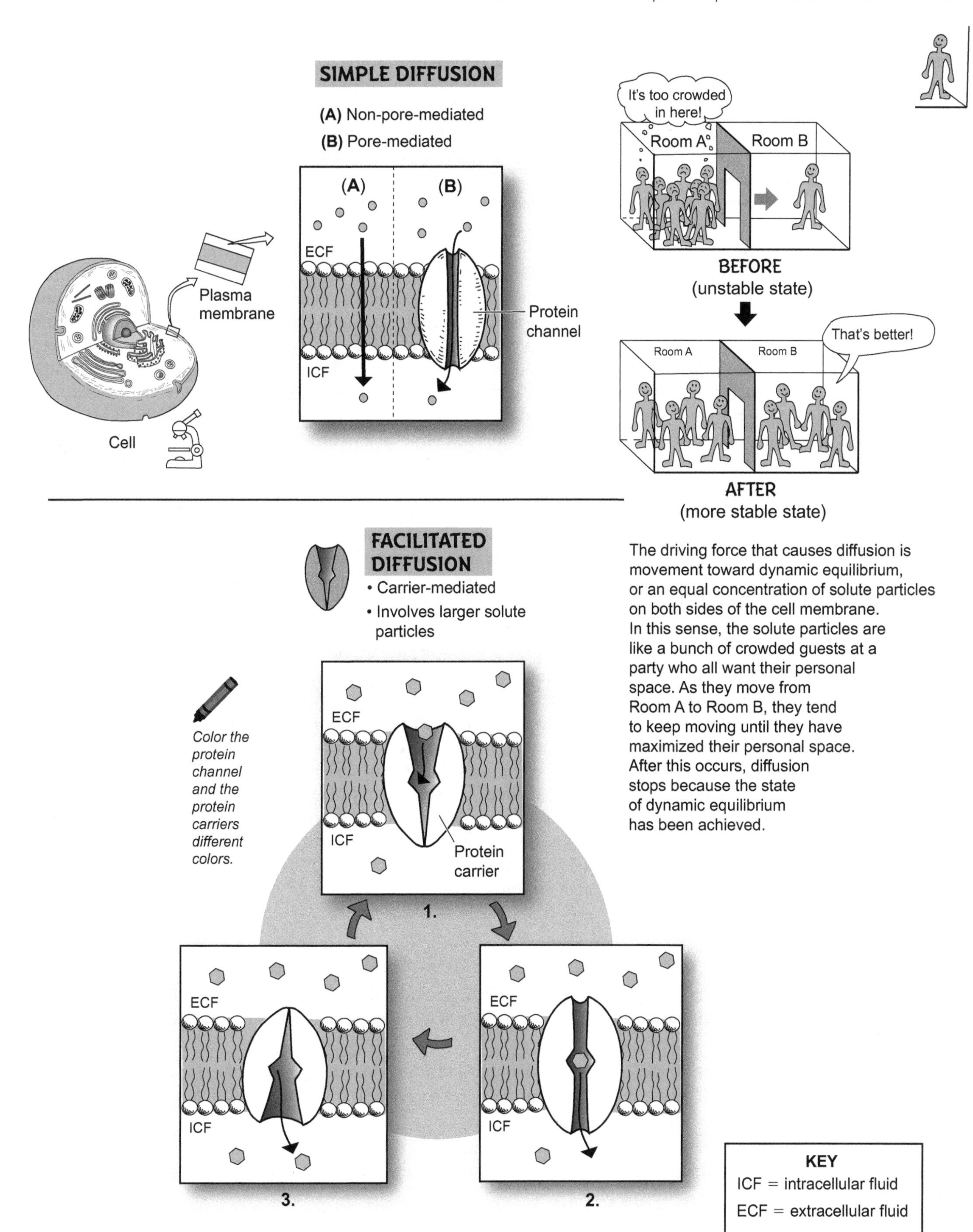
SIMPLE DIFFUSION
(A) Non-pore-mediated
(B) Pore-mediated
(A)
(B)
ECF
ICF
Protein channel
Plasma membrane
Cell
It's too crowded in here!
Room A
Room B
BEFORE
(unstable state)
That's better!
Room A
Room B
AFTER
(more stable state)
FACILITATED DIFFUSION
• Carrier-mediated
• Involves larger solute particles
Color the protein channel and the protein carriers different colors.
ECF
ICF
Protein carrier
1.
2.
3.
The driving force that causes diffusion is movement toward dynamic equilibrium, or an equal concentration of solute particles on both sides of the cell membrane. In this sense, the solute particles are like a bunch of crowded guests at a party who all want their personal space. As they move from Room A to Room B, they tend to keep moving until they have maximized their personal space. After this occurs, diffusion stops because the state of dynamic equilibrium has been achieved.
KEY
ICF = intracellular fluid
ECF = extracellular fluid

Description

In biological systems, **osmosis** is defined as the flow of **water** across a semipermeable membrane toward the solution with the higher solute concentration. This common, important process occurs when water moves in and out of plant and animal cells by crossing their cell membranes. The driving force is the tendency toward achieving dynamic equilibrium, in which there is an equal concentration of solute on both sides of the cell membrane. When examining osmosis, the solute in question is impermeable to the cell membrane. It is also essential to realize that the solution inside the cell (intracellular solution) is always compared to the solution outside the cell (extracellular solution). For example, if you told me I was tall, I might ask, "Compared to what?" Compared to a toddler, I may be tall but not when compared to a giraffe. Similarly, if you were to tell me that a solution was concentrated, I could also ask, "Compared to what?" The word "concentrated" makes sense only in the context of a relative comparison. Although your body has regulatory mechanisms to keep the concentration of the intracellular and extracellular solutions relatively stable, the extracellular solution is more likely to change in its solute concentration because of external influences. Sometimes it is more dilute, and other times it is more concentrated.

The illustration of the hollow, glass, U-shaped tube at the top of the facing page demonstrates the process of osmosis. An artificial, semipermeable membrane separates two solutions of different concentrations—one on side A and the other on side B. Notice that the solute in both solutions is a protein that is impermeable because it is too large to cross the membrane. The solution on side B is more concentrated than the solution on side A. Any time the two solutions are unequal in concentration, water always moves toward the solution with the higher solute concentration. In this case, water will move from side A to side B. Why? Think of this as water's attempt to dilute the more concentrated solution to reach dynamic equilibrium. As water moves, the volume of solution on side B increases, and the volume of solution on side A decreases. Water stops moving when the two solutions have equal solute concentrations—the state of dynamic equilibrium. At dynamic equilibrium, the net movement of water across the membrane is zero.

The illustration in the middle of the facing page shows a red blood cell in three different types of solutions: (1) **isotonic**, (2) **hypertonic**, and (3) **hypotonic.** Your body's extracellular solutions are normally **isotonic** (iso = equal) **solutions** with the same concentration of impermeable solutes as the intracellular solution. This is a stable state for body cells. A cell in this solution has already achieved dynamic equilibrium, so the net movement of water in and out of the cell is zero. A **hypertonic solution** (hyper = more, greater) is one that has a greater concentration of impermeable solutes relative to the intracellular solution. A cell placed in this solution will shrink (crenate) because of water leaving the cell. A **hypotonic** (hypo = less) **solution** is one that has a lesser concentration of impermeable solutes relative to the intracellular solution. A cell placed in this solution will swell and possibly burst (lyse) because of water rushing into the cell. By convention, biologists typically use these terms to describe the extracellular solution, though they technically can be used to describe either the intracellular or the extracellular solution as a comparison.

Applications

Three applications for osmosis are illustrated on the bottom of the facing page:

1. Using an isotonic solution is illustrated by giving a patient an intravenous (IV) saline solution at the hospital. The saline solution has to be isotonic to your red blood cells so they neither shrink nor burst.
2. Creating a hypertonic solution is illustrated by adding salt to the slimy film coating the surface of the slug. As a result, water leaves the cells of the slug and can cause it to shrivel up and die.
3. Using a hypotonic solution is illustrated by placing wilted lettuce in a bowl of water in the refrigerator overnight to refresh it. Water in the bowl enters the cells of the lettuce leaf, making it firm again.

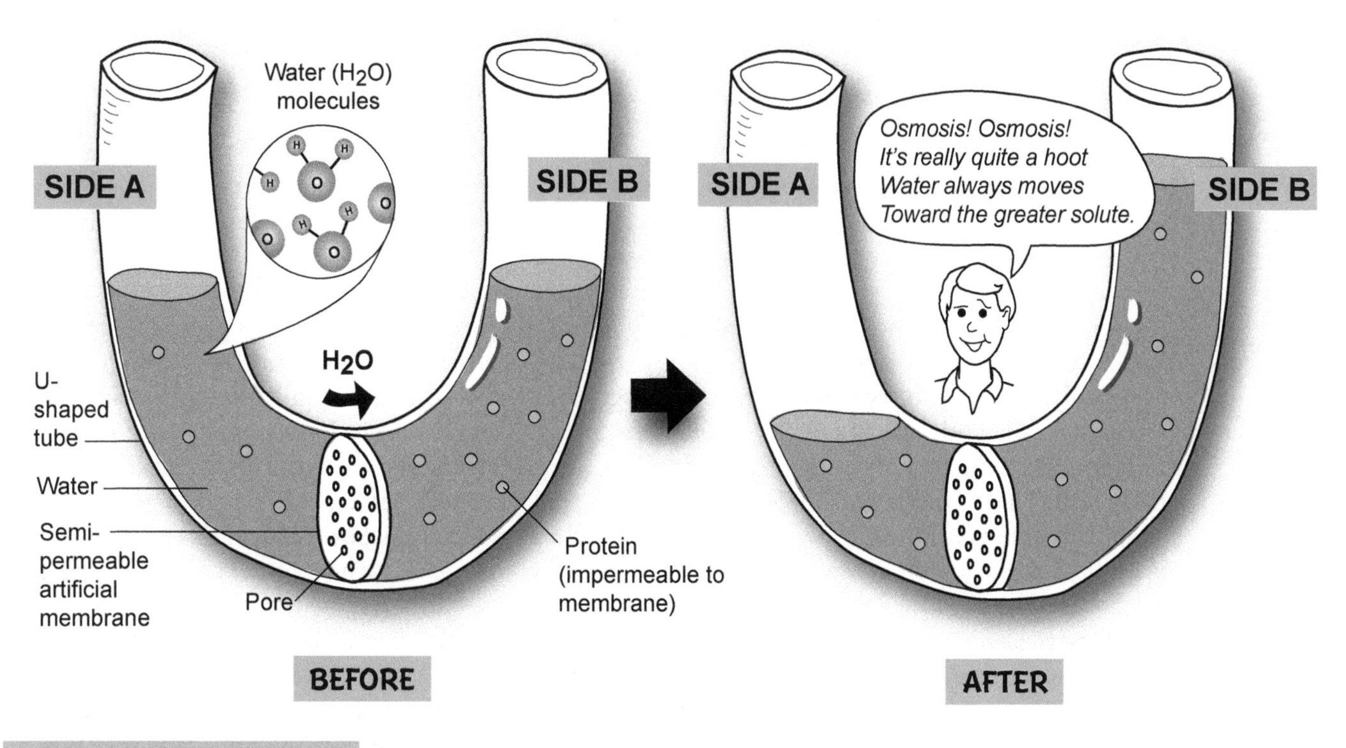

Three Possible Solutions

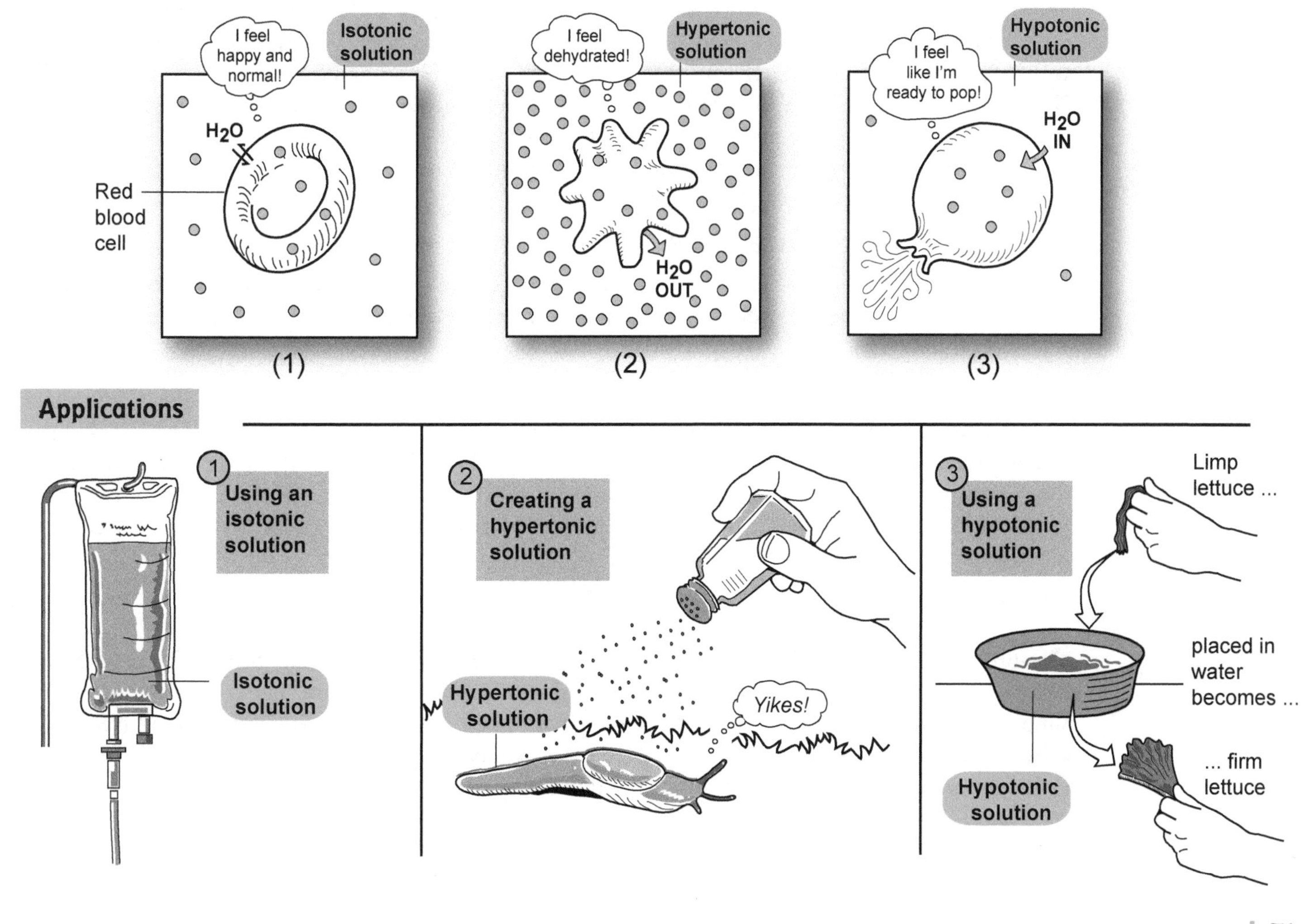

Description

Filtration falls under the category of **passive transport** and has the following features:

- Requires a force (*blood pressure*) to make it happen
- Requires a pressure gradient.
- Separates liquids (*water* and *small solutes*) from solids *(plasma proteins, red blood cells, white blood cells,* and *platelets)*.
- Produces a filtrate (*final filtered solution*)

Filtration is a common process that occurs constantly through capillary beds that link arteries to veins. Just as narrow, single-lane roads lead to larger streets and then to highways, capillaries eventually lead to the body's highways—large arteries and large veins. Capillaries are found throughout the body and are the most microscopic of the blood vessels. In fact, they are so small that red blood cells must pass through them single-file. Through filtration, they provide direct contact with body cells. Their structure is simple: The wall of a capillary is made of a single, flat layer of cells called **simple squamous epithelium.** These thin cells allow for easy filtration. Surrounding this epithelial layer is a thin **basement membrane** composed of extracellular materials, such as protein fibers, secreted by epithelial cells.

Almost all tissues have an ample supply of capillary beds—a branching network of capillaries with a higher-pressure arteriole end and a lower-pressure venule end. Like a one-way street, blood always moves from the arteriole end to the venule end. Because filtration is completely dependent on pressure, most of it occurs nearer the arteriole end. The blood pressure is the force that makes it all happen by creating a pressure gradient. As a result of the higher pressure within the arteriole end, the liquid plasma is forced through the wall of the capillary, so water and small solutes leave the blood to form a **filtrate** called **interstitial fluid.** This fluid fills the spaces within tissues called **interstitial spaces,** which include gaps between body cells as well as those between blood vessels and body cells. Simultaneously, large solids such as blood cells and large molecules such as plasma proteins remain inside the capillary, because they are too large to pass through it.

Filtration is responsible for producing all of the following:

- **interstitial fluid** (the only way we have to get fluid to tissues)
- **urine** (processed blood plasma that contains mostly water)
- **cerebrospinal fluid** (CSF)—(constantly circulates around the brain and spinal cord and has many functions: provides a protective cushion, distributes nutrients, and removes waste products)

There are three different types of capillaries (see p. 130). Some are more permeable than others, which determines how easily filtration occurs. For example, specialized capillaries in the kidneys called **glomeruli** (sing. *glomerulus*) (see p. 222) are highly permeable to allow for easy filtration of the plasma to form urine, and capillaries in the brain are the least permeable in the body to protect the brain's precious neurons from being damaged by toxic substances that may have entered the bloodstream.

Analogy

The process of filtration is the same, generally, as a coffeemaker making coffee. Compared with filtration through a capillary, the **force** is **gravity** instead of **blood pressure.** The **coffee filter** is like the **simple squamous epithelium.** The **coffee grounds** are like the **solid** materials in the blood (**red blood cells, white blood cells, plasma proteins**, etc.). The **water** is like the **plasma.** The **coffee solution** is called the *filtrate*, which is like the **interstitial fluid.**

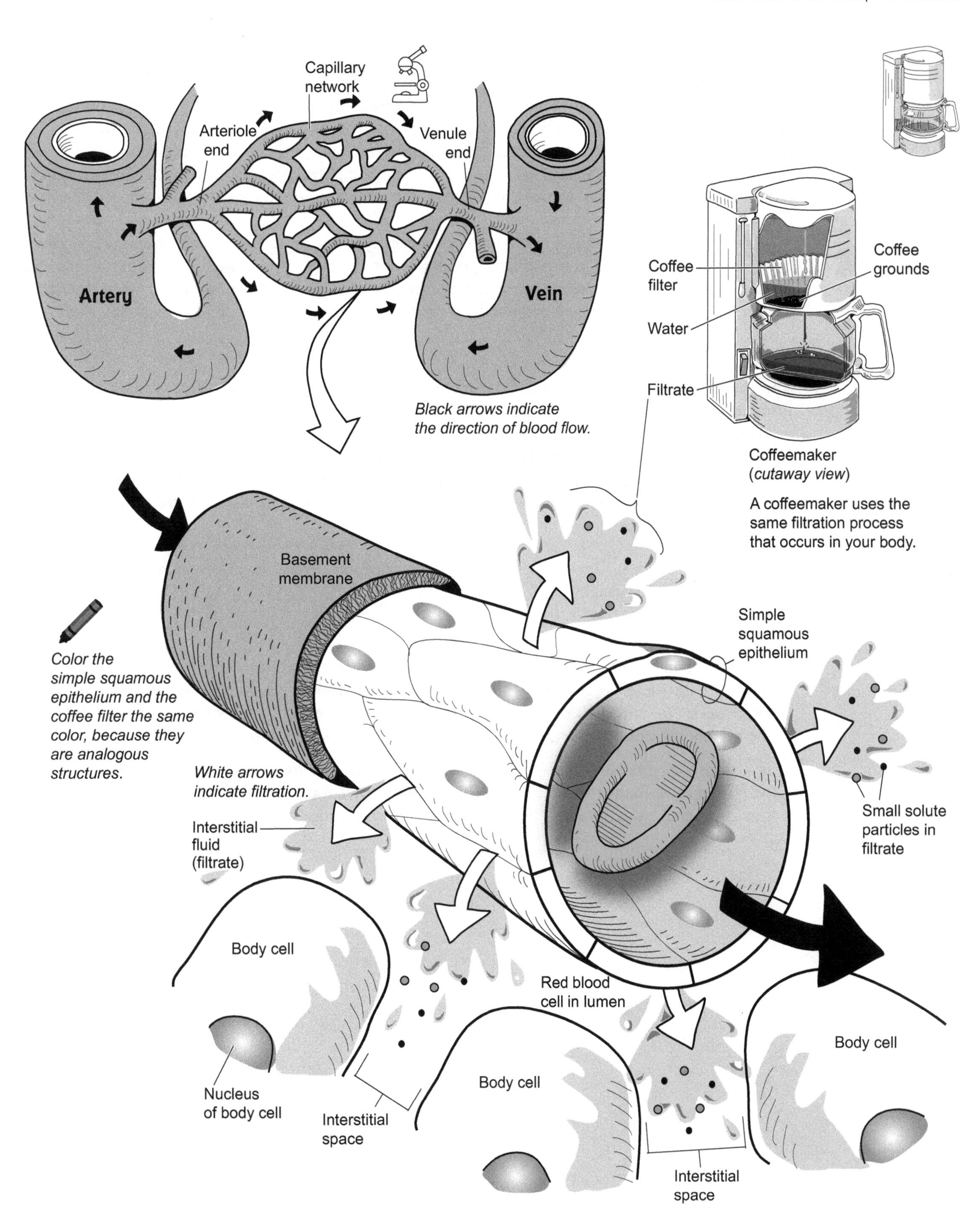
Capillary network
Arteriole end
Venule end
Artery
Vein
Black arrows indicate the direction of blood flow.
Coffee filter
Coffee grounds
Water
Filtrate
Coffeemaker (cutaway view)
A coffeemaker uses the same filtration process that occurs in your body.
Basement membrane
Color the simple squamous epithelium and the coffee filter the same color, because they are analogous structures.
White arrows indicate filtration.
Simple squamous epithelium
Interstitial fluid (filtrate)
Small solute particles in filtrate
Body cell
Red blood cell in lumen
Body cell
Body cell
Nucleus of body cell
Interstitial space
Interstitial space

Description

The plasma membrane is like the gatekeeper of the cell, because it controls the substances that enter and exit. Plasma membranes selectively allow only certain substances (*like nutrients*) in and other substances (*like waste products*) out. Not all substances are able to cross the plasma membrane. Generally, they have to be small and lipid-soluble to cross. If they meet these criteria, they can be transported by two different methods: (1) **passive transport** or (2) **active transport.** In passive transport, the cells do not have to expend any energy for the process to occur. Active transport requires that cellular energy be used. Typically, this energy is in the form of the free energy liberated from ATP hydrolysis.

The process of **active transport** falls under the category of the same name—**active transport.** It moves a solute particle against a concentration gradient with the help of a protein transporter within the plasma membrane. This protein uses the free energy from ATP hydrolysis to move the solute particle(s) against the gradient. An example is the sodium-potassium pump present in all cells (see p. 80).

Analogy

The process of **active transport** is like the action of a **sump pump** in a homeowner's basement, which prevents flooding inside the home. Both require energy to allow the pump to transport a substance against a concentration gradient. The **protein pump** uses the **free energy** from ATP hydrolysis, and the **sump pump** uses **electrical energy.** The protein pump transports a solute particle from a region of lower concentration to a region of higher concentration. Similarly, the sump pump transports *water* from a region of lower *water* concentration inside the basement to a region of higher *water* concentration outside the home.

Note: The limitation of this analogy is that active transport always involves a solute particle rather than a solvent such as water.

Study Tips

- Students often confuse the process of **active transport** and the category of **active transport** because they have the same name. Don't make this mistake. The former refers to the specific "sump pump-like process" described above, and the latter refers to the general category for any process that uses the free energy from ATP hydrolysis to transport substances into or out of the cell.
- If a transport process is active, this means that it uses the free energy from ATP hydrolysis to perform a task. To distinguish the different types of active transport (exocytosis, endocytosis, active transport, etc.), ask yourself: "What is the free energy specifically used for in each case?"

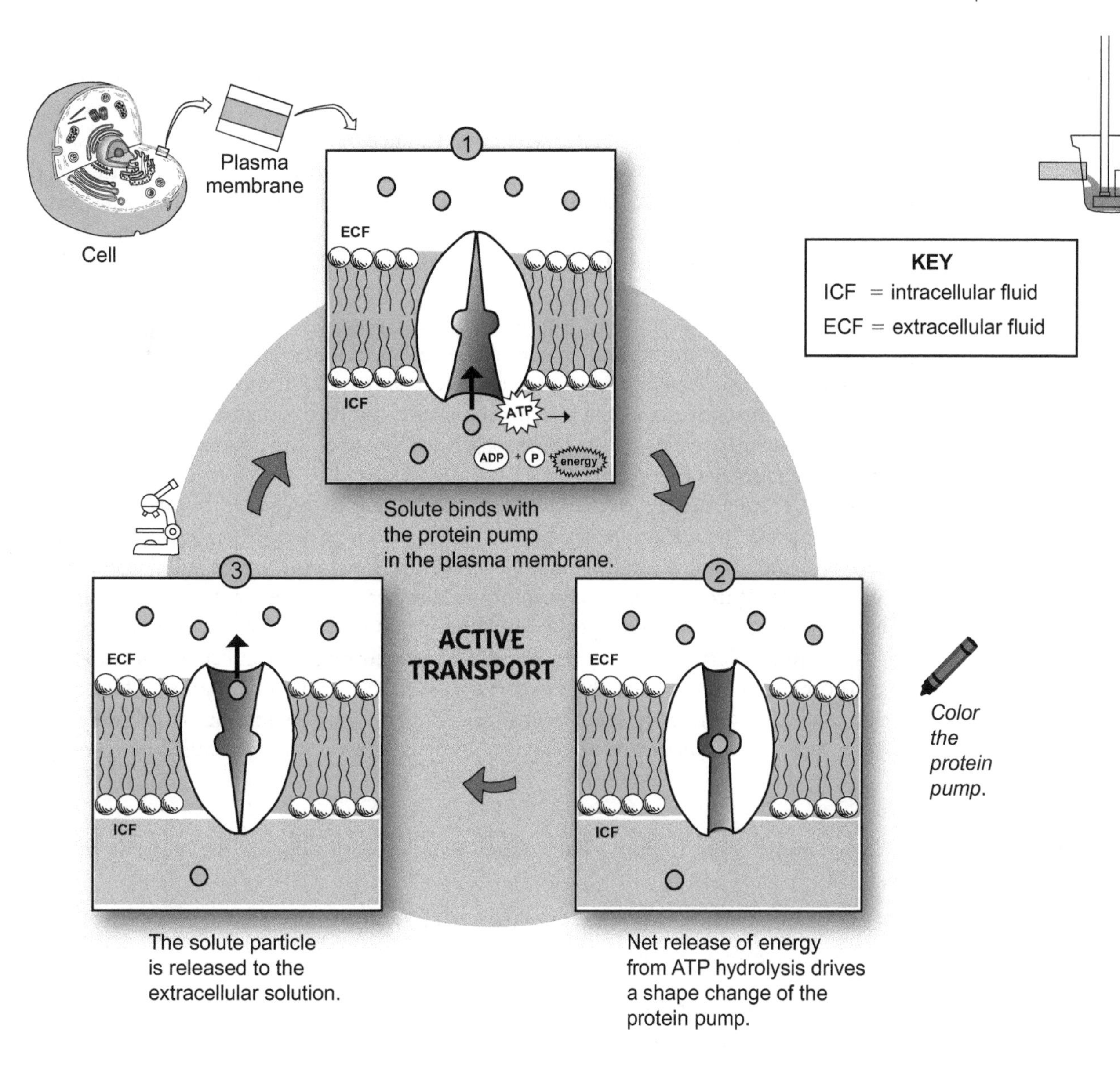

KEY
ICF = intracellular fluid
ECF = extracellular fluid

Color the protein pump.

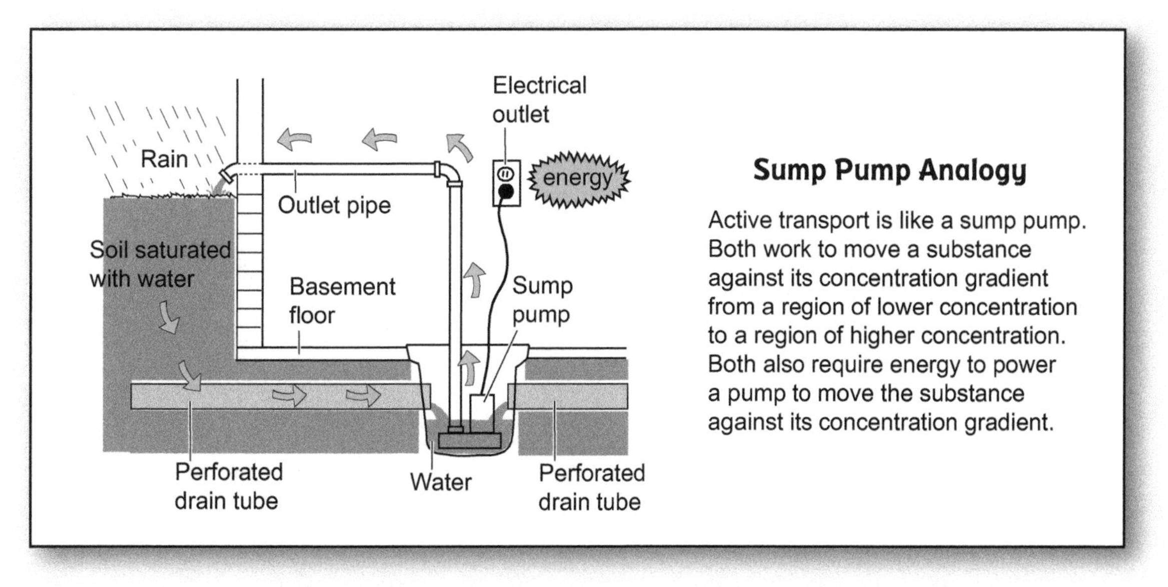

Sump Pump Analogy

Active transport is like a sump pump. Both work to move a substance against its concentration gradient from a region of lower concentration to a region of higher concentration. Both also require energy to power a pump to move the substance against its concentration gradient.

Description

The plasma membrane is like the gatekeeper of the cell because it controls which substances enter and exit. Plasma membranes selectively allow only certain substances (*like nutrients*) in and other substances (*like waste products*) out. Not all substances are able to cross the plasma membrane. Generally, they have to be small and lipid-soluble to cross. If they meet these criteria, they can be transported by two different methods: (1) **passive transport** or (2) **active transport.** In passive transport, the cell doesn't have to transport any energy for the process to occur. Active transport requires cellular energy be used. Typically, this energy is in the form of the free energy liberated from ATP hydrolysis.

Exocytosis is a type of active transport. This is the process by which substances made by the cell (e.g., a hormone) are concentrated within a vesicle. Special regulatory proteins associated with the plasma membrane use cellular energy to move the vesicle toward the plasma membrane, so the two can merge. Then the contents of the vesicle are released to the extracellular solution. This is how cells within your pancreas release the hormone insulin, and for example, how neurons release neurotransmitters such as acetylcholine.

Endocytosis is another type of active transport that is actually the reverse of exocytosis. In this process the plasma membrane invaginates to trap a solid particle (*bacteria, virus*) or some liquid (*extracellular solution)* in a small, pouch-like structure. This pouch uses cellular energy to gradually close, pinch itself off from the plasma membrane, and become a vesicle within the cell.

There are two types of endocytosis: (1) **pinocytosis** and (2) **phagocytosis.** If the material in the pouch is a liquid solution such as the extracellular solution, it is called **pinocytosis.** Cells use this process to obtain nutrients from the extracellular solution. If the material is a solid object such as a bacterium, it is called **phagocytosis.** Macrophages are a type of white blood cell that attacks foreign pathogens. Because they normally are engaged in phagocytosis, they are called *phagocytic* cells.

Study Tips

Knowing common roots, prefixes, and suffixes is important to decode the meaning of scientific terms (see p. 323).

exo- *= outer, outside* ***cyto-*** *= cell* ***-sis*** *= condition*
exocytosis *= to bring outside the cell*

endo- *= within* ***cyto-*** *= cell* ***-sis*** *= condition*
endocytosis *= to bring within the cell*

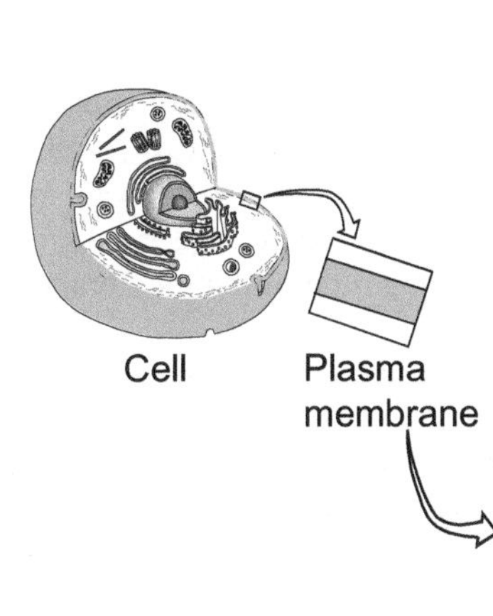

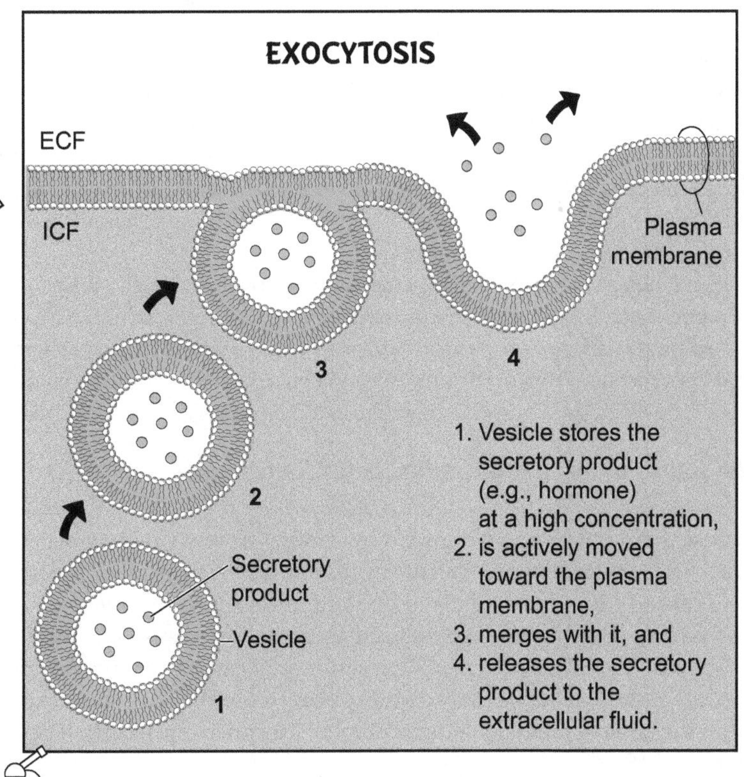

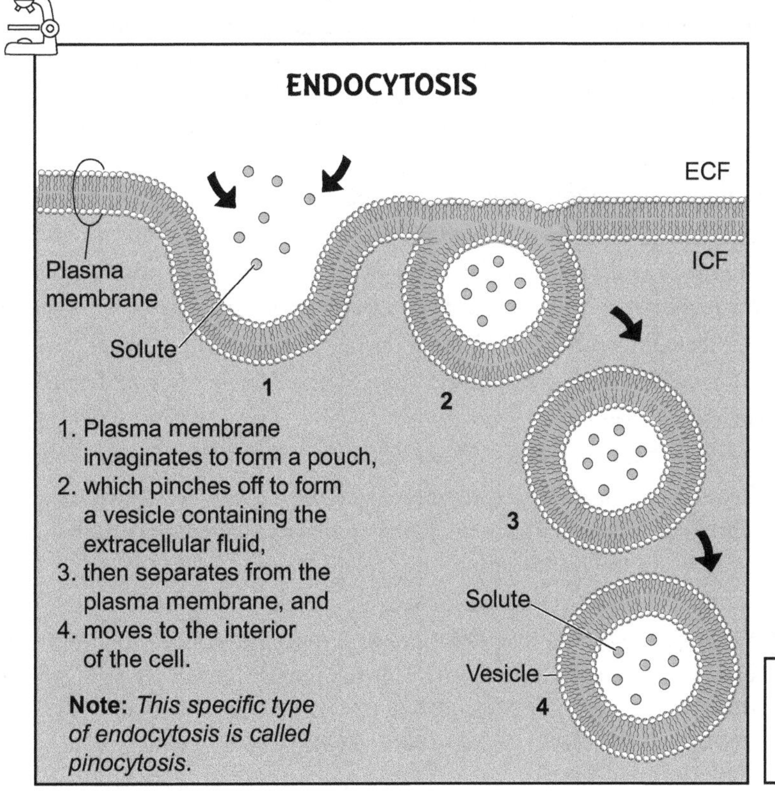

KEY

ICF = intracellular fluid

ECF = extracellular fluid

Description

A **membrane potential** (MP) is a separation of positive (+) and negative (−) charges across a **plasma membrane** that results in an *electrical potential* or a *voltage*. This voltage is measured in units called *millivolts* (mV)—much smaller than that produced by a battery. Typically, the outside of the plasma membrane is positively charged, and the inside is negatively charged. Most human body cells have an MP, but the exact value is different for different cell types and ranges between −5 mV and −100 mV. The negative sign means that the inside of the membrane is negative relative to the outside.

Let's compare the MP to a 1.5 V battery, which also has an electrical potential because of its positive (+) pole at one end and negative (−) pole at the other. The battery has **potential energy** because of its *potential* to do work.

Potential energy is like water behind a dam. Because the *position* of the water level is higher on one side of the dam and lower on the other, it also has the *potential* to do work. When the floodgate opens, this potential energy gets converted into kinetic energy or energy in motion. As the water flows through the gate, it can turn a turbine that generates electricity—hydroelectric power.

Similarly, the potential energy in the battery can be converted into a flow of electricity to power a flashlight. For the **plasma membrane**, this potential energy is stored not as water behind a dam but as a cation (+) concentration gradient. Based on the rule for simple diffusion, ions passively flow down their concentration gradients from regions of higher ion concentration to regions of lower ion concentration. This cation concentration gradient is the potential energy source for cells to do work. For example, in a neuron this gradient is used to generate a nerve impulse.

What is the source of these positive and negative charges?

The short answer to this question is that the *extracellular fluid* (ECF) and *intracellular fluid* (ICF) are electrically neutral because they contain equal numbers of *cations* (+ ions) and *anions* (− ions). The only exceptions are at the inner and outer surfaces of the **plasma membrane.** Here, there is an accumulation of positive (+) charges on the outside surface of the membrane and an accumulation of negative (−) charges on the inside. The positive charges on the outside are mainly the result of **sodium** (**Na^+**) ions—the most dominant cations in the **ECF.** The negative charges on the inside are mostly the result of negatively charged **proteins** too large to leave the cell. But it's also because of the loss of the most dominant cations in the ICF—**potassium** (**K^+**) ions.

The real situation is actually more complex (surprise, surprise!). Let's summarize the key factors that contribute to the MP:

- **Ion concentration gradients for Na^+ and K^+ across the plasma membrane**
  - **Na^+** has a greater concentration in the ECF than the ICF, so it could diffuse *into* the cell.
  - **K^+** has a greater concentration in the ICF than the ECF, so it could diffuse *out* of the cell.
  - Na^+ and K^+ can pass through gated channels only in the membranes that are usually closed but can be stimulated to open.
  - These ion concentration gradients are a source of potential energy to do work.
- **Differing permeabilities for Na^+ and K^+**
  - Membrane proteins called *leakage channels* allow both Na^+ and K^+ to diffuse across the membrane.
  - Plasma membranes are *much* more permeable to K^+ than Na^+.

    (1) K^+ leaves the cell as it diffuses down its concentration gradient.
    The result? This loss of positive charge makes the interior of the cell more negative (*see illustration*).
- **Opposite charges attract**
  - As previously mentioned, as K^+ diffuses out of the cell, this loss of positive charge makes the interior more negative.

    (2) Because opposite charges attract, some of the K^+ ions are drawn back into the negatively charged interior. The result? This relatively small gain in positive charge makes the interior less negative than it might be otherwise (*see illustration*).
- **Action of the Na^+ −K^+ pump** (see p. 80)

    (3) With each cycle, the **Na^+−K^+ pump** actively transports three Na^+ ions out for every two K^+ ions it brings into the cell. The result? This maintains the MP by preserving the charge imbalance across the membrane. How? More positive charges are moved out of the cell (+3), and fewer positive charges are brought into the cell (+2) (*see illustration*).

Study Tips

To distinguish *cations* (+ ions) from *anions* (− ions) remember this phrase: "I ***positively*** hate ***cats***."

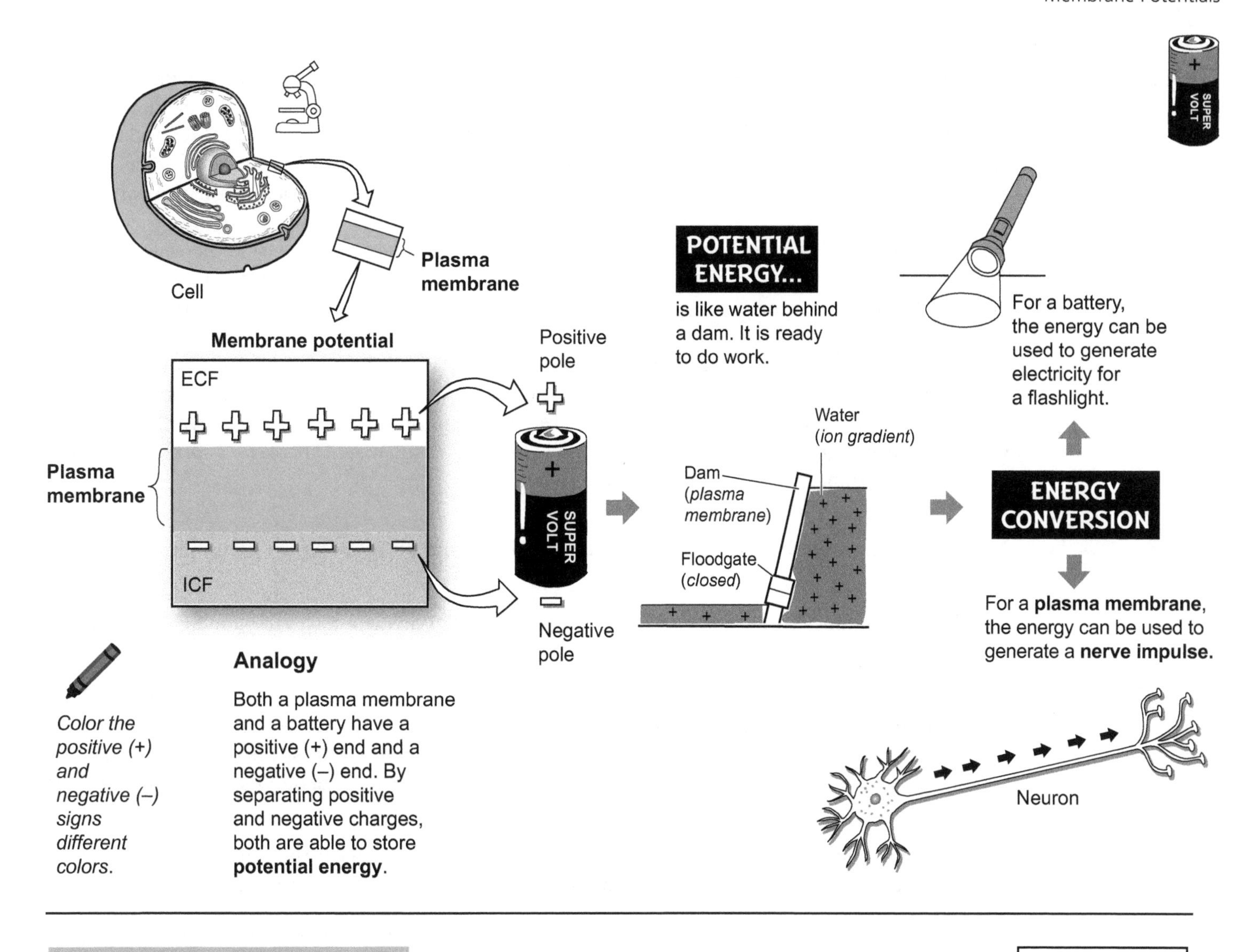

Color the positive (+) and negative (–) signs different colors.

Analogy

Both a plasma membrane and a battery have a positive (+) end and a negative (–) end. By separating positive and negative charges, both are able to store **potential energy**.

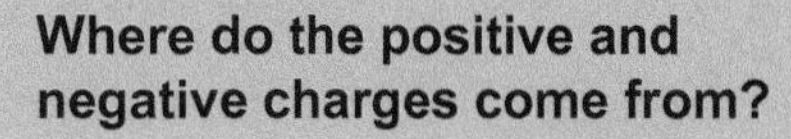

Where do the positive and negative charges come from?

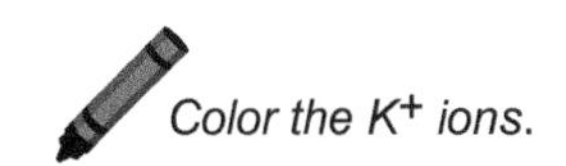

Color the K^+ ions.

KEY	
Na^+	= Sodium
K^+	= Potassium

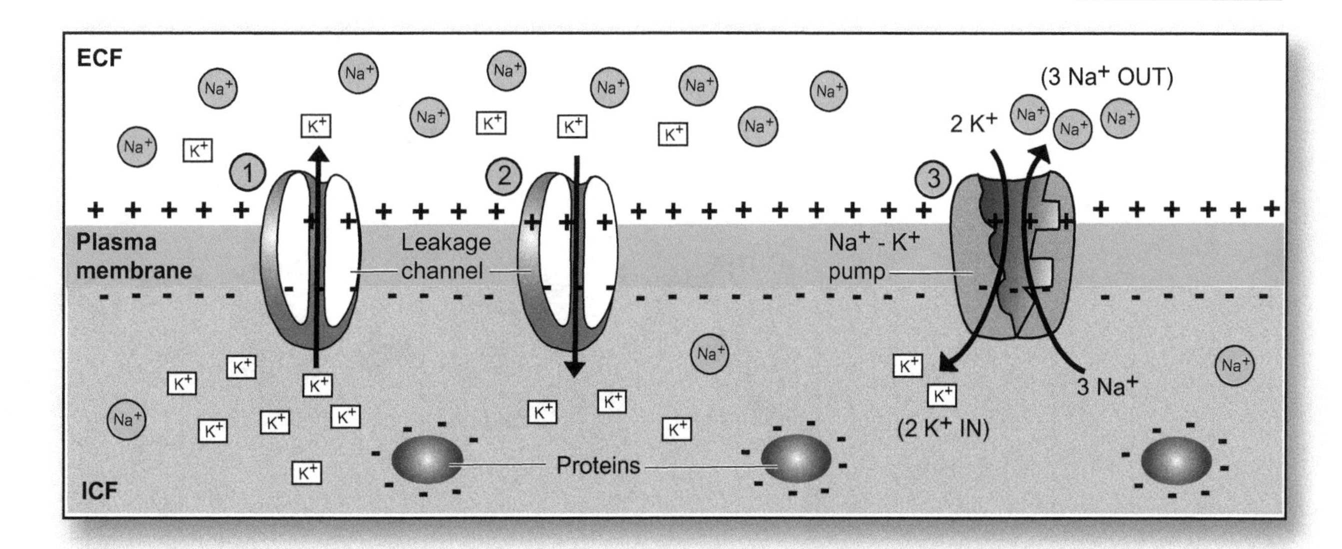

Description

The **sodium-potassium pump** is a protein found in the plasma membrane of all animal cells. It functions as an ion exchanger to actively transport sodium for potassium across the membrane. The net release of energy from the hydrolysis of one ATP molecule is used to transport three sodium ions *out* of the cell and two potassium ions *into* the cell. Since a single cell has many of these protein pumps, the net effect of all of them working together is to maintain a gradient for both sodium and potassium.

The concentration of potassium ions is normally 10–20 times greater inside the cells than out. The reverse is true for sodium. Because of this difference in concentration, sodium's tendency is to diffuse into cells, and potassium's tendency is to diffuse out of cells. This occurs through leakage channels in the membrane.

A significant amount of the cell's resting energy—about 33%—is used to keep the pump working. The activity of the pump depends on the concentration of sodium in the cytosol. The greater the concentration, the more active the pump; the lesser the concentration, the less active. If the pump were to stop working, the sodium concentration in the cytosol would increase, and nothing could prevent water from entering the cell via osmosis (see p. 70).

Cycle

In the illustration on the facing page, the sodium-potassium pump is shown going through a four-step cycle:

① **Sodium binding:** three sodium ions in the cytosol bind at their respective binding sites on the pump.

② **Shape change:** The binding of sodium causes the hydrolysis of one ATP molecule into ADP, a phosphate group and a net release of free energy. The free energy is used to bind the phosphate group to the pump, which induces a shape change in the pump. This shape change both enables the release of the three sodium ions and makes it easier for two potassium ions to bind from the extracellular solution.

③ **Potassium binding:** The binding of the two potassium ions causes the phosphate group to be released from the pump, which, in turn, causes the shape of the protein to change again.

④ **Potassium released:** The restoring of the pump to its original shape releases the two potassium ions into the cytosol. The pump is now ready to bind three new sodium ions, and the cycle repeats itself.

Analogy

Functionally, the sodium-potassium pump is like a revolving door. It takes energy to move the revolving door, just as it takes ATP energy to drive the pump. Each cycle of the pump results in three sodium ions moving *out* of the cell and two potassium ions moving *into* the cell.

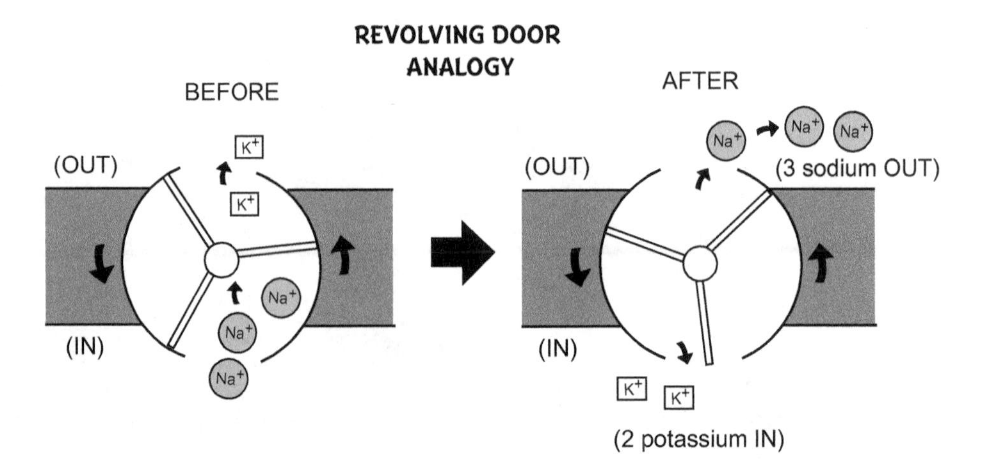

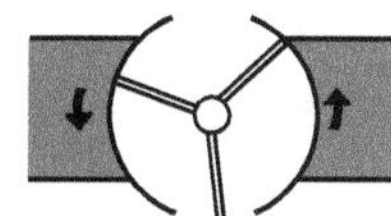

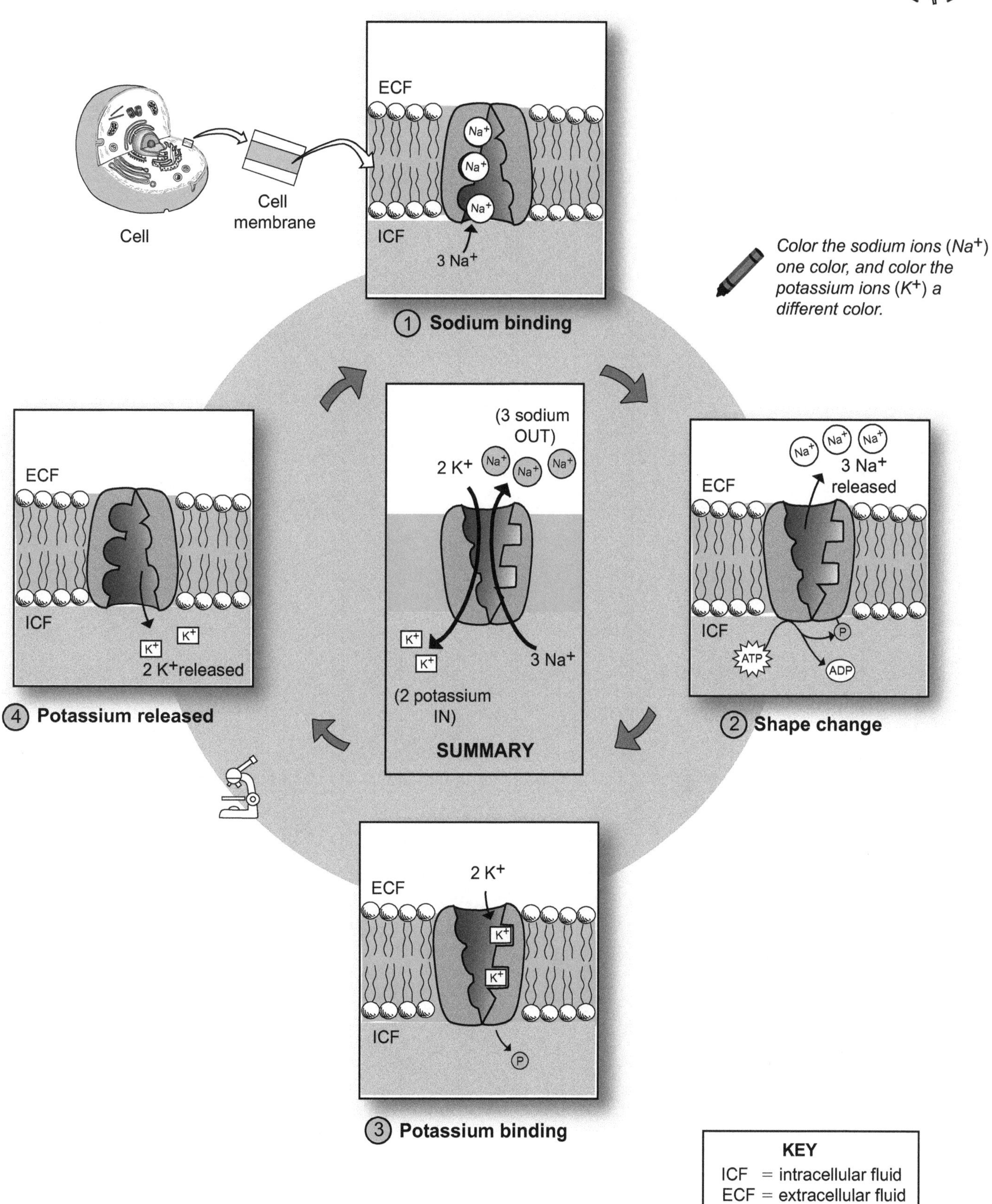

Color the sodium ions (Na^+) one color, and color the potassium ions (K^+) a different color.

KEY
ICF = intracellular fluid
ECF = extracellular fluid

Notes

NERVE AND MUSCLE INTEGRATION

Description

Your body has more than 600 skeletal muscles, which constitute most of your body mass. More than 45 miles of nerves are "wired" through your body like electrical wiring in a house. Each nerve is a collection of **neurons** (*nerve cells*), connective tissue, and blood vessels. Motor neurons connect to muscle tissue to stimulate it to contract.

Neurons are the fundamental cells in nervous tissue. Though they are of various types, they all share certain features. Surrounding the nucleus of every neuron is a region called the **cell body**, where most of the organelles are found. This is the metabolic center of the cell. Branching out from the cell—like tentacles from an octopus are two types of processes—**dendrites** or **axons**. As a rule, each neuron has only one axon per cell but may have one or more dendrites.

The neuron in the illustration is called a multipolar (*multi* = many) neuron because it has *many* dendrites branching out of the cell body. The dendrites act as the sensory receptors of the cell, and the axon conducts an action potential (*nerve impulse*) along the length of the cell like copper wires in an electrical cord. The **synaptic knobs** are doorknob-like structures located at the end of the axon that contain chemical messengers called **neurotransmitters**. Any time a neuron is stimulated, it conducts a nerve impulse along its axon and usually responds by releasing neurotransmitters.

Nerve impulses flow through a neuron in a one-way direction:

dendrite → cell body → axon → synaptic knobs

First, a stimulus must arrive at the dendrite. The type of stimulus needed to cause a response depends on the type of neuron. For example, sensory neurons have dendrites in the skin that respond to various stimuli such as temperature, touch, and pressure. When dendrites detect a stimulus, it triggers a signal to be sent to the cell body, then along the axon, and finally to the synaptic knobs.

Neurons Connect to . . .

Neurons can connect to three different things: (1) **other neurons**, (2) **glands**, and (3) **muscles**. Let's look at each, in turn. In the nervous system, neurons connect to other neurons like paper dolls in a chain. When this occurs, the orientation of one cell to another always follows a predictable "head to tail"–type pattern. More specifically, the synaptic knobs ("tail") of the first cell always connect to the dendrites ("head") of the second cell. The neurotransmitters move between the two cells in a specialized junction called the **synapse**. For example, your brain is composed of billions of neurons that connect to each other.

Sometimes neurons connect to an endocrine gland to stimulate the release of other chemical messengers called **hormones**. For example, neurons are wired to your adrenal glands, located on top of your kidneys. As part of a normal response called the *fight-or-flight* response, neurotransmitters stimulate hormone-producing cells within the gland to secrete epinephrine, or *adrenaline*. It travels through the blood to specific target organs in the body. As one example, epinephrine targets the heart, resulting in an increase in heart rate, which may be needed to help you flee a dangerous situation.

Last, neurons that connect to muscle tissue (*skeletal, cardiac,* and *smooth*) are called motor neurons. Acetylcholine (ACh) is the name of the neurotransmitter released by motor neurons that stimulates all the skeletal muscles in your body to contract. As ACh is released, it crosses a synapse and docks with receptors in the skeletal muscle cells to induce a response—contraction!

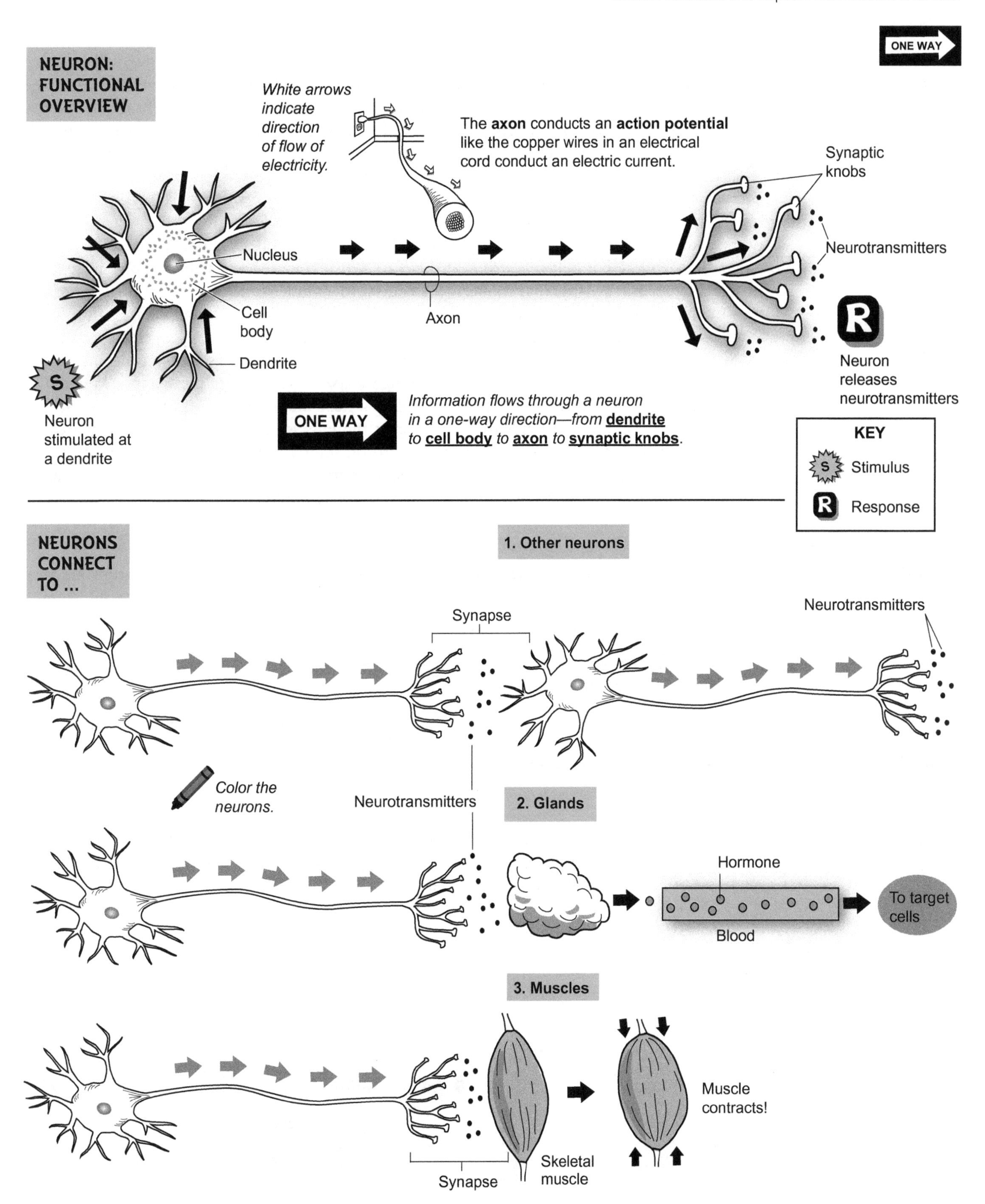

ONE WAY
NEURON: FUNCTIONAL OVERVIEW
White arrows indicate direction of flow of electricity.
The axon conducts an action potential like the copper wires in an electrical cord conduct an electric current.
Synaptic knobs
Neurotransmitters
Nucleus
Cell body
Axon
Dendrite
S
Neuron stimulated at a dendrite
R
Neuron releases neurotransmitters
ONE WAY
Information flows through a neuron in a one-way direction—from dendrite to cell body to axon to synaptic knobs.
KEY
S Stimulus
R Response
NEURONS CONNECT TO ...
1. Other neurons
Synapse
Neurotransmitters
Color the neurons.
Neurotransmitters
2. Glands
Hormone
Blood
To target cells
3. Muscles
Muscle contracts!
Synapse
Skeletal muscle

Description

There are many different structural types of neurons. For the sake of comparison, we will examine only three different types: **unipolar**, **bipolar**, and **multipolar** neurons. They are named after the number of long cellular processes (*dendrites, dendritic process, axon*) that branch off the cell body. A *uni*polar neuron has *one* very short process off the cell body that connects to a long, single axon with dendrites at one end and synaptic knobs on the other. A *bi*polar neuron has *two* separate processes—one long dendritic process that connects to the dendrites and one long axon that connects to the synaptic knobs. A *multi*polar neuron has *many* dendrites branching off the cell body. It also has a single axon that terminates in the synaptic knobs.

Every neuron has three basic parts: a **dendrite**(s), a **cell body**, and an **axon.** A neuron may have more than one dendrite but only one axon. At the end of the axon are the **synaptic knobs**. Here there is a small space called a **synapse** that connects the neuron to a muscle or a gland or another neuron.

Information flows through a neuron in a one-way direction. The starting point is always the dendrites—sensory receptors that receive various types of stimuli. With a strong enough stimulus—a threshold stimulus—a potential is generated that passes from the dendrites to the cell body. From here, an action potential is generated that travels along the axon to the end of the synaptic knobs. The synaptic knob produces and releases a chemical called a **neurotransmitter.** This chemical messenger diffuses across the synapse and binds to a receptor in the muscle cell, glandular cell, or neuron to which it is connected (see pp. 96–97). After the neurotransmitter binds to its receptor, it induces a response. For example, it may stimulate muscle tissue to contract, or cause cells in a gland to release a hormone, or stimulate a neuron to fire and generate a nervous impulse.

Location

- **Unipolar neurons**—*ex.:* sensory neurons of the peripheral nervous system. These neurons are very common.
- **Bipolar neurons**—*ex.:* photoreceptors in the retina of the eye. These neurons are rare.
- **Multipolar neurons**—*ex.:* most common type of neuron in the brain and spinal cord; motor neurons in the peripheral nervous system.

Function

All neurons conduct nerve impulses.

Study Tip

- The term *synaptic knobs* has many other names so don't let this confuse you. These include *synaptic terminals*, *axon terminals*, *synaptic end bulbs*, and *presynaptic terminals*. Phew! I think it's time to pick one and eliminate the others, don't you?

Key to Illustration

1. Unipolar neuron	D = Dendrite
2. Bipolar neuron	C = Cell body
3. Multipolar neuron	A = Axon
	SK = Synaptic knob

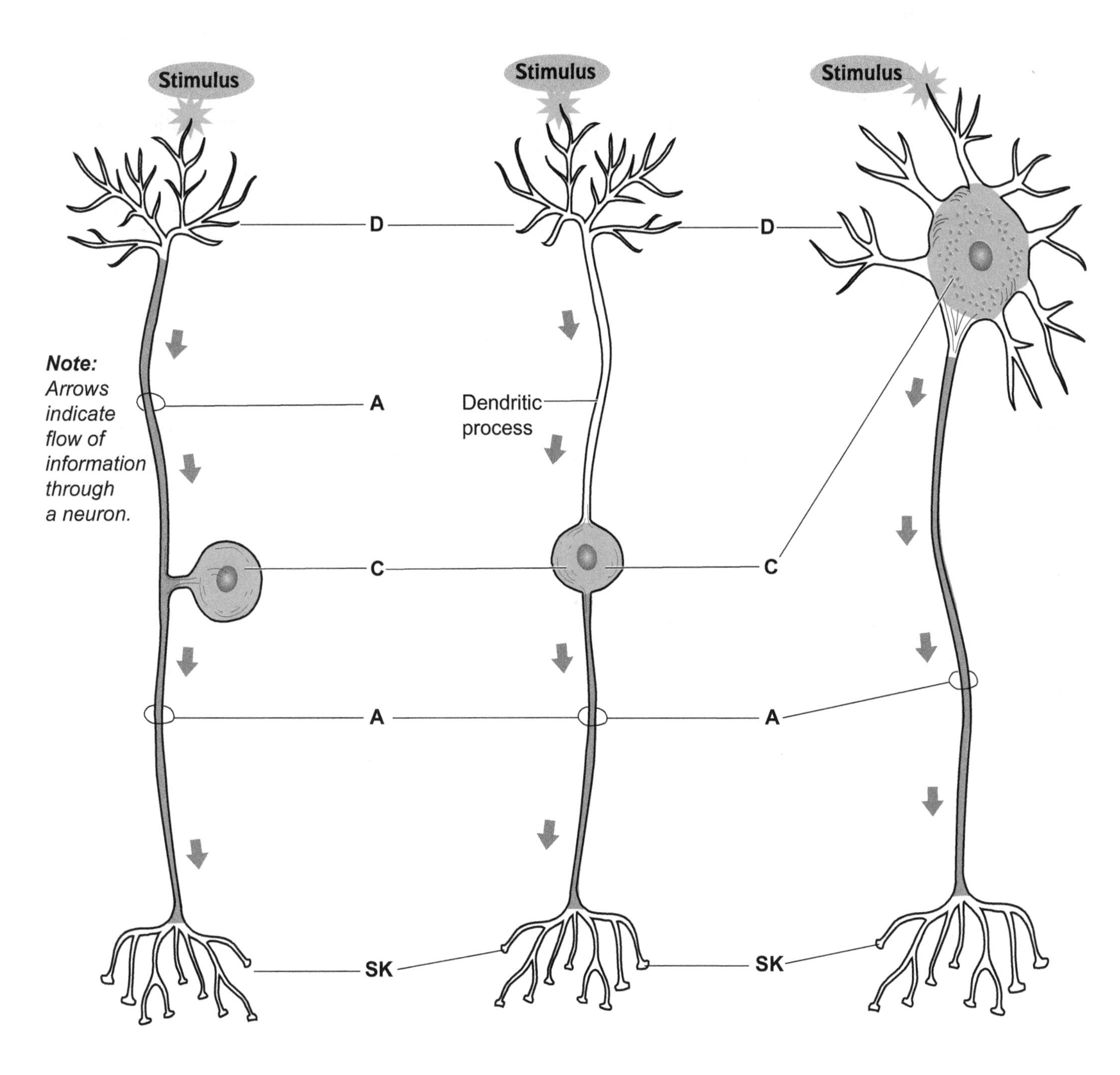

1. ____________ 2. ____________ 3. ____________

Color all the dendrites one color and the synaptic knobs a different color.

D = ____________

C = ____________

A = ____________

SK = ____________

Description

A multipolar neuron is one type of **neuron.** It has many processes called **dendrites** that respond to stimuli. The **soma** (*cell body*) is the regional area that includes the nucleus, cytoplasm, and various organelles. The rough endoplasmic reticulum is found in large clusters called **Nissl bodies.** The cell body tapers off into a funnel-shaped structure called the **axon hillock** that becomes the **axon.** Nervous impulses are conducted along the length of the axon. The axon ends in the **synaptic knob.**

Axons can be either **myelinated** or **unmyelinated.** In the process of **myelination,** the axon is wrapped in a sheath of lipid and protein that insulates the nervous impulse and speeds impulse conduction (see pp. 94–95). This process begins in the fetus and continues into late adolescence. In the central nervous system (CNS) cells called **oligodendrocytes** are responsible for the myelination process. In the peripheral nervous system (PNS), **neurolemmocytes** (*Schwann cells*) perform this task by wrapping themselves around the axon many times. These layers of plasma (*cell*) membrane constitute the **myelin sheath.** During this process, the nucleus and other organelles are pushed to the outer surface of the neurolemmocyte (*Schwann cell*). This outer layer is called the **neurilemma.** Segments of unwrapped axon between neurolemmocytes (*Schwann cells*) are called **nodes of Ranvier.**

Analogy

Each neuron has only one **axon.** The axon is like an **electrical cord.** The axon conducts a nervous impulse like the copper wires in the cord conduct electricity. The **myelin sheath** serves to insulate the axon like the **plastic casing** around the electrical cord.

Location

Multipolar neurons are the most common type of neuron in the brain and spinal cord but also occur as motor neurons in the peripheral nervous system.

Function

All neurons conduct nerve impulses.

Key to Illustration

1. Dendrite
2. Cell body (*soma*)
3. Nucleus of neuron
4. Nissl bodies
5. Axon hillock
6. Axon
7. Neurolemmocyte (*Schwann cell*)
8. Nodes of Ranvier
9. Nucleus of neurolemmocyte (*Schwann cell*)
10. Neurilemma
11. Endoneurium
12. Synaptic knob

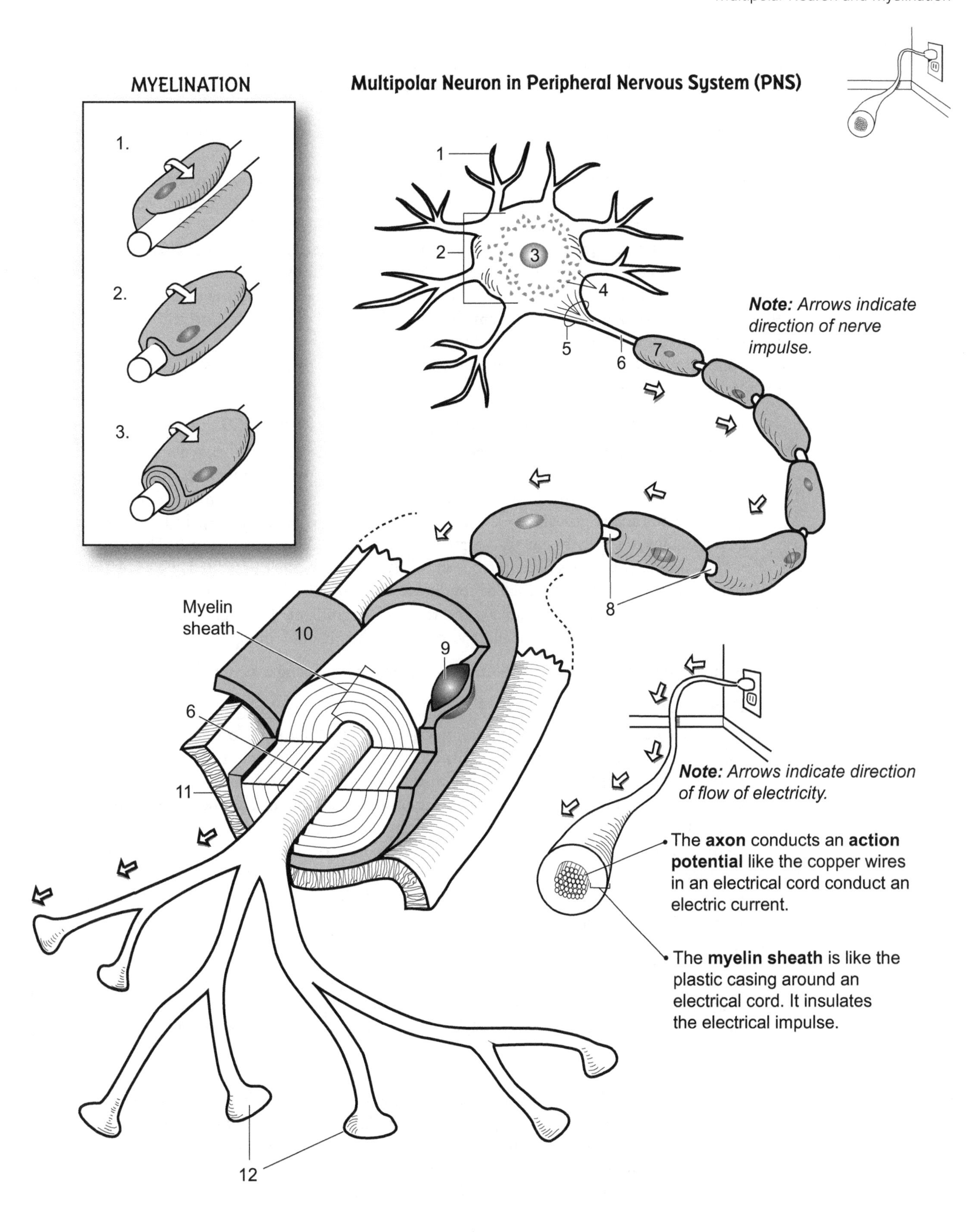
MYELINATION
1.
2.
3.
Multipolar Neuron in Peripheral Nervous System (PNS)
1
2
3
4
5
6
7
8
9
10
11
12
Myelin sheath
6
Note: Arrows indicate direction of nerve impulse.
Note: Arrows indicate direction of flow of electricity.
The axon conducts an action potential like the copper wires in an electrical cord conduct an electric current.
The myelin sheath is like the plastic casing around an electrical cord. It insulates the electrical impulse.

Description

Action potentials (*nerve impulses*) travel at great speeds along the long, slender **axons** of neurons. Like waves of electric current, they always move in a one-way direction from the cell body toward the synaptic knobs. Their purpose is to send a signal to other neurons, muscles, or glands. The axon is like a long cylinder of plasma membrane that separates the **intracellular fluid** (ICF) from the **extracellular fluid** (ECF).

The chemical composition of these two fluids differs in several ways. For example, the ECF has a greater concentration of **sodium ions** (Na^+), and the ICF has a greater concentration of **potassium ions** (K^+). Within the axon's plasma membrane are voltage-gated (V-G) ion channels. These proteins may be in one of two states—*open* or *closed*. They are triggered to open and close by a rapid change in membrane potential (voltage). Let's examine two different V-G channels: **Na^+ channels** and **K^+ channels.** Think of them like floodgates in a dam. When opened, they allow many ions to pass through the plasma membrane. V-G **Na^+ channels** only allow **Na^+** to diffuse into the axon, and V-G **K^+ channels** only allow **K^+** to diffuse out of the axon.

Conducting a nerve impulse involves three key steps: (1) *resting potential*, (2) *depolarization*, and (3) *repolarization*. Let's look at each step.

① **Resting (*membrane*) potential:** ***the state of the neuron when it is NOT conducting a nerve impulse.***

Resting potentials are also called **membrane potentials** (see p. 78). When a neuron is **not** conducting a nerve impulse, it is in this state. All living cell membranes are *polarized*. Just as a battery has a positive (+) end and a negative (−) end, the same is true for a cell membrane. The outside surface of the membrane is positive (+), and the inner surface is negative (−). Just as a battery stores a small voltage, the membrane also stores a voltage, normally about – 70 mV. This can be measured with a voltmeter that has one electrode in the ECF and the other in the ICF. This voltage is a type of potential energy that can be used to conduct a nerve impulse. This potential energy comes from two ion gradients: (1) the **Na^+ gradient** and the (2) **K^+ gradient.** These gradients are maintained by a protein pump not shown in the illustration called the **Na^+-K^+ pump** (see p. 80).

Action potential: ***the state of the axon when it is conducting a nerve impulse.***

② **Depolarization**

In order to conduct a nerve impulse, a stimulus strong enough to open a V-G NA^+ channel—called the **threshold**—must arrive at the neuron. The type of stimulus is unique to the type of neuron. For example, tactile receptors in the skin would be responsive only to touch. After this threshold is reached, it leads to the opening of V-G **Na^+ channels** in the first segment of the axon. This causes a sudden influx of **Na^+** resulting in a *depolarization*. This rapid shift of positive charges inside the axon reverses the normal polarity of the plasma membrane. In other words, the inside becomes more positive with respect to the outside. This sudden change in potential triggers the next segment of the axon to open its V-G **Na^+ channels**, and a chain reaction ensues. Like a tidal wave, depolarization moves forward along each segment of the axon.

③ **Repolarization**

In the wake of the depolarization "tidal wave," a *repolarization* occurs. The change in potential from the depolarization triggers V-G **K^+ channels** to open, and **K^+** diffuses out of the axon. This rapid outflux of positive charge helps flip the polarity almost back to what it used to be in the resting potential.

After repolarization, more K^+ diffuses out of the axon through other special membrane proteins called **leakage channels** to re-establish the original resting potential. These proteins are passive channels that do not open and close. After a short resting period called a **refractory period**, the neuron is once again ready to conduct another nerve impulse. Lastly, the excess Na^+ in the ICF is pumped out of the axon by the Na^+-K^+ pump (see p. 80).

The cycle then repeats itself: *resting potential*, *depolarization*, *repolarization*. And on and on it goes!

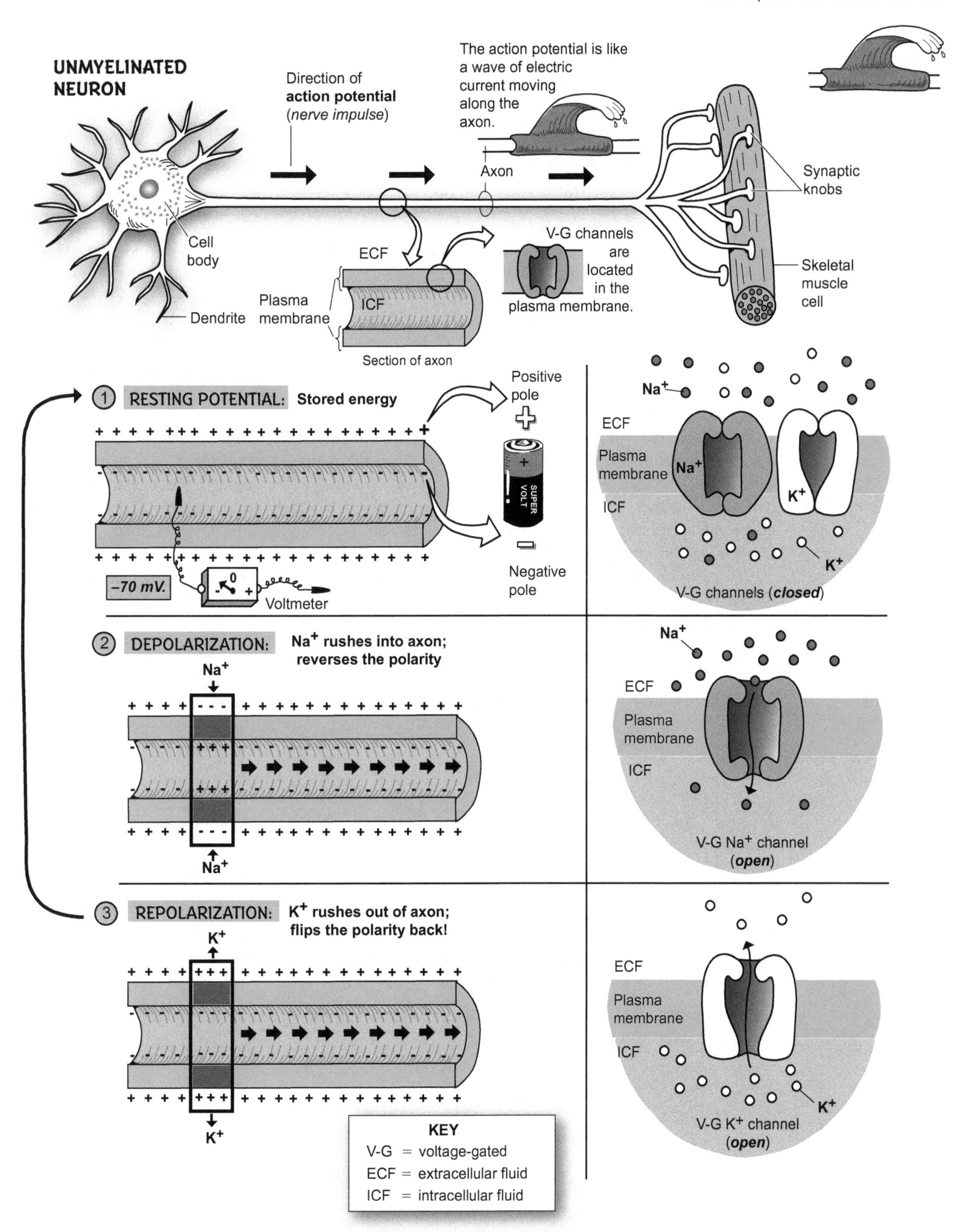
UNMYELINATED NEURON
Direction of **action potential** (*nerve impulse*)
The action potential is like a wave of electric current moving along the axon.
Axon
Synaptic knobs
Cell body
Dendrite
ECF
ICF
Plasma membrane
Section of axon
V-G channels are located in the plasma membrane.
Skeletal muscle cell
① RESTING POTENTIAL: Stored energy
Positive pole
Negative pole
SUPER VOLT
–70 mV.
Voltmeter
Na^+
ECF
Plasma membrane
ICF
K^+
V-G channels (*closed*)
② DEPOLARIZATION: Na^+ rushes into axon; reverses the polarity
V-G Na^+ channel (*open*)
③ REPOLARIZATION: K^+ rushes out of axon; flips the polarity back!
V-G K^+ channel (*open*)
KEY
V-G = voltage-gated
ECF = extracellular fluid
ICF = intracellular fluid

Description

Like an electric current travels through the copper wires in an electrical cord, nerve impulses travel along axons. The two general types of impulse conduction are: (1) continuous conduction, and (2) saltatory conduction (see p. 94). This module explains **continuous conduction**, which has the following key features:

- Occurs only in **unmyelinated neurons.**
- Is slower than saltatory conduction (about 2 m/sec).

Speed is determined mainly by myelination and diameter of the **axon**. Some axons are thick, and others are thin. The general rule is: "Thicker is quicker." Therefore, thin, unmyelinated axons are much slower than thick myelinated axons. For example, in one second, impulses in thick myelinated axons can travel longer than the length of a football field (100 yards), whereas those in thin unmyelinated would not have covered enough ground to make a first down (10 yards). That's a big difference in speed.

Unmyelinated neurons lack a myelin sheath around their axons. The **extracellular fluid** (ECF) comes into direct contact with the axon. This allows for easy exchange of ions across the plasma membrane of the axon.

Impulse conduction involves a repeated cycle of depolarization and repolarization all along the length of the axon (see p. 90). This is caused by the rapid opening and closing of two different voltage-gated (V-G) channels: (1) **Na^+ channels** and (2) **K^+ channels.** Like any gate, these channels can be in two states—either *open* or *closed*. They are triggered to open and close by a rapid change in membrane potential (voltage). First, a **threshold** stimulus triggers V-G Na^+ channels to open, and sodium ions diffuse into the **intracellular fluid** (ICF) within the axon.

Then V-G K^+ channels open, and K^+ ions diffuse out of the axon and into the **extracellular fluid** (ECF). This flux of ions across the membrane creates an action potential. With less surface area on the axon to depolarize, the impulse travels faster.

Analogy

Continuous conduction is like a domino effect in which each adjacent segment of the axon must be stimulated one after the other. The dominos are like the voltage-gated channels that must open and close in rapid succession.

Sequence of Events

The Na^+ channels are shown in the facing illustration. **Note:** K^+ channels are not shown in the illustration.

1. V-G **Na^+ channels** are stimulated to open, and Na^+ ions diffuse into the axon, resulting in a depolarization. This is followed by V-G **K^+ channels** opening to cause a repolarization.
2. In the adjacent part of the axon, V-G **Na^+ channels** again are stimulated to open, and Na^+ ions diffuse into the axon, resulting in another depolarization. This is followed by V-G **K^+ channels** opening to cause another repolarization.
3. This process of stimulating each adjacent segment of the axon continues along the entire length of the axon.

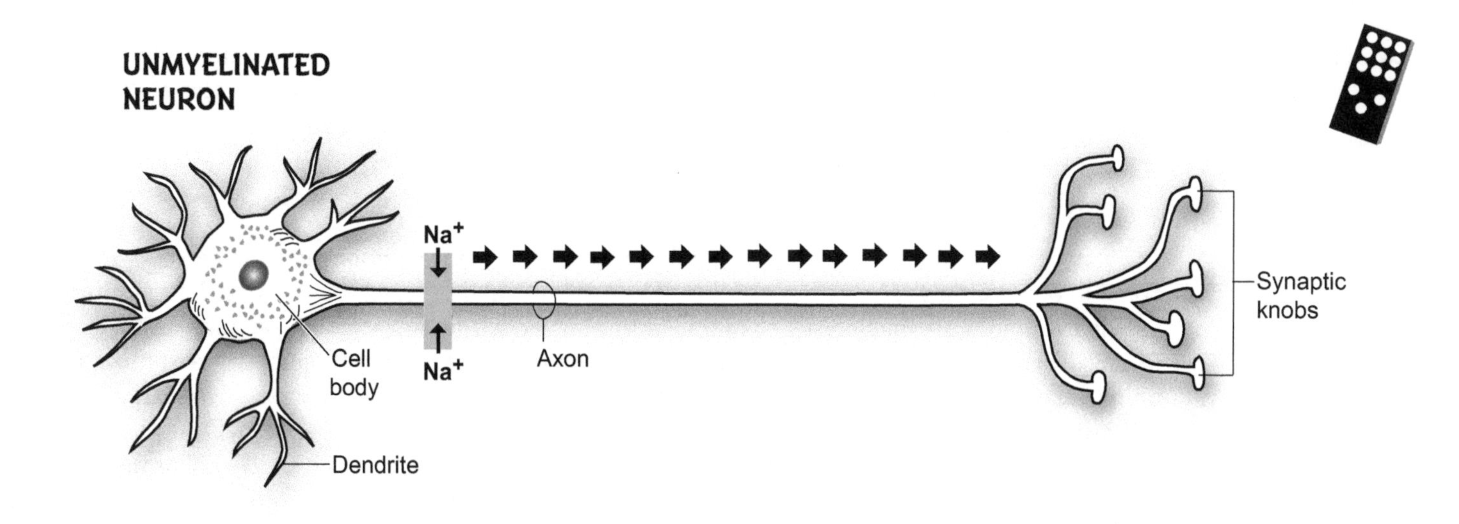

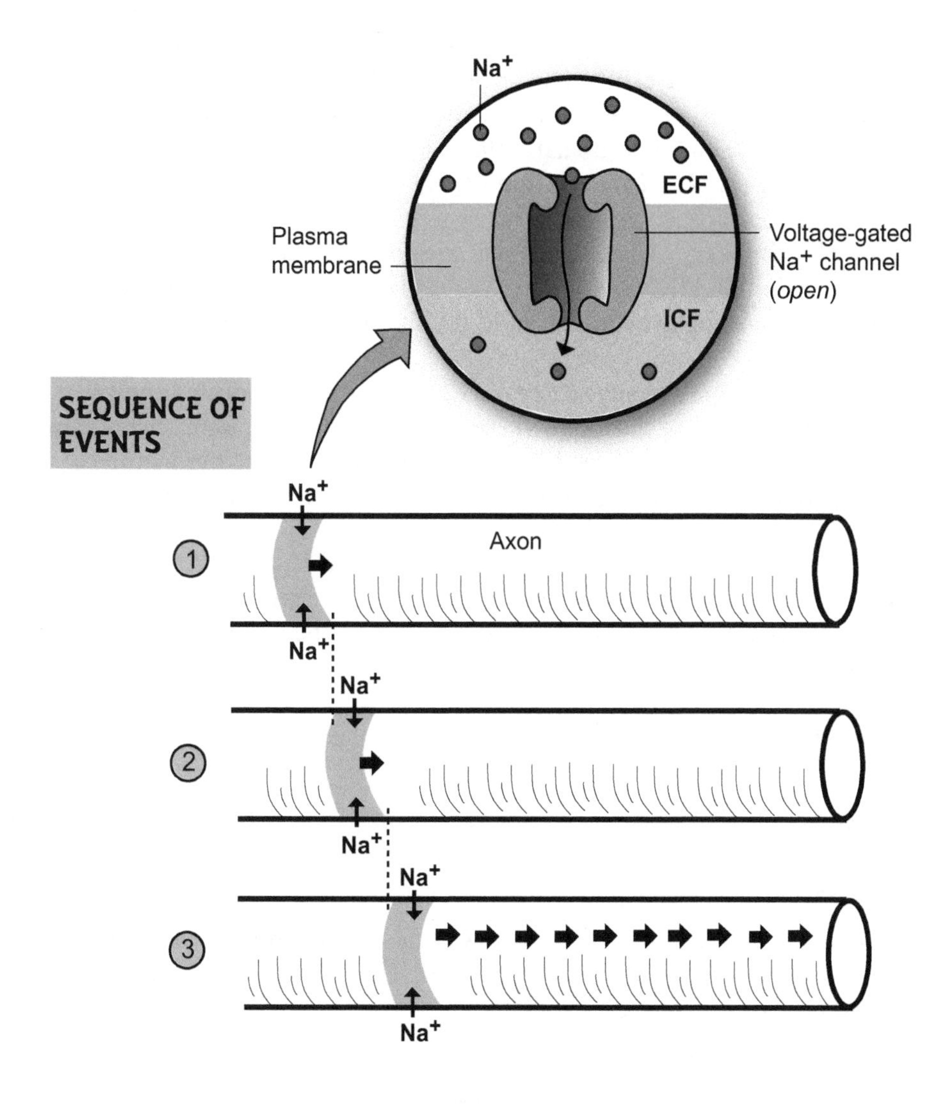

DOMINO EFFECT: Continuous conduction in an unmyelinated neuron is like a domino effect, in which each segment of the axon must be stimulated, one after the other.

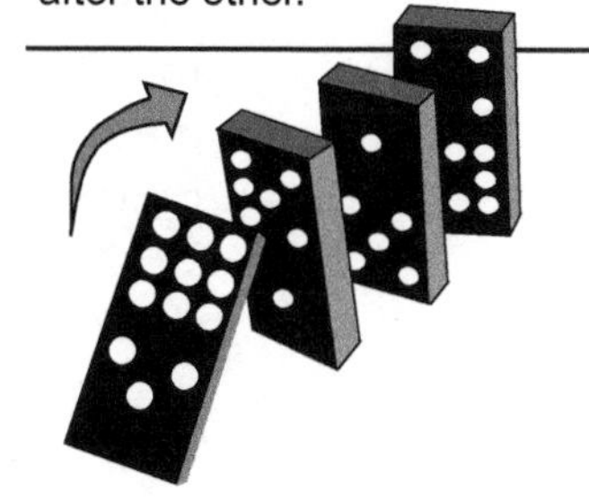

Description

The two general types of nerve impulse conduction are: (1) continuous conduction (see p. 92) and (2) saltatory conduction. This module explains the latter, **saltatory** (*saltare* = leaping) **conduction.** Saltatory conduction has the following key features:

- Occurs only in **myelinated neurons.**
- Is faster than continuous conduction (about 120 m/sec).

Speed of impulse conduction is mainly determined by myelination and diameter of the **axon.** Some axons are thick and others are thin. The general rule is: "Thicker is quicker." Therefore, thick myelinated axons are much faster than thin unmyelinated axons. For example, in one second, impulses in thick myelinated axons can travel longer than the length of a football field, whereas those in thin unmyelinated axons would not have covered enough ground to make a first down. That's a huge difference in speed.

Myelinated neurons in the peripheral nervous system have their axons covered by **neurolemmocytes** (*Schwann cells*). During development of the nervous system, these cells wrap themselves around the axon, forming layers of plasma membrane collectively referred to as the **myelin sheath.** This covering serves as insulation like the plastic coating around the copper wires in an electrical cord. Short segments of exposed axon between the neurolemmocytes are called **nodes of Ranvier.**

Impulse conduction involves a repeated cycle of depolarization and repolarization all along the length of the axon (see p. 90). This is caused by the rapid opening and closing of two different voltage-gated (V-G) channels: (1) **Na^+ channels** and (2) **K^+ channels.** Like any gate, these channels can be in two states—either *open* or *closed*. They are triggered to open and close by a rapid change in membrane potential (voltage). First, V-G Na^+ channels are stimulated to open to allow sodium ions to diffuse into the **intracellular fluid (ICF)** within the axon. Then V-G K^+ channels open, and K^+ ions diffuse out of the axon and into the **extracellular fluid** (ECF). This flux of ions across the membrane creates an action potential.

In myelinated neurons, V-G Na^+ channels and V-G K^+ channels are located mostly at nodes of Ranvier. Why? This is the only part of the axon where ions can cross the plasma membrane because ions within the ECF come in direct contact with the axon. Because ions can't cross the barrier of the myelin sheath, voltage-gated channels are not needed in segments of the axon covered by myelin.

Analogy

Saltatory conduction is like skipping a stone across the surface of the water, as the impulse seems to "jump" from node to node. Strictly speaking, the impulse does not actually jump. But this image is useful to illustrate that saltatory conduction is up to 60 times faster than continuous conduction, which is more like a domino effect.

Sequence of Events

The Na^+ channels are shown in the facing illustration. (**Note:** V-G K^+ channels are not shown in the illustration.)

① V-G **Na^+ channels** at the **first node** are stimulated to open, and sodium ions diffuse into the axon, causing depolarization. The action potential generated is shuttled through the ICF of the axon to the next node of Ranvier.

② The action potential triggers the V-G **Na^+ channels** at the **second node** to open, and sodium ions diffuse into the axon, causing another depolarization. The action potential generated is again shuttled through the axon to the next node of Ranvier.

③ The action potential triggers the V-G **Na^+ channels** at the **third node** to open, and sodium ions diffuse into the axon, causing another **depolarization.** The action potential generated is again shuttled through the axon to the next node of Ranvier. This process repeats itself until the action potential is conducted along the entire length of the axon.

Study Tip

Remember the alliteration: <u>S</u>altatory is <u>S</u>peedy like <u>S</u>kipping a <u>S</u>tone.

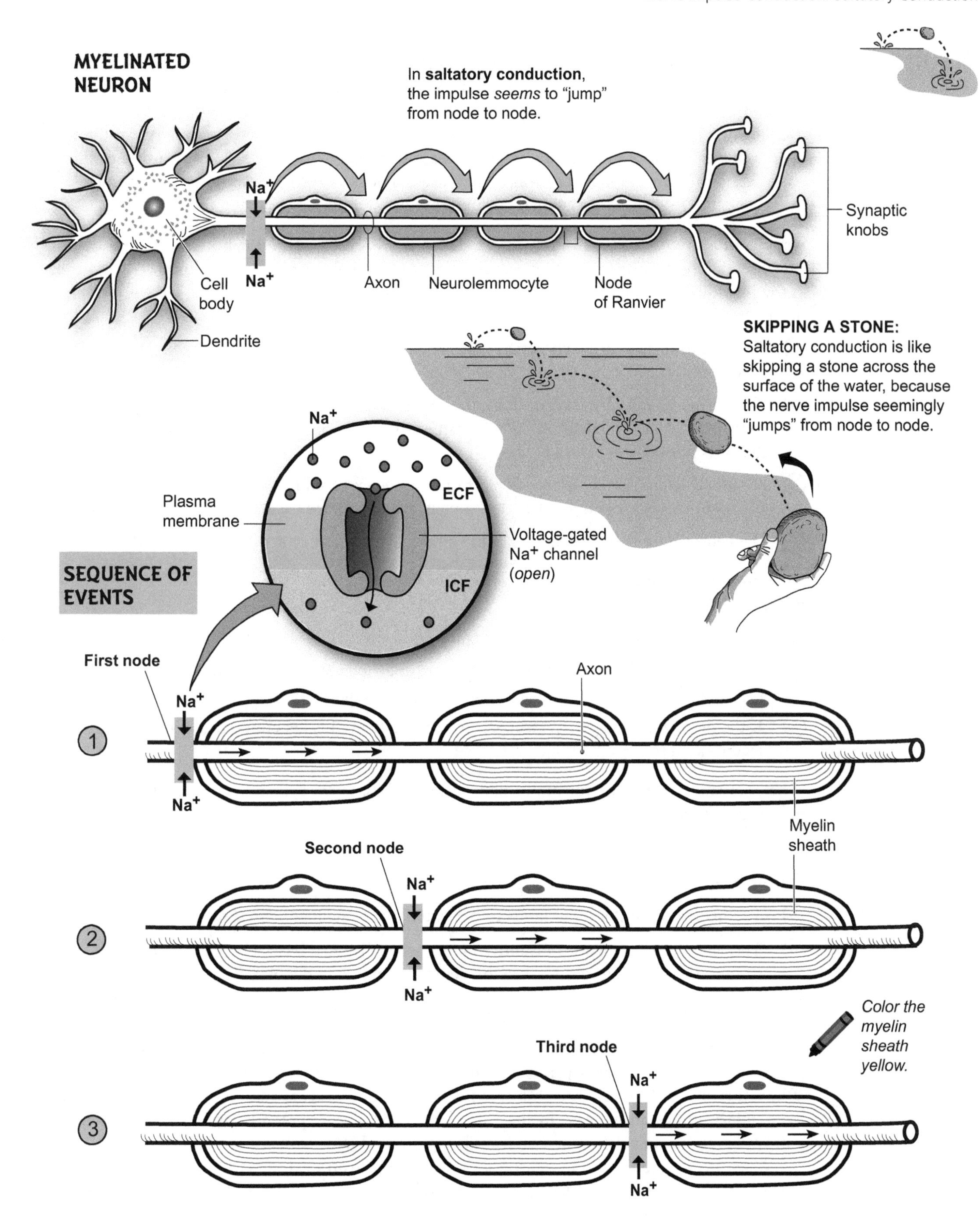
MYELINATED NEURON
In **saltatory conduction**, the impulse *seems* to "jump" from node to node.
Na+
Na+
Synaptic knobs
Cell body
Axon
Neurolemmocyte
Node of Ranvier
Dendrite
SKIPPING A STONE:
Saltatory conduction is like skipping a stone across the surface of the water, because the nerve impulse seemingly "jumps" from node to node.
Na+
ECF
Plasma membrane
Voltage-gated Na+ channel (*open*)
ICF
SEQUENCE OF EVENTS
First node
Axon
Na+
1
Na+
Myelin sheath
Second node
Na+
2
Na+
Color the myelin sheath yellow.
Third node
Na+
3
Na+

Description

Synaptic transmission is the method by which neurons communicate with other cells. The illustration shows the **synaptic knobs** of a **motor neuron** connecting to a **skeletal muscle cell** in the **neuromuscular junction** (NMJ), whereas the synaptic knobs of neuron #1 connect to the dendrites of neuron #2 in a **neuro-neuro junction** (NNJ). The **synaptic cleft** is the narrow, fluid-filled space between the synaptic knob and the muscle cell/nerve cell. In both cases, a chemical messenger called a **neurotransmitter** must cross this gap, like a ferryboat crossing a channel, and dock at the receptor in the **plasma membrane** of the other cell. This may lead to either a stimulatory or inhibitory response in the other cell.

Steps 1–4: ***Secretion of neurotransmitters.*** These steps are the same for both an NMJ and an NNJ. The only difference is in the type of neurotransmitter(s) involved. Only **acetylcholine** (**ACh**) is released in the NMJ, whereas NNJs have dozens of different possible neurotransmitters. Here is a summary of this mechanism:

① Neurotransmitters are synthesized in synaptic knobs and stored within **vesicles** at high concentrations.

② When a nerve impulse travels along the axon to the synaptic knobs, it triggers the opening of **voltage-gated (V-G) calcium (Ca^{++}) channels** because of a change in voltage. Note that these channels normally are *closed*.

③ Because calcium is at a higher concentration in the extracellular fluid, it diffuses into the synaptic knobs through the open V-G Ca^{++} channels, where it helps trigger exocytosis (see p. 76).

④ The neurotransmitter is released into the fluid-filled **synapse**, or synaptic cleft, and diffuses across it.

Steps 5–7: ***Inducing a response in either muscle or nerve tissue.***
For the **NMJ:**

⑤ **ACh** binds to the ACh-gated Na^+ channel (*closed state*).

⑥ The binding of **ACh** induces the ACh-gated Na^+ channel to *open*, allowing only sodium ions (**Na^+**) to diffuse into the muscle cell resulting in depolarization.

⑦ If the polarization is past threshold, an action potential is generated in the muscle cell that may lead to a muscle contraction.

When nerve impulses from the motor neuron cease, **ACh** in the synapse is quickly broken down, and muscle action potentials stop.

For the **NNJ:**

⑤ The **neurotransmitter** (**NT**) binds to NT-gated Na+ channel (*closed state*).

⑥ The binding of the **neurotransmitter** induces the NT-gated Na+ channel to *open*, allowing only sodium ions (**Na^+**) to diffuse into the neuron, resulting in depolarization.

⑦ If the depolarizing potential reaches threshold, it can trigger an **action potential** (*nerve impulse*).

Alternatively, the neurotransmitter may bind to a *different* NT-gated channel that is specific for either **potassium** (K^+) or chloride (Cl^-). The illustration shows a NT-gated K^+-gated channel as an example. When this channel opens, K^+ diffuses out of neuron #2. This loss of positive charge leads to **hyperpolarization**—an increase in the value of the membrane potential (see p. 78). This inhibitory pathway prevents an action potential from being induced in neuron #2.

When nerve impulses from neuron #1 cease, the neurotransmitter in the synapse has two fates: (1) like ACh, it is broken down, or (2) it is reused by being transported back into the synaptic knob or to a neighboring neuroglial cell.

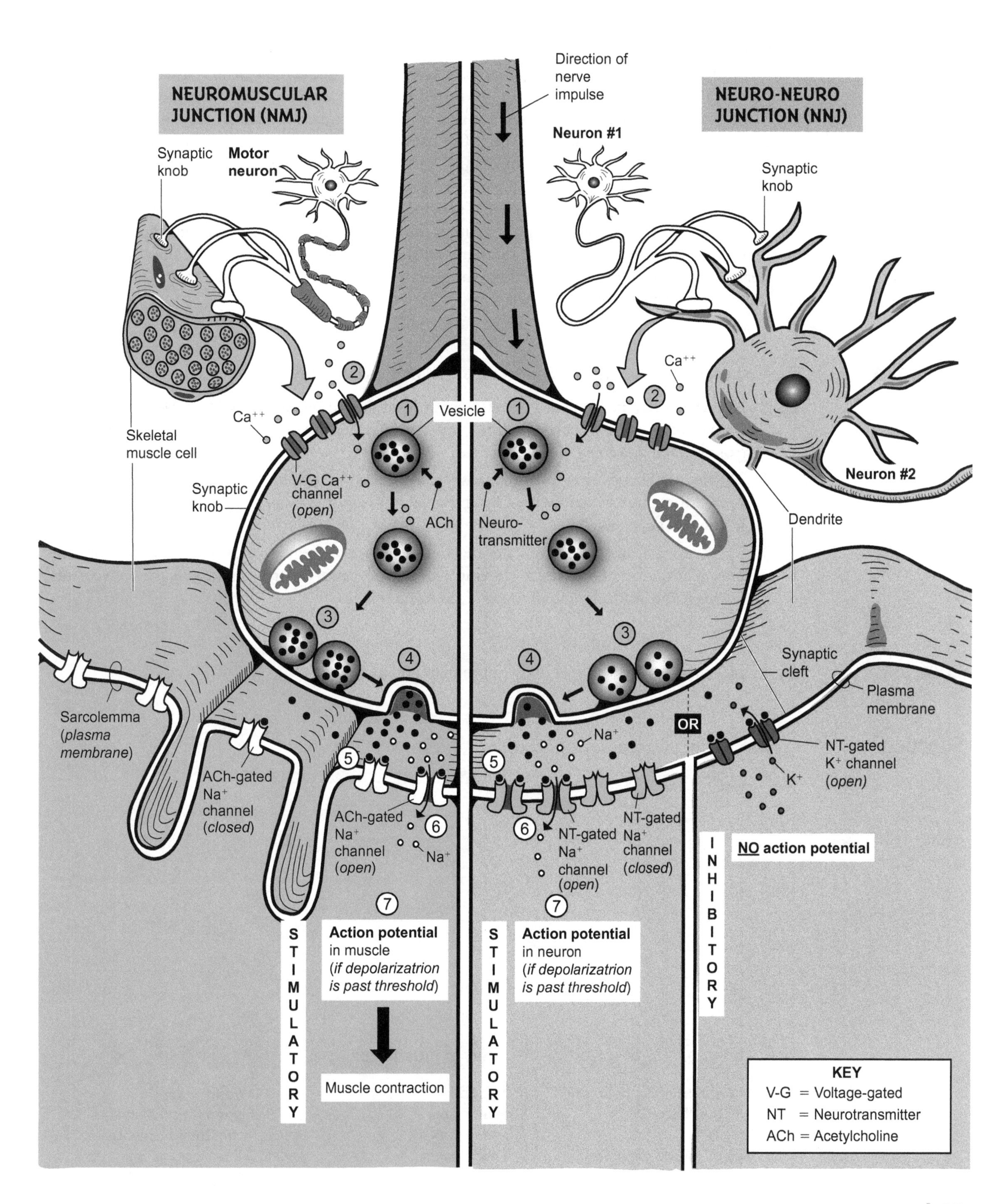
Direction of nerve impulse
NEUROMUSCULAR JUNCTION (NMJ)
NEURO-NEURO JUNCTION (NNJ)
Neuron #1
Synaptic knob
Motor neuron
Synaptic knob
Ca++
Ca++
Vesicle
Skeletal muscle cell
Neuron #2
V-G Ca++ channel (open)
Synaptic knob
ACh
Neuro-transmitter
Dendrite
Synaptic cleft
Plasma membrane
OR
Sarcolemma (plasma membrane)
Na+
NT-gated K+ channel (open)
ACh-gated Na+ channel (closed)
K+
ACh-gated Na+ channel (open)
Na+
NT-gated Na+ channel (open)
NT-gated Na+ channel (closed)
NO action potential
INHIBITORY
STIMULATORY
Action potential in muscle (if depolarizatrion is past threshold)
STIMULATORY
Action potential in neuron (if depolarizatrion is past threshold)
Muscle contraction
KEY
V-G = Voltage-gated
NT = Neurotransmitter
ACh = Acetylcholine

Description

A whole skeletal muscle is packaged like a series of tubes within other tubes. A muscle is first divided into bundles of long tubes called **fascicles.** Each fascicle is a bundle of skeletal muscle cells or fibers. Each skeletal muscle cell is a bundle of **myofibrils.** Each myofibril is composed primarily of myofilaments made of two key protein filaments—myosin and actin. These proteins are part of a repeated unit called a **sarcomere** (see pp. 100–101), the structural and functional unit for muscle contraction. The ends of a sarcomere are defined by the **Z-lines,** made of protein.

All of these different bundles of tubelike structures are held together with connective tissue. The **epimysium** (*epi* = upon, *mys* = muscle) is a tough, fibrous, connective tissue that completely surrounds the outside of a whole muscle. Within the whole muscle, the **perimysium** (*peri* = around, *mys* = muscle) fills the space around and between fascicles. Surrounding each skeletal muscle cell is another connective tissue called the **endomysium** (*endo* = within, *mys* = muscle) that mainly serves to bind one skeletal muscle cell to another.

Analogies

Three analogies are given for structures at the level of the sarcomere.

1. An **actin,** or thin, filament compares with a **double-stranded chain of pearls.** Each **pearl** is equivalent to **one molecule of actin.** (Note that this analogy does not include the troponin and tropomyosin proteins.)
2. The myosin filament has **myosin heads** (cross bridges) branching off of it, which later attach to actin during muscle contraction. From the lateral view, these heads appear angled like the **tail feathers in an arrow.**
3. The **myosin heads** attach to actin and pull on it with a regular movement. To visualize this movement, imagine the **heads moving** like a **boat rower's oars.** Unlike the oar movements, however, the heads do not all move at the same time.

Location

All skeletal muscles in the human body (more than 600 in all).

Function

Contraction for the purpose of movement.

Key to Illustration

1. Whole muscle (biceps brachii from the illustration)
2. Fascicle
3. Skeletal muscle cell (fiber)
4. Myofibril
5. Epimysium
6. Perimysium
7. Endomysium
8. Myosin heads
9. Myosin (*thick*) filaments
10. Actin (*thin*) filaments
11. Z-line

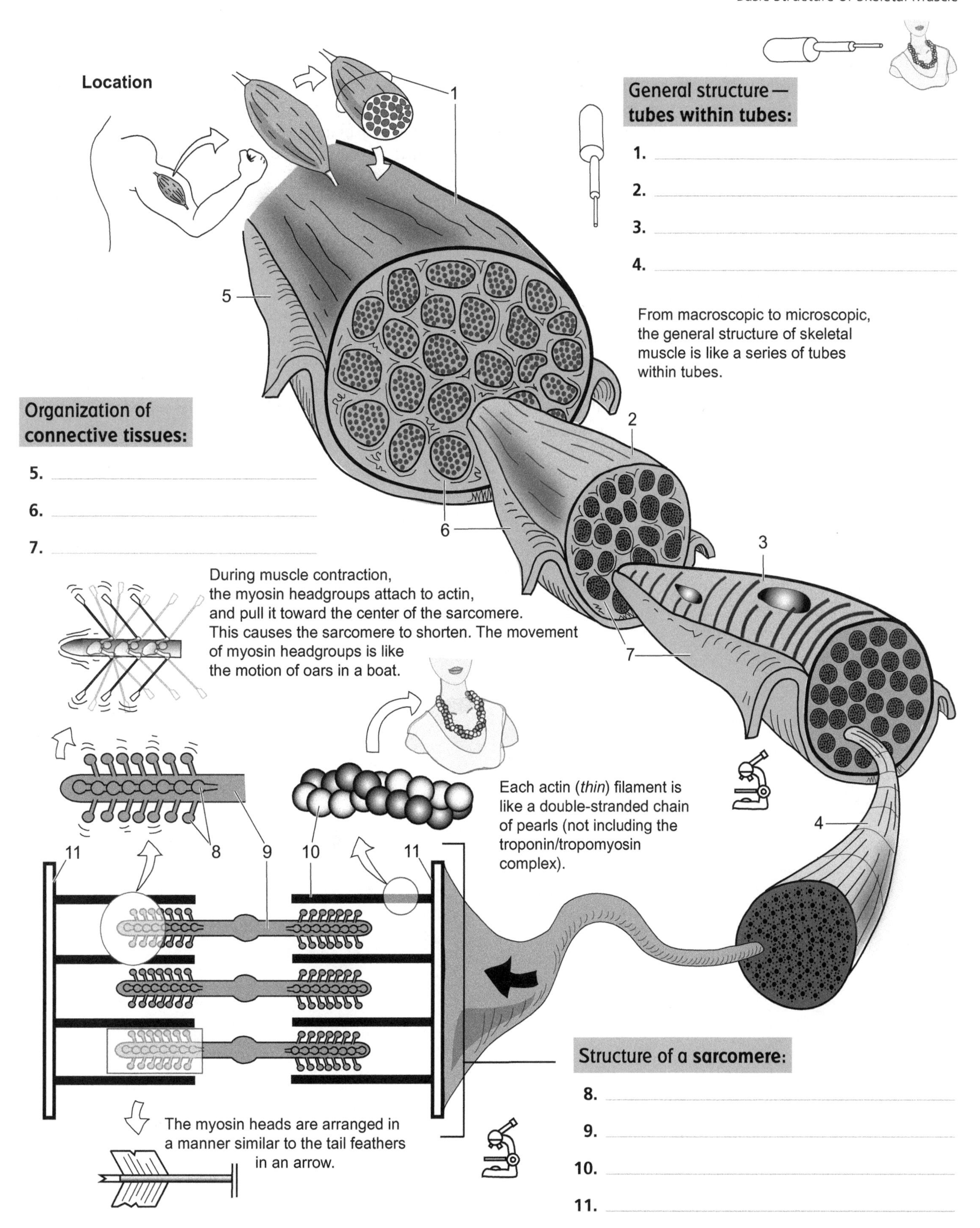
Location
General structure — tubes within tubes:
1.
2.
3.
4.
From macroscopic to microscopic, the general structure of skeletal muscle is like a series of tubes within tubes.
Organization of connective tissues:
5.
6.
7.
During muscle contraction, the myosin headgroups attach to actin, and pull it toward the center of the sarcomere. This causes the sarcomere to shorten. The movement of myosin headgroups is like the motion of oars in a boat.
Each actin (thin) filament is like a double-stranded chain of pearls (not including the troponin/tropomyosin complex).
Structure of a sarcomere:
8.
9.
10.
11.
The myosin heads are arranged in a manner similar to the tail feathers in an arrow.
1
2
3
4
5
6
7
8
9
10
11
11

Description

In the **sliding filament mechanism** of muscle contraction, the stimulus for skeletal muscle contraction is always a nerve impulse—no nerve impulse, no contraction. The **sarcomere** is the functional unit of muscle contraction, and its ends are marked by proteins called **Z-lines**. In the middle are the two proteins needed for muscle contraction: (1) **actin** (*thin*) **filaments**, and (2) **myosin** (*thick*) **filaments**. The myosin filaments have oar-like structures called **myosin heads** that stick out. Their function is to attach to and pull on actin.

During contraction, all the myosin heads pull on actin together, causing actin to slide past myosin as the actin is pulled toward the center of the sarcomere. The end result is that the sarcomere shortens. This shortening can be visualized most easily by noting that the Z-lines move toward each other. Also, note that neither the length of the myosin nor the actin changes during contraction. The thing that does change is the amount of *overlap* between the myosin and actin, which increases.

As goes the sarcomere, so goes the skeletal muscle cells and the whole muscle. When the sarcomeres contract, this causes the skeletal muscle cells to shorten, which shortens the whole muscle.

Relaxation occurs when nerve impulses stop. Then the linkages between the myosin heads and the actin filaments are broken, allowing the actin to slide past the myosin. This lengthening of the sarcomere continues until it has returned to its original position. These cycles of contraction and relaxation are repeated as needed.

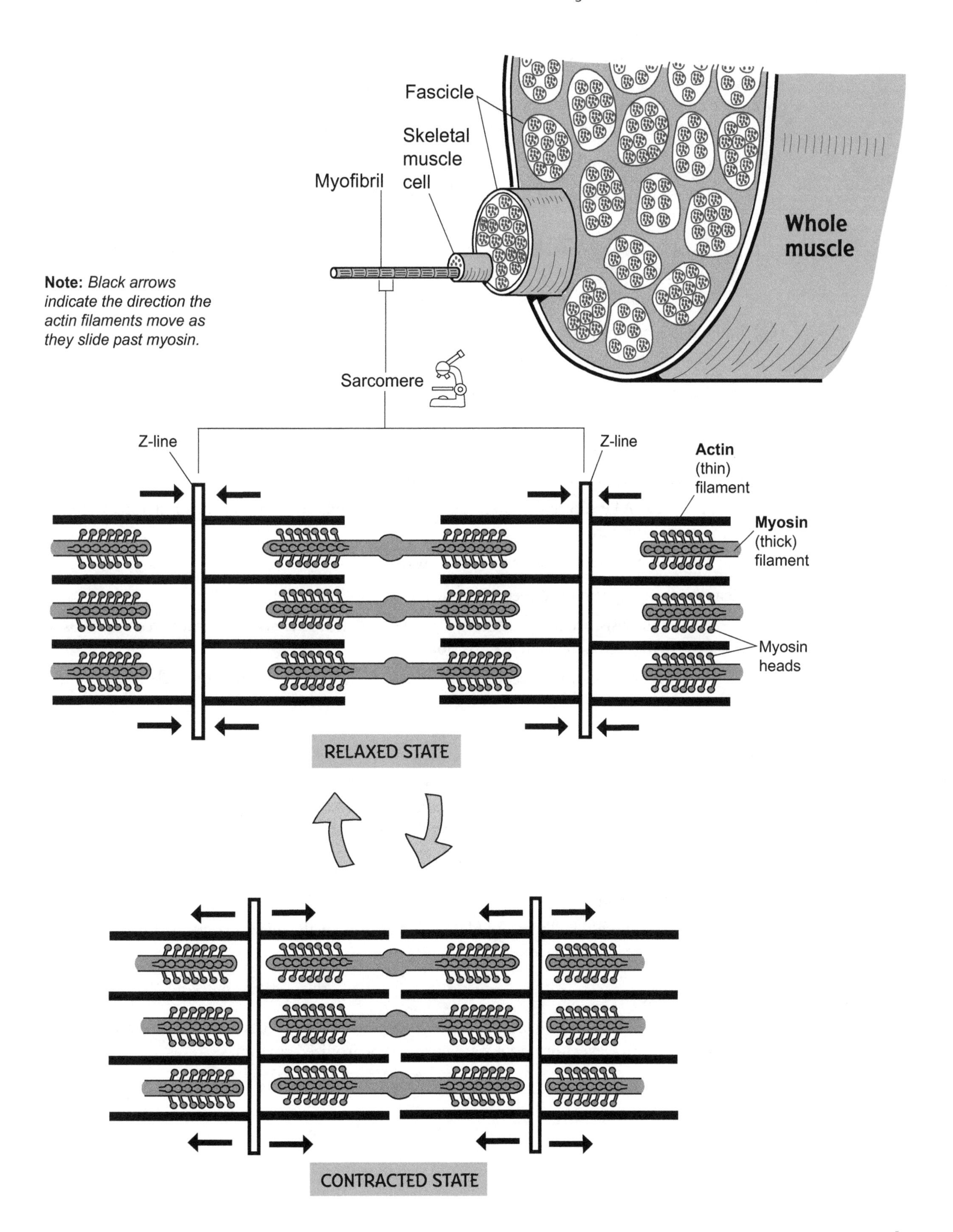
Fascicle
Skeletal muscle cell
Myofibril
Whole muscle
Note: Black arrows indicate the direction the actin filaments move as they slide past myosin.
Sarcomere
Z-line
Z-line
Actin (thin) filament
Myosin (thick) filament
Myosin heads
RELAXED STATE
CONTRACTED STATE

Description

All muscles can perform only one major function—contraction. At the microscopic level, it is the sarcomeres that shorten or contract. The process involves actin filaments sliding past myosin in a four-step contraction cycle: (1) **attachment**, (2) **pulling**, (3) **detachment**, and (4) **reactivation.**

A step-by-step summary of the contraction cycle is as follows:

① **Attachment:** Myosin head attaches to actin.

- The regulatory proteins **troponin** and **tropomyosin** already have shifted to expose the **myosin binding sites.**
- The products of **ATP hydrolysis**, **ADP** and **P**, are still bound to the myosin head from the previous cycle.
- The myosin head attaches to the actin molecule (*peg in a hole analogy*).

② **Pulling:** Myosin head pulls on actin.

- The myosin head bends inward, which releases **ADP** and **P.**
- The myosin molecule bends, thereby pulling the actin filament toward the center of the sarcomere.
- This pulling action is called a *power stroke.*

③ **Detachment:** Myosin head releases from actin.

- As a new **ATP** molecule enters the cycle and binds to a myosin head, it forces the head to detach from actin.
- The link between myosin and actin is broken.

④ **Reactivation:** Myosin head reenergizes itself for another cycle.

- ATP hydrolysis occurs, producing a net release of free energy that activates the myosin head by moving it to a high energy position.
- The myosin head now is ready to attach to actin, and the contraction cycle is repeated.

Analogies

- **Peg in a hole analogy**

  The peg fits in the hole. In this case, the actin binding site on the myosin head is like the peg, and the myosin binding site on the actin molecule is like the hole. In reality, it is a chemical bond rather than a peg and hole connection that allows the myosin head to bind to actin. But this analogy helps visualize the links formed between myosin and actin.

- **Cotton swab analogy**

  Each myosin molecule is like a double-headed cotton swab (*with one end cut off*) that is able to bend like a hinge in two places.

- **Rower analogy**

  As the myosin heads move through their four-step cycle, they somewhat resemble the movement of oars in a boat. However, they do not move in a synchronized rhythm.

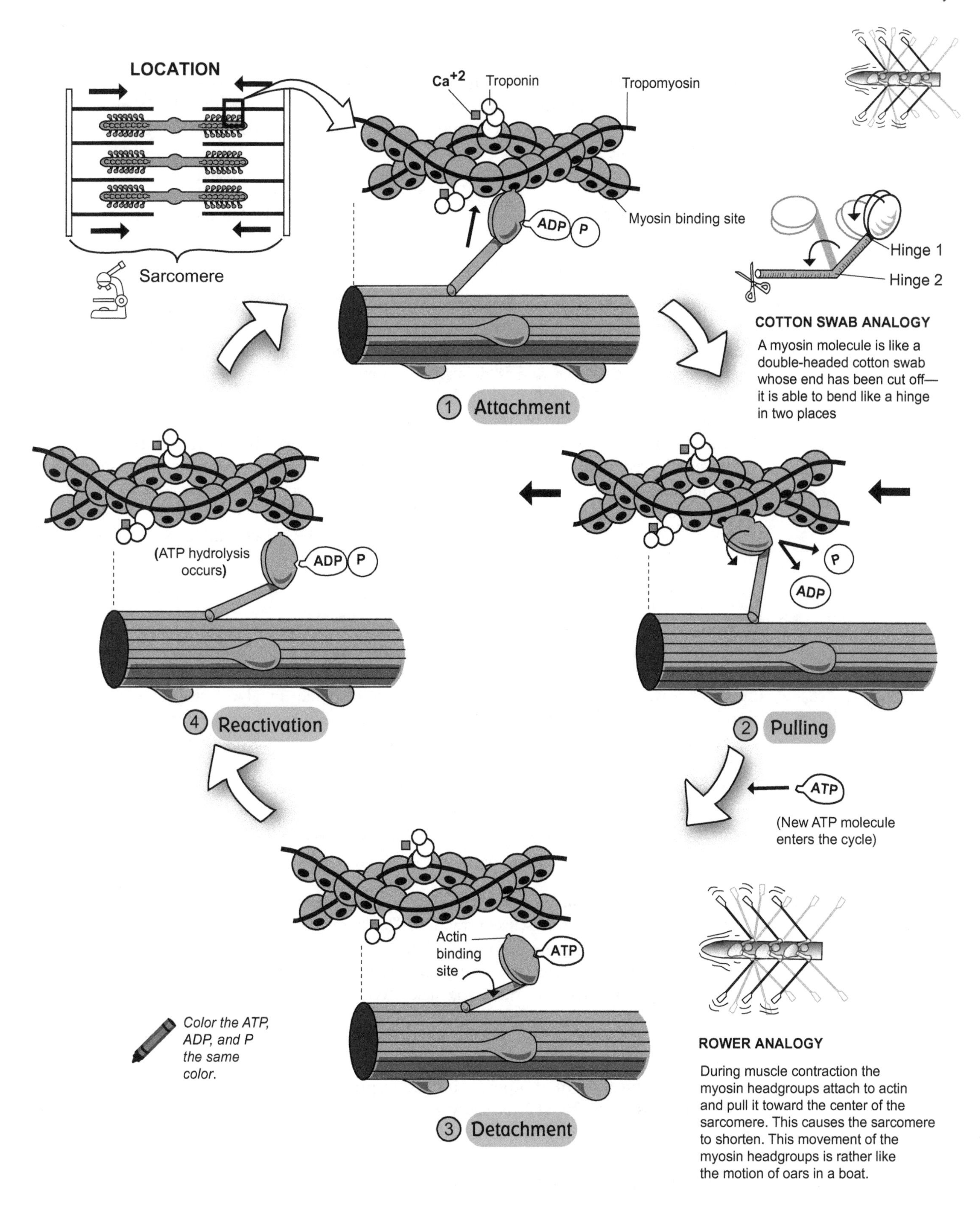
LOCATION
Sarcomere
Ca+2
Troponin
Tropomyosin
Myosin binding site
ADP
P
Hinge 1
Hinge 2
COTTON SWAB ANALOGY
A myosin molecule is like a double-headed cotton swab whose end has been cut off—it is able to bend like a hinge in two places
1 Attachment
(ATP hydrolysis occurs)
ADP
P
4 Reactivation
ADP
P
2 Pulling
ATP
(New ATP molecule enters the cycle)
Actin binding site
ATP
3 Detachment
Color the ATP, ADP, and P the same color.
ROWER ANALOGY
During muscle contraction the myosin headgroups attach to actin and pull it toward the center of the sarcomere. This causes the sarcomere to shorten. This movement of the myosin headgroups is rather like the motion of oars in a boat.

Description

ATP is the common currency that fuels cellular activities and the primary energy molecule used by all cells. Muscle contraction demands enormous amounts of ATP—even those that last only a few seconds. For example, a contracting muscle cell uses about 2 million molecules of ATP every second! Wow! Without a constant supply of ATP, muscle contraction ceases. This begs the question: Where does all of this ATP come from?

Here is a summary of the four major sources of energy for muscle contraction:

1. **ATP hydrolysis** (see p. 64): Muscle cells store a local supply of ATP, but it does not provide much energy.
   - Time allowed for muscle contraction: about 5 seconds.

2. **Phosphorylation from creatine phosphate:** Whereas ATP is the *primary* energy molecule used by cells, **creatine phosphate (CP)** is the *secondary* energy molecule found *only* in muscle cells. Creating it is a two-part process. First, the small chemical **creatine** (C) is produced by various organs, including the liver, and is delivered to muscle cells. Then, within the muscle cells, a phosphate group is added to C with the help of an enzyme.

   The function of CP is to transfer its phosphate group directly onto ADP to create more ATP. In a sense, CP "recharges" ADP to make more ATP. One CP yields one ATP. During the relaxed state, muscle cells make excess ATP, which quickly undergoes ATP hydrolysis, supplying a phosphate group to transform C into CP.

   In the contracting state, the levels of ADP are rising within the muscle cells, so CP transfers its phosphate group onto the excess ADP to form more ATP, which can be used to power more muscle contractions. But this process doesn't last long.
   - Time allowed for contraction: about 15 seconds.

3. **Fermentaton** (see p. 198): Fermentation is the production of two **lactate** molecules from the breakdown of a single **glucose** molecule when no oxygen is present in the cell. This process occurs in the cytosol and yields a net gain of **two ATP. Glucose** is provided by either the blood or the breakdown of local **glycogen** within the muscle tissue. In **glycolysis**, a glucose molecule normally is converted into pyruvate. When oxygen is present, pyruvate enters the mitochondria to begin aerobic respiration. But under conditions of no oxygen, the pyruvate is converted into lactate instead.

   Lactate is a waste product that accumulates in muscle cells, then diffuses out and is transported into liver cells. Although two ATP is not much for each cycle of fermentation, it doubles the time of contraction compared with the previous two mechanisms.
   - Time allowed for muscle contraction: about 30 seconds.

4. **Aerobic respiration** (see p. 188): Aerobic respiration is the essential mechanism used to deliver a constant supply of ATP for muscle contraction. Without it, we couldn't sustain muscle contraction for long periods of time. In aerobic respiration, oxygen is required to produce more ATP through the **citric acid cycle** and **electron transport systems** inside the **mitochondrion.** The **pyruvate** produced in glycolysis enters the citric acid cycle in the mitochondrion instead of getting converted into lactate. The glucose needed for glycolysis comes from the same sources it did in fermentation.

   In addition to glucose, energy sources include **fatty acids** from adipose tissue and **amino acids** from the breakdown of proteins. Instead of producing only two ATP (as in fermentation), the new total is about **36 ATP** for every glucose molecule broken down. As an example, if you plan to run a marathon, you will be relying on aerobic respiration to fuel all those muscle contractions for running!
   - Time allowed for muscle contraction: hours.

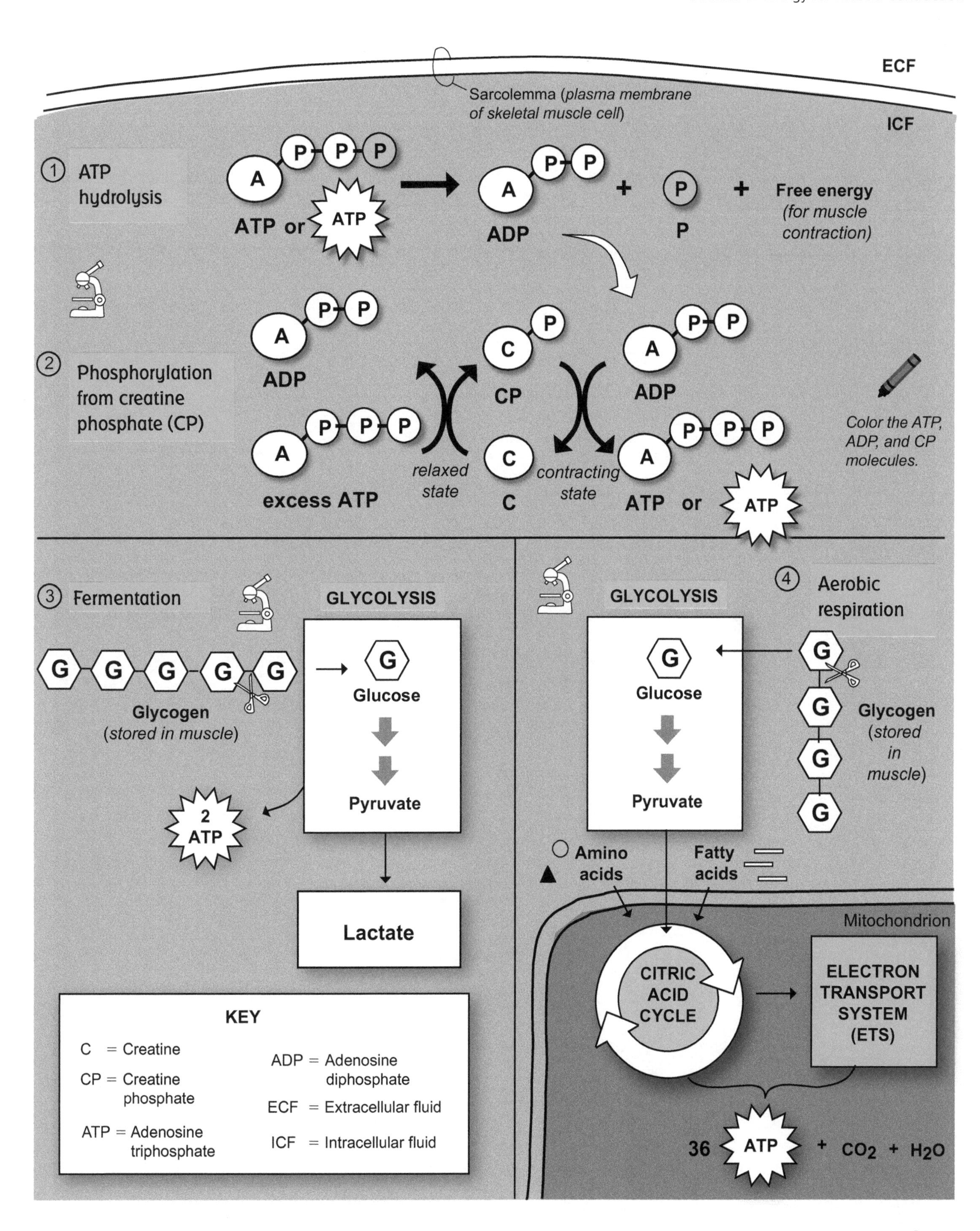
ECF
Sarcolemma (*plasma membrane of skeletal muscle cell*)
ICF
① ATP hydrolysis
A P P P
ATP or ATP
A P P
ADP
+ P
P
+ Free energy
(*for muscle contraction*)
② Phosphorylation from creatine phosphate (CP)
ADP
C P
CP
ADP
excess ATP
relaxed state
C
contracting state
ATP or ATP
Color the ATP, ADP, and CP molecules.
③ Fermentation
GLYCOLYSIS
G G G G G
Glycogen
(*stored in muscle*)
Glucose
Pyruvate
2 ATP
Lactate
KEY
C = Creatine
CP = Creatine phosphate
ATP = Adenosine triphosphate
ADP = Adenosine diphosphate
ECF = Extracellular fluid
ICF = Intracellular fluid
④ Aerobic respiration
GLYCOLYSIS
Glucose
Pyruvate
Glycogen
(*stored in muscle*)
Amino acids
Fatty acids
Mitochondrion
CITRIC ACID CYCLE
ELECTRON TRANSPORT SYSTEM (ETS)
36 ATP + CO_2 + H_2O

Notes

CARDIOVASCULAR SYSTEM

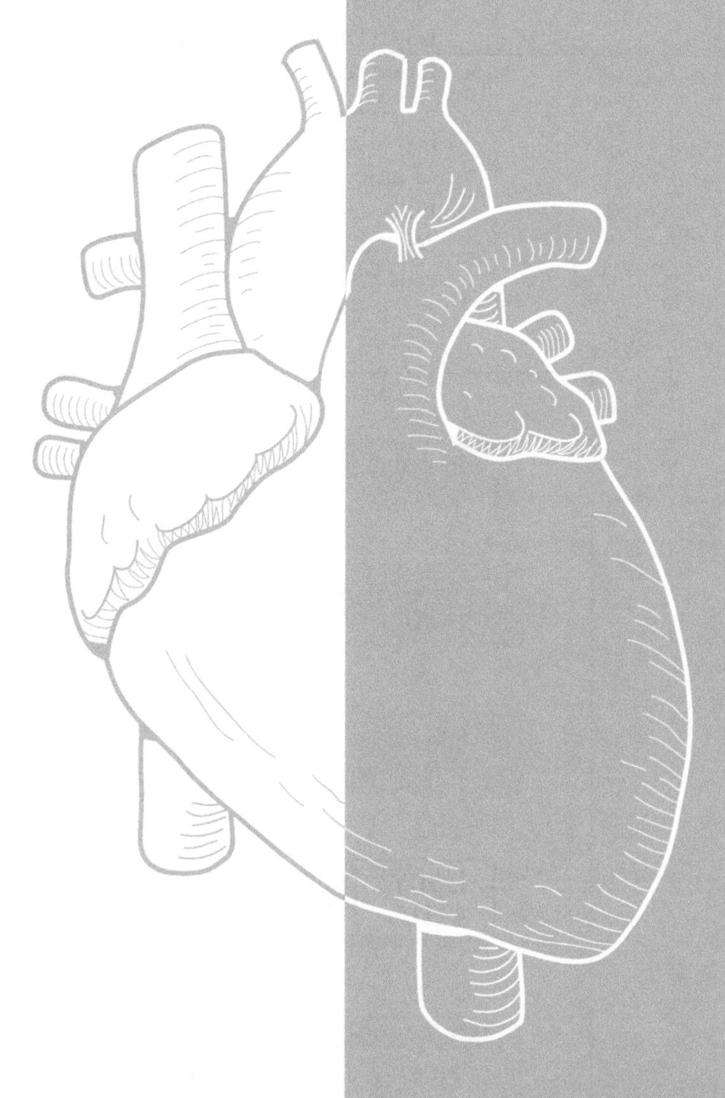

Description

The **cardiovascular system** consists of the **heart** and all the **blood vessels.** Functionally, the heart is like a double pump with each pump connected to the other through a long series of blood vessels. It consists of two receiving chambers called **atria** and two pumping chambers called **ventricles.** The left side of the heart always pumps oxygenated blood, and the right side always pumps deoxygenated blood.

The illustration on the facing page shows blood flow through the heart, through the **pulmonary circuit,** and through the **systemic circuit.** The pulmonary circuit refers to all the blood vessels that take deoxygenated blood from the right ventricle of the heart to the lungs, and then return oxygenated blood to the left atrium. After this oxygenated blood is pumped from the left atrium to the left ventricle, it is pumped out to the rest of the body. The blood vessels that transport this oxygenated blood to the body are part of the systemic circuit.

All gas exchange occurs within **capillaries.** Capillaries are microscopic blood vessels only one cell layer thick. Their wall is made of **simple squamous epithelium.** These flat cells easily permit the diffusion of gases, such as **oxygen** (O_2) and **carbon dioxide** (CO_2). Oxygen diffuses out of the blood and into body cells to be used in the process of cellular respiration. Carbon dioxide is a normal by-product of cellular respiration and gradually builds up within body cells. Carbon dioxide diffuses from the body cells into the capillary.

Beginning in the right atrium, this is a flowchart for the blood flow:

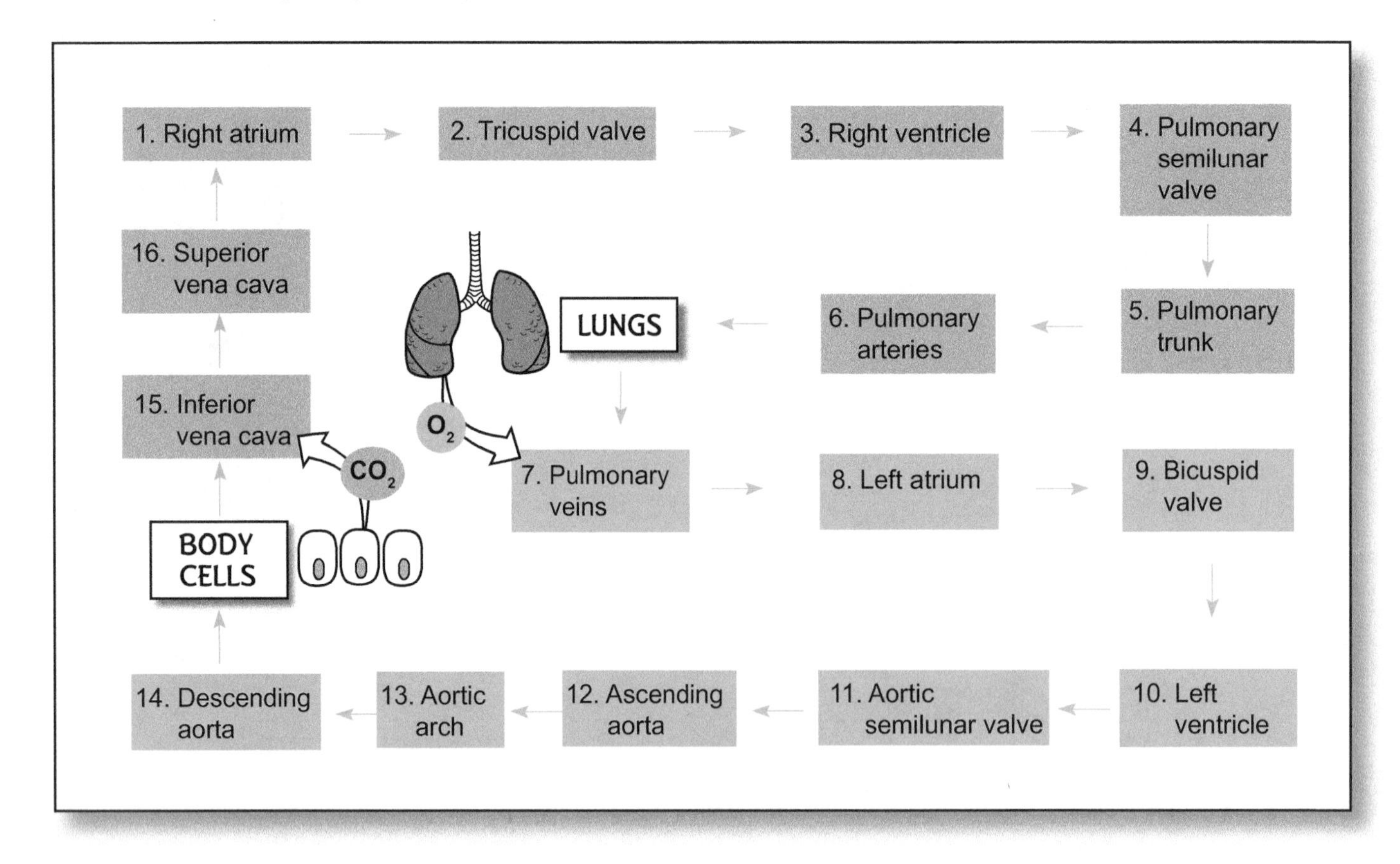

Study Tip

In this pathway that begins and ends with the right atrium, here is a handy tip to recall which atrioventricular valve—bicuspid or tricuspid—comes first, and which comes second, in the pathway.

You ride your **tri**cycle (**tri**cuspid) before your **bi**cycle (**bi**cuspid).

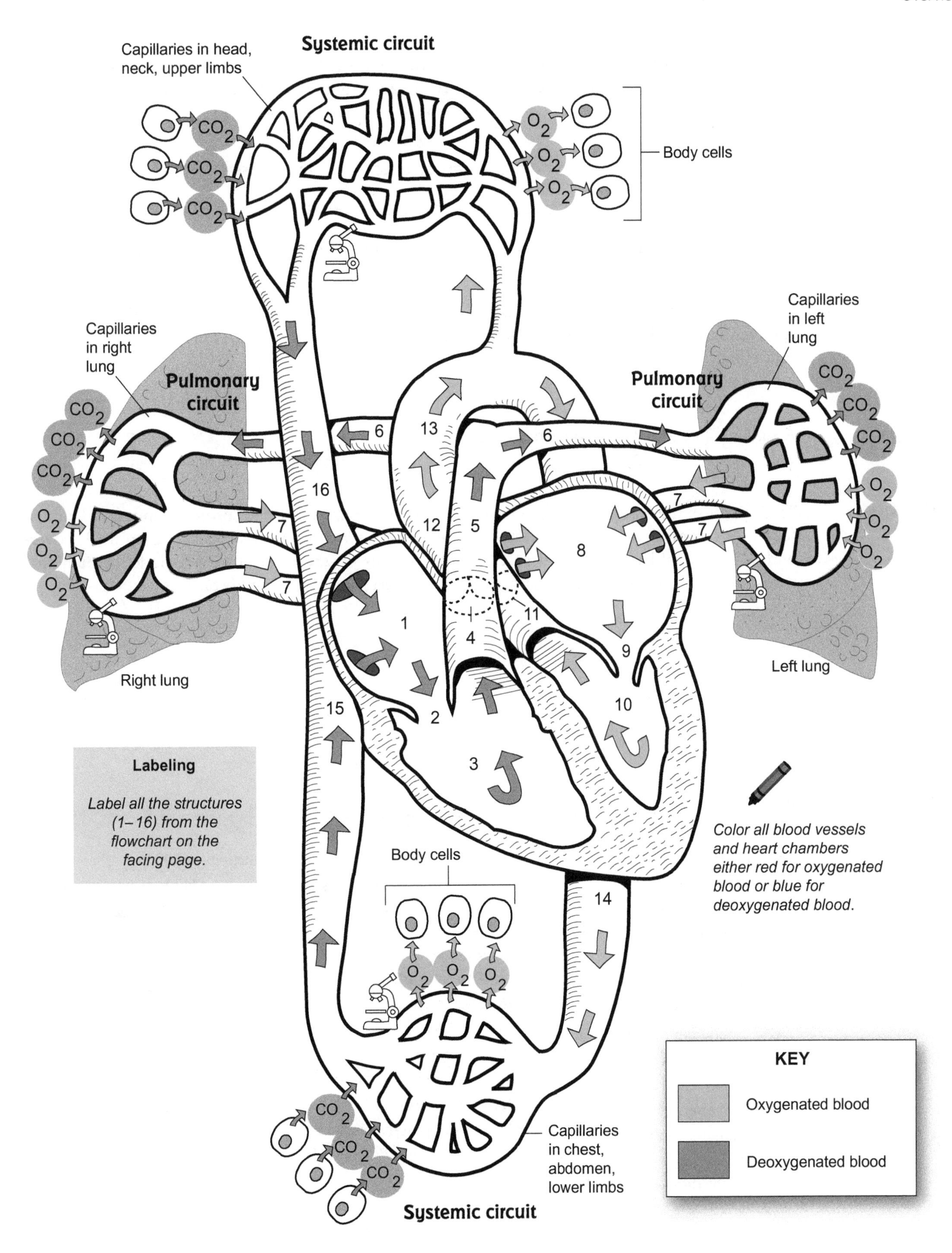

Labeling

Label all the structures (1–16) from the flowchart on the facing page.

Color all blood vessels and heart chambers either red for oxygenated blood or blue for deoxygenated blood.

Description

Coronary circulation refers to the blood supply to the heart. The **coronary arteries** supply oxygenated blood to the heart, and the **cardiac veins** carry deoxygenated blood back to the heart. The following flowchart summarizes coronary circulation through the blood vessels.

Base of aorta → left and right coronary arteries → branches of coronary arteries

(circumflex a., anterior interventricular a., marginal a., posterior interventricular a.) →

coronary capillaries → cardiac veins → coronary sinus → right atrium

When **coronary arteries** become blocked, the blood supply to the heart is reduced. This deprives cardiac muscle cells of oxygen. If this blockage persists over many years, it may lead to a **myocardial infarction** (*heart attack*).

Analogy

A **sulcus** is like a **shallow groove** or **gulley.**

Study Tip

To distinguish the anterior from the posterior view of the heart, use the **coronary sinus** as a landmark for the posterior view. It is often difficult to see on a dissected specimen because it is normally covered with a horizontal band of fatty tissue. Good landmarks for the anterior view include the **pulmonary trunk, anterior interventricular artery, circumflex artery,** and **ascending aorta.**

Key to Illustration

Blood vessels (B)

- B1. Superior vena cava
- B2. Inferior vena cava
- B3. Ascending aorta
- B4. Aortic arch
- B5. Descending aorta
- B6. Brachiocephalic trunk
- B7. L. common carotid a.
- B8. L. subclavian a.
- B9. Pulmonary trunk
- B10. Pulmonary arteries
- B11. Pulmonary veins
- B12. R. coronary a. (*in r. anterior atrioventricular groove*)
- B13. L. coronary a.
- B14. Circumflex a.
- B15. Anterior interventricular a. (*in anterior interventricular sulcus*)
- B16. Marginal a.
- B17. Anterior cardiac v.
- B18. Great cardiac v.
- B19. Small cardiac v.
- B20. Coronary sinus
- B21. Posterior v. of l. ventricle
- B22. Middle cardiac v.
- B23. Posterior interventricular a. (*in posterior interventricular sulcus*)

Structures (S)

- S1. Ligamentum arteriosum
- S2. Apex (tip) of heart
- S3. Auricle of r. atrium
- S4. Auricle of l. atrium

Chambers (C)

- C1. R. atrium
- C2. L. atrium
- C3. R. ventricle
- C4. L. ventricle

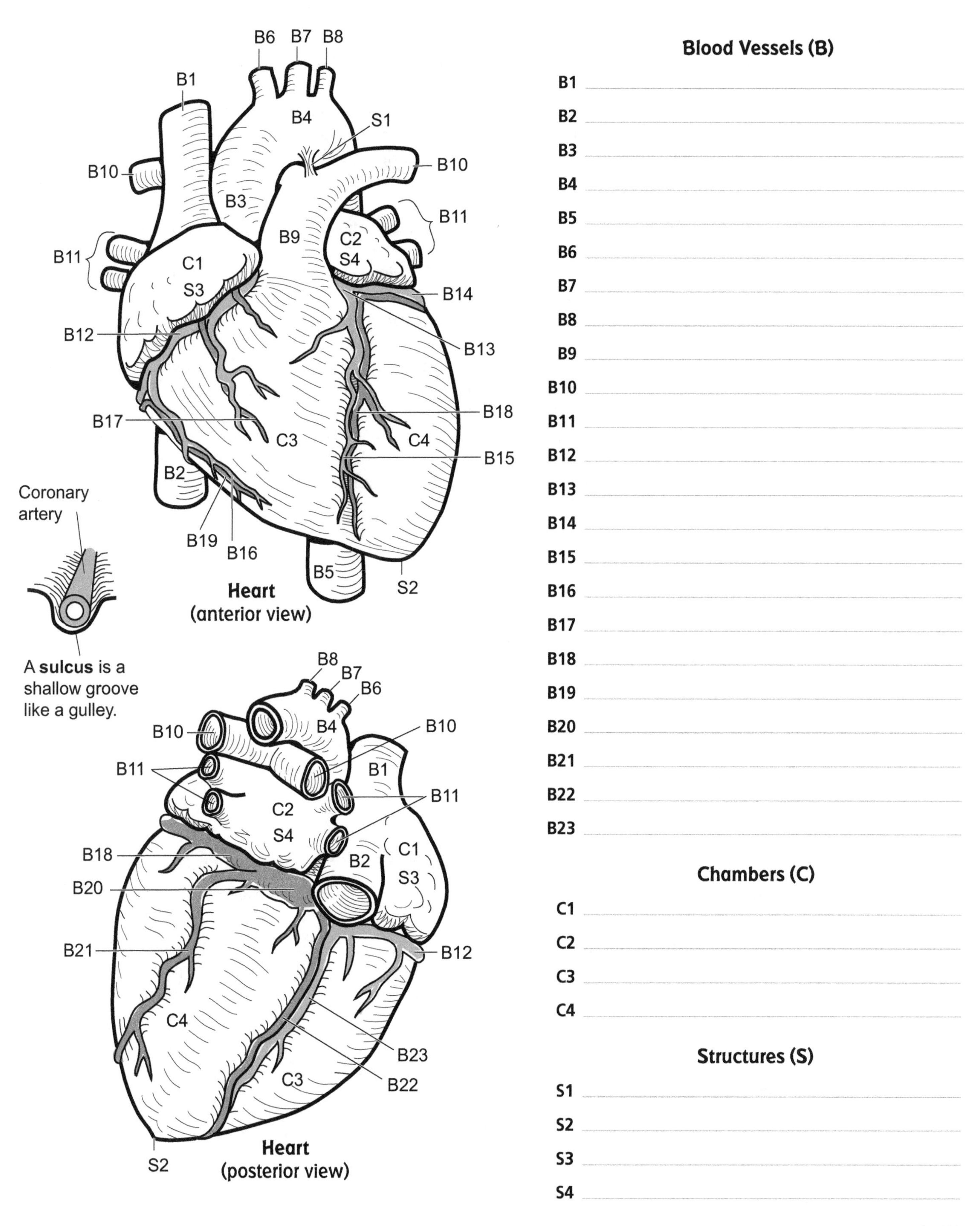
B6 B7 B8
B1
B4
S1
B10
B3
B9
C2
S4
B11
C1
S3
B14
B12
B13
B17
B18
C3
C4
B15
B2
Coronary artery
B19
B16
B5
S2
Heart
(anterior view)
A **sulcus** is a shallow groove like a gulley.
B20
B21
B22
B23
Heart
(posterior view)
Blood Vessels (B)
B1
B2
B3
B4
B5
B6
B7
B8
B9
B10
B11
B12
B13
B14
B15
B16
B17
B18
B19
B20
B21
B22
B23
Chambers (C)
C1
C2
C3
C4
Structures (S)
S1
S2
S3
S4

Description

The heart is divided into left and right halves and has four chambers, two atria and two ventricles. The **atria** are the first chambers to receive blood from the body. They fill with blood, contract, and transfer blood to the pumping chambers, or **ventricles.** The **right ventricle** pumps deoxygenated blood to the lungs, and the **left ventricle** pumps oxygenated blood to the rest of the body. The heart has two different types of valves: **atrioventricular** (AV) **valves** and **semilunar valves.** The AV valves are located between the atria and the ventricles. The one on the right side of the heart has three valve flaps, so it is called the **tricuspid valve**, and the one on the left side has two valve flaps, so it is called the **bicuspid** *(mitral)* **valve.** These valves permit a one-way flow of blood from atria to ventricles.

Long, fibrous, cord-like structures called **chordae tendineae** anchor the valve flaps to the **papillary muscles**—long, cone-shaped, muscular extensions of the inner ventricles. The chordae tendineae and papillary muscles help keep the AV valves closed during ventricular contraction. The semilunar valves are located at the base of each major artery that leaves each ventricle.

On the right side is the pulmonary semilunar valve, and on the left is the aortic semilunar valve. These valves prevent backflow of blood into the ventricles. From outermost to innermost, the wall of the heart is made of three layers: epicardium, myocardium, and endocardium. The **epicardium** (*visceral pericardium*) is made of fibrous connective tissue and is the innermost layer of the pericardial sac that surrounds the heart. The **myocardium** is composed of multiple layers of cardiac muscle and many blood vessels and nerves. The **endocardium** is made of simple squamous epithelium and lines the inside of all the heart chambers and valves. It is continuous with the endothelium of blood vessels that enter and exit the heart, such as the aorta and the pulmonary veins.

Key to Illustration

Blood Vessels (B)

- B1. Superior vena cava
- B2. Inferior vena cava
- B3. Ascending aorta
- B4. Aortic arch
- B5. Descending aorta
- B6. Brachiocephalic trunk
- B7. L. common carotid a.
- B8. L. subclavian a.
- B9. Pulmonary trunk
- B10. Pulmonary arteries
- B11. Pulmonary veins

Chambers (C)

- C1. R. atrium
- C2. L. atrium
- C3. R. ventricle
- C4. L. ventricle

Structures (S)

- S1. Ligamentum arteriosum
- S2. Interventricular septum
- S3. Chordae tendineae
- S4. Papillary muscle
- S5. Apex (tip)

Valves (V)

- V1. Pulmonary semilunar
- V2. Aortic semilunar
- V3. Tricuspid leaflet
- V4. Bicuspid leaflet

Wall Layers (W)

- W1. Epicardium
- W2. Myocardium
- W3. Endocardium

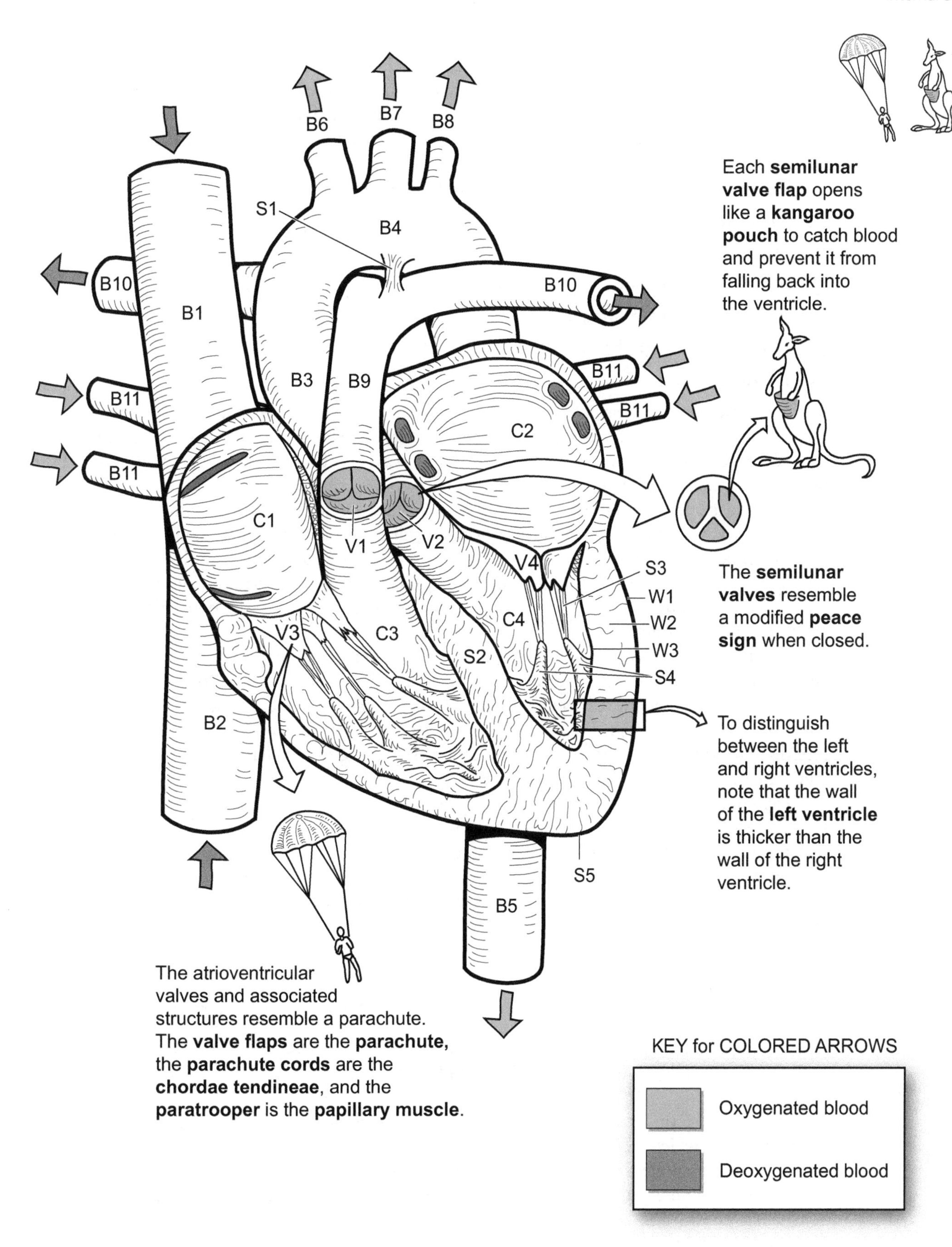
B6
B7
B8
S1
B4
B10
B1
B10
B11
B3
B9
B11
B11
C2
B11
C1
V1
V2
V4
S3
W1
W2
W3
S4
C4
V3
C3
S2
B2
S5
B5
Each semilunar valve flap opens like a kangaroo pouch to catch blood and prevent it from falling back into the ventricle.
The semilunar valves resemble a modified peace sign when closed.
To distinguish between the left and right ventricles, note that the wall of the left ventricle is thicker than the wall of the right ventricle.
The atrioventricular valves and associated structures resemble a parachute. The valve flaps are the parachute, the parachute cords are the chordae tendineae, and the paratrooper is the papillary muscle.
KEY for COLORED ARROWS
Oxygenated blood
Deoxygenated blood

Description

The **heart** has its own internal regulation system to achieve two important functions: (1) triggering the heartbeat, and (2) coordinating the timing between contraction of the atria and contraction of the ventricles. This control system is referred to as the **intrinsic conduction system.** Without this control system to ensure that the heart chambers completely fill with blood before contracting, the heart would be a very inefficient pump.

The intrinsic conduction system consists of the following six structures:

① **Sinoatrial (SA) node**

② **Internodal pathway**

③ **Atrioventricular node**

④ **Atrioventricular (AV) bundle**

⑤ **Bundle branches (right and left)**

⑥ **Purkinje fibers**

This system is a network of interconnecting cardiac muscle cells that spread through the atria and the ventricles. Notice that this system can be structurally divided into two main parts: the **nodes** (SA and AV nodes) and the **pathway** (internodal pathway, AV bundle, bundle branches, and Purkinje fibers). Let's examine each, in turn.

The nodes consist of clusters of specialized cardiac muscle cells that contain very few contractile proteins—myosin and actin—found in normal cardiac muscle cells. These cells are the body's only autorhythmic muscle cells. Instead of contracting, they serve as a kind of "spark plug," or stimulus, to establish the heartbeat. The **SA node** is located within the wall of the right atrium and is referred to as the *primary pacemaker*. The **AV node** is located in the septum between the two atria and is referred to as the *secondary pacemaker*.

A long network of specialized cardiac muscle cells—the pathway—ensures that the heart chambers are stimulated to contract in a coordinated manner. Like smooth muscle cells, cardiac muscle cells can stimulate adjacent cells. Think of the pathway like a series of dominoes. Knocking over the first domino will cause all the others to fall over. Similarly, because cardiac muscle cells interdigitate with each other like pieces of a jigsaw puzzle, stimulating the first cell will quickly stimulate all the others in the pathway.

The **internodal pathway** radiates from the SA node and extends to the left and right atria and AV node. The **AV bundle** is a relatively short segment of cells that extends from the AV node and penetrates into the top of the interventricular septum. It's the only electrical connection between the atria and the ventricles. From here, it splits into two pathways—the right and left bundle branches, which extend through the interventricular septum toward the apex of the heart. The bundle branches extend into the walls of their respective ventricles and into the papillary muscles to become the Purkinje fibers.

Impulse Pathway

The flow of the impulse for contraction always moves in the following sequence:

SA node → internodal pathway → AV node → AV bundle → bundle branches → Purkinje fibers

Let's examine this sequence with respect to the coordination of the filling and contracting of the atria and the ventricles. First, the autorhythmic cells within the SA node trigger the impulse to spread to the left and right atria through the internodal pathway. Simultaneously, the atria fill with blood and expand. The impulse causes the atria to contract, which forces blood into the ventricles. A delay in ventricular contraction is needed to allow the ventricles to fill with blood. This delay comes in the form of the time it takes to stimulate the AV node and send the impulse down the AV bundle and bundle branches. By the time the impulse has spread to the Purkinje fibers, the ventricles have finished filling with blood, and the ventricles are stimulated to contract.

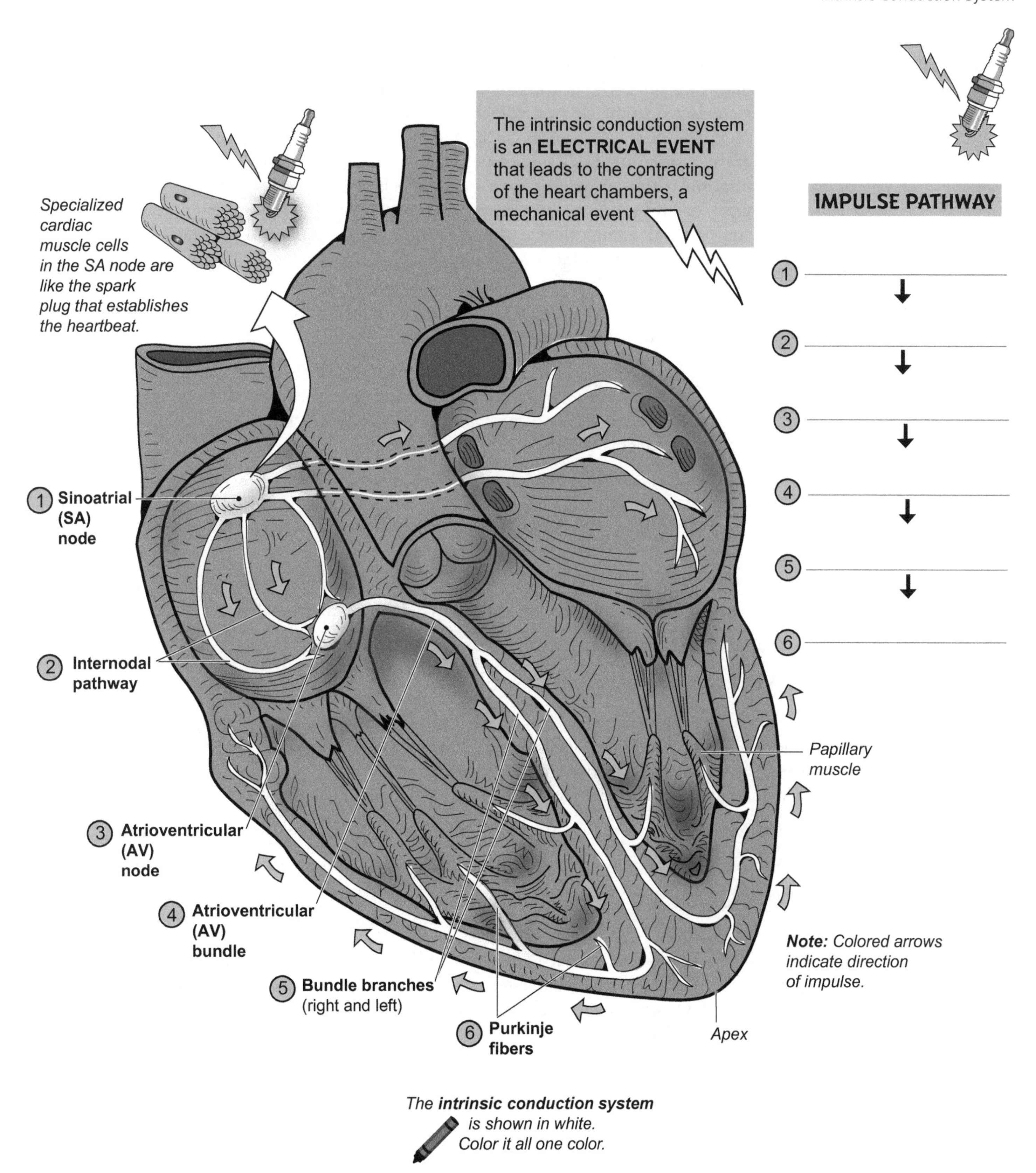

The ***intrinsic conduction system*** *is shown in white. Color it all one color.*

Description

An **electrocardiogram** (ECG or **EKG**) is a graph of the heart's electrical activity as expressed in **millivolts** (**mV**) over time. The instrument used to obtain an ECG is called an **electrocardiograph.** A person is connected to the electrocardiograph with electrodes placed on the arms and legs (*limb leads*) and along the chest (*chest leads*). Each of the limb leads gives a slightly different picture of the heart's electrical activity. An ECG is used to detect if the electrical conduction pathway within the heart is normal and if any damage has been done to the heart.

In a typical lead II recording, three different waves appear: **P**, **QRS complex**, and **T**. Each wave represents an electrical event called a *depolarization* or a *repolarization*. These electrical events stimulate cardiac muscle within the heart wall to either contract or relax. Consequently, these events lead to the contraction and relaxation of the heart chambers—atria and ventricles.

- **P wave:** *atrial depolarization*—at the end of the P wave, both atria have depolarized, which causes the atria to contract.
- **QRS complex:** *ventricular depolarization*—at the end of the QRS complex, both ventricles have depolarized, which causes the ventricles to contract. **Note:** Atrial repolarization also occurs during this period, but it is masked by the ventricular depolarization.
- **T wave:** *ventricular repolarization*—at the end of the T wave, both ventricles have repolarized, which causes the ventricles to relax.

Reading an ECG is an art that requires a lot of expertise. Even so, we can explain the general way in which they are interpreted. Two types of variations may signal abnormalities:

1. Variation in *wave height* (they may be *elevated* or *depressed*)
2. Variation in normal *time intervals*

Let's consider changes in wave height. For example, if a P wave is elevated, it may indicate atrial enlargement. If the QRS complex is elevated, it may indicate ventricular enlargement. A tall and pointed T wave may indicate myocardial ischemia.

The following three **normal time intervals** are examined on any ECG:

- **P–R interval** = time from beginning of P wave to start of QRS complex; 0.2 sec.
- **S–T segment** = time from end of S wave to beginning of T wave; 0.1 sec.
- **Q–T segment** = time from beginning of QRS complex to end of T wave; 0.4 sec.

Any time interval longer or shorter than normal also may indicate an abnormality. For example, a P–Q interval much longer than normal indicates a blockage in the normal conduction pathway. This could be caused by scar tissue resulting from a childhood infection of rheumatic fever. Another example is a longer than normal Q–T segment, which may indicate myocardial damage.

Normal Electrocardiogram or ECG (Lead II)

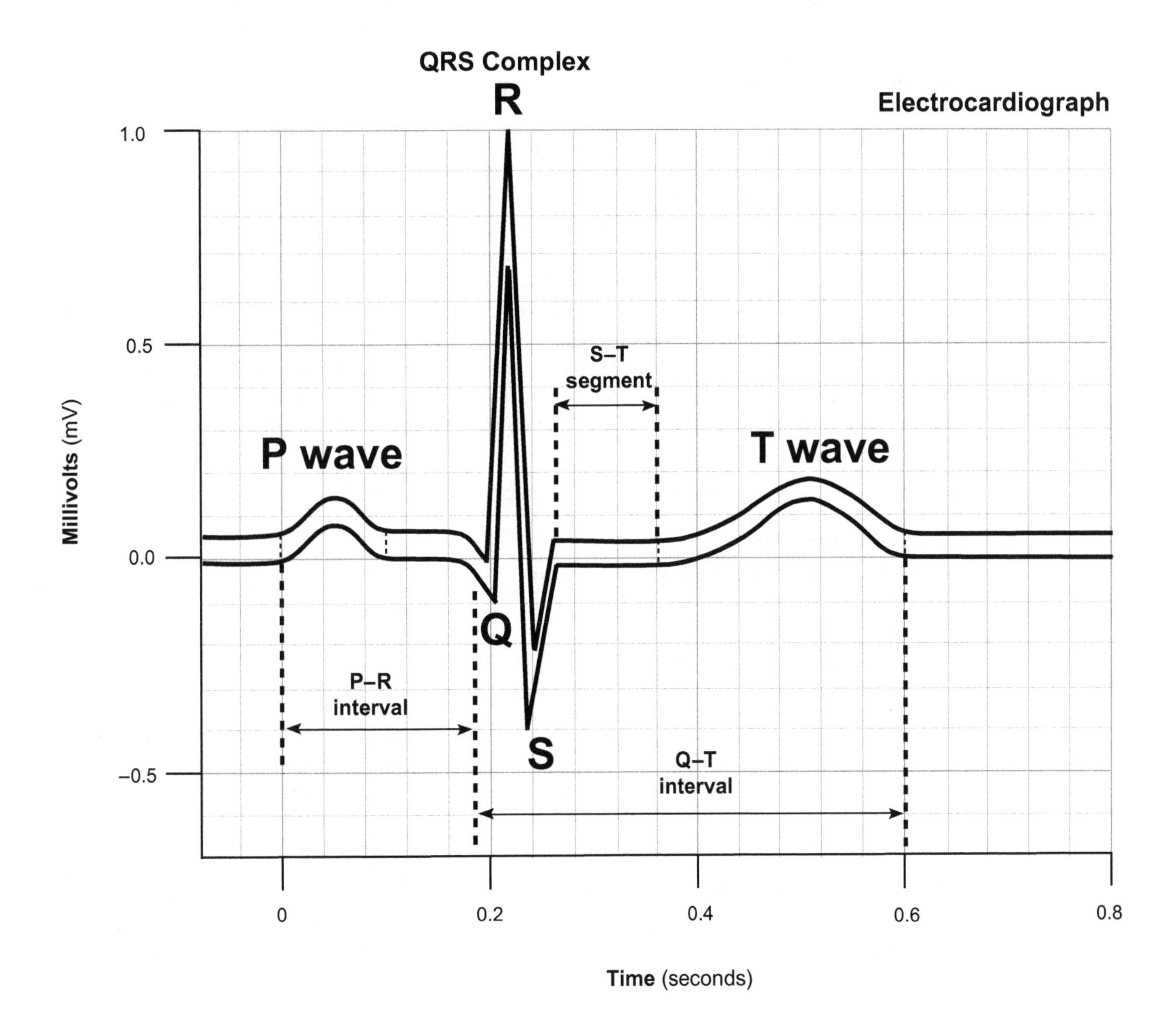

Color the P wave, QRS complex, and T wave different colors.

Description

The heart is essentially two pumps that work together as one synchronized unit. The left side pumps only oxygenated blood, and the right side pumps only deoxygenated blood. The **cardiac cycle** refers to all the pumping actions that occur within the heart during one entire heartbeat. It consists of both the atria and the ventricles filling with blood and then contracting. It begins with contraction of the atria and ends with refilling of the atria. On average, this continuous cycle takes about 800 msec. to complete in an adult.

The atria and the ventricles have repeated patterns of contraction (systole) and relaxation (diastole). The atria function as the receiving chambers of blood from the body. After they fill with blood, they contract and force blood downward into the true pumps—the more powerful ventricles. Because these chambers have the more difficult task of pumping the blood to the body, they have a thicker layer of cardiac muscle in their walls *and a larger volume*. Recall that the heart has two pairs of valves: atrioventricular (AV) valves and semilunar valves. The AV valves function as one-way valves to permit blood to flow from the atria to the ventricles. The closing of the semilunar valves prevents blood from flowing back into the ventricles after being ejected.

The electrocardiogram (ECG) is related to the cardiac cycle. An ECG measures electrical changes in the heart muscle that induce the contractions of the chambers. **These contractions lead to pressure changes within the chambers, which, in turn, induce the heart valves to either open or close.**

Steps

To understand the sequence of events, the cardiac cycle has been divided into five steps, as follows:

① Atrial contraction

- **ECG connection:** from P wave to Q wave
- **AV valves** *open*
- **Semilunar valves** *closed*
- **Description:** The left and right atria contract simultaneously, causing the atrial pressure to increase. This forces blood through the AV valves and into the ventricles. The ventricles are relaxed and filling with blood. The semilunar valves are closed because the pressure in the ventricles is too low to force them open.

② Isovolumetric ventricular contraction

- **ECG connection:** begins with R wave
- **AV valves** *closed*
- **Semilunar valves** *closed*
- **Description:** The ventricles begin contracting in this phase, causing the ventricular pressure to increase. Higher ventricular pressure relative to the atria closes the AV valves. The volume of blood in the ventricles remains constant (isovolumetric). All heart valves are closed.

③ Ventricular ejection

- **ECG connection:** from S wave to T wave
- **AV valves** *closed*
- **Semilunar valves** *open*
- **Description:** The ventricles continue contracting during this phase. Like wringing out a wet rag, the ventricles wring the blood out, beginning at the apex and moving upward. This allows for the maximum volume of blood to be ejected—about 70 mL per ventricle. The high ventricular pressure forces the semilunar valves to open, and blood is ejected into the pulmonary arteries and the aorta.

④ Isovolumetric ventricular relaxation

- **ECG connection:** begins at end of T wave
- **AV valves** *closed*
- **Semilunar valves** *closed*
- **Description:** The ventricles are relaxing, and the heart valves are all closed. The volume of blood in the ventricles remains constant (isovolumetric). Because ventricular pressures are higher than atrial pressures, no blood flows into the ventricles. Ventricular pressure is quickly decreasing during this phase.

⑤ Passive ventricular filling

- **ECG connection:** after T wave to next P wave
- **AV valves** *open*
- **Semilunar valves** *closed*
- **Description:** As blood flows into the atria, the atrial pressure increases until it exceeds the ventricular pressure. This forces the AV valves open, and blood fills the ventricles. This is the main way the ventricles are filled. At the end of this phase, the ventricles will be about 70–80% filled.

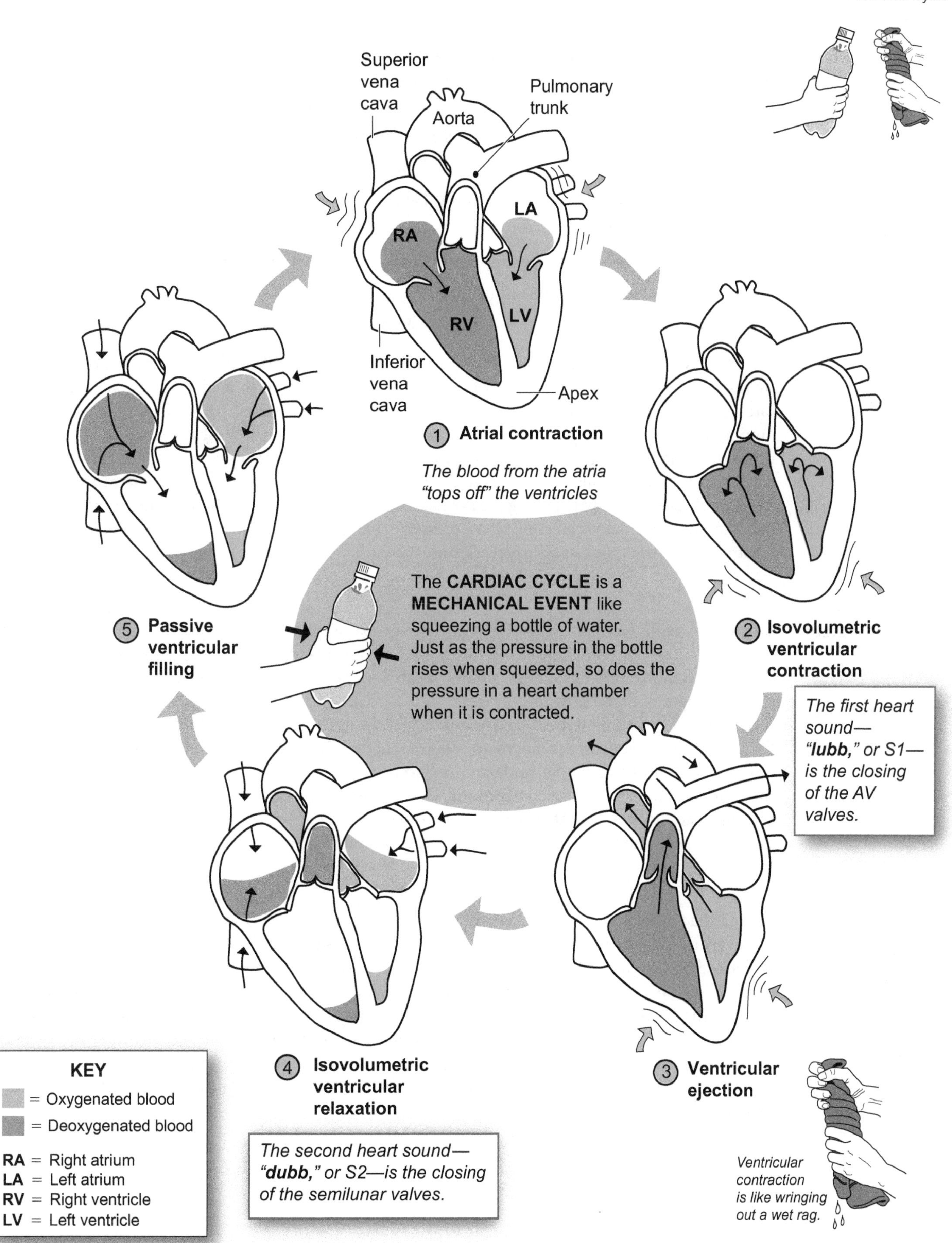
Superior vena cava
Aorta
Pulmonary trunk
LA
RA
RV
LV
Inferior vena cava
Apex
1 Atrial contraction
The blood from the atria "tops off" the ventricles
The CARDIAC CYCLE is a MECHANICAL EVENT like squeezing a bottle of water. Just as the pressure in the bottle rises when squeezed, so does the pressure in a heart chamber when it is contracted.
5 Passive ventricular filling
2 Isovolumetric ventricular contraction
The first heart sound—"lubb," or S1—is the closing of the AV valves.
4 Isovolumetric ventricular relaxation
The second heart sound—"dubb," or S2—is the closing of the semilunar valves.
3 Ventricular ejection
Ventricular contraction is like wringing out a wet rag.
KEY
= Oxygenated blood
= Deoxygenated blood
RA = Right atrium
LA = Left atrium
RV = Right ventricle
LV = Left ventricle

Description

In the central nervous system (CNS), the **cardiovascular (CV) center** in the **medulla oblongata** is the command-and-control center for regulating heart function. It uses reflex pathways to control heart rate. The three major peripheral sensory receptors that provide input to the CV center are:

1. **Proprioceptors**—measure tension changes in muscles and joints
2. **Chemoreceptors**—detect changes in blood acidity by sensing changes in CO_2 and H^+ levels
3. **Baroreceptors**—located in the carotid sinus, aortic arch, and other arteries; detect changes in blood pressure

The CV center sends its motor output to the heart through two different nerves:

1. **Vagus nerve**—to decrease heart rate
2. **Cardiac accelerator nerve**—to increase heart rate

The internal pacemakers within the heart—the **SA node** and **AV node**—help set its normal rate and rhythm. But the heart has to be able to respond to various stimuli so it can increase or decrease heart rate as needed. For example, during exercise, the heart rate increases, which, in turn, elevates cardiac output, thereby providing more oxygen and glucose to skeletal muscle tissue. As you exercise, proprioceptors detect increased tension in skeletal muscles, and chemoreceptors detect increased levels of CO_2 in the blood. This triggers nerve impulses to be sent along sensory neurons to the CV center. Motor output then is carried along the cardiac accelerator nerve to stimulate the heart rate to increase.

Now let's use the variable of blood pressure changes to follow the reflex pathway shown in the illustration for both the parasympathetic and sympathetic divisions.

Parasympathetic Control

The **vagus nerve** (cranial nerve X) is part of the parasympathetic division of the ANS and sends impulses to the heart to decrease the heart rate. It is a mixed nerve—meaning that it contains both sensory and motor neurons. Baroreceptors, located in the carotid sinus, aortic arch, and other arteries, detect changes in blood pressure.

As blood pressure increases above normal levels, impulses are sent along sensory neurons to the CV center and then carried along motor neurons within the vagus nerve to the heart. The vagus nerve innervates the heart at the SA node and AV node. It functions to inhibit the heart, thereby reducing the heart rate and strength of contractions. This, in turn, decreases cardiac output, which lowers blood pressure back to normal. At the same time, impulses also are sent out along vasomotor nerves (not shown in illustration) that stimulate vasodilation, thereby further lowering the blood pressure.

Sympathetic Control

The **cardiac accelerator nerve** is part of the sympathetic division of the ANS and sends impulses to the heart to increase the heart rate. As blood pressure decreases below normal levels, this is detected by baroreceptors that send impulses along a neural network over to the CV center, then to the spinal cord and to the heart. The final nerve in this pathway—the **cardiac accelerator nerve**—innervates the heart at the SA node, AV node, and cardiac muscle in the ventricular wall (ventricular myocardium). The response is that both the heart rate and strength of contractions increase. This, in turn, increases cardiac output, which increases blood pressure. At the same time, impulses are sent out along vasomotor nerves (not shown in illustration) that stimulate vasoconstriction and thereby further increase blood pressure.

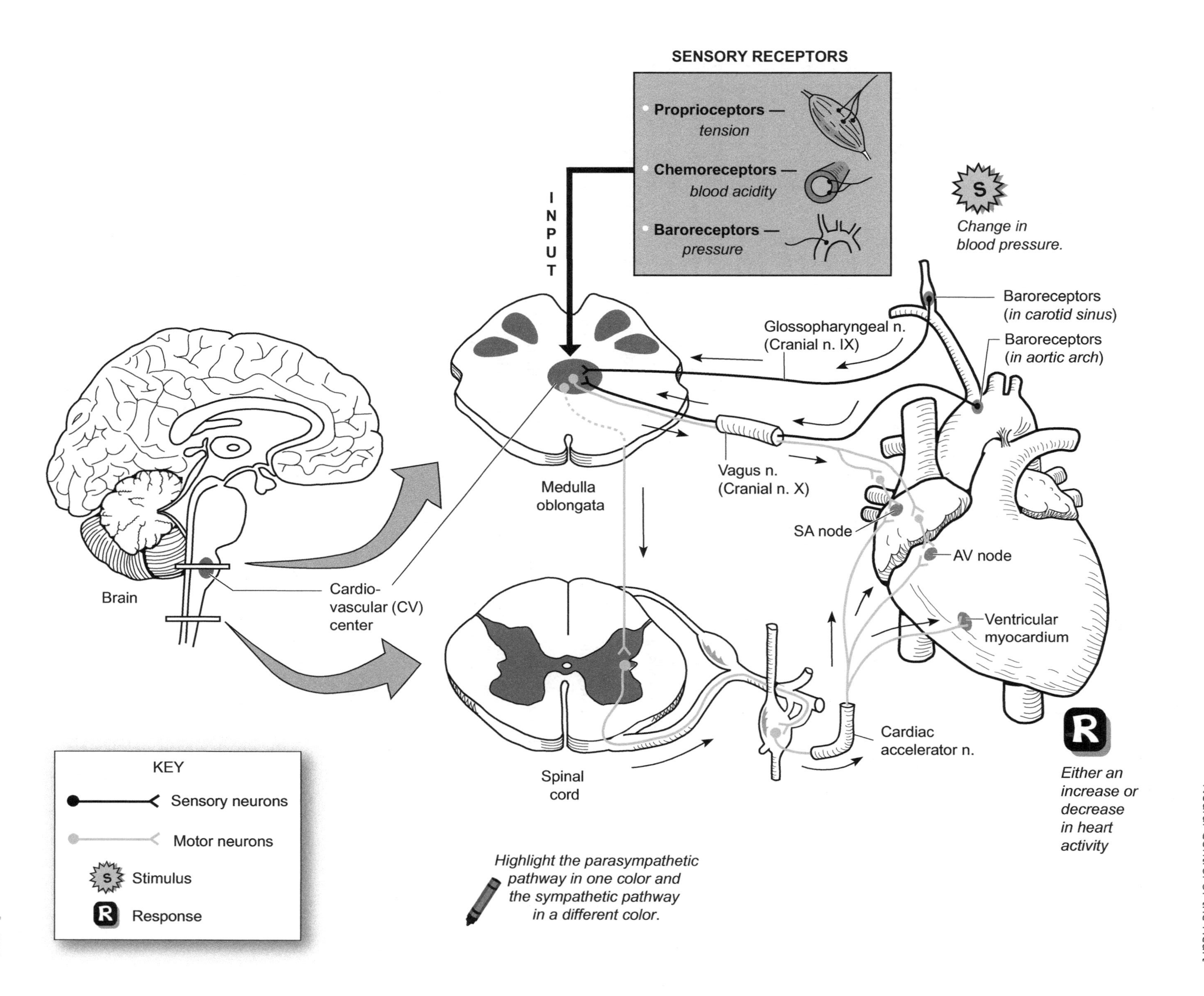
SENSORY RECEPTORS
Proprioceptors — tension
Chemoreceptors — blood acidity
Baroreceptors — pressure
INPUT
S
Change in blood pressure.
Baroreceptors (in carotid sinus)
Baroreceptors (in aortic arch)
Glossopharyngeal n. (Cranial n. IX)
Vagus n. (Cranial n. X)
Medulla oblongata
SA node
AV node
Ventricular myocardium
Brain
Cardio-vascular (CV) center
Spinal cord
Cardiac accelerator n.
R
Either an increase or decrease in heart activity
KEY
Sensory neurons
Motor neurons
Stimulus
Response
Highlight the parasympathetic pathway in one color and the sympathetic pathway in a different color.

Description

The body has five fundamental types of blood vessels: *arteries, arterioles, capillaries, venules,* and *veins*. All of them connect in the following predictable pattern:

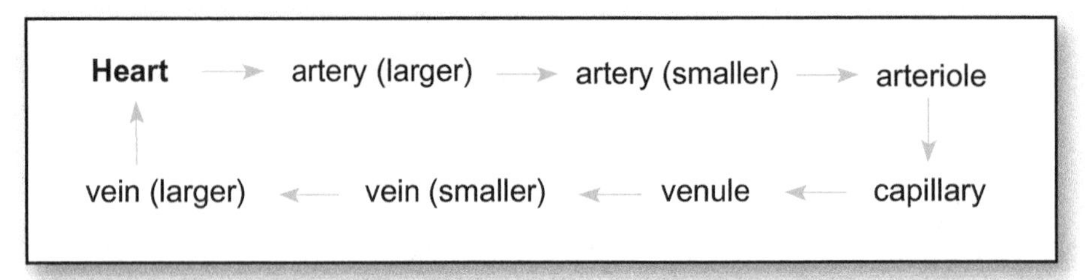

Arteries always carry blood away from the heart. They are thicker-walled than veins because the blood within them is at a higher pressure. All **veins** always carry blood back to the heart. Because the pressure within them is lower, they are thinner-walled. Larger veins contain valves—similar to one-way semilunar valves in the heart—that assist the low-pressure venous blood return to the heart. When a person is standing, gravity works against venous blood flow. Like a staircase breaks up the distance between the first and second floors into smaller steps, valves break up larger veins into smaller segments. Arteries and veins connect at the microscopic level by capillary networks. **Capillaries** are the smallest blood vessels in the body and are important functionally because gas and fluid exchange occurs here. Their entire wall is often a single cell layer in thickness. **Arterioles** and **venules** are microscopic vessels that feed and drain capillaries, respectively. Nearest the capillaries, they are structurally similar to a capillary except they have small amounts of smooth muscle around them. The smooth muscle around an arteriole can be stimulated to either constrict (*vasoconstriction*) or relax (*vasodilation*). This, in turn, causes changes in blood flow to a capillary and systemic blood pressure. Vasoconstriction of arterioles decreases blood flow and increases blood pressure, and vasodilation does just the opposite.

Arteries and veins have three major layers in their walls: *tunica externa, tunica media,* and *tunica interna*:

- **Tunica externa**—connective tissue layer made mostly of collagen fibers
- **Tunica media**—layers of smooth muscle and some elastic fibers
- **Tunica interna**—endothelial layer (simple squamous epithelium) with underlying loose connective tissue

Study Tips

- Don't confuse yourself by trying to distinguish between arteries and veins as to whether they carry oxygenated blood or deoxygenated blood. *This does not work* because some arteries/veins carry oxygenated blood and others carry deoxygenated blood.
- To recall the general function of arteries, use the phrase: ***Arteries Away!*** **Arteries** always carry blood *away* from the heart.
- To recall one structural difference between arteries and veins: ***Veins have Valves*** (arteries do not have valves).
- To distinguish an artery from a vein under the microscope, remember that **arteries** always have a **thicker tunica media.**

Key to Illustration

1. Tunica interna (*tunica intima*)
2. Tunica media
3. Tunica externa (*tunica adventitia*)
4. Lumen of blood vessel

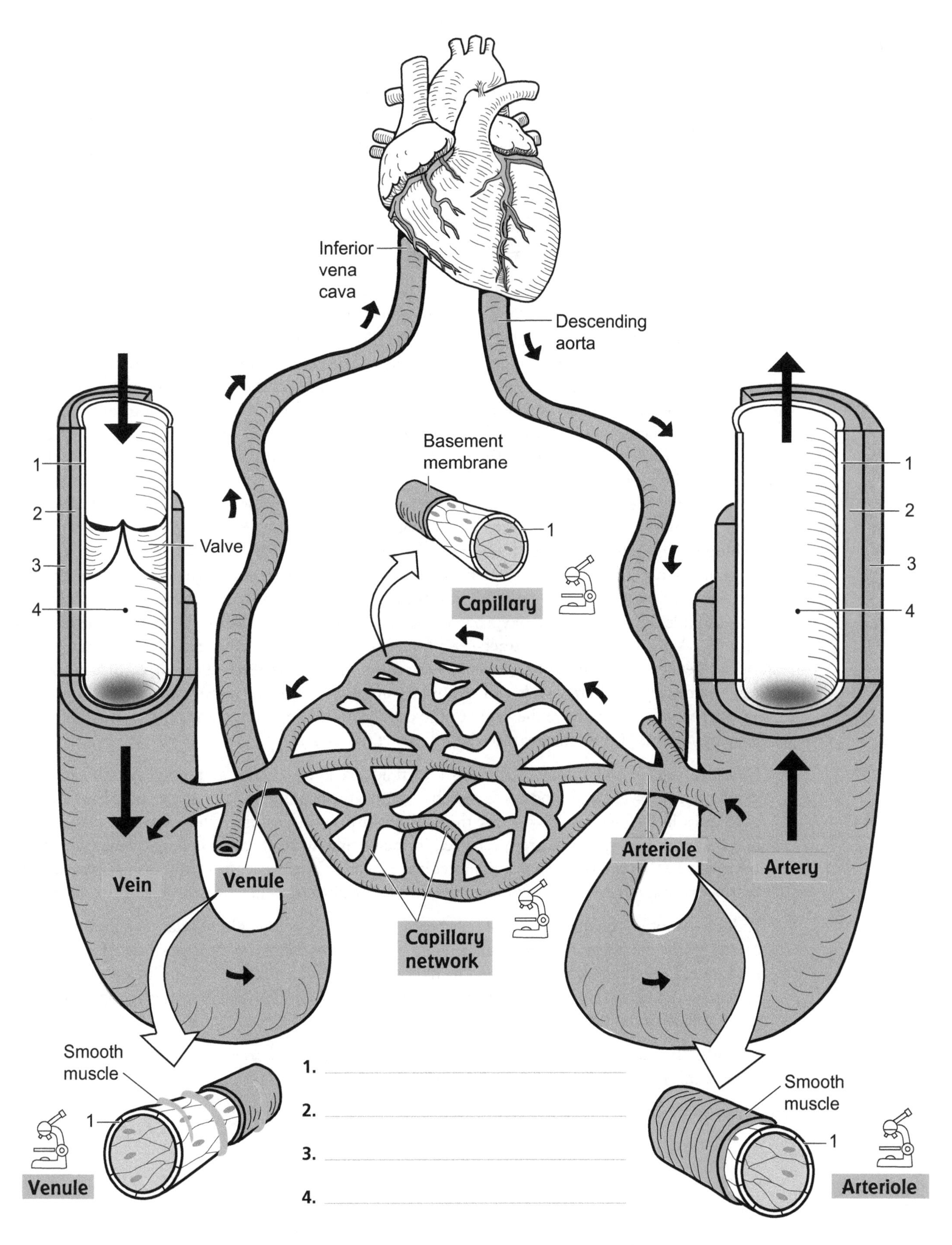
Inferior vena cava
Descending aorta
Basement membrane
1
Capillary
1
2
3
4
Valve
Vein
Venule
Capillary network
Arteriole
Artery
Smooth muscle
1
Venule
Smooth muscle
1
Arteriole

1. ______________

2. ______________

3. ______________

4. ______________

Description

The illustration gives an overview of the general pattern of circulation. Blood always follows a predictable circuit through blood vessels. There are five fundamental types of blood vessels in the body: *arteries, arterioles, capillaries, venules,* and *veins*. All of them connect in the following pattern:

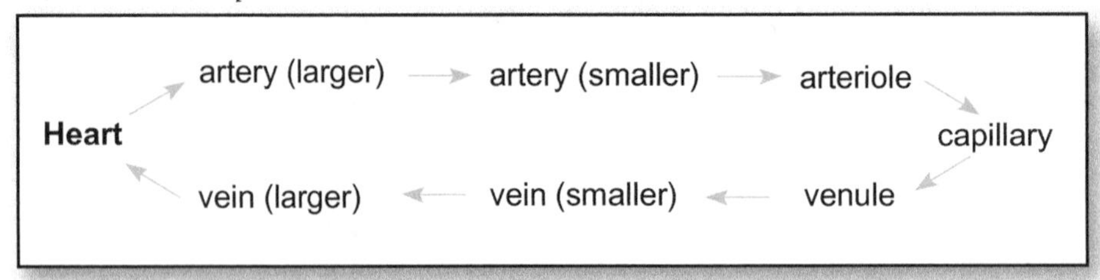

The schematic illustration on the facing page shows three of these five: *arteries, veins,* and *capillaries*. **Arteries** always carry blood away from the heart. They are thicker-walled than veins because the blood within them is at a higher pressure. **As distance from the heart increases, blood pressure decreases.** All **veins** always carry blood back to the heart. Because the pressure within them is lower, they are thinner-walled. Arteries and veins connect together at the microscopic level by capillary networks. **Capillaries** are the smallest blood vessels in the body and are very important functionally because gas exchange and fluid exchange occurs here. Oxygen exits the blood to be used by body cells, and carbon dioxide enters the blood from cells. The liquid plasma is filtered out of the blood to become **interstitial fluid** (*tissue fluid*).

Let's follow the general pattern of circulation. Veins carrying low-pressure, deoxygenated blood drain into the **vena cava**, which drains into the heart's **right atrium** (**RA**). This receiving chamber fills with blood, contracts, and forces blood into the **right ventricle** (**RV**). All this deoxygenated blood is then pumped out of the right ventricle to go to the lungs to get oxygenated. In the lungs, the pulmonary capillaries function only for gas exchange, in which oxygen diffuses into the blood and carbon dioxide diffuses out. The oxygenated blood is then transported through veins to the **left atrium** (**LA**). The LA fills with blood, contracts, and forces blood into the **left ventricle** (**LV**). This oxygenated blood is then pumped out to the body via the **aorta.** The heart feeds its own cardiac muscle first through **coronary capillaries** so it can continue pumping blood every minute of every day. Arteries carry oxygenated blood above the heart to the capillaries in the brain, trunk, and upper limbs. Other arteries carry blood below the heart to the following major areas:

- **Digestive organs and spleen**—After gas and fluid exchange occurs at the **splenic** and **mesenteric capillaries**, deoxygenated blood is carried by veins to the **hepatic portal system** in the liver. Note that capillaries in this system are not for the typical purpose of gas and fluid exchange. Instead, these highly permeable capillaries are specialized for delivering nutrients absorbed by the digestive tract to liver cells before entering the general circulation. The liver cells serve as special processing centers that perform many functions. For example, they store glucose as glycogen and detoxify alcohols.
- **Kidney**—Another unique group of permeable capillaries is the **glomerular capillaries.** Like the capillaries in the hepatic portal system, these are also not for the typical purpose of gas and fluid exchange. Instead, they are specialized to filter only the blood plasma, place it in a separate tubular system, and process this liquid into urine. These capillaries lead into the **peritubular capillaries,** where gas and fluid exchange does occur.
- **Gonads**—In the male, gas and fluid exchange occurs at capillaries in the testes, whereas in the female, this occurs at capillaries in the ovaries.
- **Liver, lower limbs**

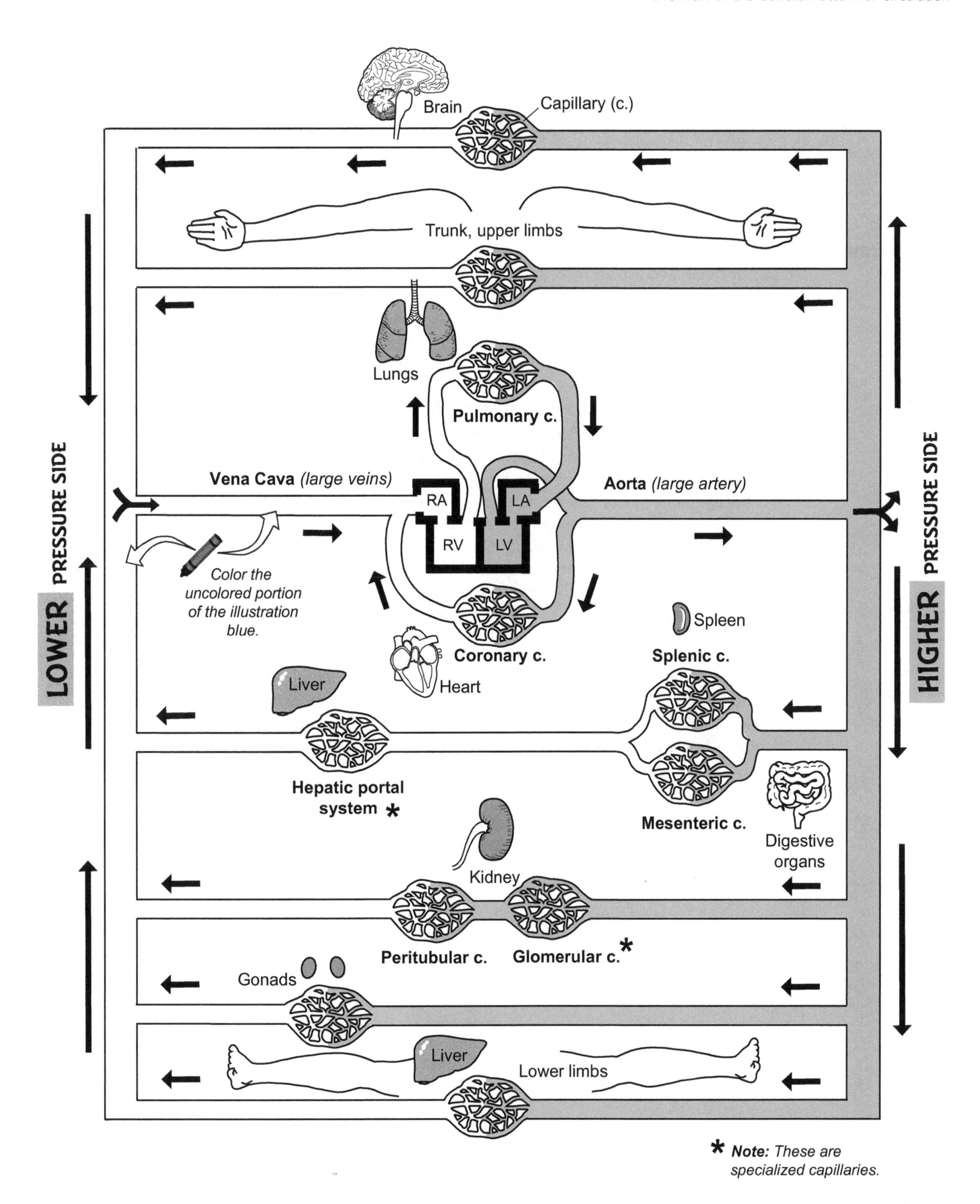

* **Note:** *These are specialized capillaries.*

Description

Fluid flow through any tube—whether water through a garden hose or blood through a vessel—follows a pattern. To understand **blood flow**, you must think of the fluid not as a single unit but as moving in concentric layers. The two types of flow patterns are: (1) **laminar flow**, and (2) **turbulent flow**. As another clinical example, turbulent blood flow is heard moving through a defective heart valve as a "murmur."

Types of Blood Flow

① **Laminar** (*lamina* = layer) **flow** is the normal flow pattern in healthy blood vessels.

The concentric layers of fluid flow together, but not at the same rate. As shown in the illustration, the outer layers travel more slowly than the inner layers. Why? As the fluid in the outermost layer rubs against the vessel wall, it is hindered by the force of friction and moves the slowest. In contrast, the fluid in the center of the tube is interfered with the least, so it moves the fastest. Normal, healthy blood vessels are smooth on their inner surface. Like the smooth surface on the inside of a garden hose or a drainpipe, this allows the fluid to flow through most easily.

② **Turbulent flow** is the result of any disruption in the normal, laminar flow.

Some low levels of turbulent flow are normal as blood flows through all the vessels in the body, but high levels may indicate an abnormality. What causes it? Two things: (1) physical change in a vessel (*ex.:* constriction, sharp turn, or narrowing of a vessel) or (2) Disease state (*ex.:* atherosclerosis). Because large blood vessels connect to smaller ones, some narrowing is normal, but it is a gradual change that doesn't usually result in turbulent flow.

Let's consider an abnormality like the constriction shown in the illustration. As fluid strikes the wall of the constriction, its normal, linear path is disrupted. The result is the generation of swirling little currents—like backflows.

As a clinical example, the carotid arteries are the major vessels that deliver oxygenated blood to the brain. If a person has atherosclerosis, these arteries may become blocked because of plaque deposits. This can lead to a stroke or even death. Because turbulent flow produces its own unique sounds, a physician can detect this problem by placing a stethoscope on the patient's neck. This problem can be compounded because the swirling currents of turbulent flow may cause part of the plaque to break off—becoming a mobile fragment called an embolus. The embolus then may travel downstream and cause a blockage in some part of the brain.

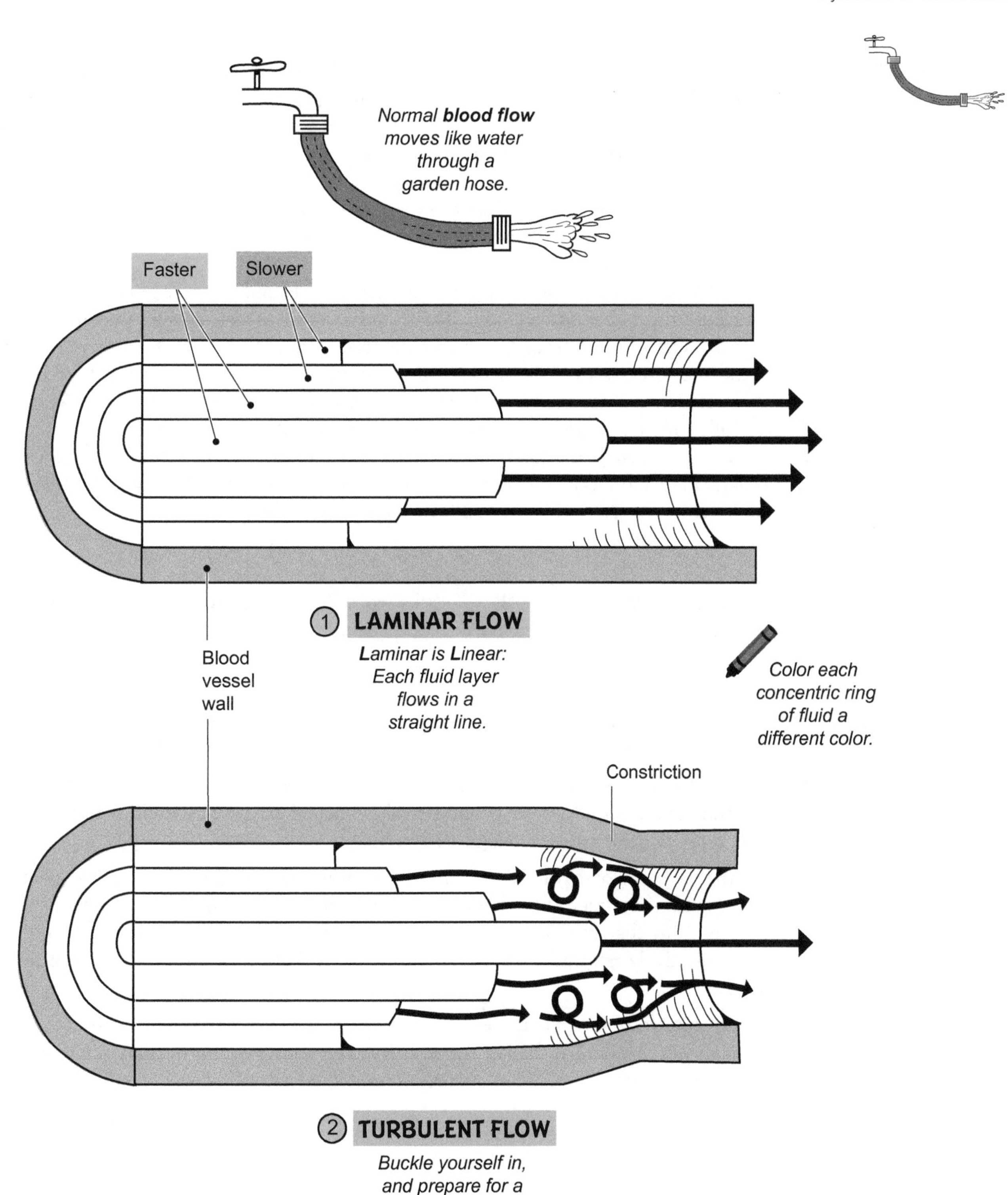

① LAMINAR FLOW

Laminar is Linear: Each fluid layer flows in a straight line.

Color each concentric ring of fluid a different color.

② TURBULENT FLOW

Buckle yourself in, and prepare for a bumpy ride!

Description

The heart generates the blood pressure by ejecting blood into the arterial system. **Blood pressure** (BP) is a type of **hydrostatic pressure**—or force of a fluid against the wall of a tube. Arteries carry blood away from the heart, and veins carry it back to the heart. **As distance from the heart increases, blood pressure decreases.** Therefore, on average, arteries have much higher pressure than veins. As the ventricles contract and force blood out into the arteries, these vessels expand, and the pressure rises to a *maximum* pressure. As the ventricles relax, and no more blood is ejected into the arteries, the arteries recoil, and pressure falls to a *minimum* pressure. This constant cycle of rising and falling blood pressure is the source of our pulse.

When blood pressure is taken, it is measured in mm Hg and expressed numerically as the maximum pressure over the minimum pressure. This is referred to as the **systolic pressure** over the **diastolic pressure**. For example, a normal blood pressure reading might be 120/70, read as "120 over 70."

Blood pressure has to be homeostatically maintained at a normal level. If blood pressure exceeds normal levels, it can cause smaller vessels to rupture, leading to anything from a stroke even to blindness. If levels fall too low, this also can lead to numerous problems. For example, filtration is totally dependent on pressure. The kidneys constantly filter the blood to remove waste products, and the capillaries in all our organs filter the blood to form tissue fluid. No pressure, no filtration. The body has numerous physiological mechanisms to increase BP when it falls too low.

Analogy

Like water pressure in a garden hose results from the force of water against the inside of the hose, blood pressure results from the force of blood against the wall of the blood vessel.

Measuring BP

① When a blood pressure cuff is wrapped around the upper arm, the intended purpose is to compress the **brachial artery**, which delivers oxygenated blood to the arm.

② The valve is closed on the blood pressure instrument, then the bulb is squeezed until the cuff pressure exceeds the systolic pressure in the brachial artery. The result is that the artery closes so blood temporarily stops flowing through it.

③ As the valve is slowly opened, the pressure within the cuff decreases. Consequently, the brachial artery begins to open. Blood squirts through the constricted vessel, resulting in turbulent flow (see p. 126) that can be detected by a stethoscope. These characteristic "whooshing" sounds are collectively referred to as the **sounds of Korotkoff**, and the pressure at onset is the systolic pressure.

④ Turbulent blood flow continues until the brachial artery has expanded back to its normal size. The pressure reading at the last sound before the restoration of normal blood flow represents the diastolic pressure.

Blood pressure (BP)
is a type of **hydrostatic pressure (HP)** or fluid force against the wall of a tube.

As **water pressure** in a garden hose results from the force of water against the inside of the hose, **blood pressure** results from the force of blood against the wall of a blood vessel.

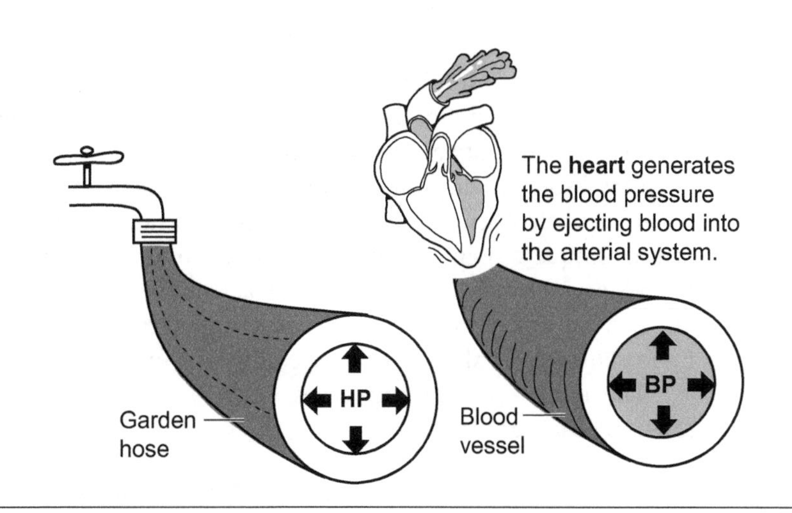

Measuring Blood Pressure

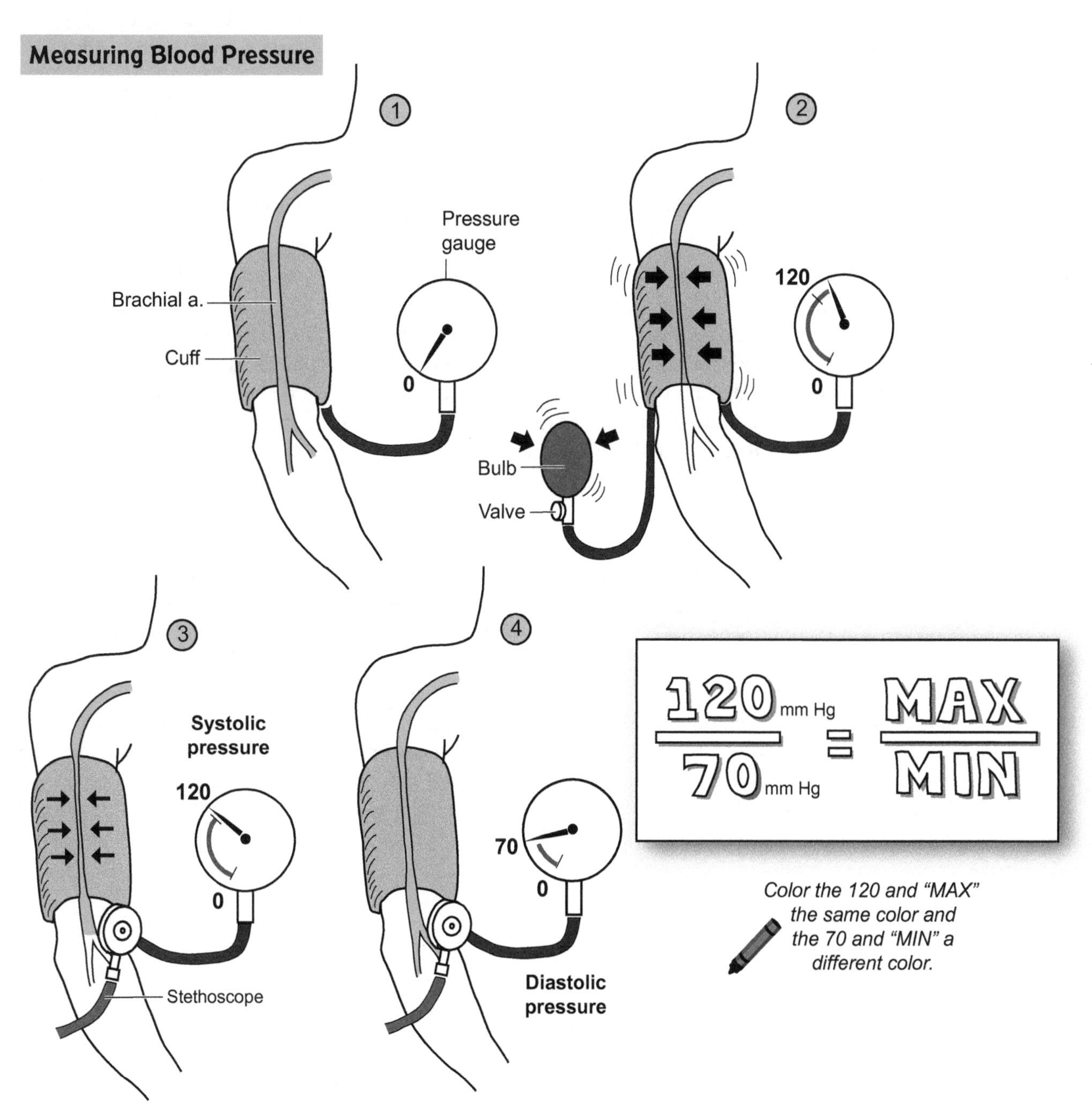

Color the 120 and "MAX" the same color and the 70 and "MIN" a different color.

Description

Capillaries, the most microscopic blood vessels in the body, join **arterioles** to **venules**. Their diameter is so small that red blood cells must pass through single file. These vessels form an interconnecting network or mesh of vessels called capillary beds. All the various tissues in the body depend on these capillary beds to remove wastes and to deliver nutrients and fluid.

The tubelike structure of each capillary is simple. The wall of the vessel is made up of endothelial cells or simple squamous epithelial cells. Each of these cells is flat to allow for easy diffusion of solutes from blood to tissue cells and vice versa. Wrapped around the outside of these cells like a thin blanket are protein fibers called the basement membrane. Gaps between adjacent endothelial cells called intercellular clefts allow for passage of fluid and small solutes.

The three types of capillaries are: (1) continuous capillaries, (2) fenestrated capillaries, and (3) sinusoidal capillaries.

Type	Comments/Permeability	Location in Body
① **Continuous capillary**	• Most common • *Least permeable*	Skin, connective tissues, skeletal muscle, smooth muscle, and lungs
② **Fenestrated capillary**	• *Most permeable*. Endothelial cells contain small holes called fenestrations that increase permeability	Kidneys, small intestine, choroid plexuses in brain, and some endocrine glands
③ **Sinusoidal capillary**	• *Most permeable*—has the largest fenestrations and the largest intercellular clefts	Liver, red bone marrow, spleen, and some endocrine glands

Blood flow through a capillary bed is controlled by bands of smooth muscle called precapillary sphincters. When they are closed, blood bypasses the capillary bed by going through the **shunt**. When they are open, blood moves through the entire capillary bed (as illustrated). This allows the body to respond to different situations. For example, during exercise, the sphincters have to be opened in the capillaries in skeletal muscles to meet the muscle's increased oxygen demand and deal with increased production of carbon dioxide. During the digestion of a heavy meal, more blood has to be shunted to the capillaries in the digestive tract rather than to the skeletal muscles.

Solute Diffusion

Capillary beds are the sites of nutrient and waste exchange between the blood and body cells. Nutrients move from the blood into tissue cells, and wastes move from tissue cells into the blood. For example, the respiratory gases—oxygen and carbon dioxide—**diffuse** across the wall of the capillary. Oxygen is a vital nutrient that diffuses out of the blood and into tissue cells, and carbon dioxide is a waste product that diffuses from tissue cells into the blood.

Recall that diffusion depends on a gradient (see pp. 68–68). The rule for simple diffusion is that a solute moves from a region of higher solute concentration to a region of lower solute concentration. Oxygen is carried in the blood primarily by the protein hemoglobin. As cells consume oxygen to metabolize glucose in the cellular respiration process, this ensures that the gradient for oxygen is maintained. Carbon dioxide is a by-product of cellular respiration, so it maintains a gradient as it constantly accumulates within body cells.

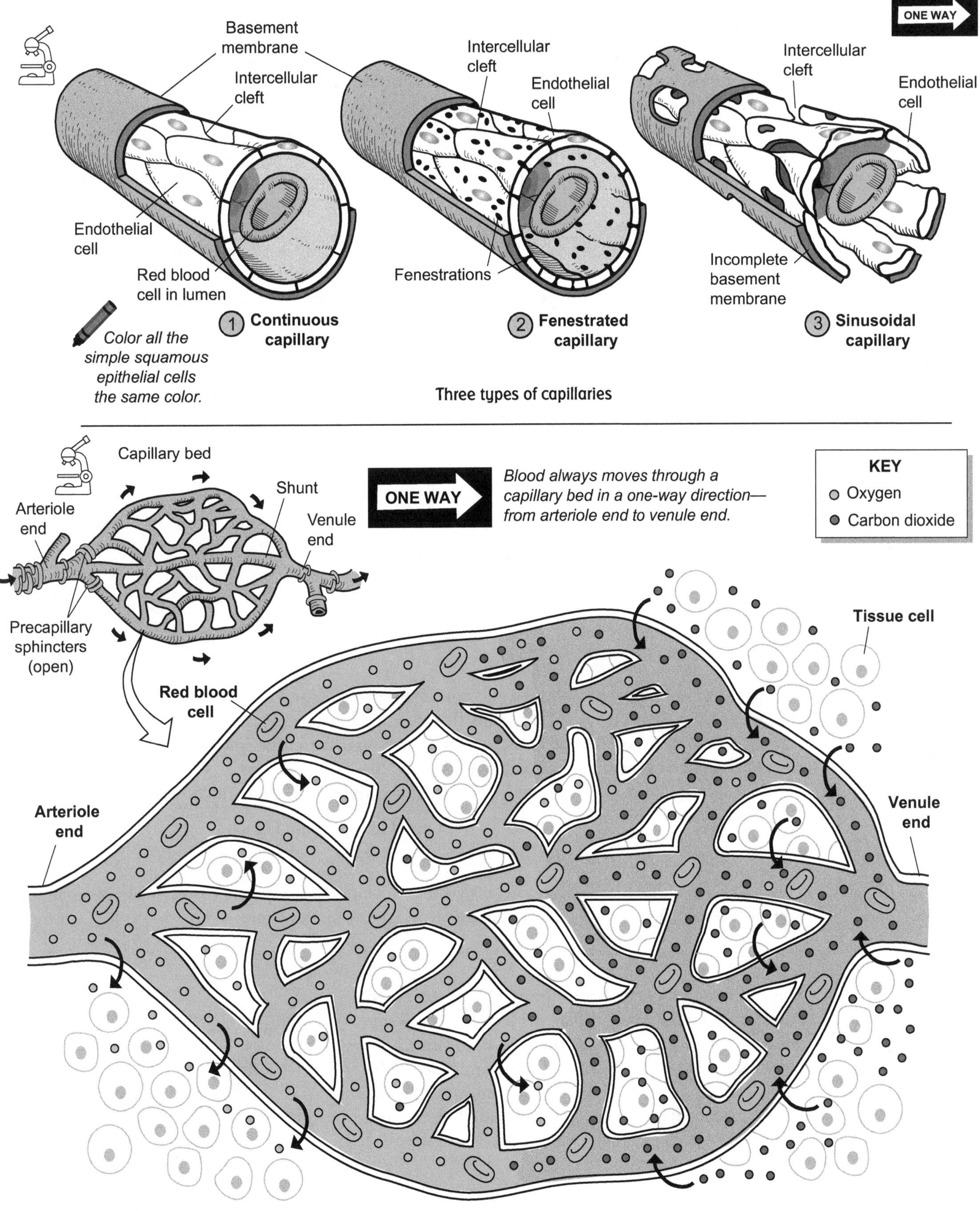

Three types of capillaries

Diffusion of oxygen and carbon dioxide across a capillary bed

Description

Gas exchange occurs across capillary beds. This is how oxygen is delivered to your cells and how carbon dioxide is removed from tissues. In addition to diffusion of gases, two processes are occurring simultaneously with diffusion: **filtration** and **reabsorption.** Let's examine each, in turn.

The filtration process is completely dependent on a force. No force, no filtration. In capillaries, this force is the blood pressure inside the capillaries and is called the **capillary hydrostatic pressure (CHP)**. It drives the filtration process and is measured in millimeters of mercury (Hg). Across a capillary bed, the CHP quickly drops as blood moves from the *higher* pressure **arteriole end** (35 mm Hg) to the *lower* pressure **venule end** (18 mm Hg). Working against the **CHP** is a counterforce called the **blood colloidal osmotic pressure (BCOP)**.

Focus on the last two words of this term: osmotic pressure. **Osmotic pressure** represents the force from water's tendency to move across a semipermeable membrane toward the solution with the greater concentration of nonpenetrating solutes (primarily proteins). The osmotic pressure is proportional to the solute concentration. The higher the solute concentration, the greater is the osmotic force. Because the concentration of nonpenetrating solutes in the blood is greater than the tissue fluid within the interstitial spaces, water's tendency is to move back into the blood.

Unlike the CHP, the BCOP remains constant across the capillary bed at 25 mm Hg. Why? Because the concentration of nonpenetrating solutes does not change from one end of the capillary bed to the other.

Formula

Filtration occurs *only* at the **arteriole end** of the capillary bed. Here is the formula to calculate the net filtration pressure (NFP):

$$\text{NFP} = \text{CHP} - \text{BCOP}$$

substituting the numbers:

$$\text{NFP} = 35\ \text{mm Hg} - 25\ \text{mm Hg} = +10\ \text{mm Hg}$$

This positive (+) number indicates a net movement of water and small solutes (such as sodium and chloride ions) out of the blood and into the interstitial spaces. This process of filtering the blood plasma is the only way the body has to create interstitial fluid. The interstitial fluid is then drained into the lymphatic system to become lymph (see pp. 206–207).

Reabsorption of water occurs *only* at the venule end. Let's calculate the NFP again:

$$\text{NFP} = 18\ \text{mm Hg} - 25\ \text{mm Hg} = -7\ \text{mm Hg}$$

The negative (−) number indicates net movement of water back into the blood. In short, some water is always reabsorbed at the venule end of the capillary because of osmosis.

Analogy

The capillary hydrostatic pressure (CHP) is the blood pressure inside the capillary. The CHP is like the water pressure inside a garden hose that has been turned on. Now imagine that you punctured some tiny holes in the hose. The fluid squirting out of the hose is like the solution being filtered out of the blood to create the tissue fluid within the interstitial spaces.

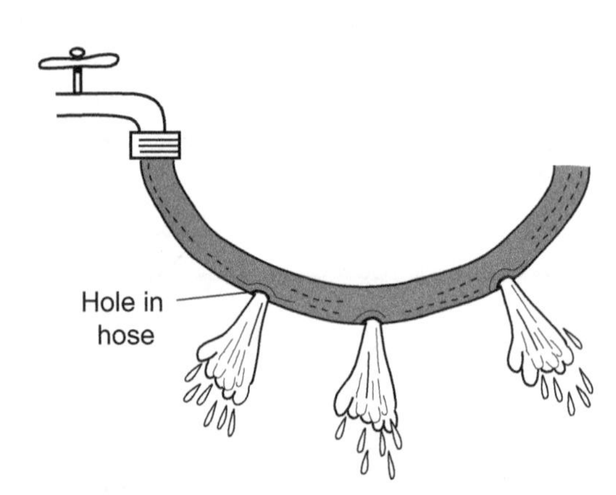

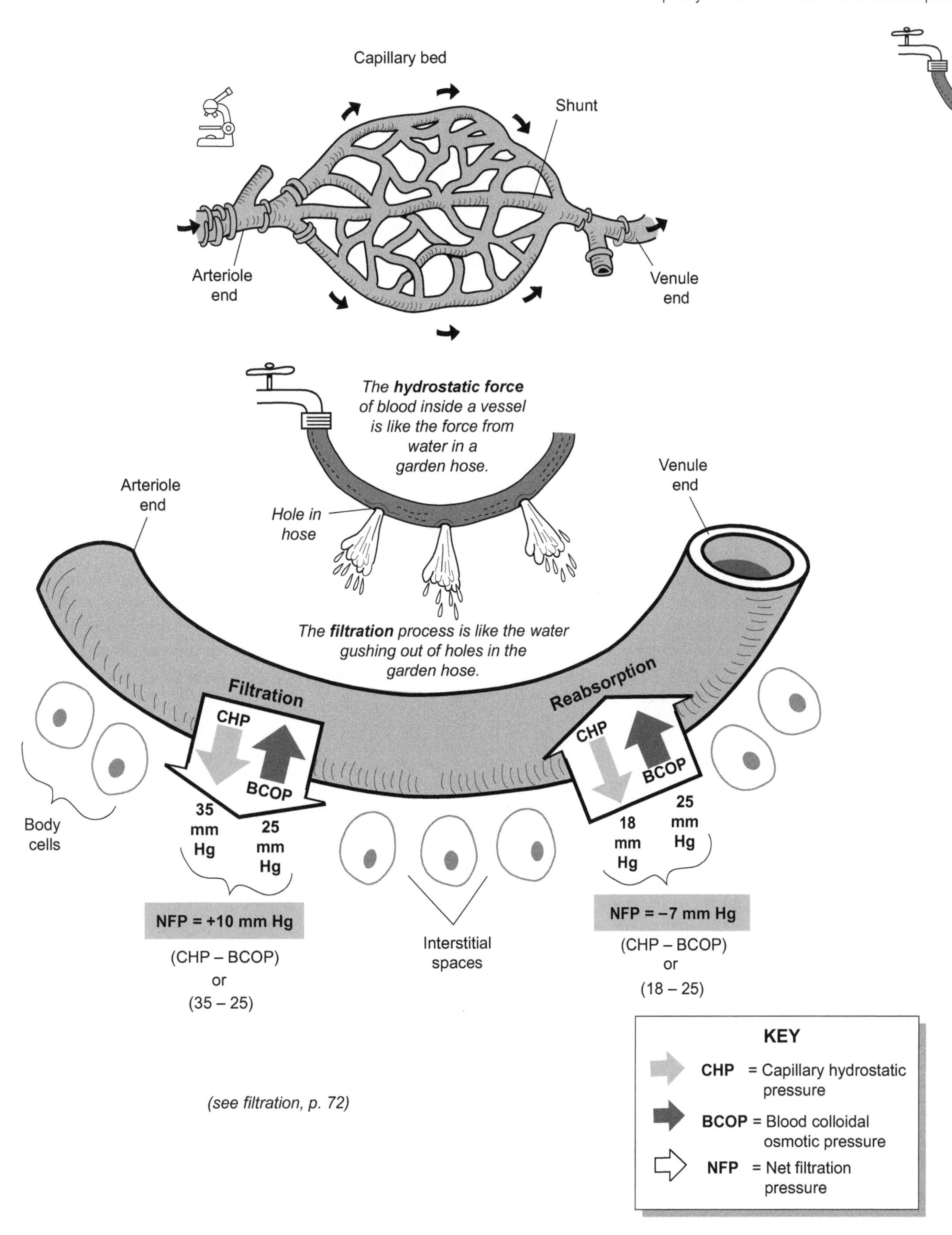

(see filtration, p. 72)

Description

***Venous return** refers to the volume of blood returning to the heart through the veins.* Veins are under much lower pressure than arteries, which makes it difficult to return venous blood to the heart. For example, the average pressure in the venous system is about 16 mm Hg, and the average pressure in the arterial system is about 60 mm Hg. Veins contain semilunar valves at regular intervals, which allow for blood flow in one direction and prevent backflow. Venous return is easier when a person is lying down because the force of gravity is not a factor. But when someone is standing up, consider how long a journey it is for the venous blood in your foot to return to the heart while moving against gravity. Although the heart's normal pumping action helps move the venous blood back to the heart, the body has two other mechanisms we will refer to collectively as venous "pumps": (1) **skeletal muscle "pump,"** and (2) **respiratory "pump."**

① Skeletal muscle "pump"

When you are walking, the normal muscle contractions in your legs greatly aid in venous return. They actually squeeze blood up through the peripheral veins toward the heart in an action called "milking." Here is how it works: The illustration shows a short segment of a vein with two semilunar valves. The **proximal valve** is nearer the heart, and the **distal valve** is farther from the heart. As the skeletal muscle around this vein contracts, the vessel is squeezed, thereby increasing the pressure inside the vein. This increased pressure forces the proximal valve to open, but the distal valve remains closed because of back pressure. The end result of this action is that some venous blood is moved up into the next section of the vein. By repeating this pattern, venous blood is given an extra push on its journey back to the heart.

If you have to stand for a long time, it is good to flex your gastrocnemius muscles (calf muscles) to prevent fainting. Fainting is caused by a decrease in venous return, which leads to decreased cardiac output, which causes a decrease in the blood supply to the brain. This means that less oxygen and glucose are being delivered to neurons in the brain, which causes fainting. By flexing your gastrocnemius muscles after standing still for a long time, you will put your skeletal muscle "pump" to work for you.

② Respiratory "pump"

Normal breathing facilitates venous return. The deeper the breathing, the greater is the assistance in this regard. As shown in the illustration, normal breathing involves a repeating cycle of contraction and relaxation of the **diaphragm.** When this muscle contracts for a normal inspiration, it moves downward and compresses the contents of the **abdominal cavity**, which increases the abdominal pressure. Simultaneously, the volume in the **thoracic cavity** increases, causing a decrease in pressure. This pressure gradient forces blood out of the various veins in the abdominal region and into the inferior vena cava (**IVC**) to allow blood to move back into the right atrium of the heart.

Venous "pumps"

The BIG SQUEEZE!

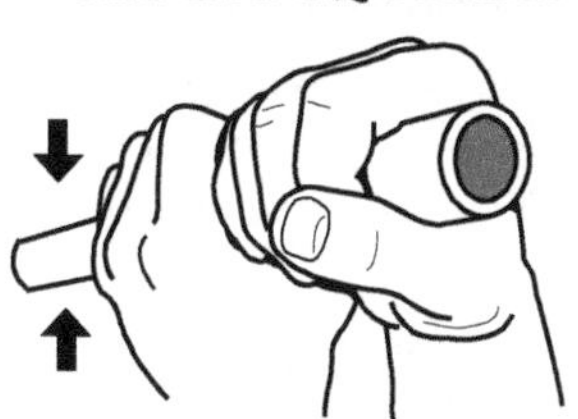

The venous "pumps" are external forces that act on veins like squeezing a hollow rubber tube with your hands.

① Skeletal muscle "pump"

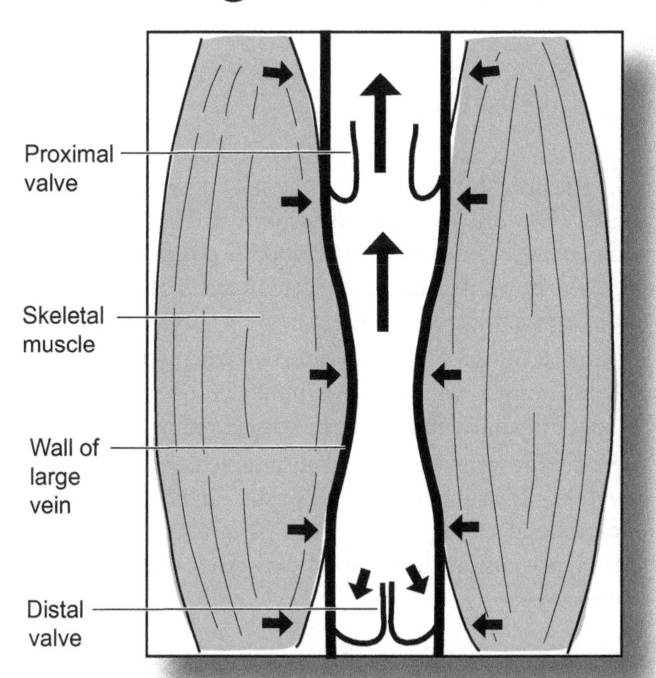

Color the blood in the large vein blue.

② Respiratory "pump"

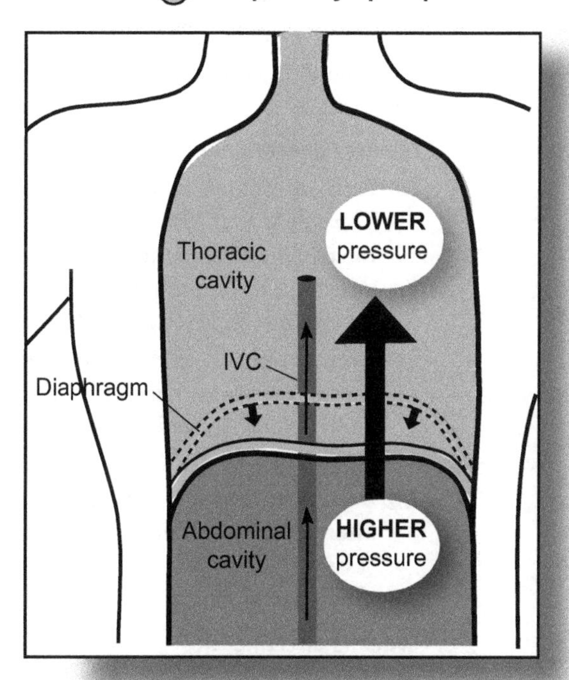

Which way does a pressure gradient go? It always moves air or fluids from HIGH to LOW.

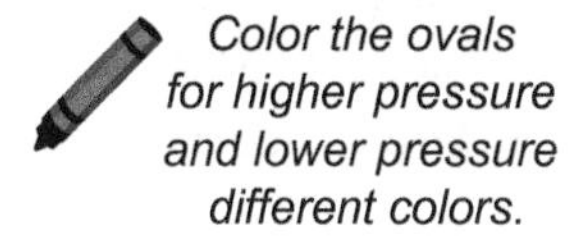

Color the ovals for higher pressure and lower pressure different colors.

Description

Cardiac output (**CO**) is the amount of blood pumped out of a ventricle in 1 minute and is expressed as milliliters of blood per minute. It is mathematically defined as the product of the **stroke volume** (**SV**) times the **heart rate** (**HR**). This is the mathematical equation: **CO = SV × HR**

A normal stroke volume is the amount of blood pumped out of each ventricle in one beat—about 70 mL for an adult heart. A healthy adult male has a resting heart rate of about 70 beats per minute (bpm). By plugging these numbers into the formula given above, we can calculate a normal CO. As shown in the equation at the bottom of the illustration, it calculates to approximately 4,900 mL of blood per minute. If converted to liters, it is 4.9 L per minute.

The illustration shows this by assuming that we could collect blood continually (from only one ventricle) in a 6-liter glass beaker. At the end of 1 minute, it would have filled the beaker to 4.9 L of blood. The CO can be measured by taking a person's blood pressure and pulse, then doing some calculations.

For example, let's say your blood pressure was measured as 118/80 and pulse was 70 bpm. From this, calculate the pulse pressure (systolic pressure minus diastolic pressure). In this case, $118 - 80 = 36$. Then plug the numbers into this equation:

$$\begin{aligned} CO &= 2\text{ mL} \times \text{pulse pressure} \times HR \\ &= 2 \times 36 \times 70 \\ &= \mathbf{5{,}320\ mL/min.} \end{aligned}$$

In general, factors that effect SV also affect HR. For example, *during exercise, both SV and HR increase.* Even so, let's consider the key factors that govern each of these two variables.

Factors affecting SV

- **Preload:** defined as the amount of tension in the ventricular cardiac muscle cells prior to contracting. The rule is that *the greater the tension on the cells, the more forcefully they contract.* This relationship is referred to as the **Frank-Starling Law.** For example, during exercise, more blood is returned to the heart through the venous return. As more blood enters the heart, it stretches cardiac muscles cells, resulting in a more forceful ejection of blood. For this reason, exercise increases SV.
- **Contractility:** defined as the degree to which cardiac muscle cells can shorten when stimulated by a specific chemical substance. For example, calcium ions have a positive influence on cardiac muscle contraction. Abnormally low levels of calcium result in an irregular heartbeat, decreasing the SV.
- **Afterload:** defined as the amount of force needed from ventricular cardiac muscle cells to eject blood from the ventricles and past the semilunar valves. Anything that impedes blood flow can increase afterload. For example, any blockage in the peripheral blood vessels (such as atherosclerosis) would restrict blood flow and increase the afterload. Hypertension (high blood pressure) also increases the afterload. As the afterload increases, the SV decreases.

Factors affecting HR

- **Age:** HR gradually decreases from childhood to adulthood. A newborn may have an HR of 120 bpm, whereas an adult male may have 70 bpm. In the elderly, the HR increases again relative to that of a young adult.
- **Sex:** On average, females have slightly higher resting HRs than males. The difference is about 5 bpm.
- **State of activity:** During certain phases of the sleep cycle, the HR decreases. But during exercise the HR temporarily increases.
- **Endurance training:** Marathon runners may have a resting HR of 50 bpm. This type of training increases heart size as well as SV. This allows for a normal CO with a lower HR.
- **Stress, anxiety:** Stress and anxiety increase HR.

The CO does not remain constant. It regularly rises and falls. Maintaining the CO in a normal range is primarily the job of the cardiovascular (CV) center in the medulla oblongata of the brain. Using reflex pathways in the autonomic nervous system, the CV center regulates the rhythm and force of the heart rate. Hormones also help to control cardiac activity. For example, epinephrine and norepinephrine are powerful cardiac stimulators. In short, the nervous system and hormonal regulators work together to regulate heart activity, thereby indirectly ensuring that a normal CO is achieved.

Cardiac output (CO) = **stroke volume** (SV) × **heart rate** (HR)

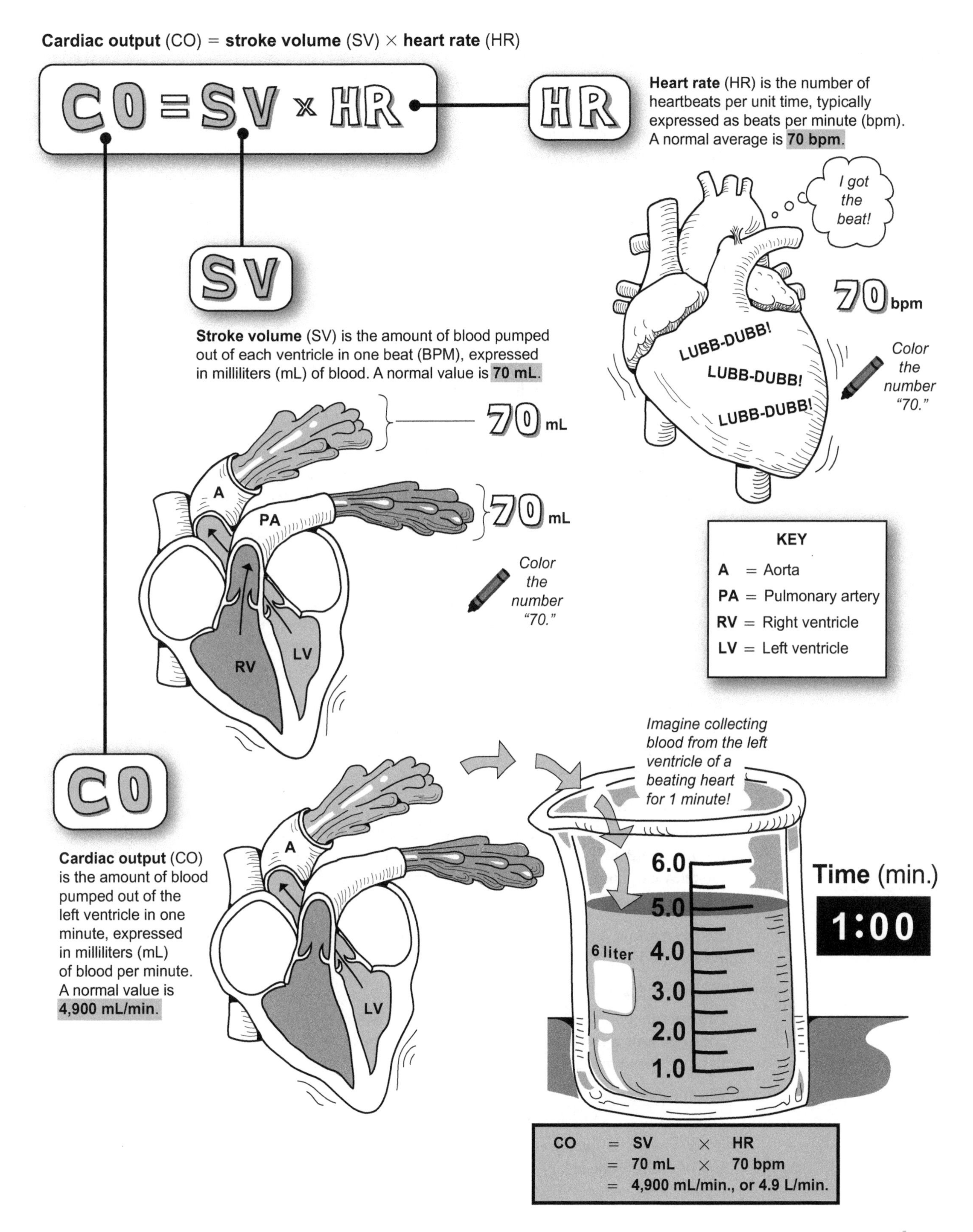

CO	=	SV	×	HR
	=	70 mL	×	70 bpm
	=	4,900 mL/min., or 4.9 L/min.		

Description

The following four regulatory mechanisms are used to control **blood pressure** (**BP**): cardiovascular center, neural regulation, hormonal regulation, and autoregulation.

① **Neural regulation through the cardiovascular center**

Neural regulation uses reflex pathways to regulate normal blood pressure. As its name implies, the **cardiovascular center** (**CV**), in the **medulla oblongata**, is the nervous system's command-and-control center for regulating the heart rate and vasoconstriction. It receives peripheral sensory input from three major sources: **proprioceptors** (*detect changes in muscle tension*), **chemoreceptors** (*detect changes in blood acidity*), and **baroreceptors** (*detect changes in blood pressure*). The CV responds by sending the appropriate motor output to the heart and blood vessels. Depending on the stimulus, the **heart rate** (**HR**) either increases or decreases, and blood vessels are stimulated to constrict. The end result is that the BP either increases or decreases back to normal levels. For details of these reflex pathways, see p. 120.

② **Hormonal regulation**

Hormonal regulation involves the use of chemical messengers called **hormones** that bind to receptors at a target tissue in order to induce a response.

- Renin-angiotensin-aldosterone (**RAA**) system.

  When blood pressure decreases, it stimulates the kidney to release the enzyme renin into the blood. This leads to the production of a powerful vasoconstrictor called angiotensin II. This, in turn, stimulates the adrenal cortex to produce the hormone aldosterone. The net result is an increase in blood pressure. For details of this mechanism, see p. 242.

- Epinephrine (**EPI**)/norepinephrine (**NE**)

  EPI and **NE** are produced by cells in the adrenal medulla in response to emergency or stressful situations. Two of the organs they target are the heart and blood vessels. The result is that heart rate increases and blood vessels constrict, respectively. Together, these two responses lead to an increase in blood pressure. For details of this mechanism, see p. 294.

- Antidiuretic hormone (**ADH**)

  Water loss from the blood, like what occurs during excessive sweating, stimulates the posterior lobe of the pituitary gland to produce **ADH**. This hormone targets the nephrons in the kidneys and causes more water to be reabsorbed into the blood, thus restoring normal blood pressure. For the details of this mechanism, see p. 236.

- Atrial natriuretic peptide (**ANP**)

  As blood volume increases, this stretches the atria in the heart, which stimulates cells in the atria to produce **ANP**. By increasing vasodilation, sodium ion excretion, and urine production and blocking the release of hormones like ADH and aldosterone, ANP decreases BP.

③ **Autoregulation**—the automatic control of blood flow to a tissue

These mechanisms are normally not as significant as the neural and hormonal mechanisms mentioned above in 1 and 2.

- **Physical changes**

  Changes in body temperature affect BP. For example, cold causes superficial blood vessels to constrict, which increases BP. In contrast, heat causes blood vessels to dilate, leading to a decrease in blood pressure.

- **Chemical changes**

  Different types of body cells, such as smooth muscles cells, endothelial cells, and macrophages, produce various chemicals that signal blood vessels to either dilate or constrict. For example, smooth cells produce lactic acid, which causes blood vessels to dilate. Other vasodilating chemicals include nitric oxide (NO), H^+, and K^+.

GOAL: Maintain Normal Blood Pressure

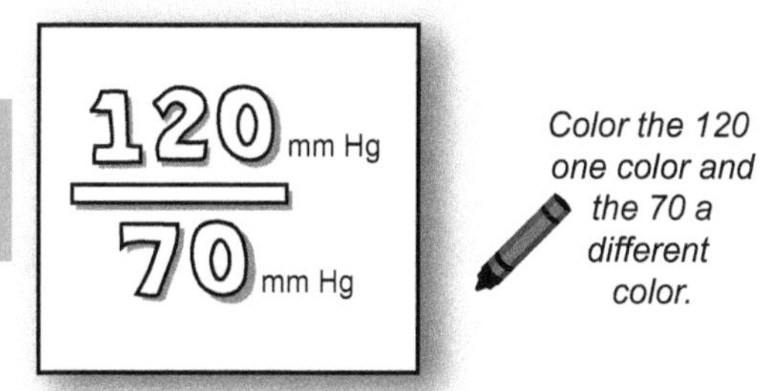

Color the 120 one color and the 70 a different color.

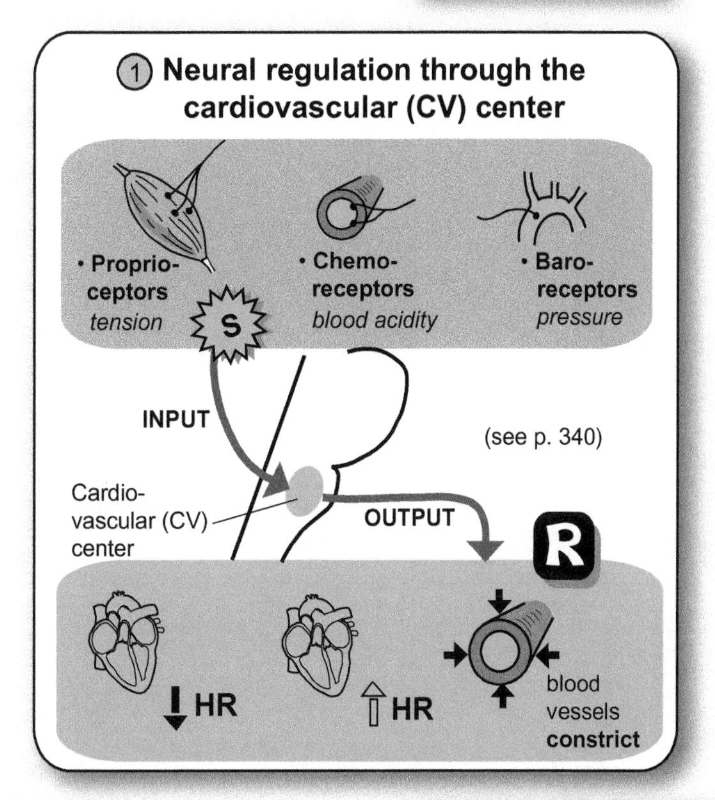

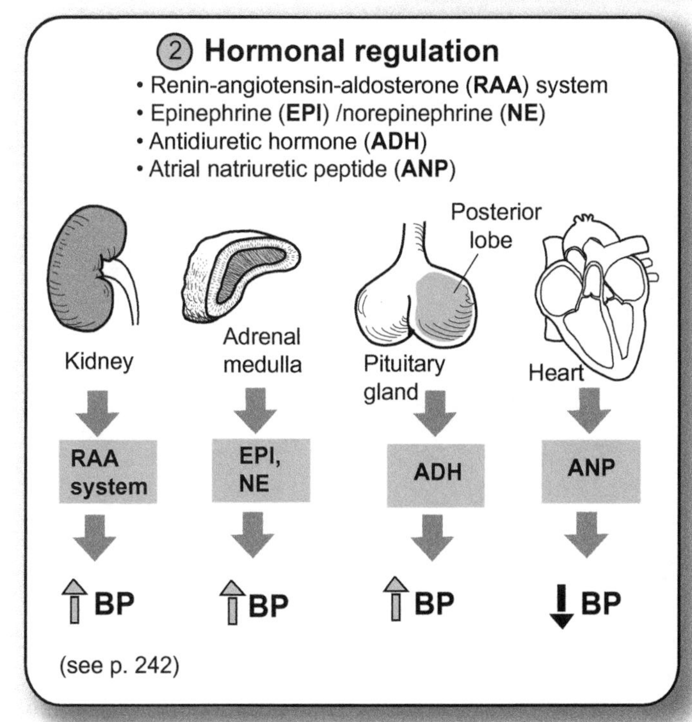

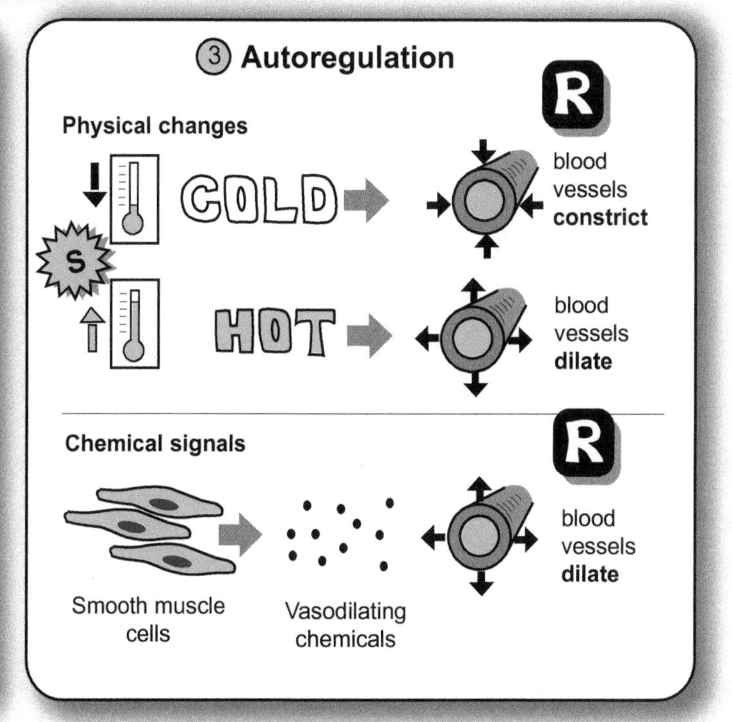

Notes

RESPIRATORY SYSTEM

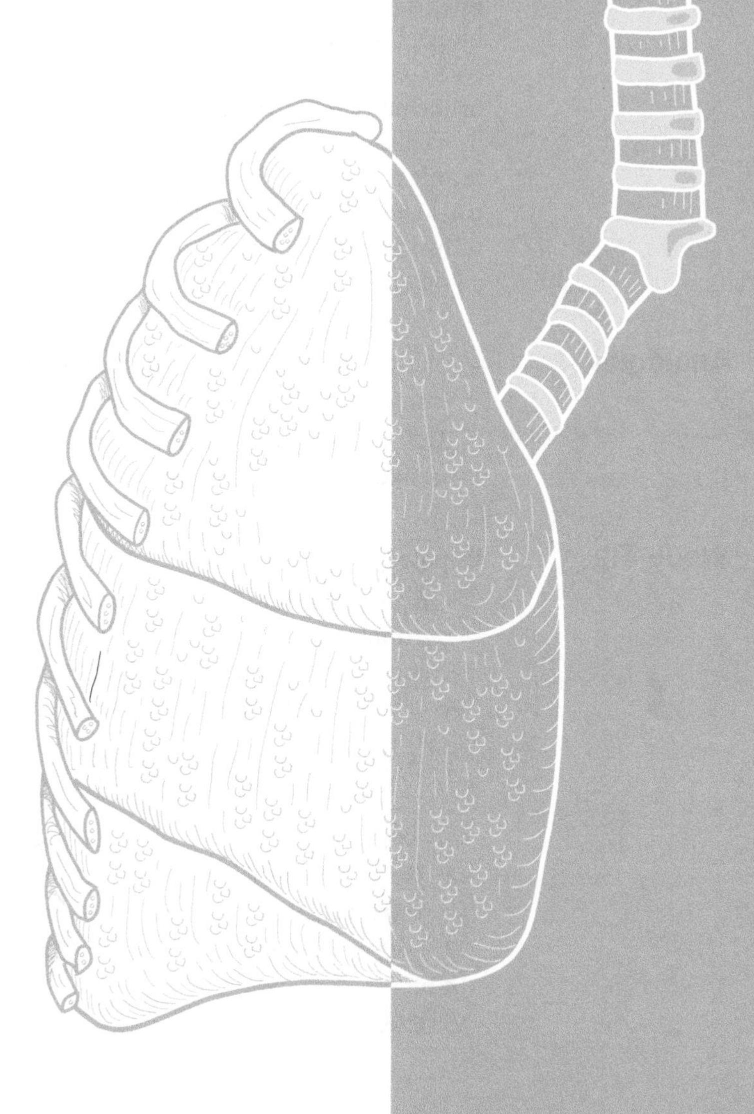

Description

The respiratory system is divided into two major divisions: *upper respiratory system* and *lower respiratory system*. The **upper respiratory system** consists of the nose, nasal cavity, paranasal sinuses, and pharynx. The **lower respiratory system** consists of the larynx, trachea, bronchi, and lungs.

Let's trace the pathway of a molecule of oxygen (O_2) through the respiratory system to its final destination at a body cell. The O_2 molecule enters the **nasal cavity** through the **external nares** *(nostrils)*. As it passes to the back of this moist chamber, it enters the **nasopharynx**, then the **oropharynx**, and finally the **laryngopharynx**.

After passing the rigid, flap-like structure called the **epiglottis**, it enters the **larynx**, then passes through the slit-like opening between the vocal cords called the **glottis**. Next it moves through the long, rigid tube of the **trachea** until it reaches a split in this passageway. Following the passageway branching into the left lung, it enters the **left primary bronchus**, then enters the next split in the passageway, the narrower **secondary bronchus**. The next branch is the even narrower **tertiary bronchus**. Finally, the O_2 molecule enters a microscopic tube called a **bronchiole**.

This continues to branch into a **terminal bronchiole**, then a **respiratory bronchiole**, and finally terminates in an air sac called an **alveolus**. This delicate air sac is the end of the bronchial tree in the lungs. The wall of each alveolus is made of **simple squamous epithelium**, which allows for easy diffusion of the O_2 molecule out of the alveolus and into the bloodstream, where it will be delivered to a body cell.

Analogies

- The **larynx** looks like the **head of a snapping turtle**. The **turtle's head** is the **thyroid cartilage**, and the **turtle's lower jaw** is the **cricoid cartilage**. The **neck of the turtle** is the **trachea**.
- The **clusters of alveoli** are like a **wad of bubble wrap** used in packaging. Both are small, saclike structures filled with air.

Study Tip

Palpate (*feel by touch*):

On average, the larynx is slightly larger in males than in females, but you can easily feel a portion of it in either gender. The main structure you feel beneath the skin is the thyroid cartilage, commonly called the *Adam's apple*. The portion of the Adam's apple that protrudes most anteriorly is called the laryngeal prominence.

Key to Illustration

Larynx (L)
- L1. Thyrohyoid membrane
- L2. Thyroid cartilage
- L3. Laryngeal prominence
- L4. Cricothyroid ligament
- L5. Cricoid cartilage

Upper Respiratory Tract (UP)
- UP1. Frontal sinus
- UP2. Sphenoidal sinus
- UP3. External nares (*nostrils*)
- UP4. Nasal cavity
- UP5. Superior concha
- UP6. Middle concha
- UP7. Inferior concha
- UP8. Pharyngeal tonsil
- UP9. Nasopharynx
- UP10. Palatine tonsil
- UP11. Oropharynx
- UP12. Lingual tonsil
- UP13. Epiglottis
- UP14. Laryngopharynx

Lower Respiratory Tract (LO)
- LO1. Vocal fold
- LO2. Trachea
- LO3. Tracheal rings
- LO4. Parietal pleura
- LO5. Visceral pleura
- LO6. Location of carina (*internal ridge*)
- LO7. Primary bronchus
- LO8. Secondary bronchus
- LO9. Tertiary bronchus
- LO10. Terminal bronchiole
- LO11. Respiratory bronchiole
- LO12. Alveoli
- LO13. Simple squamous epithelium

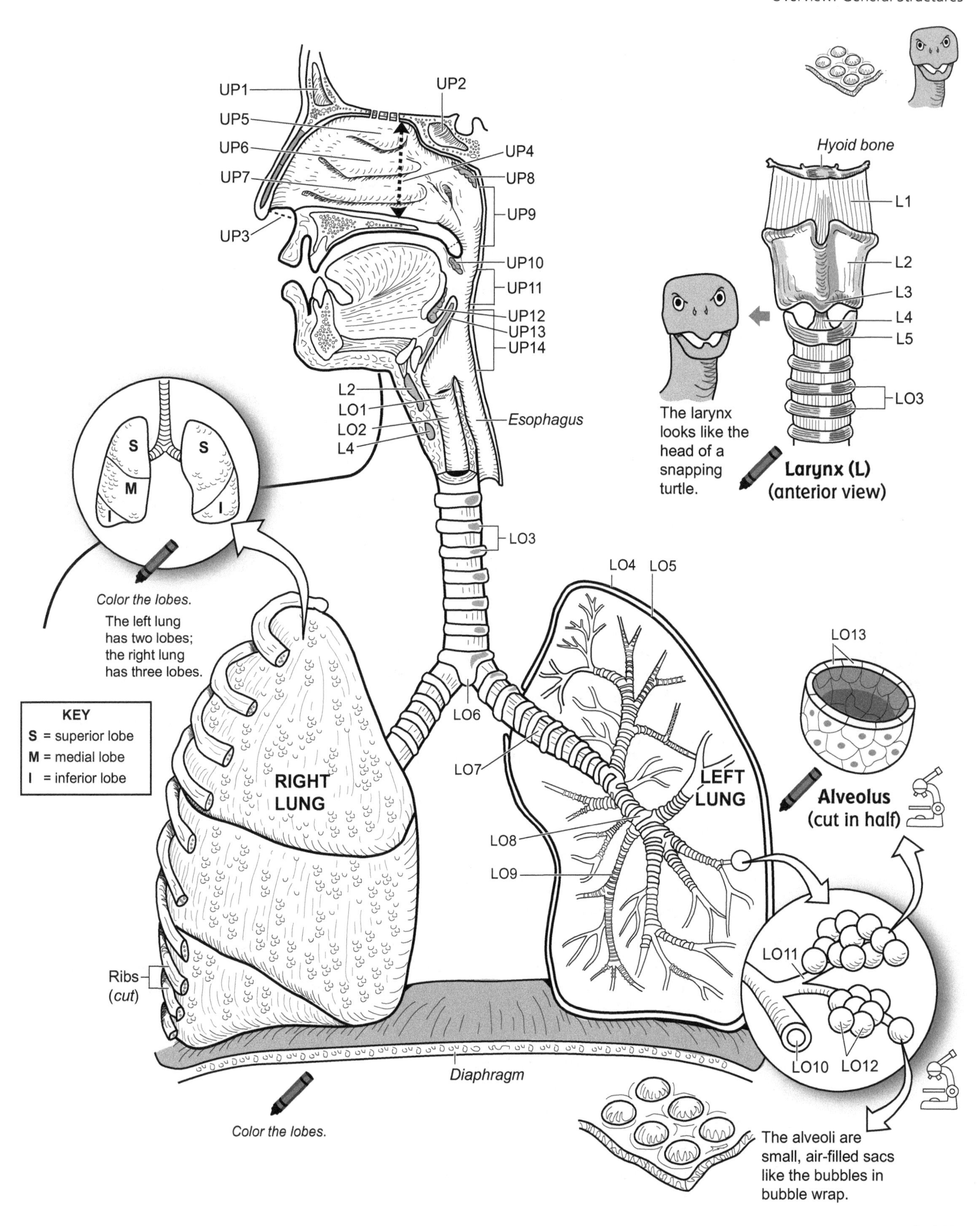
UP1
UP2
UP5
UP6
UP4
UP7
UP8
UP9
UP3
UP10
UP11
UP12
UP13
UP14
L2
LO1
LO2
L4
Esophagus
LO3
Hyoid bone
L1
L2
L3
L4
L5
LO3
The larynx looks like the head of a snapping turtle.
Larynx (L)
(anterior view)
S
S
M
I
I
Color the lobes.
The left lung has two lobes; the right lung has three lobes.
KEY
S = superior lobe
M = medial lobe
I = inferior lobe
LO4
LO5
LO13
LO6
RIGHT LUNG
LO7
LEFT LUNG
Alveolus
(cut in half)
LO8
LO9
LO11
Ribs
(cut)
LO10
LO12
Diaphragm
Color the lobes.
The alveoli are small, air-filled sacs like the bubbles in bubble wrap.

Concept 1: Atmospheric Pressure (AP)

Atmospheric pressure (**AP**) is a force that combines the weight of all the gases in the air we breathe. Of the many gases, the most abundant are: **nitrogen** (N_2), **oxygen** (O_2), **carbon dioxide** (CO_2), and **water** vapor (H_2O). Each of these gases has a different mass.

AP is the force on any surface that comes in contact with air. In the illustration, AP is indicated by the black arrows pushing against the external surface of the little boy's body. Even though he can't actually feel its presence, it is there.

The other illustration shows a **barometer**—a device used to measure the AP. A glass beaker is shown filled with liquid **mercury** (Hg). As the AP pushes down on the surface of this liquid, it forces the mercury into the open end of the glass tube so it rises to a specific height. At sea level, the standard measurement is **760 mm Hg**. This number changes as the elevation changes. *The higher the elevation, the lower the AP*. For example, a mountain climber and a person strolling in the park at sea level experience different AP. The mountain climber has less atmosphere above him, so he has less AP than the person in the park.

Concept 2: Pressure and Volume Law (Boyle's Law)

The pressure and volume law (*Boyle's law*) is a gas law stating that *volume and pressure have an inversely proportional relationship for a gas held at a constant temperature*. The illustration shows a clear, hollow sphere containing five gas molecules. These molecules have kinetic energy, so they randomly bounce around against the wall of the sphere. This is the source of the force of pressure inside the sphere.

Consider what would happen if the small sphere were to increase in size. Has the number of gas molecules changed? No. Five gas molecules are still present. The only thing that has changed is the volume of the sphere. If the volume increases, what do you predict will happen to the pressure inside? That's right. It will decrease.

If we refer to the volume and pressure in the small sphere as V_1 and P_1 and use V_2 and P_2 for the enlarged sphere, the pressure and volume law can be expressed as this **equation:** $V_1 \times P_1 = V_2 \times P_2$. The product of the volume and the pressure in the larger sphere should be equivalent to the product of the volume and the pressure in the smaller sphere. Fill in the given numbers, and plug them into the formula to see if this relationship holds true.

Application: Mechanics of Breathing

In addition to knowing atmospheric pressure and Boyle's law, we need to understand the terms **interpleural pressure** and **alveolar pressure**, so we can understand the mechanics of normal breathing. Interpleural pressure is the pressure inside the small, liquid-filled, pleural space around the lungs. At rest, the normal value is **756** mm Hg. The pressure inside the lungs at any given moment is called the alveolar pressure. Its normal value at rest is the same as the normal AP, or **760** mm Hg. Normal, quiet breathing is a repeated cycle of **inhalations** and **exhalations**. Inhalation is an *active* process because it requires the contraction of muscles such as the **diaphragm** and the **external intercostal muscles** between the ribs. In contrast, an exhalation is a passive process because it mainly involves recoil of the elastic connective tissue in the lungs and thoracic wall. Let's summarize the steps involved in a normal inhalation and a normal exhalation.

Steps in a normal inhalation:

1. Diaphragm and external intercostal muscles contract
2. Thoracic cavity expands; lung volume increases
3. Alveolar pressure drops from 760 to 758 mm Hg
4. Air flows into lungs down its pressure gradient from **760** to **758** mm Hg

Note: There is no such thing as "suction." We do not "suck" air into the lungs. Instead, air flows from a region of higher pressure to a region of lower pressure. This continues until the alveolar pressure is equal to the AP (760).

Steps in a normal exhalation:

1. Diaphragm and external intercostals relax
2. Thoracic cavity contracts; lung volume decreases
3. Recoil effect from lungs and thoracic wall
4. Alveolar pressure increases from 760 to 762 mm Hg
5. Air flows out of the lungs down its pressure gradient from **762** to **760** mm Hg. This continues until the alveolar pressure is equal to the AP (**760** mm Hg).

CONCEPT 1: ATMOSPHERIC PRESSURE (AP)

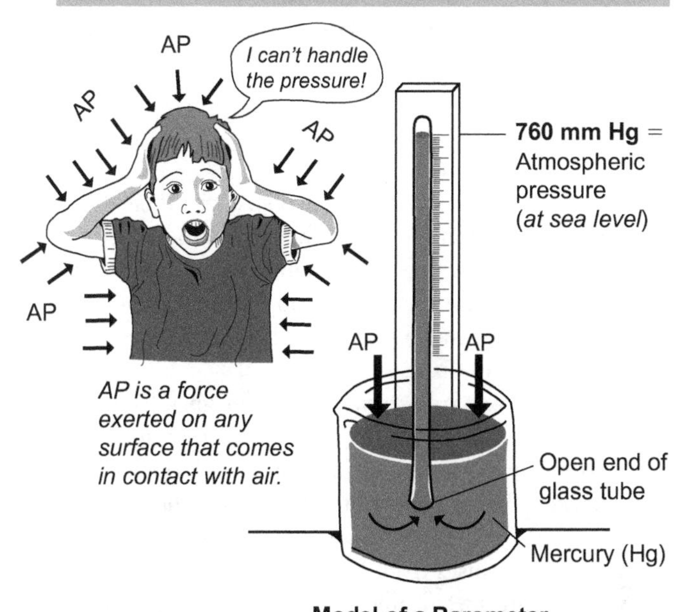

Model of a Barometer

CONCEPT 2: PRESSURE AND VOLUME LAW

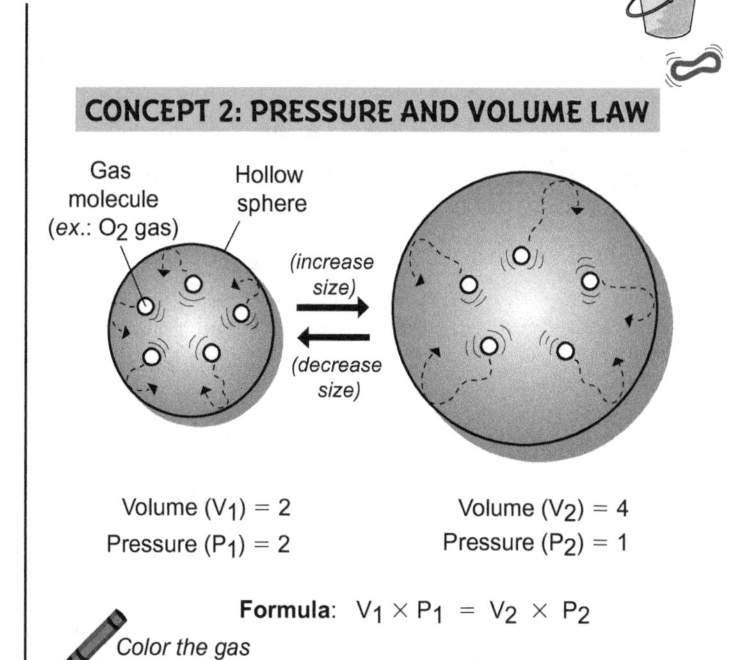

Formula: $V_1 \times P_1 = V_2 \times P_2$

Color the gas molecules.

___ ___ = ___ ___

APPLICATION: MECHANICS OF BREATHING

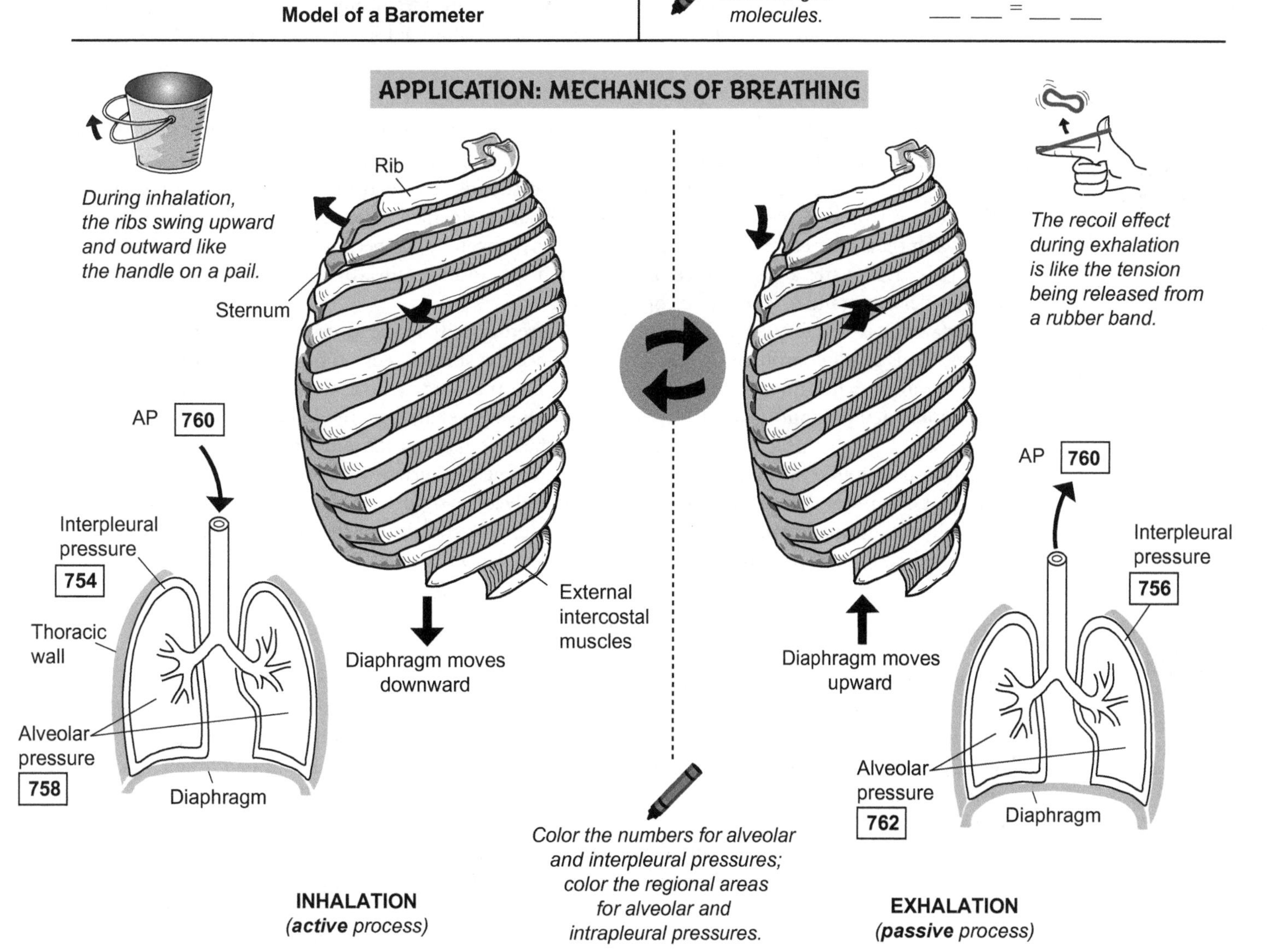

Color the numbers for alveolar and interpleural pressures; color the regional areas for alveolar and intrapleural pressures.

INHALATION
(***active*** process)

EXHALATION
(***passive*** process)

Alveolar Structure and the Respiratory Membrane

The lungs contain about 300 million alveoli to increase surface area for exchange of respiratory gases. Like the bubbles in bubble wrap, the **alveoli** (sing. *alveolus*) are the delicate, microscopic air sacs where gas exchange occurs between the lungs and the blood. Each alveolus is composed of three different types of cells:

1. **Simple squamous epithelial cells** (*Type I cells*)
2. **Surfactant-secreting cells** (*Type II cells*)
3. **Alveolar macrophages** (*dust cells*)

The numerous simple squamous epithelial cells make up the wall of each alveolus, and their flat shape allows for better diffusion of respiratory gases (oxygen and carbon dioxide). The surfactant-secreting cells are fewer in number and are scattered within the alveoli. They secrete **surfactant**—an oily fluid film that lines the inside of the alveoli like a soap bubble and serves to reduce surface tension to prevent collapse. The alveolar macrophages are the "housekeepers" of the alveoli. They move around and engulf microorganisms, dust particles, and other debris to keep the lungs clean and free of disease.

Recall that the blood flows to each lung through the **pulmonary arteries** (left and right). Each of these vessels branches and becomes smaller until it finally reaches numerous capillary beds that surround the alveolar sacs—clusters of alveoli. This structural relationship is like a plastic mesh bag (blood capillaries) surrounding a cluster of grapes (alveolar sac) at the market. The **respiratory membrane** is called the "blood/air barrier" and is very thin to allow for easy diffusion of oxygen and carbon dioxide. This is the site where oxygen diffuses from alveolus to blood and carbon dioxide diffuses from blood to alveolus. This structure consists of the fusion of the alveolar and capillary walls. More specifically, it is where the **simple squamous epithelium** of the alveolus meets the **simple squamous epithelium** of the capillary with basement membranes in between.

Surfactant and Surface Tension

Surface tension is both a force and a property of water, attributed to the fact that water molecules are **polar**. Like a magnet, the oxygen end of any water molecule has a partially negative charge and the hydrogen end has a partially positive charge. When the negative end of one water molecule aligns itself with the positive end of another, a **hydrogen bond** is formed. In fact, water molecules are always rearranging themselves to maximize the number of hydrogen bonds because this is the most stable state for them. This gives water a high degree of surface tension and explains, by the way, why the insect called a **water strider** can stand on the surface of the water.

Surfactant is a fluid secreted by **surfactant-secreting cells** (Type II cells) that contains a mixture of phospholipids and lipoproteins. It functions to reduce the surface tension on alveoli to prevent them from collapsing after exhalation. If pure water were to line the inside of the alveoli, the surface tension in the water would tend to pull inward on the alveoli. During exhalation, the delicate alveoli would collapse because of the high surface tension on them. Surfactant reduces the amount of hydrogen bonds in the water normally found inside the alveoli, thereby reducing the surface tension. This keeps the alveoli partly inflated at all times. During normal breathing, the alveoli would look like slowly pulsating spheres that expand in size during inhalation and shrink in size during exhalation. This process is more energy-efficient than if we had to refill collapsed alveoli with every inhalation.

Lung Compliance

Note: Lung compliance is the only topic not illustrated on the facing page.

Lung compliance is a measure of the degree of effort needed to expand, or ease of expanding, the lungs and thoracic wall. This is determined primarily by two factors: (1) elasticity of lung tissue, and (2) alveolar surface tension. Two states of compliance are mentioned—high and low. Normal, healthy lungs have a *high* degree of compliance because of the elastic connective tissue within the lungs and the reduced surface tension in the lungs because of surfactant. States of *low* compliance may occur with some respiratory disorders.

For example, a premature baby may lack the normal surfactant found in a newborn. This increases surface tension on the alveoli. Emphysema also leads to lower compliance because of destruction of the walls of the alveoli. In short, any disease state that increases alveolar surface tension and/or damages normal elastic connective tissue in the lungs will result in lowered compliance.

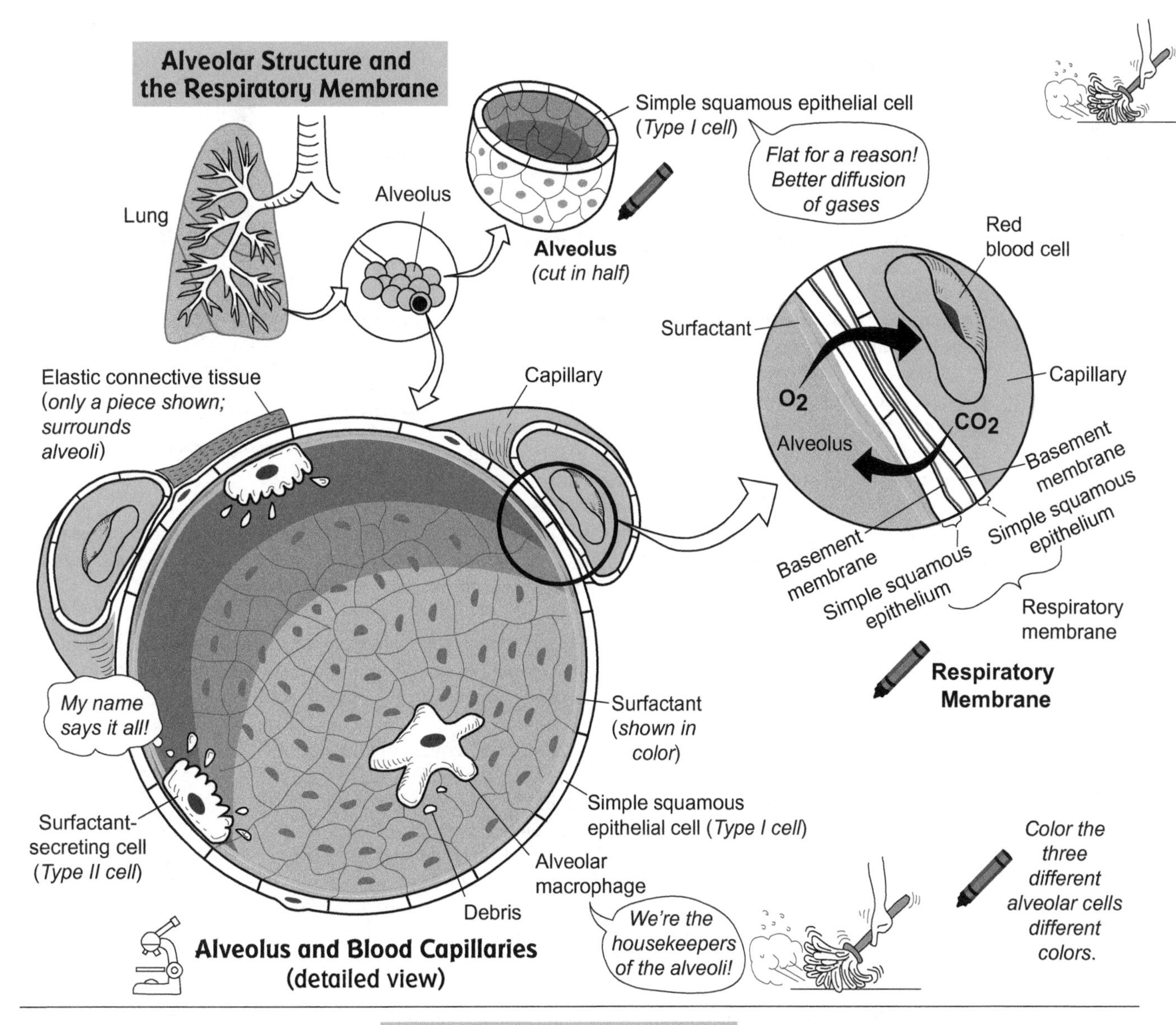

Surfactant and Surface Tension

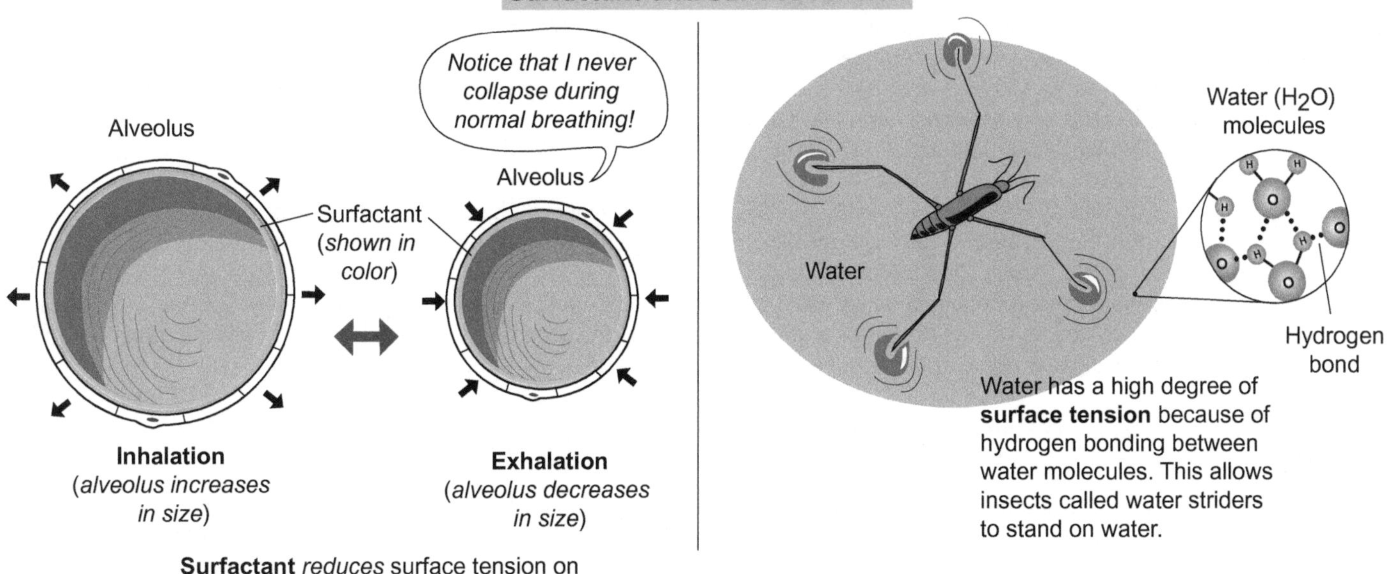

Surfactant *reduces* surface tension on alveoli, which prevents them from collapsing

Water has a high degree of **surface tension** because of hydrogen bonding between water molecules. This allows insects called water striders to stand on water.

This module deals with the following concepts:

- **Dalton's law**
- **Henry's law**
- **Diffusion of O_2 and CO_2 into and out of the blood**
- **External respiration and internal respiration**

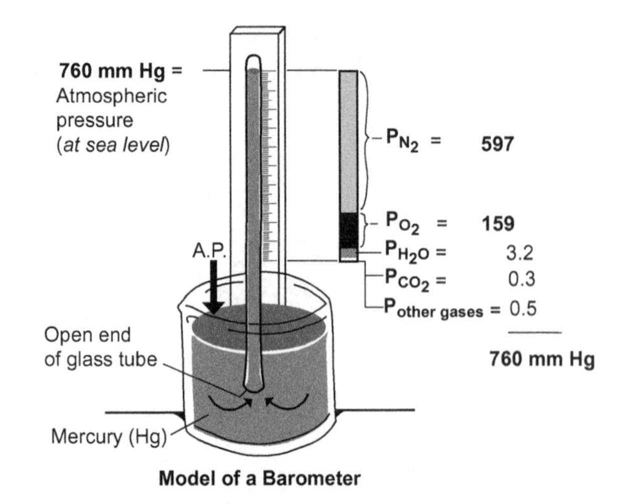

Model of a Barometer

Dalton's Law

Dalton's **law** is a gas law that deals with **partial pressures** of gases. It states that *in a mixture of gases (like the air we breathe), each gas exerts its own partial pressure.* Therefore, the sum of all the partial pressures is the total pressure exerted by the gas mixture.

Consider a barometer that measures atmospheric pressure (AP). At sea level, the normal AP is 760 mm Hg. Air contains the following gases: nitrogen (78%), oxygen (21%), water vapor (0.4%), carbon dioxide (0.04%), and other gases (0.06%). The partial pressure of nitrogen gas (P_{N_2}) is 78% of 760 or **597** mm Hg. The partial pressure of oxygen (P_{O_2}) is 21% of 760, or **159** mm Hg. Therefore, the combined partial pressures from nitrogen and oxygen account for **99%** of atmospheric pressure (756 of the 760 total). In the body, we have to consider the partial pressures of physiologically important gases such as O_2 and CO_2 in the alveolar air and the blood.

Henry's Law

Another gas law, **Henry's law**, states that at a given temperature, the amount of gas dissolved in a liquid is directly proportional to the partial pressure of that gas. Gases normally move in and out of solution, with some more soluble in liquids than others. For example, carbon dioxide (CO_2) is highly soluble in water—much more so than oxygen (O_2).

As an example of a gas moving out of solution, let's consider soda pop. All sodas contain carbonated water—CO_2 dissolved in water. Although much of the CO_2 remains in solution, some is found in the air inside the bottle. If you open a soda and allow it to sit out at room temperature for a long time, all the CO_2 eventually will move out of solution, and the soda will go "flat." If you shake a fresh bottle, you can see the CO_2 bubble out of solution, which builds up pressure inside the bottle. This increase in pressure causes the soda to shoot out when the bottle is opened. Similarly, in the lungs, CO_2 moves out of solution from the blood capillaries and into the alveolar air.

Shake it up! CO_2 bubbles out of solution.

The increase in pressure causes the soda to burst out.

Diffusion of O_2 and CO_2 into and out of the Blood

The **blood P_{CO_2} level** is about 45 mm Hg as it approaches the lungs, and the P_{CO_2} in the alveoli is about 40 mm Hg. Because of the small gradient, CO_2 slowly diffuses out of the blood and into the alveoli. As the blood leaves the lungs, the P_{CO_2} level has dropped slightly, to about 40 mm Hg. Because

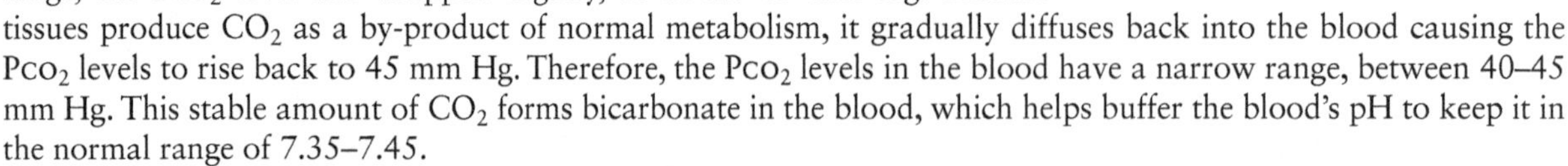

tissues produce CO_2 as a by-product of normal metabolism, it gradually diffuses back into the blood causing the P_{CO_2} levels to rise back to 45 mm Hg. Therefore, the P_{CO_2} levels in the blood have a narrow range, between 40–45 mm Hg. This stable amount of CO_2 forms bicarbonate in the blood, which helps buffer the blood's pH to keep it in the normal range of 7.35–7.45.

The **blood P_{O_2} level** is about 40 mm Hg as it approaches the lungs, while the P_{O_2} of the alveolar air is about 105 mm Hg. Because of the large gradient, O_2 quickly diffuses into the blood. Because the hemoglobin in the red blood cells is especially efficient at binding O_2, the blood P_{O_2} levels rise to 100 mm Hg—the typical level for oxygenated blood. When the oxygenated blood approaches tissues low in oxygen, it rapidly diffuses into tissues. In deoxygenated blood, the P_{O_2} drops to about 40 mm Hg. In short, the blood P_{O_2} levels vary widely, from 40 mm Hg in deoxygenated blood to 100 mm Hg in oxygenated blood.

External Respiration and Internal Respiration

- **External respiration** occurs between the alveoli in the lungs and the blood capillaries; O_2 moves from the alveoli into the blood, and CO_2 moves from the blood into the alveoli.
- **Internal respiration** occurs between the blood capillaries and body cells; O_2 moves from the blood into body cells, and CO_2 moves from the body cells (where it is produced as a waste product) into the blood.

Once oxygen is inside body cells, it is used in the important process of **cellular respiration** (see p. 454).

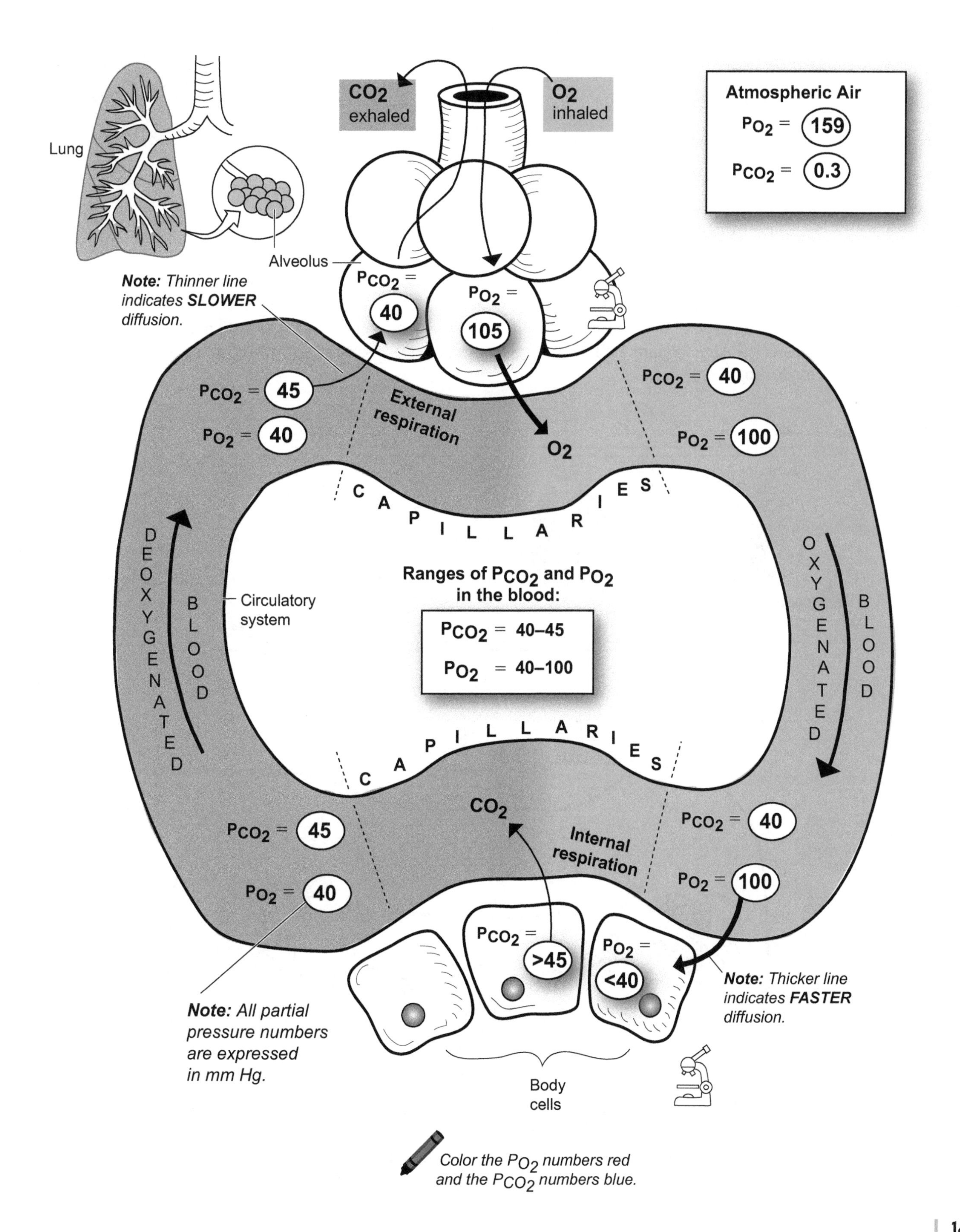
Lung
Alveolus
CO2 exhaled
O2 inhaled
Atmospheric Air
PO2 = 159
PCO2 = 0.3
PCO2 = 40
PO2 = 105
Note: Thinner line indicates SLOWER diffusion.
PCO2 = 45
PO2 = 40
External respiration
O2
PCO2 = 40
PO2 = 100
CAPILLARIES
DEOXYGENATED BLOOD
Circulatory system
Ranges of PCO2 and PO2 in the blood:
PCO2 = 40–45
PO2 = 40–100
OXYGENATED BLOOD
CAPILLARIES
CO2
Internal respiration
PCO2 = 45
PO2 = 40
PCO2 = 40
PO2 = 100
PCO2 = >45
PO2 = <40
Note: Thicker line indicates FASTER diffusion.
Note: All partial pressure numbers are expressed in mm Hg.
Body cells
Color the PO2 numbers red and the PCO2 numbers blue.

Description

Hemoglobin (Hb) is a protein pigment found in red blood cells. Each Hb molecule is composed of four subunits: alpha 1, alpha 2, beta 1, and beta 2. Each subunit has an iron-containing heme group where a molecule of oxygen is able to bind.

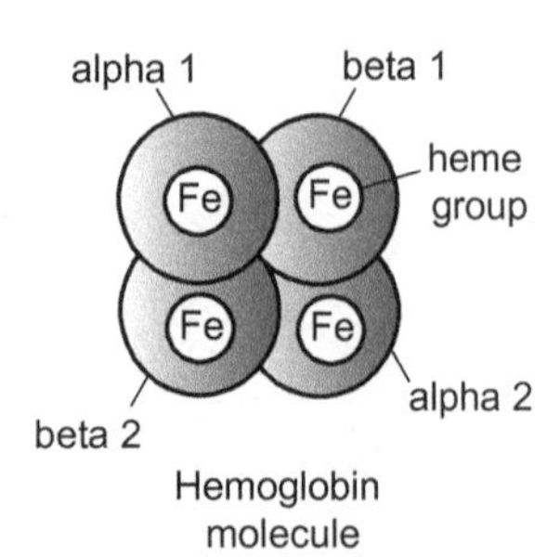

Function

The primary function of Hb is to transport oxygen from the lungs to the body cells, but it also transports some carbon dioxide from the body cells to the lungs. It is a pigment because its color changes depending on whether oxygen is bound to the molecule. It is bright red in its oxygenated form but changes to dark red in its deoxygenated form.

Two Stages for Hemoglobin

With respect to oxygen transport, hemoglobin exists in two states: **oxyhemoglobin** or **deoxyhemoglobin.** Deoxyhemoglobin is the state in which no oxygen is bound to Hb. Once inside the capillaries in the lungs, each Hb molecule binds one molecule of oxygen at a time to hold a maximum of four O_2 molecules. After binding its fourth oxygen molecule, it is saturated with oxygen and said to be in its oxyhemoglobin form.

Deoxyhemoglobin and Oxyhemoglobin

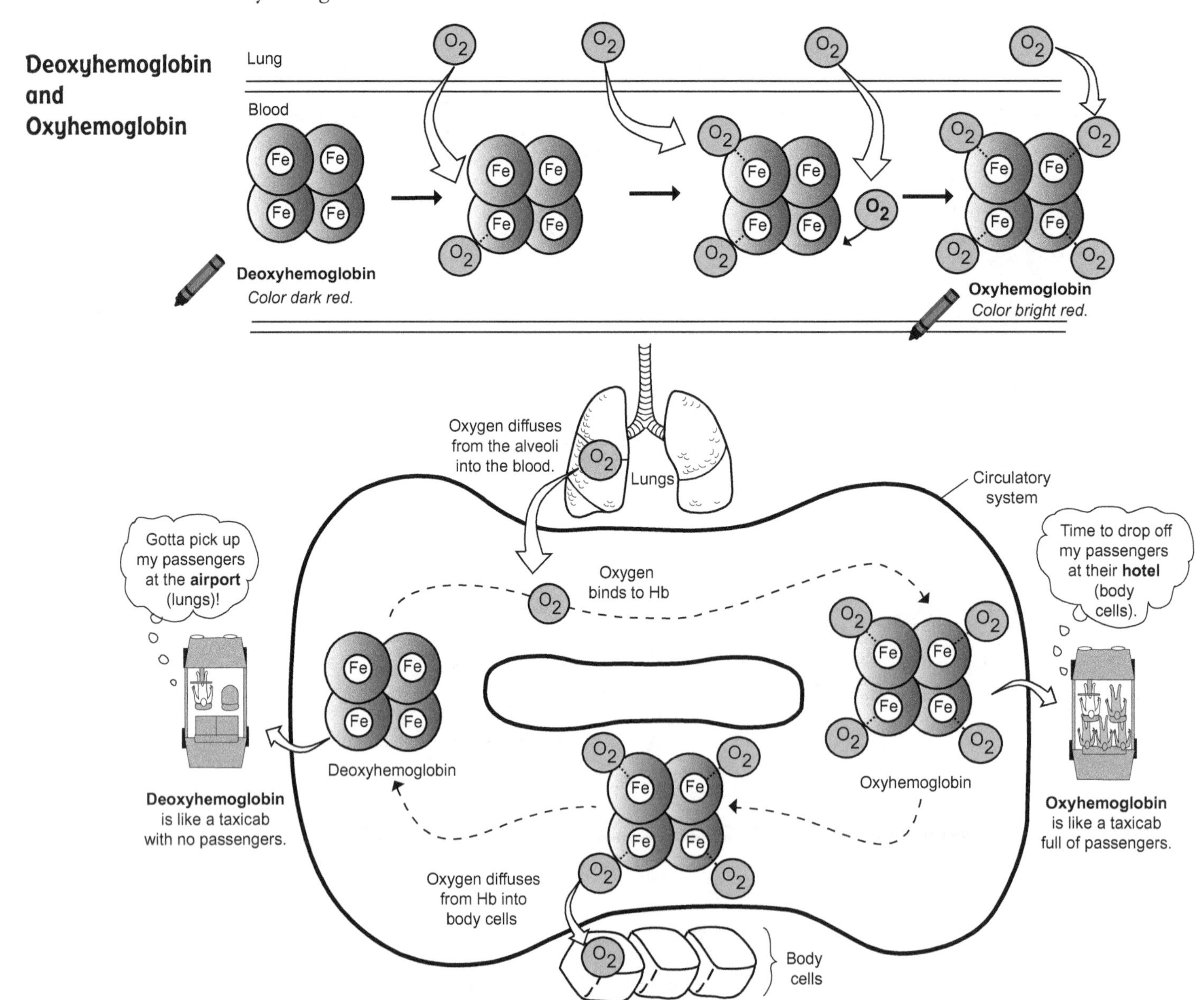

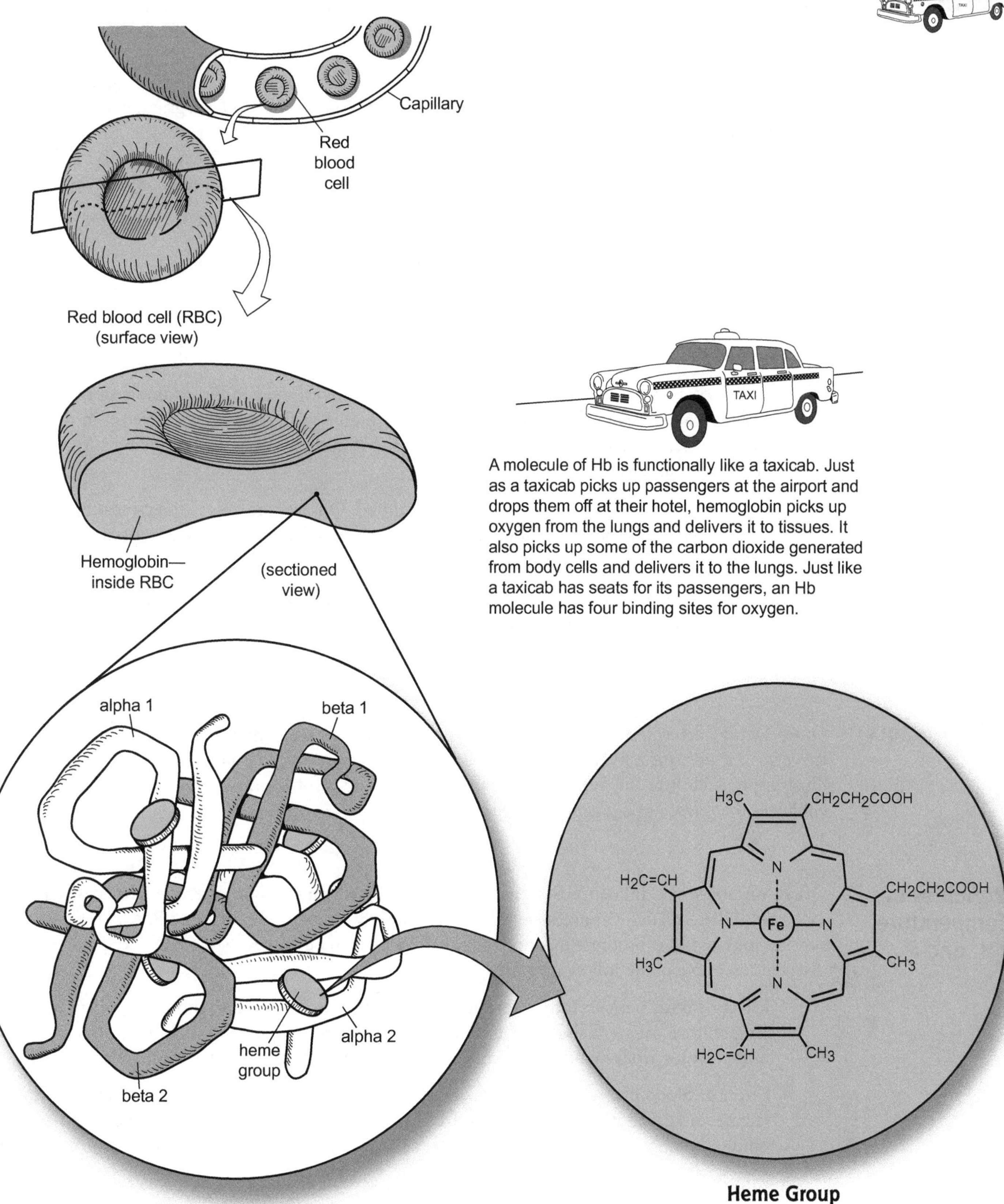

A molecule of Hb is functionally like a taxicab. Just as a taxicab picks up passengers at the airport and drops them off at their hotel, hemoglobin picks up oxygen from the lungs and delivers it to tissues. It also picks up some of the carbon dioxide generated from body cells and delivers it to the lungs. Just like a taxicab has seats for its passengers, an Hb molecule has four binding sites for oxygen.

Heme Group

Hemoglobin (Hb) Molecule

Description

About 98% of all the oxygen (O_2) is transported through the blood on **hemoglobin** (**Hb**). The remaining 2% is dissolved in the plasma. Because all tissues need oxygen to survive, it is vital to understand factors that aid in oxygen binding to, and releasing from, hemoglobin. The binding of O_2 to Hb depends primarily on the *partial pressure created by oxygen* (Po_2) in the blood (see p. 148). When the percent saturation of Hb is plotted versus the blood Po_2, it yields a graph called the **oxygen-hemoglobin dissociation curve.**

Deoxyhemoglobin is like a taxicab without any oxygen "passengers." When the taxi is filled with four passengers, it is **oxyhemoglobin** (see p. 150). This cycle can be represented by the reversible reaction given below:

Hb	+	O_2	↔	Hb-O_2
deoxyhemoglobin				oxyhemoglobin

Examples

Let's interpret the curve according to the three different points numbered 1–3:

① At a Po_2 of **100**, hemoglobin is about 98% saturated with oxygen. An example of this would be the blood capillaries in the lungs. Because of the high Po_2 in the air sacs (alveoli) in the lungs, a large amount of oxygen diffuses rapidly into the blood capillaries surrounding the alveoli and immediately binds to hemoglobin.

② At a Po_2 of **40**, hemoglobin is about 75% saturated. On average, this is the typical blood Po_2 for tissues at rest. This means 25% of the available O_2 is released from hemoglobin to be used by tissues at rest.

③ At a Po_2 of **20**, hemoglobin is about 35% saturated. This is a typical blood Po_2 for active tissues such as contracting muscle tissues during exercise. Notice that a large release of the available O_2—about 65%—comes with only a small change in Po_2 (20 mm Hg).

Other Factors

Other factors determine the **affinity** of hemoglobin for oxygen. These factors can *shift the normal curve to the left* (*higher affinity*) or *to the right* (*lower affinity*), as indicated by the arrows in the illustration. A shift to the *left* enhances the *binding* of oxygen to hemoglobin, and a shift to the *right* enhances the *release* of oxygen from hemoglobin.

Variables of Temperature, pH, and P_{CO_2}

Other variables also influence the binding of oxygen to hemoglobin. Exercise provides a practical example of three related variables we will examine: body temperature, pH, and Pco_2. During exercise, skeletal muscle tissue becomes more metabolically active and changes the normal oxygen-hemoglobin dissociation curve in the following ways:

- **Effect of body temperature**—Contracting skeletal muscles generate more heat. This increase in temperature causes a shift to the right (lower affinity for oxygen), making it easier to release more oxygen from hemoglobin and deliver more oxygen to the active muscle tissue.
- **Effect of blood pH**—pH within muscle cells becomes more acidic because of the production of lactate and carbonic acid. As a result, free hydrogen ion (H^+) concentration increases. Some H^+ bind to hemoglobin. This induces a subtle shape change that causes another shift to the right.
- **Effect of P_{CO_2}**—All metabolically active tissues produce more CO_2 as a waste product. As with H^+, some of the CO_2 also binds to hemoglobin, inducing another shape change that causes a shift to the right.

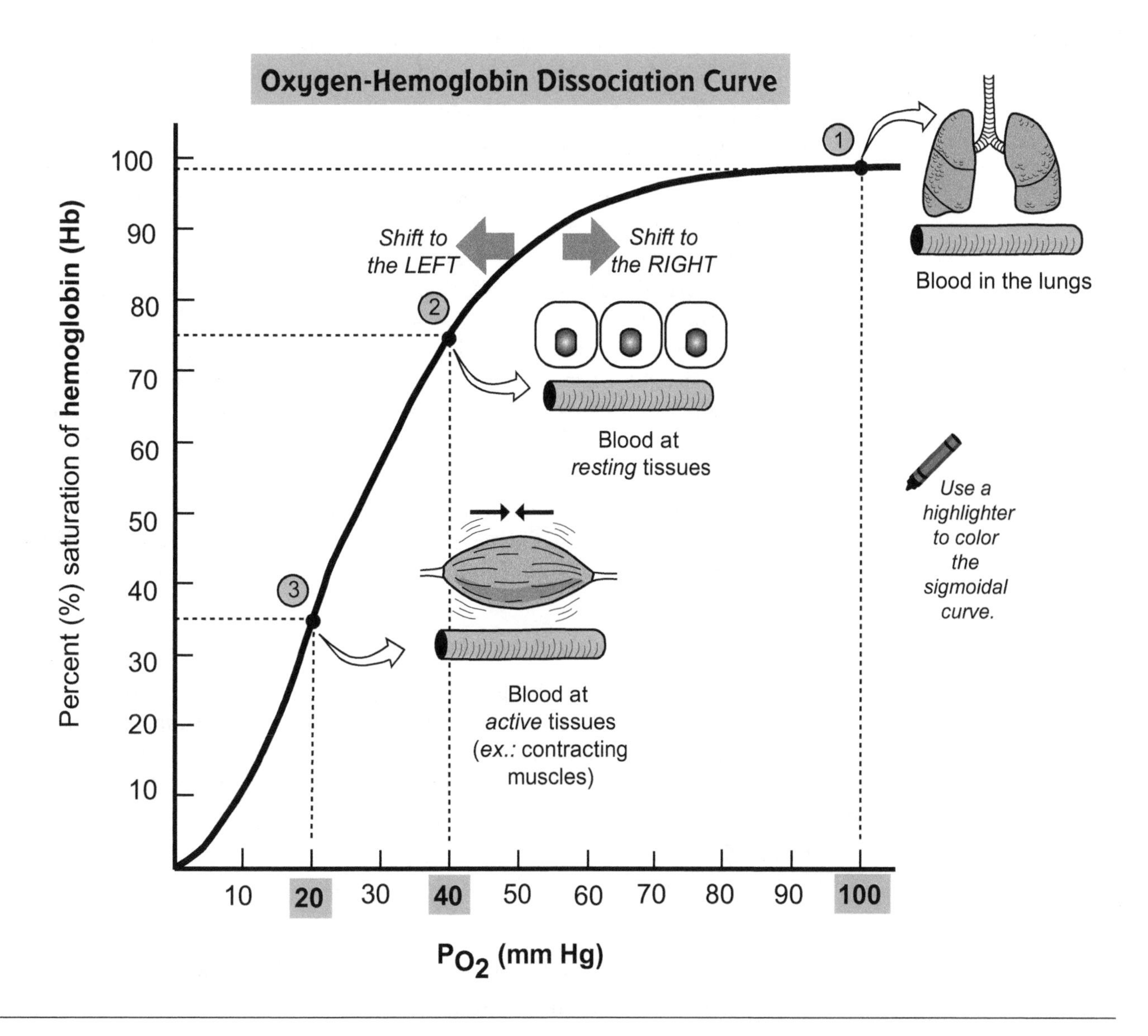
Oxygen-Hemoglobin Dissociation Curve
Percent (%) saturation of hemoglobin (Hb)
100
90
80
70
60
50
40
30
20
10
1
2
3
Shift to the LEFT
Shift to the RIGHT
Blood in the lungs
Blood at *resting* tissues
Blood at *active* tissues (*ex.:* contracting muscles)
Use a highlighter to color the sigmoidal curve.
10 20 30 40 50 60 70 80 90 100
P_{O_2} (mm Hg)

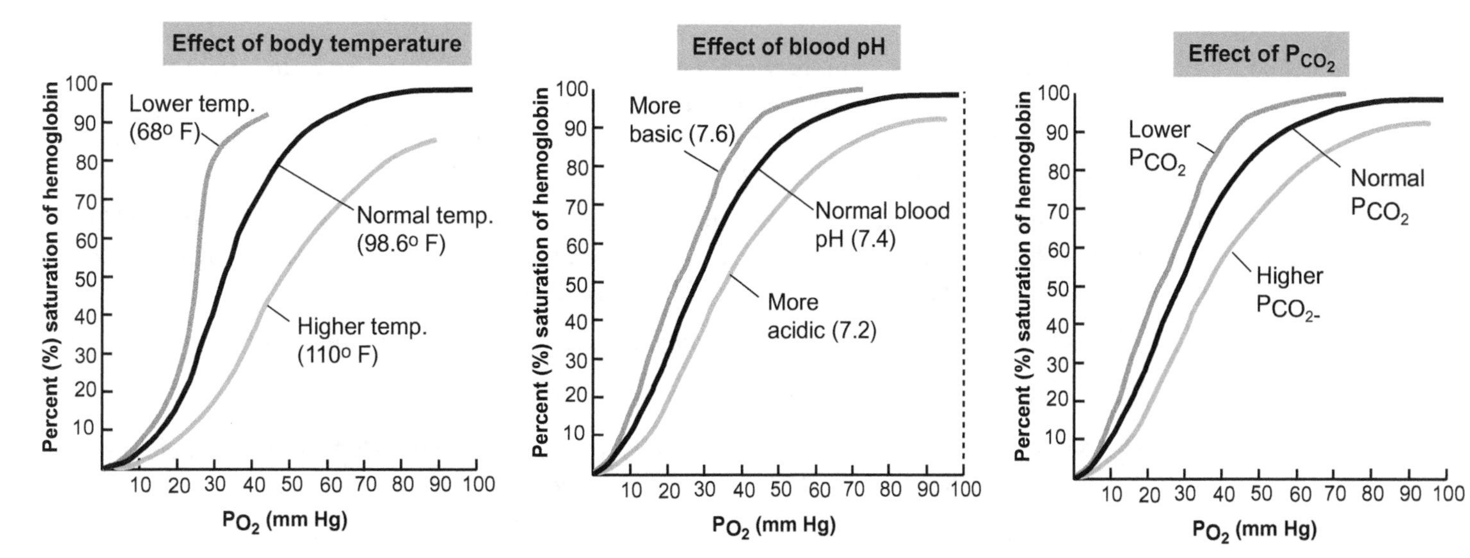
Effect of body temperature
Percent (%) saturation of hemoglobin
Lower temp. (68° F)
Normal temp. (98.6° F)
Higher temp. (110° F)
P_{O_2} (mm Hg)
Effect of blood pH
More basic (7.6)
Normal blood pH (7.4)
More acidic (7.2)
Effect of P_{CO_2}
Lower P_{CO_2}
Normal P_{CO_2}
Higher P_{CO_2}-
10 20 30 40 50 60 70 80 90 100

Description

CO_2 and O_2 are transported in the blood in different ways, as described below.

At the tissues: CO_2 is loaded into the blood, and O_2 is unloaded into tissue cells.

- CO_2 is transported in three different ways:

  ① About 7% remains dissolved in the plasma as CO_2.

  ② The majority—about **70%**—is transported in the form of bicarbonate ions (HCO_3^-). How? After CO_2 diffuses into **red blood cells (RBCs)**, it undergoes a chemical reaction with water that produces an unstable intermediate product called carbonic acid (H_2CO_3). This reaction is fast, aided by the enzyme **carbonic anhydrase.** Because carbonic acid is unstable, it immediately decomposes into bicarbonate ions (HCO_3^-) and hydrogen ions (H^+).

  The levels of HCO_3^- quickly build up inside the RBC. The excess HCO_3^- has to be moved out into the plasma so more CO_2 can be converted into HCO_3^-. This is accomplished by the **chloride shift**, in which a membrane protein exchanges a HCO_3^- ion for a Cl^- ion. Each ion has the same charge, so this does not result in any change in total charge across the membrane.

  ③ About 23% is transported as **carbaminohemoglobin** (Hb-CO_2) when CO_2 loosely binds to hemoglobin (Hb). Because the main function of hemoglobin is to transport O_2, CO_2 *binds to a different binding site on the molecule than O_2 so the two gases are not competing for the same binding site.*

- O_2 is unloaded to tissue cells in two different ways:

  ① Slightly more than 98% of the O_2 is transported in the form of **oxyhemoglobin** (Hb-O_2) when O_2 loosely binds to hemoglobin (Hb). Then oxyhemoglobin releases the O_2, and it diffuses tissue cells. Here it is used in cellular respiration. Some of the excess hydrogen ions (H^+) already present in the RBC temporarily bind to Hb to form H-Hb.

  ② Slightly under 2% of the O_2 is dissolved in the plasma as O_2. Some of this diffuses into tissue cells.

In the lungs—CO_2 is unloaded into the alveoli, and O_2 is loaded into the blood.

- CO_2 is transported in reverse from how it was at the tissues.

  ① Some of the CO_2 dissolved in the plasma diffuses into the alveoli.

  ② The chemical reaction between CO_2 and water runs in reverse. In other words, the bicarbonate ions (HCO_3^-) and hydrogen ions recombine to form carbonic acid ($H_2CO_3^-$) again. Then, carbonic acid decomposes into CO_2 and water. The CO_2 liberated from this process diffuses into the **alveoli.**

The **chloride shift** is reversed in the lungs. In other words, a membrane protein transports a HCO_3^- into the RBC in exchange for moving a Cl^- into the plasma. This supplies lots of HCO_3^- for the purpose of continuing to run the reaction in the reverse direction.

  ③ CO_2 is released from **carbaminohemoglobin**, and it diffuses into the alveoli.

- O_2 is transported in reverse from how it was at the tissues.

  ① More than 98% of the O_2 diffuses into red blood cells and loosely binds to **hemoglobin** (Hb) to form **oxyhemoglobin** (Hb-O_2). Some hydrogen ions (H^+) are produced as a by-product.

  ② Less than 2% of the O_2 diffuses into the blood and remains dissolved in the plasma as O_2.

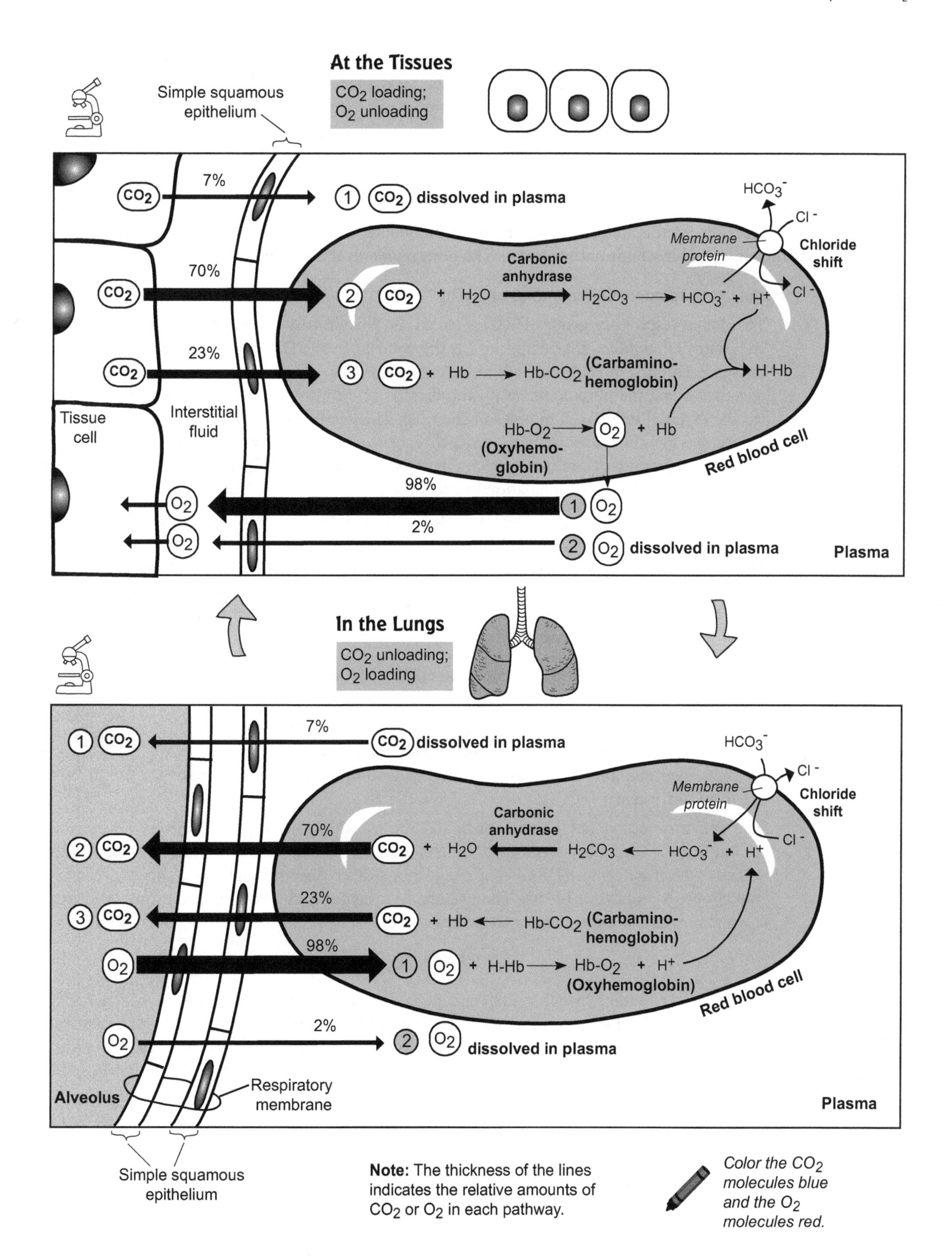

Note: The thickness of the lines indicates the relative amounts of CO_2 or O_2 in each pathway.

Color the CO_2 molecules blue and the O_2 molecules red.

Description

We take the rhythm and coordination of normal breathing for granted. The pattern of inhaling and exhaling constantly repeats itself. At rest, we take an average of 12–16 breaths per minute. This deceivingly simple process actually is complex and is regulated by the nervous system. The latest research has revealed that this mechanism is even more complex than previously thought.

Classic Explanation

In the classic explanation, clusters of specific neurons constitute respiratory control centers located in two regions of the brainstem—the **medulla oblongata** and the **pons**. The medulla oblongata sets the rate and rhythm of normal breathing. The pons regulates the rate and depth of breathing.

Medullary Respiratory Centers: DRG and VRG

- The **dorsal respiratory group** (DRG) is called the "inspiratory center" because it stimulates inhalations. Like a sparkplug, the neurons in the DRG fire in regular bursts. Each burst lasts about 2 seconds. As this happens, it simultaneously sends impulses along the **phrenic nerve** (to the diaphragm) and the **intercostal nerves** (to the **external intercostal muscles** of the ribs). This stimulates these muscles to contract, which increases the size of the thoracic cavity, and air rushes into the lungs. A 3-second delay between consecutive firings of the DRG allows time for a passive exhalation to follow each inhalation.
- The **ventral respiratory center** (**VRG**) is called the "expiratory center," but it is active *only* during a *forced* exhalation. During normal breathing, the VRG is inactive because exhalation is a passive, recoil process that does not require stimulation of muscles. During a *forced* exhalation, though, the DRG sends impulses to stimulate the VRG. The VRG responds by sending impulses to the internal intercostal muscles and abdominal muscles, which then contract, decreasing the size of the thoracic cavity for a forced exhalation.

Pons Respiratory Centers: Pneumotaxic Center (PC) and Apneustic Center (AC)

- The **pneumotaxic center** (PC) is the "regulator." It is like a traffic cop coordinating the flow of traffic to keep it running smoothly. Similarly, the pons coordinates the transition between inhalation and exhalation. It also prevents overinflation of the lungs by always sending inhibitory impulses to the inspiratory center (DRG).

- The **apneustic center** (AC) also coordinates the transition between inhalation and exhalation by fine-tuning the medullary respiratory centers. It accomplishes this by sending stimulatory impulses to the inspiratory center (DRG) that result in a slower, deeper inhalation. This is necessary when you choose to hold your breath. The pneumotaxic center is able to inhibit the apneustic center so a normal breath is not too slow or too deep.

Changes in Naming and Conceptual Understanding

The latest understanding brings a slight change in naming and conceptual understanding. Let's look at each, in turn. As for the naming, a new term has been introduced: the *respiration regulatory center*. This consists of all the aforementioned control centers (DRG, VRG, PC, and AC). But instead of considering them as individual units, we now think of them as being strung together in a loop or a circuit. This makes the classic definition of each control center a little more murky.

Conceptually, the latest thinking is that the respiration control center acts as a central pattern generator, or CPG. Like an electrical circuit, CPGs are neural circuits that generate periodic motor commands for rhythmic movements. The CPGs function automatically most of the time, but also allow voluntary control to override them.

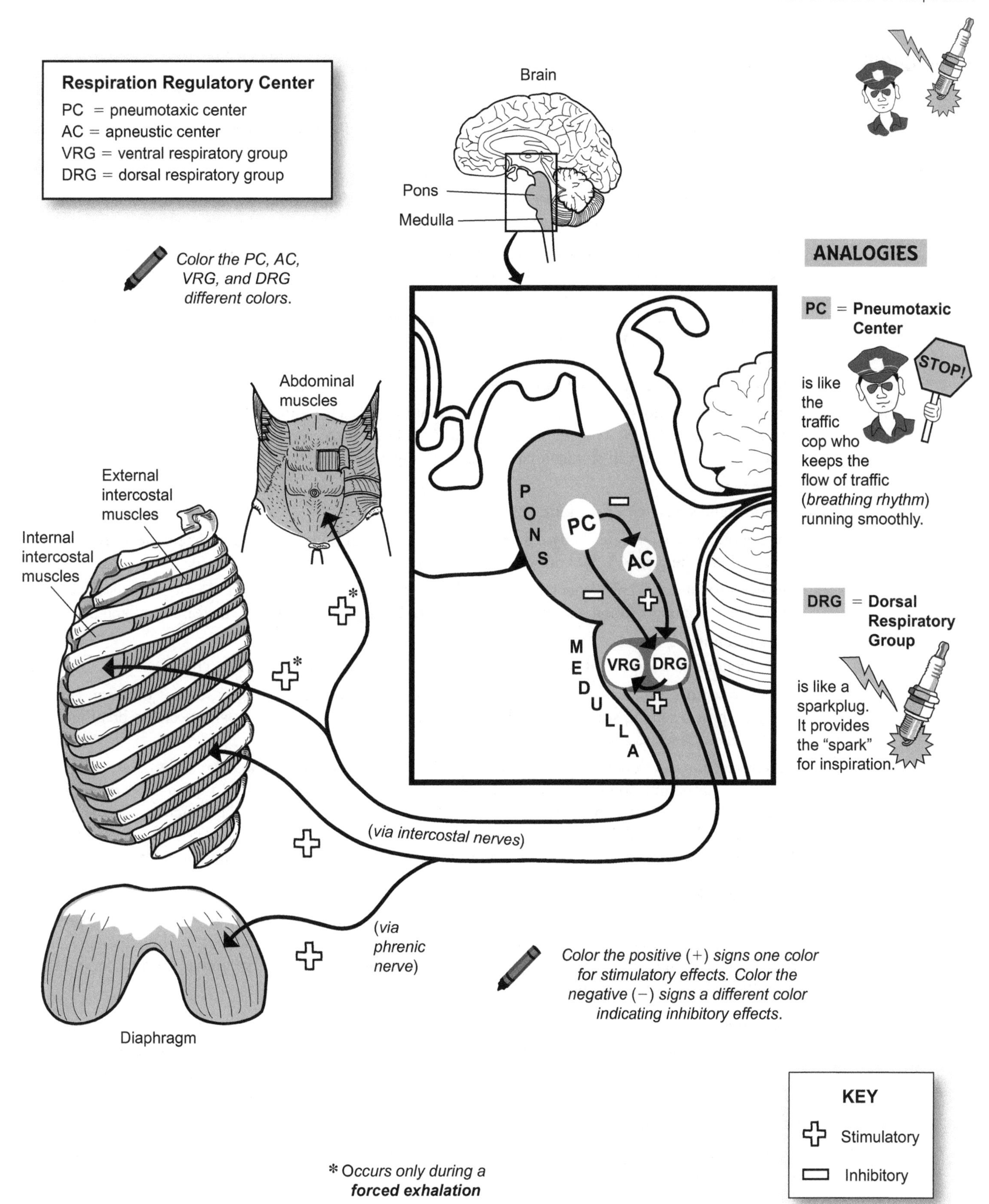
Respiration Regulatory Center
PC = pneumotaxic center
AC = apneustic center
VRG = ventral respiratory group
DRG = dorsal respiratory group
Color the PC, AC, VRG, and DRG different colors.
Brain
Pons
Medulla
PONS
MEDULLA
PC
AC
VRG
DRG
Abdominal muscles
External intercostal muscles
Internal intercostal muscles
(via intercostal nerves)
(via phrenic nerve)
Diaphragm
Color the positive (+) signs one color for stimulatory effects. Color the negative (−) signs a different color indicating inhibitory effects.
ANALOGIES
PC = Pneumotaxic Center
STOP!
is like the traffic cop who keeps the flow of traffic (breathing rhythm) running smoothly.
DRG = Dorsal Respiratory Group
is like a sparkplug. It provides the "spark" for inspiration.
KEY
Stimulatory
Inhibitory
* Occurs only during a forced exhalation

Description

Even though the medulla oblongata and the pons control the typical rate and rhythm of normal breathing, they must be able to increase or decrease the breathing rate when needed. To do so, these respiratory centers in the brainstem receive various inputs from different parts of the body. This module summarizes the different types of input received and indicates whether it has a stimulatory or an inhibitory influence on normal breathing.

Functions

Chemoreceptors

Chemoreceptors are sensory neurons sensitive to shifts in specific chemicals in the blood, such as CO_2, H^+, and O_2. Changes in CO_2 levels are the *primary* stimulus, and changes in O_2 levels act as a *secondary* stimulus.

- **Peripheral chemoreceptors** are located in the aortic arch and the common carotid arteries. Consider what happens during exercise: Po_2 levels fall as skeletal muscle cells consume more oxygen; Pco_2 levels increase as skeletal muscle cells become more metabolically active; and H^+ levels increase, which makes the blood more acidic. Why? CO_2 diffuses into the blood. In the presence of the enzyme carbonic anhydrase, it chemically reacts with water (H_2O) to form carbonic acid (H_2CO_3), which dissociates into H^+ and bicarbonate (HCO_3^-). In this way, an increase in CO_2 results in an increase in H^+. All of these conditions stimulate an increase in respiration rate to supply the skeletal muscle cells with more O_2 and rid the body of excess CO_2.
- **Central chemoreceptors** are located in the medulla oblongata. They monitor Pco_2 levels and the pH of the cerebrospinal fluid (CSF). As mentioned above, as CO_2 levels increase, this also causes the pH of the CSF to become more acidic, which stimulates an increase in respiration rate.

Hypothalamic Controls

The **hypothalamus** has many functions. For example:

- The hypothalamus regulates body temperature. An increase in body temperature from a fever results in an increased breathing rate. A decrease in body temperature decreases breathing rate.
- The hypothalamus has connections with the limbic system, our "emotional brain." In times of emotional stress such as fear or anxiety, stimulatory impulses are sent to the inspiratory centers to increase breathing rate.

Pulmonary Controls

The lungs contain two important types of receptors: **irritant receptors** and **stretch receptors.**

- **Irritant receptors** are located in the bronchioles. They detect airborne particles such as dust and cigarette smoke, which results in stimulating reflex pathways to constrict bronchioles and decrease breathing rate. Other reflexes, such as coughing and sneezing, may be triggered also.
- **Stretch receptors** (*baroreceptors*) are located in the walls of the bronchi and bronchioles. As they detect stretching during inflation of the lungs, they send inhibitory impulses directly to the medullary inspiratory center. This results in an expiration, and the lungs relax. In this relaxed state, the stretch receptors no longer are stimulated, so the inhibition is lost and a new inspiration begins. This reflex is called the **inflation** (*Hering-Breuer*) **reflex.** The purpose seems to be a protective mechanism to prevent overinflation of the lungs.

Voluntary Controls: Cerebral Cortex

The **cerebral cortex** is the voluntary control over the body's activities. This can lead to either an increase or a decrease in respiration rate. For example, we may consciously choose to hold our breath. This stimulates motor neurons to send stimulatory impulses directly to our respiratory muscles for breathing. We may choose to go into a deep meditative state that indirectly leads to a decrease in respiration.

Proprioceptors

Proprioceptors are located in muscles, tendons, and joints. They monitor the tension in muscles and the movement and position of joints and deliver the information to the brain. For example, at the beginning of exercise, as tension in muscles and movement in joints increases, it stimulates an increase in respiration rate.

Other Influences on Respiration

- **Pain:** Sudden, acute pain may trigger **apnea** (absence of breathing), and extended somatic pain increases respiration rate.
- **Blood pressure (BP):** A quick increase in BP decreases respiration rate, and a drop in BP increases respiration rate.

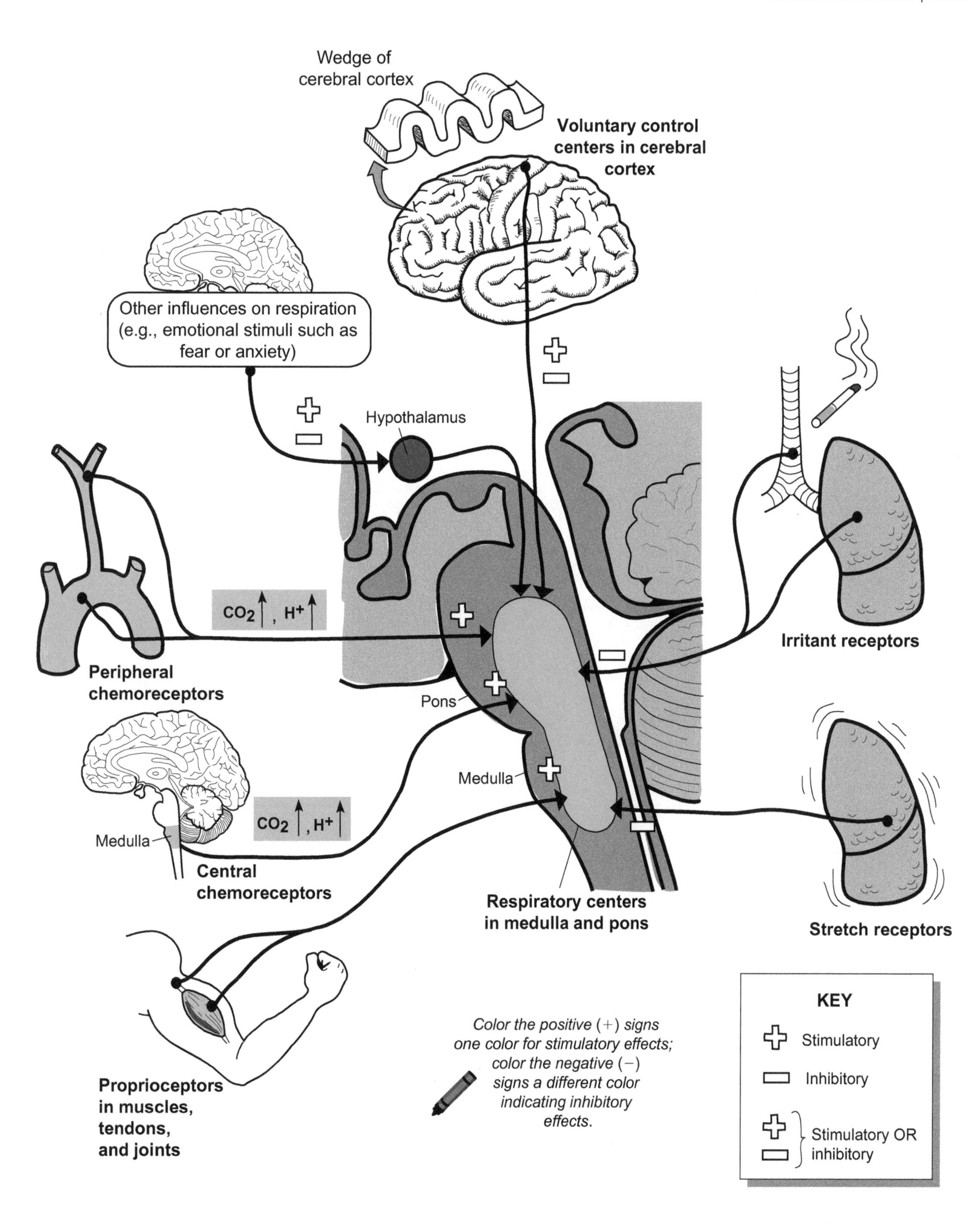
Wedge of cerebral cortex
Voluntary control centers in cerebral cortex
Other influences on respiration (e.g., emotional stimuli such as fear or anxiety)
Hypothalamus
CO2 ↑, H+ ↑
Peripheral chemoreceptors
Pons
Medulla
Medulla
CO2 ↑, H+ ↑
Central chemoreceptors
Respiratory centers in medulla and pons
Irritant receptors
Stretch receptors
Proprioceptors in muscles, tendons, and joints
Color the positive (+) signs one color for stimulatory effects; color the negative (−) signs a different color indicating inhibitory effects.
KEY
Stimulatory
Inhibitory
Stimulatory OR inhibitory

Notes

DIGESTIVE SYSTEM

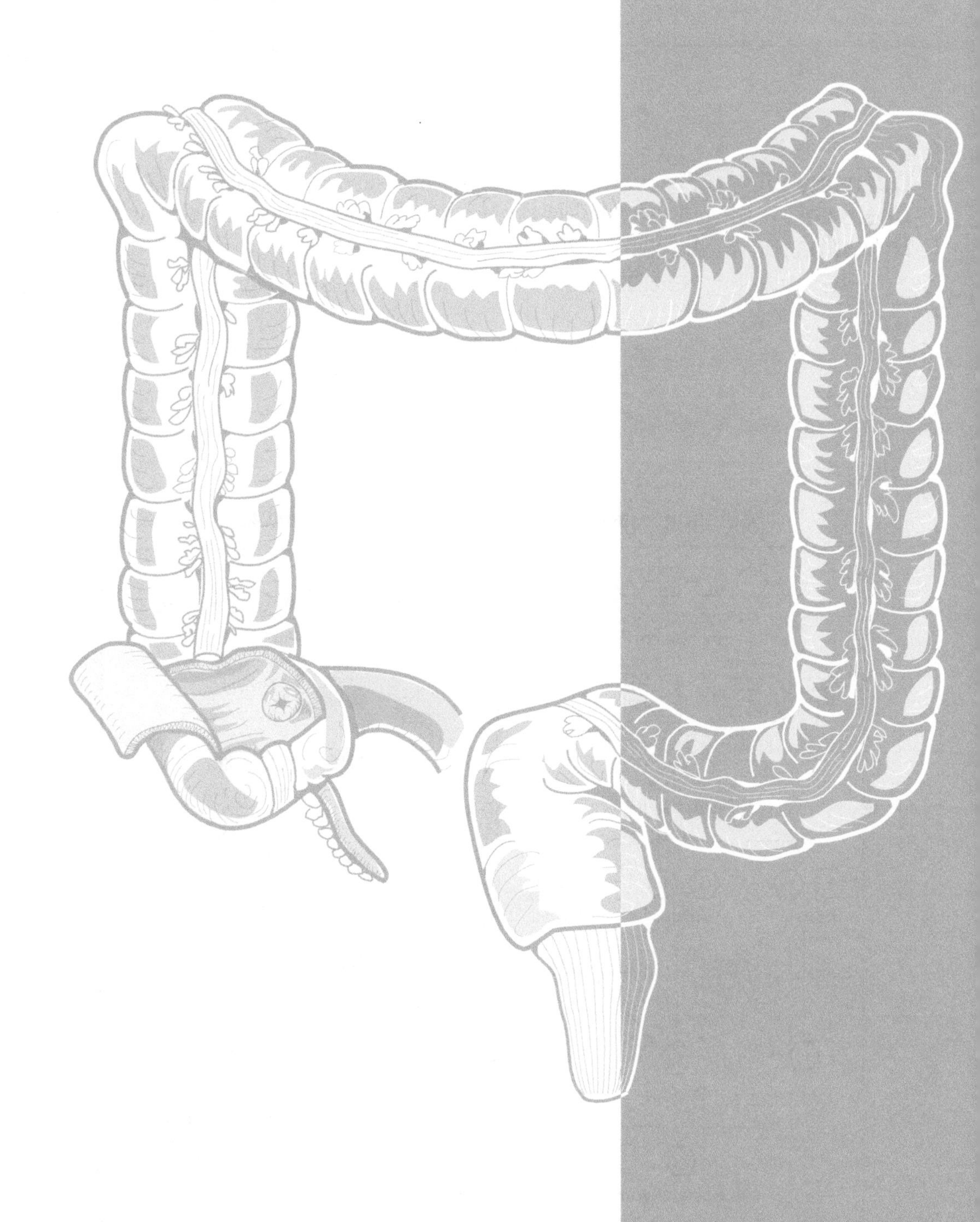

Overview of the Digestion Process

Let's look at the digestion process by examining what happens when someone eats a ham sandwich. The process begins in the mouth with **mechanical digestion**—the physical breakdown of food from chewing. The sandwich contains many different nutrients, including starch—the **complex carbohydrate** in the bread, **protein** in the ham, and **lipids** (*ex.:* triglycerides) in the mayonnaise. Each of these macromolecules has to be broken down by **chemical digestion** into its fundamental building blocks so they can be used by body cells:

Complex carbohydrates	→	Monosaccharides
Proteins	→	Amino acids
Lipids	→	Monoglycerides and fatty acids

Overview of Chemical Digestion

The process is a series of catabolic reactions that breaks chemical bonds in the macromolecules with the help of digestive enzymes. Different enzymes are required for different macromolecules, shown on the illustration.

Note: The enzymes are shown in *italics*, and the icons indicate the location in the digestive system.

= mouth = stomach = small intestine

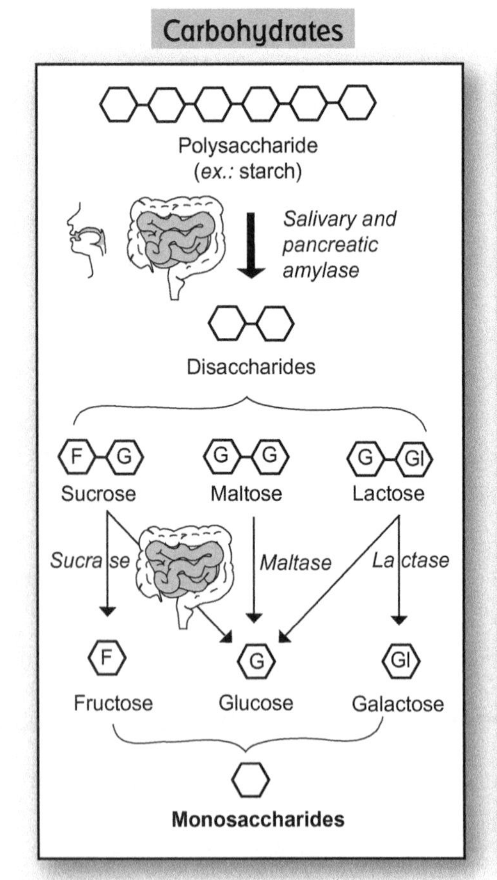

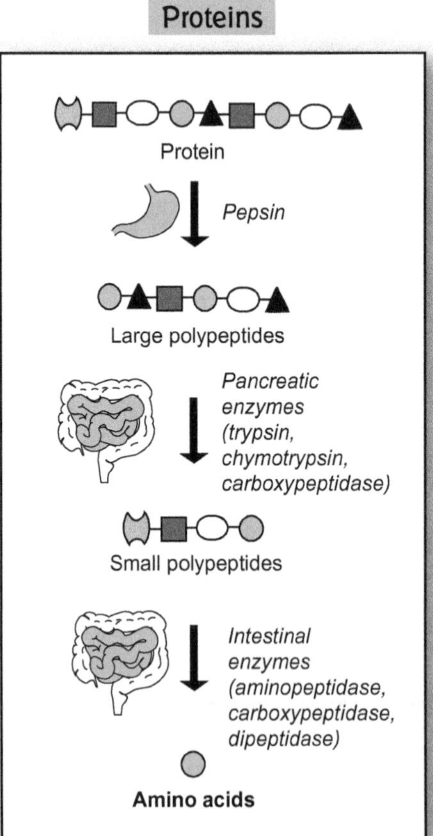

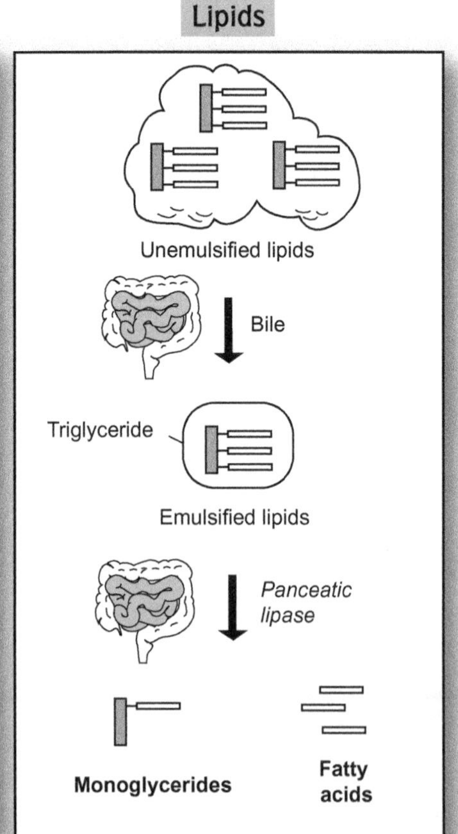

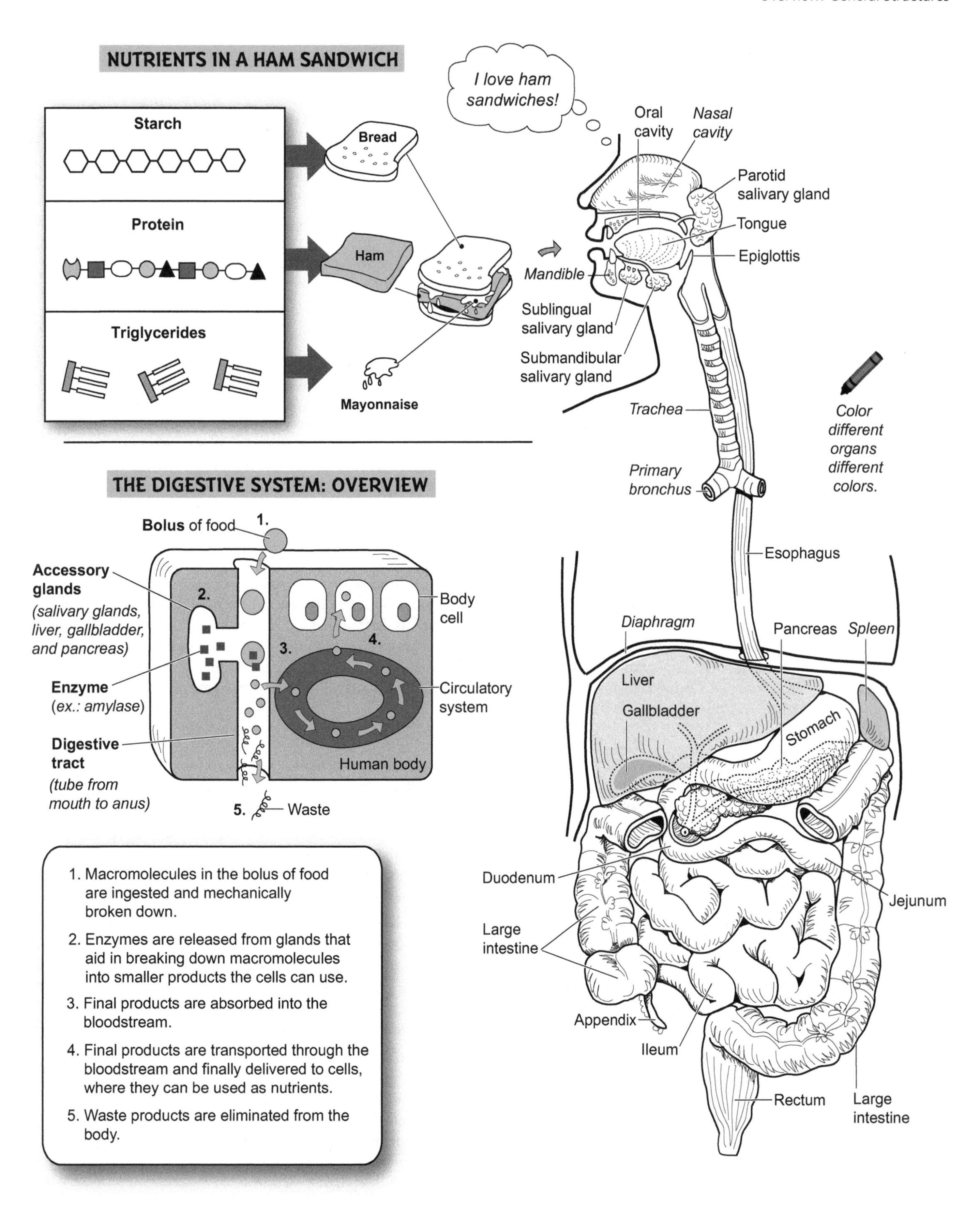
NUTRIENTS IN A HAM SANDWICH
Starch
Protein
Triglycerides
Bread
Ham
Mayonnaise
I love ham sandwiches!
Oral cavity
Nasal cavity
Parotid salivary gland
Tongue
Epiglottis
Mandible
Sublingual salivary gland
Submandibular salivary gland
Trachea
Primary bronchus
Color different organs different colors.
Esophagus
Diaphragm
Pancreas
Spleen
Liver
Gallbladder
Stomach
Duodenum
Jejunum
Large intestine
Appendix
Ileum
Rectum
Large intestine
THE DIGESTIVE SYSTEM: OVERVIEW
Bolus of food
1.
Accessory glands
(salivary glands, liver, gallbladder, and pancreas)
2.
Enzyme
(ex.: amylase)
Digestive tract
(tube from mouth to anus)
3.
4.
Body cell
Circulatory system
Human body
5.
Waste
1. Macromolecules in the bolus of food are ingested and mechanically broken down.
2. Enzymes are released from glands that aid in breaking down macromolecules into smaller products the cells can use.
3. Final products are absorbed into the bloodstream.
4. Final products are transported through the bloodstream and finally delivered to cells, where they can be used as nutrients.
5. Waste products are eliminated from the body.

Description

Digestion in the mouth deals with producing saliva from salivary glands, chewing, and swallowing. Each of these processes is summarized here.

Salivary Glands

The body contains three pairs of **salivary glands: parotid, submandibular,** and **sublingual.** Of these three, the parotid is the largest. The submandibular gland produces the most saliva, and the sublingual gland produces the least. Salivary glands are classified as exocrine glands partly because they all have ducts that connect them to the oral cavity.

At the microscopic level, salivary glands contain two types of secretory cells: **mucous cells** and **serous cells.** The mucous cells secrete a protein called **mucin** that forms mucus. The serous cells release a watery secretion containing electrolytes and salivary amylase. These collective secretions form saliva and release it into the oral cavity. Because secretion of saliva is controlled by the autonomic nervous system, salivary glands are innervated by the parasympathetic and the sympathetic divisions.

Saliva

Saliva is a mildly acidic (pH 6.4–6.8), watery secretion that has multiple functions. It contains mostly water (95%), as well as some important dissolved solutes (0.5%), including electrolytes such as sodium (Na^+), potassium (K^+), chloride (Cl^-), and others. **Mucus** is produced to lubricate the bolus to make swallowing easier. **Lysozyme** is an antibacterial agent that inhibits the growth of bacteria normally found in the oral cavity. **Antibodies** such as immunoglobulin A also control bacterial growth. The enzyme **salivary amylase** catalyzes the breakdown of complex carbohydrates into disaccharides. Like water spraying on cars in a carwash, saliva cleans oral cavity structures such as the teeth by constantly washing over them. Last, saliva serves as a fluid medium in which food molecules can dissolve to allow for chemical detection by taste buds. Without saliva, it is difficult to taste anything!

Chewing

Chewing (*mastication*) is the physical breakdown of food and the mixing of food with saliva. This is accomplished with the help of the teeth and the tongue. The four different types of teeth are incisors, canines (*cuspids*), premolars (*bicuspids*), and molars. Each type is best suited for a specific task. Incisors have a flat, chiseled edge for cutting food like biting into an apple. Cuspids have a pointed edge for puncturing, tearing, and shredding food. Bicuspids and molars have larger, flattened surfaces, best for grinding and crushing food. The tongue moves during chewing to mix the pulverized food with saliva to form a pasty, compressed mass called a **bolus.** This mixing prepares the **bolus** for swallowing.

Swallowing

Swallowing (*deglutition*) is a complex process that requires many different muscles to move ingested substances from the oral cavity to the stomach. It typically is divided into three stages: oral, pharyngeal, and esophageal. Note that the same process applies to ingested liquids and solids. The stages in the process are:

① **Oral** (or *buccal*) **phase**—a voluntary phase that begins when substances are ingested. The tongue mixes these materials with saliva to form a moistened mass called a bolus. This bolus is compressed against the hard and soft palates in the roof of the mouth by upward movements of the tongue. Then the tongue pushes the bolus toward the oropharynx, which marks the end of the oral phase.

② **Pharyngeal phase**—an involuntary process. It begins as the bolus pushes against the oropharynx, where it stimulates tactile receptors to send signals to the swallowing control center in the medulla oblongata. Three key events occur: (1) the soft palate and uvula move upward to prevent the bolus from entering the nasal cavity; (2) the bolus moves into the oropharynx; and (3) muscles raise the larynx forward and upward so the epiglottis closes over the glottis to prevent the bolus from entering the larynx or trachea. The pharyngeal phase ends after pharyngeal muscles constrict and force the bolus out of the pharynx and into the esophagus.

③ **Esophageal phase**—involuntary like the pharyngeal phase. The esophagus is a long, muscular tube that connects the pharynx to the stomach. This phase begins as the upper esophageal sphincter relaxes to allow the bolus to enter the esophagus. The rigid epiglottis springs back to its normal, upright position. Then the upper esophageal sphincter contracts to propel the bolus deeper into the esophagus. The muscular contractions of peristalsis (see p. 168) quickly move the bolus into the stomach. This entire phase takes less than 8 seconds!

DISCOVER for YOURSELF

Place your fingertips on the front of your larynx as you swallow. You can feel the larynx move forward and upward.

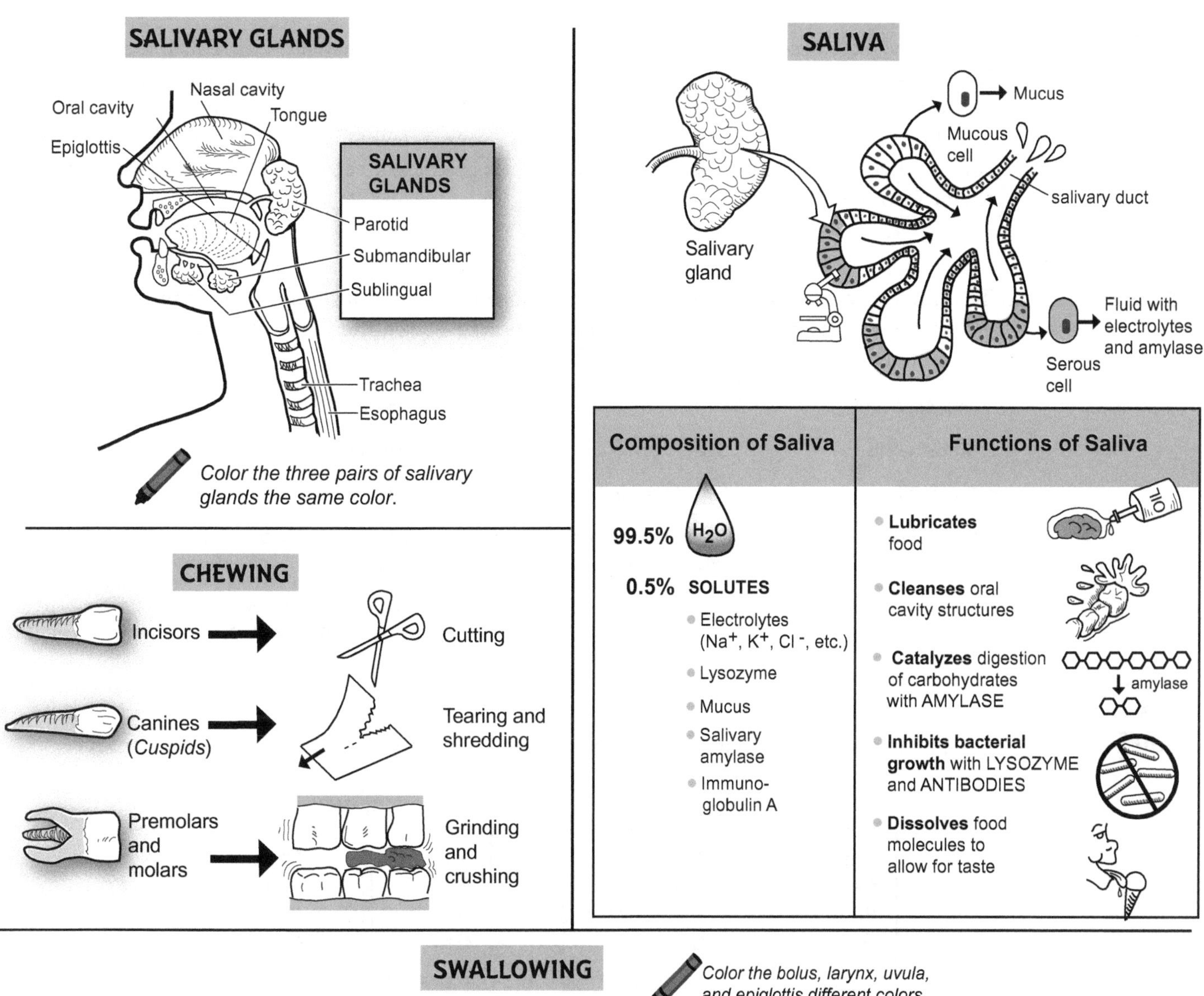

SWALLOWING

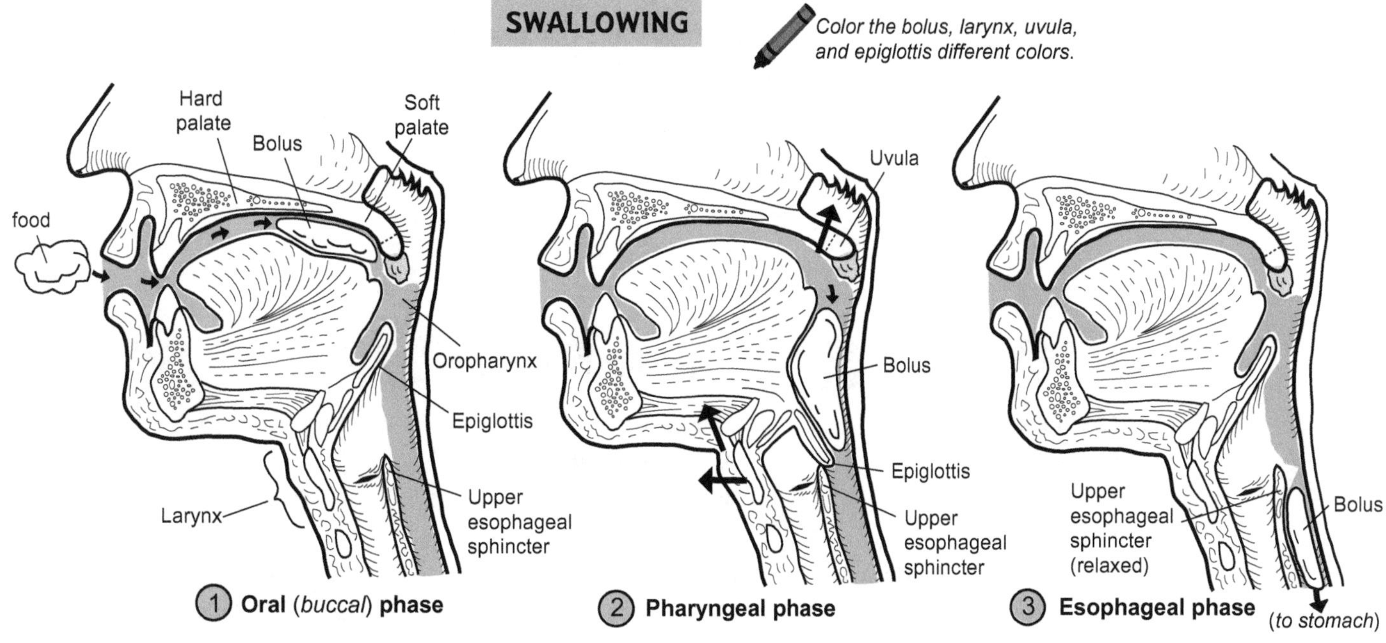

Note: *The colored region shows the pathway ingested substances follow.*

Description

The stomach connects the esophagus to the first portion of the small intestine, called the duodenum. It is a specialized sac that contains three layers of smooth muscle within its wall. The innermost lining of the stomach, called the **mucosa**, is coated with a protective layer of alkaline mucus. Folds of mucosa called **rugae** allow the stomach to increase surface area to maximize **gastric juice**. This acidic mixture contains hydrochloric acid and enzymes. When this gastric juice combines with food, a mixture called **chyme** is formed.

Function

To release gastric juice for the purpose of chemical digestion of food, the mucosa manufactures an enzyme called **pepsin** that initiates protein digestion. Within the mucosal lining of the stomach are four different types of cells that have the following functions:

Mucosal Cell	Function
Mucous cells	Secrete an alkaline **mucus** to protect the stomach lining from acidic gastric juice
Chief cells	Secrete the inactive enzyme **pepsinogen**
Parietal cells	Secrete hydrochloric acid (HCl); this helps convert pepsinogen into the active enzyme **pepsin**
G cells	Produce and secrete the hormone **gastrin**, which increases secretion from parietal and chief cells. It also induces smooth muscle contraction in the stomach wall.

Study Tip

Use the following mnemonics to recall the mucosal cell types and the substances they produce:

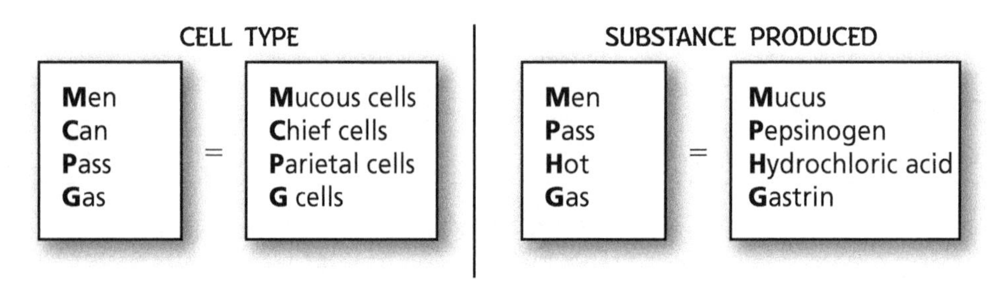

Key to Illustration

1. Esophagus
2. Cardia
3. Fundus
4. Serosa
5. Longitudinal muscle layer
6. Circular muscle layer
7. Oblique muscle layer
8. Mucosa
9. Lesser curvature
10. Body
11. Greater curvature
12. Rugae
13. Pyloric sphincter
14. Gastric pit
15. Gastric gland
16. Mucous cells
17. Parietal cells
18. Chief cells
19. G cells
20. Muscularis mucosa

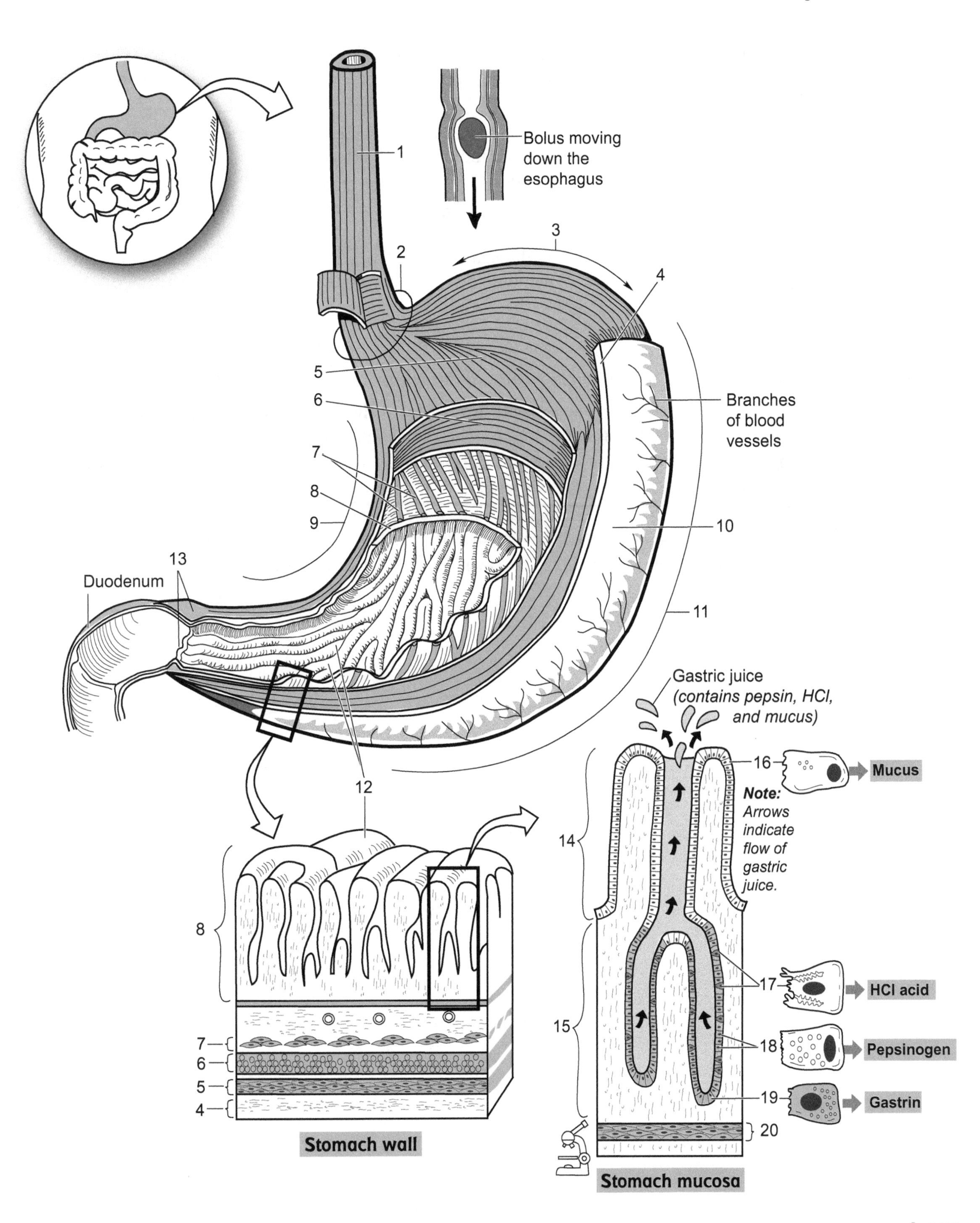
Bolus moving down the esophagus
1
2
3
4
5
6
7
8
9
10
11
12
13
Duodenum
Branches of blood vessels
Gastric juice (contains pepsin, HCl, and mucus)
14
15
16
17
18
19
20
Mucus
Note: Arrows indicate flow of gastric juice.
HCl acid
Pepsinogen
Gastrin
8
7
6
5
4
Stomach wall
Stomach mucosa

Gastric Motility

Gastric motility refers to smooth muscle contractions in the wall of the stomach. This process is regulated by the autonomic nervous system and by the endocrine system with hormones such as **gastrin.** Waves of **peristalsis** move through the stomach regularly.

① They begin as gentle muscular contractions near the **lower esophageal sphincter** (LES) and continue down the stomach toward the **pyloric sphincter.**

② As the contractile waves near the distal end of the stomach, they become much stronger and more forceful. This results in thoroughly mixing **chyme** before it passes through the pyloric sphincter.

③ As the peristaltic wave passes through the partly opened pyloric sphincter, it causes the chyme to move through it in a back-and-forth fashion. This serves to break up the larger materials left in the chyme.

Gastric Emptying

Gastric emptying refers to the movement of **chyme** out of the stomach and into the duodenum.

① Various stimuli, such as caffeine, alcohol, partially digested proteins, and distension of the stomach wall, all trigger the process of gastric emptying to begin.

② These stimuli cause two things to occur: (1) an increase in parasympathetic impulses sent along the vagus nerve to the stomach, and (2) an increase in secretion of gastrin from the stomach.

③ Gastrin and vagus nerve impulses have the combined effects of closing the LES, opening the pyloric valve, and increasing gastric motility.

④ The final result is emptying of the chyme in the stomach into the duodenum.

GASTRIC MOTILITY

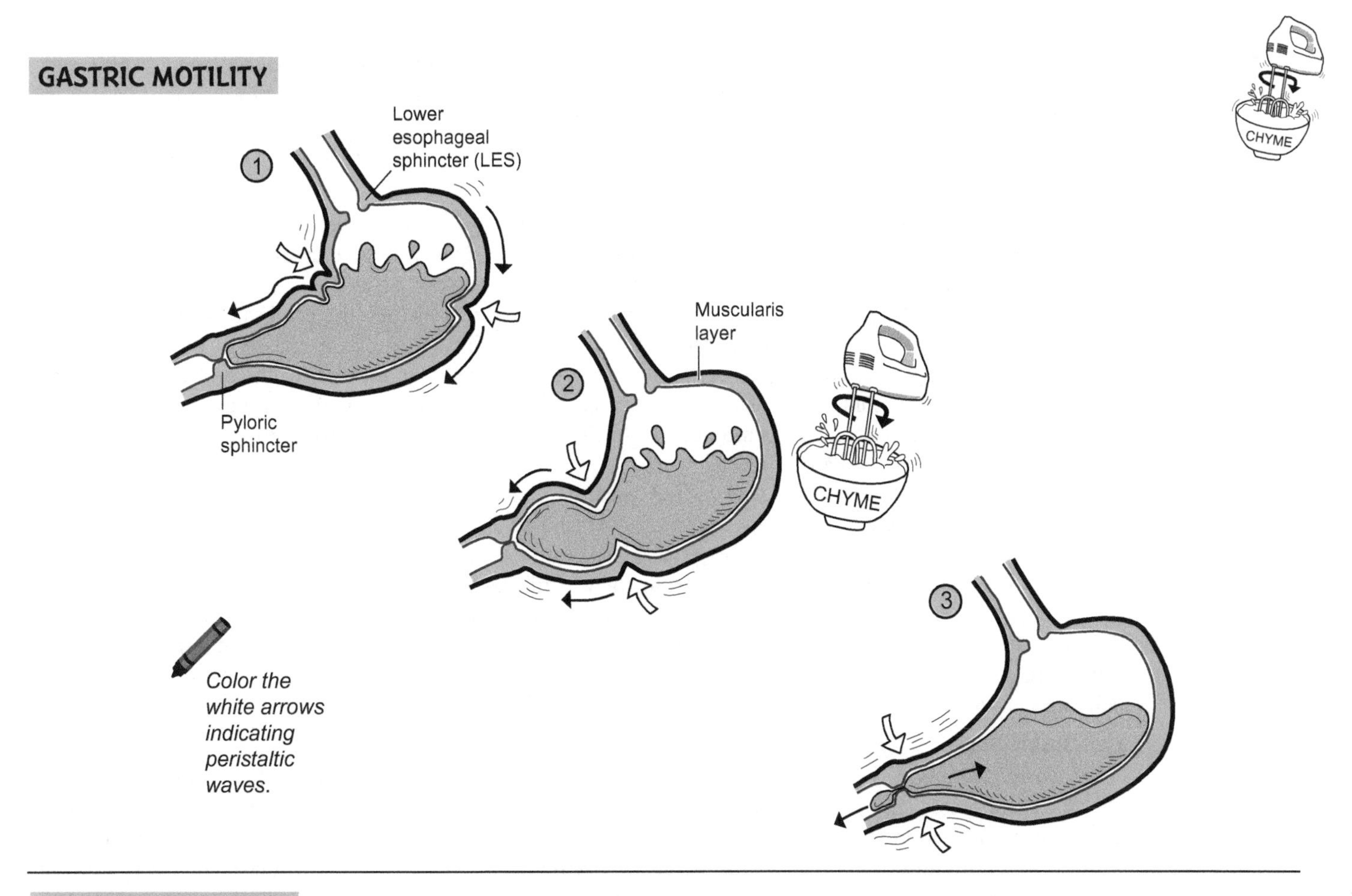

Color the white arrows indicating peristaltic waves.

GASTRIC EMPTYING

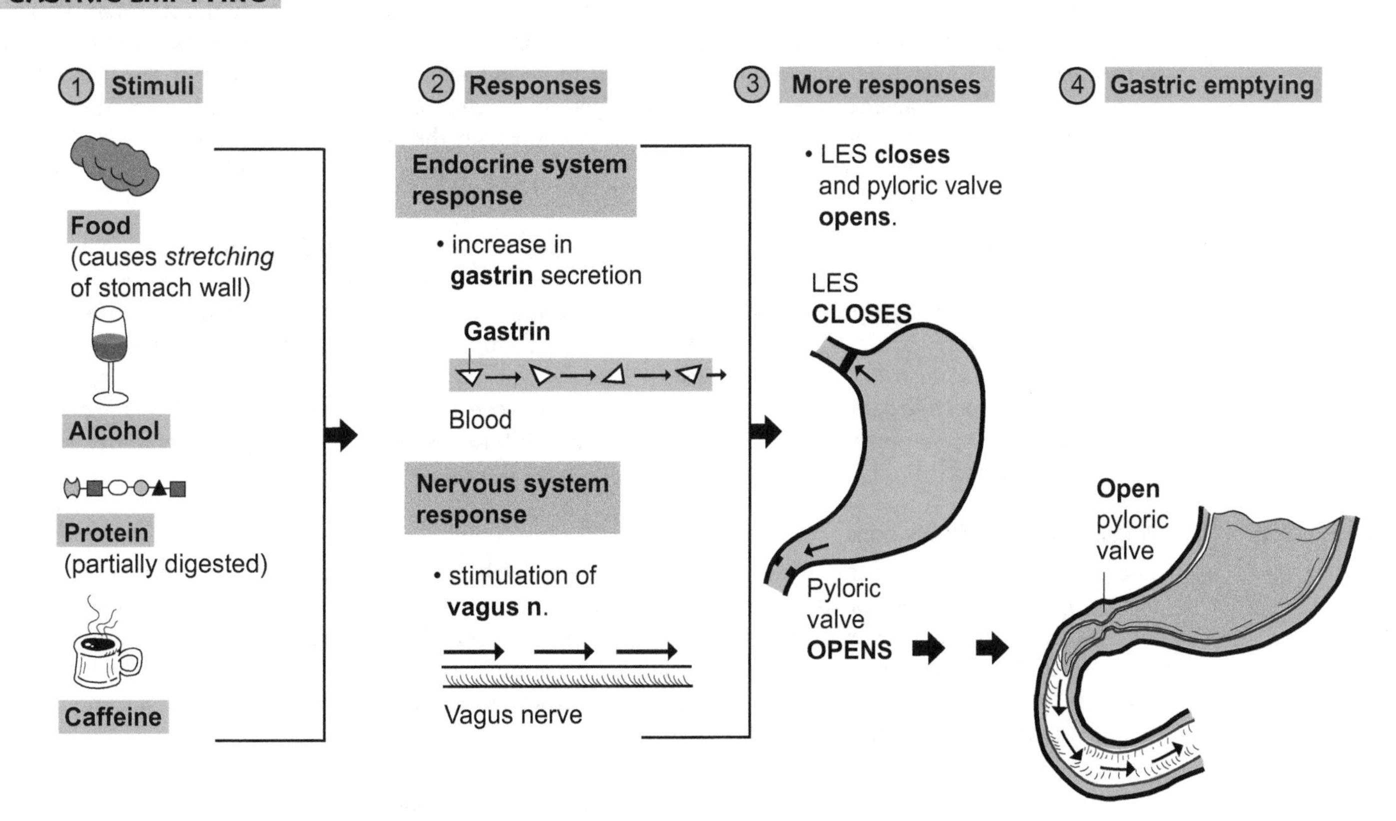

Description

The regulation of digestion is controlled simultaneously by the autonomic nervous system (ANS) and by hormones. This module focuses on the role of the nervous system in regulating digestion.

Phases

Control of digestion can be divided into three phases: (1) cephalic phase, (2) gastric phase, and (3) intestinal phase.

① Cephalic Phase

The **cephalic phase** begins before any food enters the oral cavity. The sight, smell, thought, or first taste of food stimulates reflexes to increase the production of saliva, gastric secretions, and pancreatic secretions. The collective purpose of this response is to prepare the digestive tract to receive food. This phase produces about 10% of all gastric secretions. Key digestive control areas in the nervous system include the **cerebral cortex, hypothalamus,** and **medulla oblongata.** Various stimuli (from food) are sent through nerve impulses to the cerebral cortex, where they are interpreted. Then the cerebral cortex responds by sending impulses to the medulla.

The **hypothalamus** contains the feeding center. Specific stimuli, such as smells and tastes of food, stimulate nerve impulses to be sent to the feeding center in the hypothalamus. This causes the hypothalamus to send nerve impulses to the medulla oblongata.

As stimulatory impulses are sent to the medulla oblongata from both the cerebral cortex and the hypothalamus, the medulla responds by sending out nerve impulses that produce three key responses: (1) increased saliva production, (2) increased gastric secretions, and (3) increased pancreatic secretions (small amount). More specifically, impulses sent from the medulla oblongata down the **vagus nerve** stimulate both gastric and pancreatic secretions. The gastric secretions contain pepsinogen, hydrochloric acid, and mucus. The pancreatic secretions are rich in pancreatic enzymes that aid in carbohydrate digestion, triglyceride digestion, protein digestion, and nucleic acid digestion. Cranial nerves other than the vagus stimulate the salivary glands to produce saliva.

② Gastric Phase

The **gastric phase** begins as food enters the stomach. This triggers reflex pathways that stimulate about 80% of all gastric secretions. The two types of sensory receptors in the stomach wall are stretch receptors and chemoreceptors. Upon entering the stomach, food stretches the stomach wall and triggers stretch receptors to send a nerve impulse along the vagus nerve to the medulla oblongata. The same response occurs when chemoreceptors detect a change in the pH of the gastric secretions.

After proteins in food enter the stomach, they buffer some of the normal acidity in the stomach. This leads to a more alkaline pH, which triggers the chemoreceptors to send out nerve impulses. These impulses are sent to the medulla oblongata, which responds by sending impulses back down the vagus nerve to the stomach. This further stimulates the gastric secretions of pepsinogen, hydrochloric acid, and mucus. It also stimulates the muscular contractions of peristalsis to move the chyme through the stomach and continue peristalsis in the esophagus. Finally, the hormone gastrin secreted during the gastric phase further stimulates gastric secretions.

③ Intestinal Phase

The **intestinal phase** begins when chyme enters the **duodenum** after passing through the pyloric valve. Within the wall of the duodenum are the same two important sensory receptors as in the stomach: stretch receptors and chemoreceptors. Activation of these receptors stimulates reflex pathways through the medulla that initially lead to more gastric secretion, gastric motility, and emptying of the stomach. This phase accounts for about 10% of total gastric secretion and ensures that the digestion process is fully completed. Later, when the duodenum becomes filled with fatty chyme, the hormones secretin and cholecystokinin (CCK) are produced to inhibit gastric secretion and stomach emptying. This inhibitory response is needed to signal the end of the passage of a meal.

Divisions

Parasympathetic (P) Divisions

The glands and organs of the digestive system are extensively innervated by both the sympathetic and the parasympathetic divisions of the ANS. Consequently, secretions and motility are under unconscious control by the nervous system. The **parasympathetic division (P)** controls stimulation of secretions and motility throughout the digestive tract primarily through action of the **vagus nerve** (cranial nerve X). The vagus nerve connects to all the major digestive system organs and structures except the salivary glands. It is a mixed nerve, meaning that it carries both sensory and motor information within it.

Sympathetic (S) Division

The **sympathetic division** (S) inhibits secretion and motility in the digestive system.

① CEPHALIC PHASE

The **cephalic phase** begins before any food enters the oral cavity. The sight, smell, thought, or first taste of food stimulates nerve impulses to increase the production of saliva, gastric secretions, and pancreatic secretions.

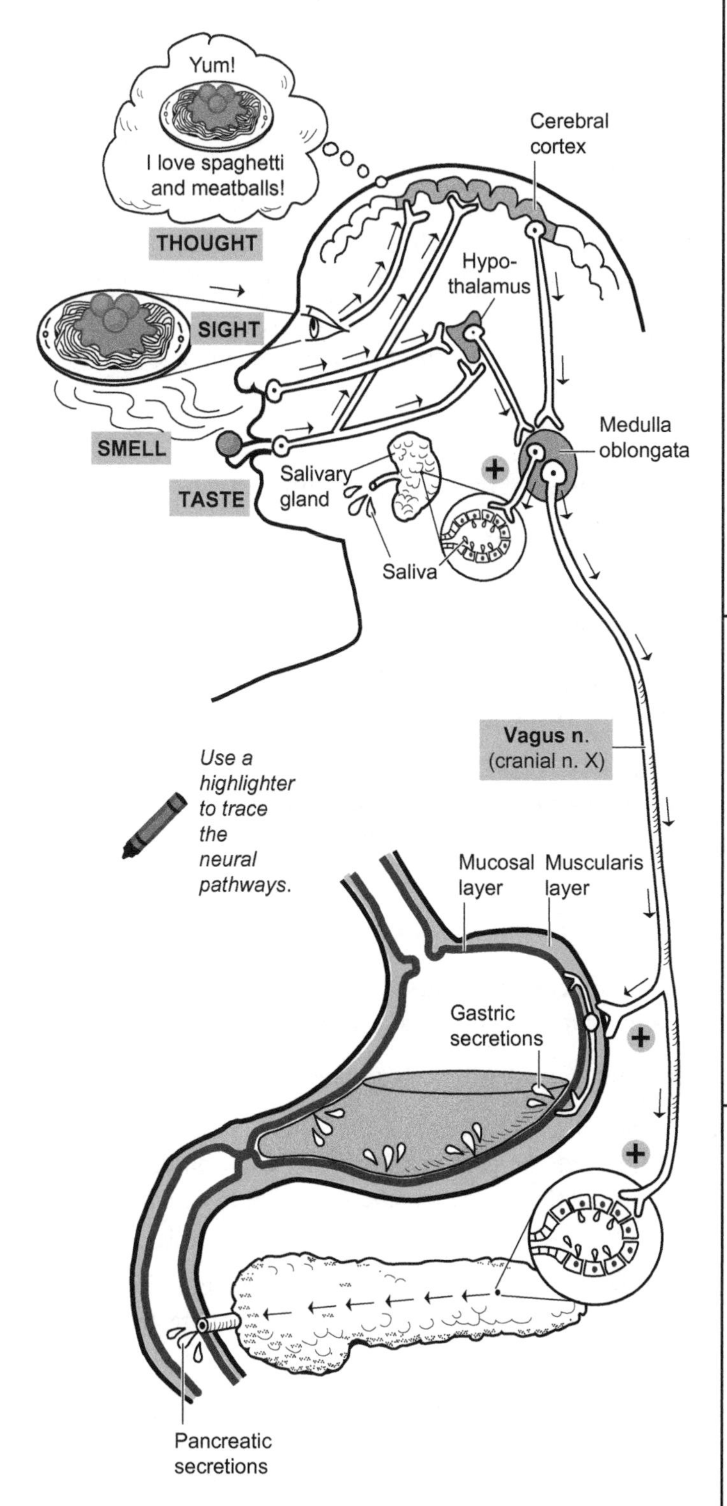

Use a highlighter to trace the neural pathways.

SYMPATHETIC (S) and PARASYMPATHETIC (P) CONTROL

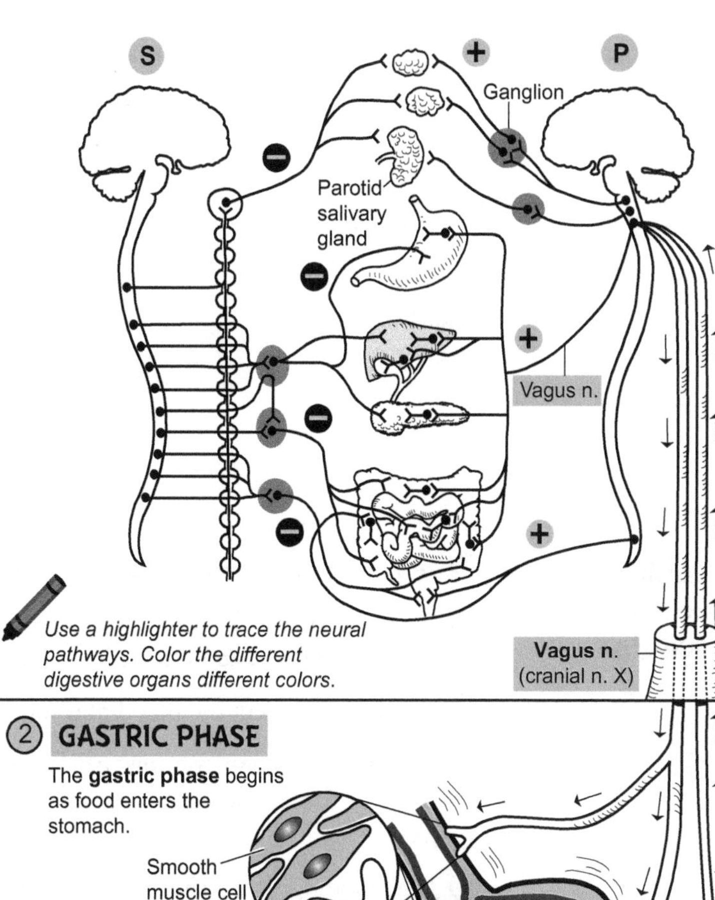

Use a highlighter to trace the neural pathways. Color the different digestive organs different colors.

② GASTRIC PHASE

The **gastric phase** begins as food enters the stomach.

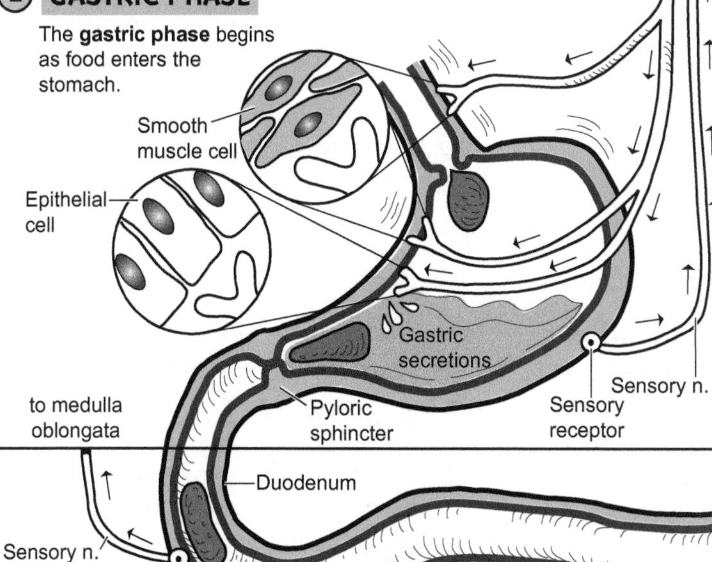

③ INTESTINAL PHASE

The **intestinal phase** begins when chyme enters the duodenum after passing through the pyloric sphincter.

General Description

The hormonal regulation of digestion is controlled mainly by three major hormones: (1) **gastrin,** (2) **secretin,** and (3) **cholecystokinin** (CCK). The table below gives an overview of each hormone:

Hormone/Symbol	Stimulus	Site of Secretion	Major Actions
Gastrin	Expansion of stomach wall by ingested materials; protein and caffeine in stomach; alkaline chyme	**G cells** within the mucosa of the stomach	Stimulates secretion of gastric juice (HCl, mucus, and pepsinogen)
Secretin	Acidic chyme in the duodenum	**S cells** within the mucosa of the duodenum	Stimulates secretion of pancreatic juice rich in bicarbonate ions (HCO_3^-) to help neutralize acid in chyme
Cholecystokinin (CCK)	Triclycerides, fatty acids, and amino acids in the duodenum	**CCK cells** in the mucosa of the duodenum	Simulates production of pancreatic juice rich in digestive enzymes such as lipase; stimulates muscle contraction in wall of gallbladder to expel stored bile; stimlates relaxation of muscle to open hepatopancreatic sphincter

Regulation of Gastric Secretion

1. A bolus moves through the lower esophageal sphincter of the esophagus and into the stomach.
2. Partially digested proteins and caffeine from food stimulate G cells within the mucosal lining to produce gastrin and release it into the blood.
3. Gastrin travels in the blood and spreads to all body organs.
4. Gastrin binds only to target cells that contain a receptor for gastrin. Because the various secretory cells in the epithelial mucosa contain the gastrin receptor, gastrin binds to them and induces a response.
5. The binding of gastrin to the gastrin receptor causes a chemical chain reaction within the cell that leads to stimulation of the production of more gastric juices (hydrochloric acid, mucus, and pepsinogen).

Regulation of Pancreas and Gallbladder Secretions

1. Acidic chyme stimulates S cells in the duodenal mucosa to manufacture secretin; fatty acids and triglycerides in chyme stimulate CCK cells in the duodenal mucosa to produce CCK.
2. Secretin and CCK are released into the blood and travel to all body organs.
3. Secretin binds to secretin receptors in pancreatic cells. This stimulates the cells to make pancreatic juice rich in bicarbonate ions (HCO_3^-) to neutralize the acid in the chyme. CCK binds to CCK receptors in pancreatic cells to stimulate the production of pancreatic juice rich in enzymes such as lipase.
4. CCK also targets the gallbladder. CCK binds to its receptors in the smooth muscle in the gallbladder wall. This stimulates smooth muscle contraction that results in expelling bile from the gallbladder and eventually into the duodenum.
5. CCK also targets the hepatopancreatic sphincter, but with a response opposite from the gallbladder. As CCK binds to its receptors in the smooth muscle cells of the sphincter, it triggers a different chemical chain reaction within the target cells. This causes the smooth muscle to relax, opening the sphincter and allowing both pancreatic secretions and bile to be released into the duodenum.

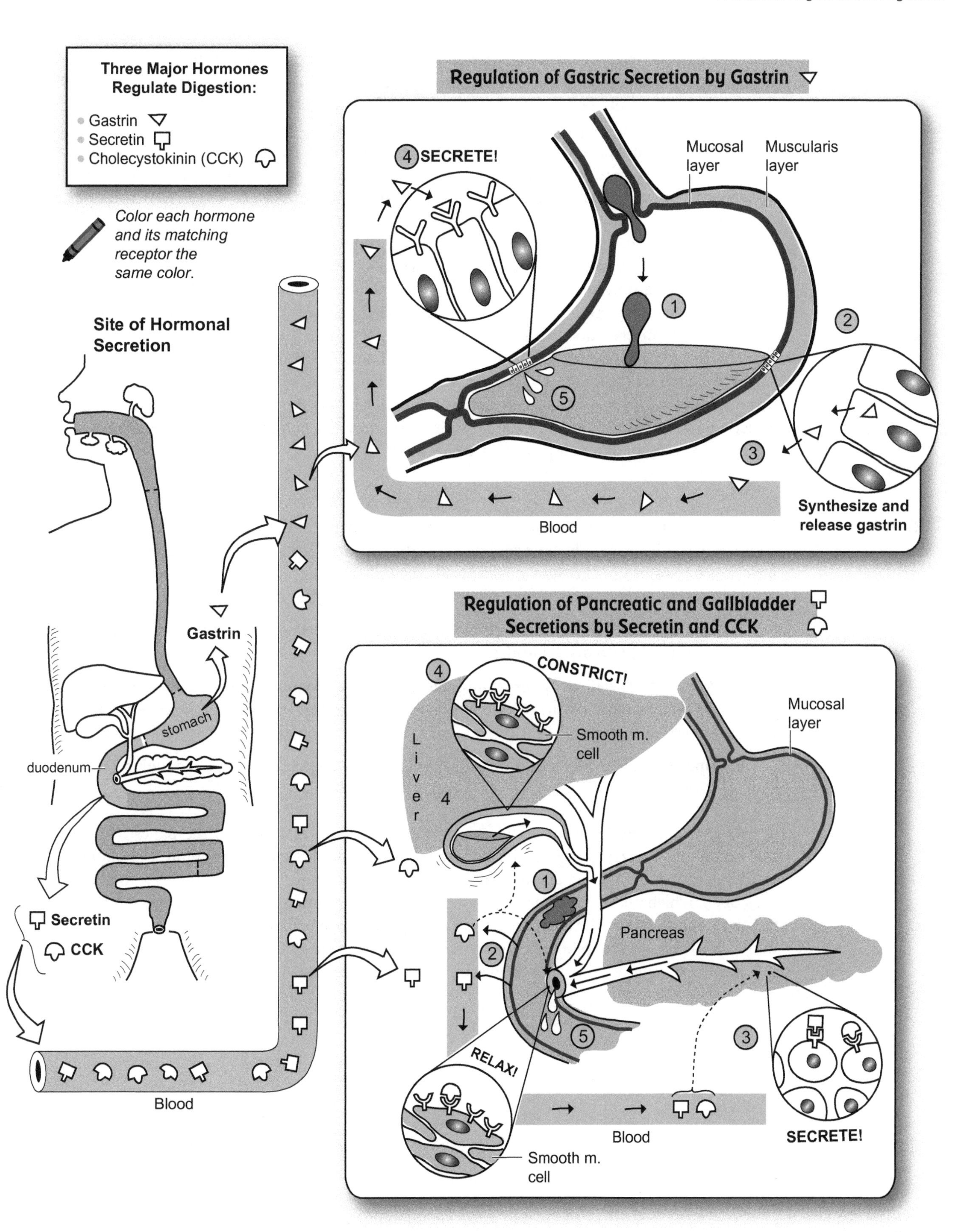
Three Major Hormones Regulate Digestion:
Gastrin
Secretin
Cholecystokinin (CCK)
Color each hormone and its matching receptor the same color.
Site of Hormonal Secretion
Gastrin
stomach
duodenum
Secretin
CCK
Blood
Regulation of Gastric Secretion by Gastrin
4 SECRETE!
Mucosal layer
Muscularis layer
1
2
3
5
Blood
Synthesize and release gastrin
Regulation of Pancreatic and Gallbladder Secretions by Secretin and CCK
4
CONSTRICT!
Liver
4
Smooth m. cell
Mucosal layer
1
2
Pancreas
3
5
RELAX!
Blood
SECRETE!
Smooth m. cell

Description

The **pancreas** is vital to chemical digestion because it produces so many digestive enzymes. It is an elongated, pinkish-colored gland divided into three main sections: *head, body,* and *tail.* The head is the widest portion; the body is the central region; and the tail marks the blunt, tapering end of the gland. The surface is bumpy and nodular. The individual nodes are called **lobules.**

Scattered within the gland are many glandular epithelial cells. Also inside is a long tube called the pancreatic duct that runs through the middle of the gland and terminates in an opening to the duodenum. Smooth muscle surrounds this opening to form a valve called the **hepatopancreatic sphincter** (*sphincter of Oddi*).

The pancreas has a dual function as both an **endocrine** and an **exocrine** gland. Making it an exocrine gland, the vast majority of the epithelial cells (99%) form clusters called **acini.** The cells in the acini produce watery secretions including bicarbonate ions and many different digestive enzymes to aid in chemical digestion. These are released into the **pancreatic duct,** through the hepatopancreatic sphincter, and into the duodenum.

Making it an endocrine gland, the remaining 1% of cells are arranged in separate cell clusters called called **pancreatic islets** (*islets of Langerhans*). These various cells produce four different hormones: *insulin, glucagon, somatostatin,* and *pancreatic polypeptide* (PP). These hormones diffuse directly into the bloodstream, travel to various target organs, and induce responses in those organs to regulate a variety of different processes in the body.

Function

Pancreatic acini produce many different enzymes for chemical digestion. These include **amylase** (for complex carbohydrates), **lipase** (for lipids), and **proteases** (for proteins). In addition, the pancreas produces **nucleases** (for digesting nucleic acids such as DNA).

Let's examine the role of each of the first three enzyme types mentioned.

1. **Amylase** catalyzes the initial step in carbohydrate digestion, the conversion of a polysaccharide (*ex.:* starch) into disaccharides.
2. **Proteases:** The three different proteases—**trypsin, chymotrypsin,** and **carboxypeptidase**—all are used to speed up the middle step in protein digestion, conversion of large polypeptides into small polypeptides.
3. **Lipase** catalyzes the breakdown of lipids such as triglycerides into **monoglycerides** and **fatty acids.**

As chyme passes from the stomach into the duodenum, it stimulates the pancreas to make and release digestive enzymes. This process is regulated by the hormones **secretion** and **cholecystokinin (CCK)** (see p. 172). The hepatopancreatic sphincter normally is closed, but the hormone CCK causes it to open, which allows the release of pancreatic secretions into the duodenum. These secretions then mix with the chyme so the enzymes can do their jobs. As the hepatopancreatic sphincter opens, it also allows for the release of bile from the common bile duct. **Bile** mediates the first step in lipid digestion. It is needed to emulsify the lipids in the chyme so the lipases can work.

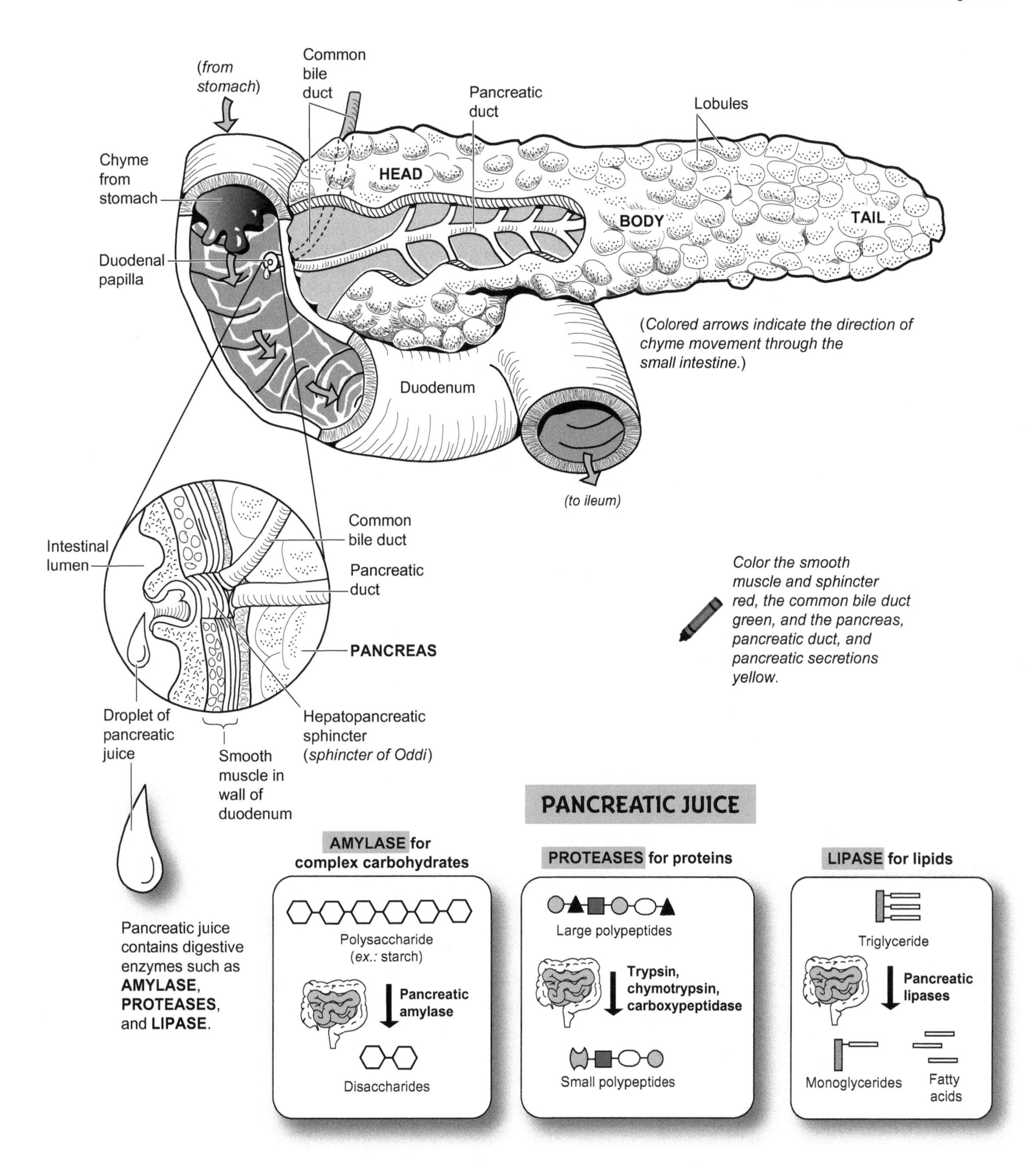

Pancreatic juice contains digestive enzymes such as **AMYLASE**, **PROTEASES**, and **LIPASE**.

Description

The **liver** is the largest abdominal organ and is located below the diaphragm in the abdominal cavity. It is divided into two major lobes—left and right—separated by a sheet of connective tissue called the **falciform ligament.** Along the bottom edge of the falciform ligament is a fibrous, cable-like structure called the **round ligament** *(ligamentum teres)* derived from the fetal **umbilical vein.** The liver is anchored to the diaphragm by the **coronary ligament.** The liver also has two minor lobes—**caudate** and **quadrate**—visible in the posterior view.

Microscopically, each lobe of the liver is composed of more than 100,000 individual hexagon-shaped units called **hepatic lobules.** These structures are the fundamental functional units of the liver. At each corner of a lobule is a cluster of three vessels called a **hepatic triad** that consists of a hepatic artery, a branch of the hepatic portal vein, and a bile duct. At the center of each lobule is a long vessel called the **central vein.**

Radiating out from the central vein like spokes from a wheel are the very permeable, blood-filled capillaries called **sinusoids.** Fixed to the inner lining of the sinusoids are macrophages, also called Kupffer's cells. The **hepatic artery** and **hepatic portal vein** deliver blood into the sinusoids, and it travels toward the central vein, then into the hepatic veins, and into the **inferior vena cava.** Most **hepatocytes** (*liver cells*) are located close to the blood in the sinusoids, which makes it easier to deposit products into the blood or screen materials out of it.

Bile is produced by hepatocytes and drains into vessels called **bile canaliculi,** then into the larger **bile ducts,** which transport the bile away from the liver and into the **gallbladder** for storage. Note that the direction of bile flow in the bile canaliculi is opposite to that of the blood in the sinusoids.

Functions

The liver has more than 200 different functions. Major functions of the liver are:

- **Synthesis** *ex.:* plasma proteins, clotting factors, bile, cholesterol
- **Storage** *ex.:* iron (Fe), glycogen, blood, fat-soluble vitamins
- **Metabolism** *ex.:* convert glucose to glycogen and glycogen to glucose; convert carbohydrates to lipids, maintain normal blood glucose levels
- **Detoxification** *ex.:* alcohol and other drugs
- **Conjugation**—this means making substances water soluble
  ex.: bilirubin is converted to a water soluble form to make it easier to be excreted

Analogy

In cross section, each **hepatic lobule** is like a simple **Ferris wheel** with six swinging chairs. Each **swinging chair** is a **hepatic triad.** The **hub** of the Ferris wheel is the **central vein,** and the **spokes** of the Ferris wheel are like the **sinusoids.**

Key to Illustration

Liver Lobes (L)

- L1 Left major
- L2 Right major

Ligaments (L)

- RL Round ligament (*ligamentum teres*)
- FL Falciform ligament
- CL Coronary ligament

Lobule Structures

- CV Central vein
- HT Hepatic triad
- BC Bile canaliculus
- BD Bile duct
- HA Branch of hepatic artery
- HPV Branch of hepatic portal vein
- S Sinusoid
- H Hepatocyte
- K Kupffer's cell (*macrophage*)

Other

- GB Gallbladder

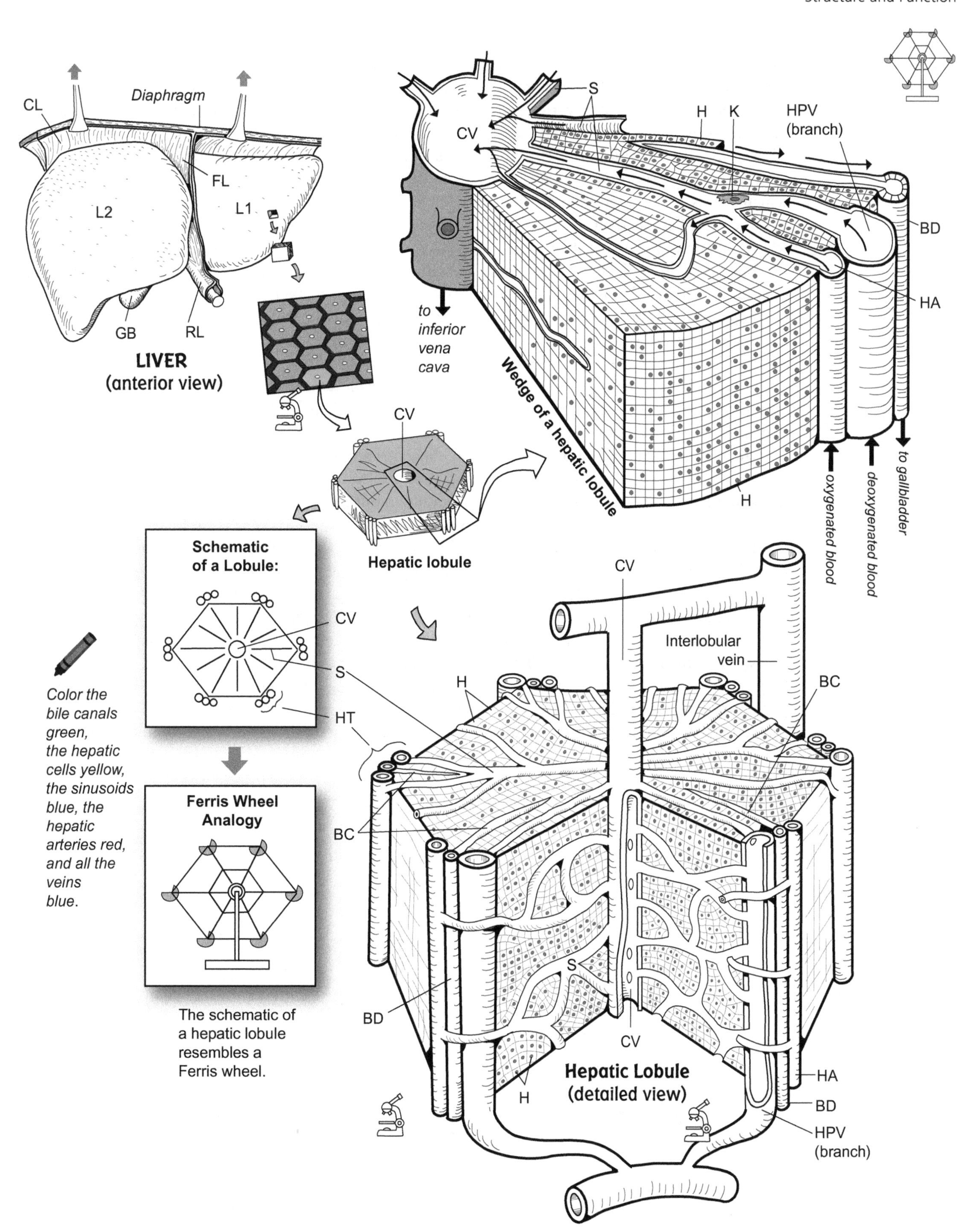
CL
Diaphragm
FL
L2
L1
GB
RL
LIVER
(anterior view)
CV
S
H
K
HPV
(branch)
BD
HA
to
inferior
vena
cava
Wedge of a hepatic lobule
H
oxygenated blood
deoxygenated blood
to gallbladder
CV
Hepatic lobule
Schematic
of a Lobule:
CV
S
HT
Ferris Wheel
Analogy
Color the
bile canals
green,
the hepatic
cells yellow,
the sinusoids
blue, the
hepatic
arteries red,
and all the
veins
blue.
The schematic of
a hepatic lobule
resembles a
Ferris wheel.
CV
Interlobular
vein
BC
H
BC
S
BD
CV
H
Hepatic Lobule
(detailed view)
HA
BD
HPV
(branch)

Description

Bile is a thick, alkaline solution produced by **hepatocytes** (*liver cells*), transported through bile ducts, stored and concentrated in the **gallbladder**, and released into the lumen of the **duodenum.** It is composed mostly of water, various ions, bile pigments such as **bilirubin** (orange or yellow pigment derived from hemoglobin catabolism), and various lipids such as **cholesterol, phospholipids**, and **triglycerides.** In the gallbladder the color of bile varies from yellow to brown or green.

After being produced, it drains away from the hepatic lobule and enters a microscopic **bile duct**, which drains into a larger **hepatic duct.** The **left hepatic duct** drains bile from the left lobe of the liver, and the **right hepatic duct** drains bile from the right lobe of the liver. The two hepatic ducts then fuse to form a **common hepatic duct.**

Bile flows down the common hepatic duct and into the **common bile duct**, which terminates in an opening leading to the duodenum. Around this opening is a thickened band of smooth muscle called the **hepatopancreatic sphincter**. Normally the smooth muscle around this sphincter is contracted, causing it to be closed.

Bile is stored in the **gallbladder**, located under the right lobe of the liver. To fill the gallbladder with bile, the hepatopancreatic sphincter must remain closed. As bile drains from the liver, it backs up into the common bile duct and enters the **cystic duct**, which leads to the gallbladder. Hard, solid deposits called **gallstones** can form in either the gallbladder or bile ducts. There are two types based on which component in the bile became too concentrated—either cholesterol or bilirubin. Gallstones vary widely in size and can become too large to exit the gallbladder, which can lead to *cholecystitis*—inflammation of the gallbladder. If the gallstones damage the lining of the gallbladder, a surgical procedure called a *cholecystectomy*—removal of the gallbladder—may need to be performed.

Emptying of the gallbladder is regulated by a hormone called **cholecystokinin**, or **CCK.** As lipid-rich **chyme** (food and gastric juice) passes through the **mucosa** (inner lining) of the duodenum, it stimulates specific cells in the duodenum to release CCK. This hormone then travels through the bloodstream and targets the smooth muscle around the gallbladder.

These smooth muscle cells have receptors for CCK. Once CCK binds to these receptors, it induces the smooth muscle to contract. The result is that bile is expelled forcefully from the gallbladder. Simultaneously, CCK targets the smooth muscle of the hepatopancreatic sphincter and causes it to relax, opening the sphincter. With the sphincter open, bile moves out of the gallbladder, down the cystic duct, down the common bile duct, past the hepatopancreatic sphincter, and into the duodenum.

Function

Emulsification of lipids (fats) is the first step in the chemical digestion of lipids. Rather than breaking chemical bonds, bile physically breaks large masses of lipids into smaller lipid droplets. This makes it easier for enzymes such as **lipase** to facilitate chemical digestion.

Key to Illustration

1. Right hepatic duct
2. Left hepatic duct
3. Common hepatic duct
4. Cystic duct
5. Common bile duct
6. Pancreatic duct

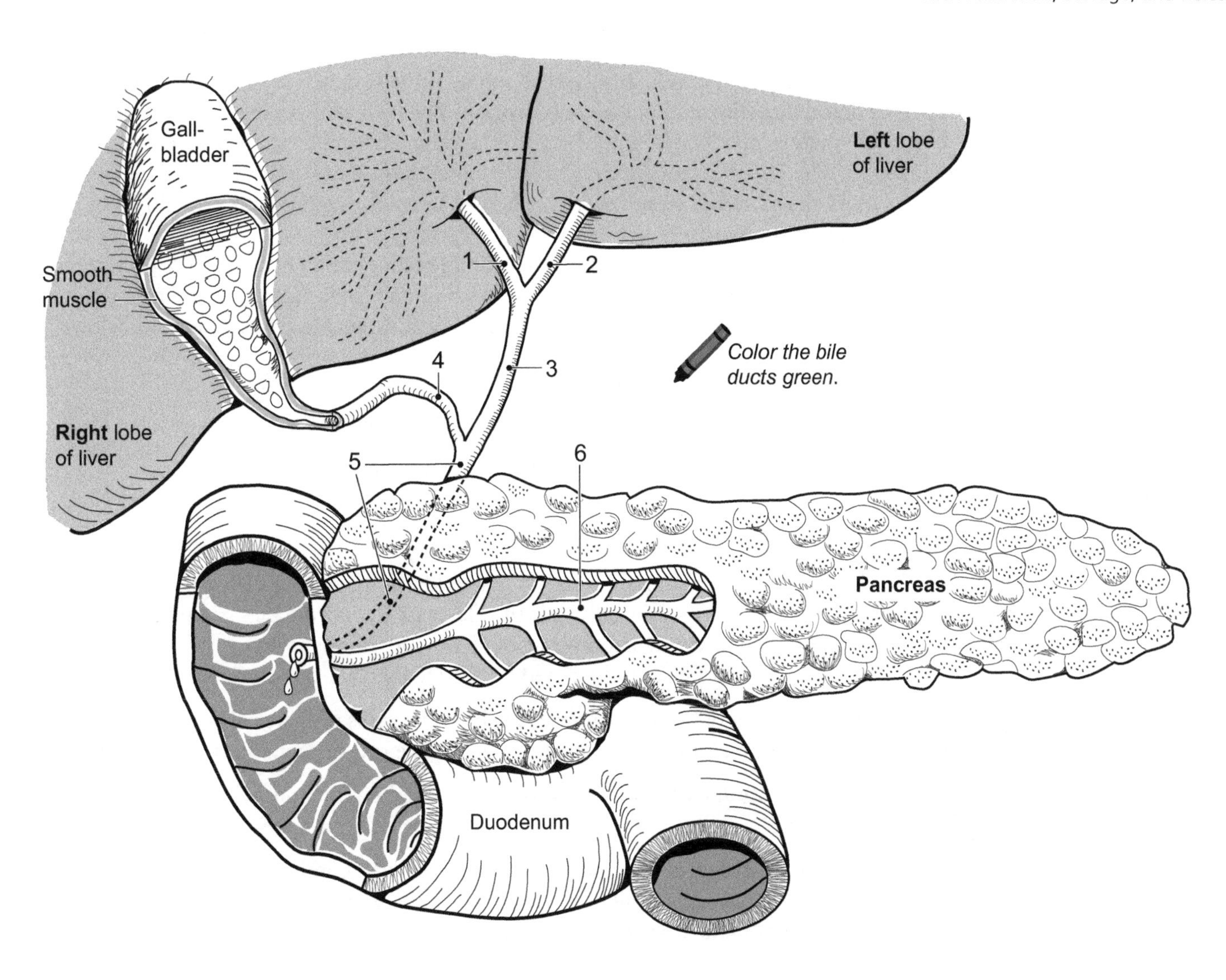
Gall-
bladder
Left lobe
of liver
Smooth
muscle
1
2
3
4
5
6
Color the bile
ducts green.
Right lobe
of liver
Pancreas
Duodenum

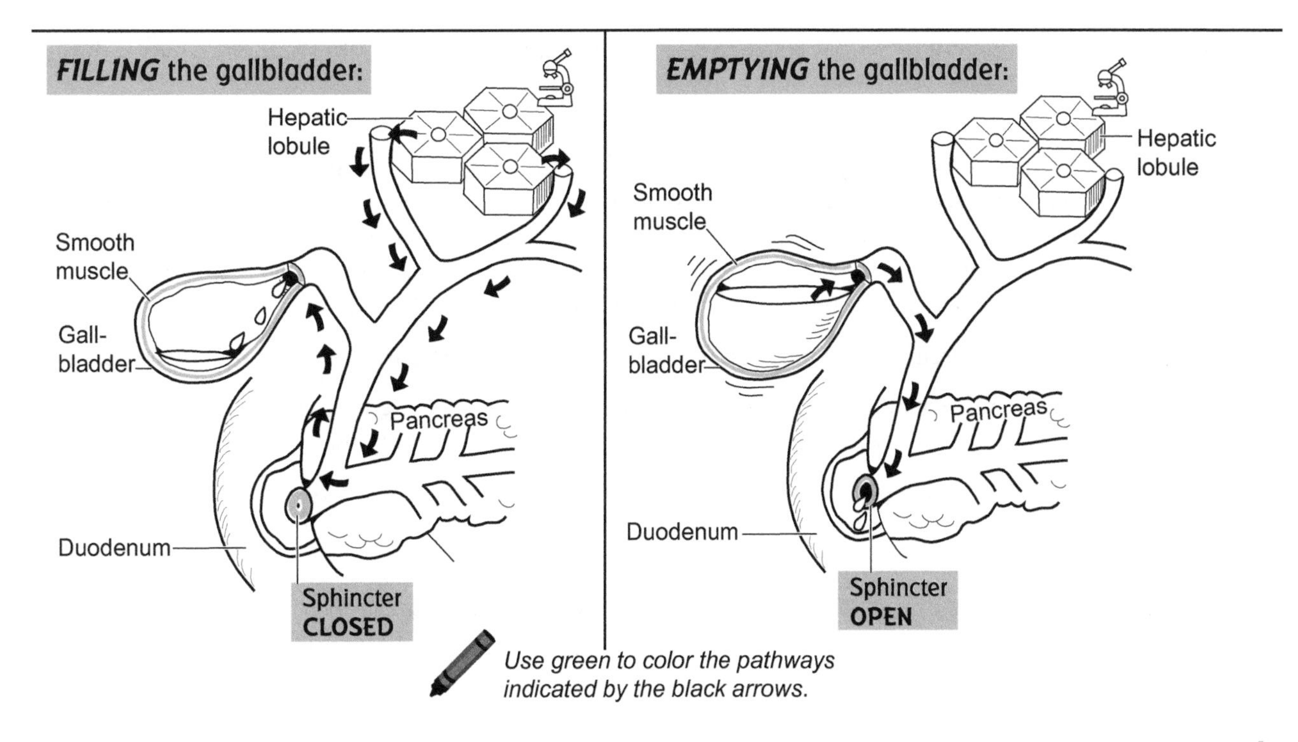
FILLING the gallbladder:
Hepatic
lobule
Smooth
muscle
Gall-
bladder
Pancreas
Duodenum
Sphincter
CLOSED
EMPTYING the gallbladder:
Hepatic
lobule
Smooth
muscle
Gall-
bladder
Pancreas
Duodenum
Sphincter
OPEN
Use green to color the pathways
indicated by the black arrows.

Description

The small intestine is a 20-foot muscular tube that connects the stomach to the large intestine. It is the major site of **chemical digestion** and nutrient **absorption.** The wall of this organ contains a muscularis layer of smooth muscle subdivided into two thin layers: an inner circular layer and an outer longitudinal layer. When contracted, the circular layer pinches the small intestine, and the longitudinal layer shortens it. All the materials from the stomach must move through this tube by different types of coordinated muscle contractions.

The small intestine uses the processes of **peristalsis** and **segmentation** to move undigested materials along its length. These processes are not unique to the small intestine; they occur in other parts of the digestive tract as well. The sites of action for these muscular movements are summarized below:

Type of Muscular Movement	Site of Action
1. Peristalsis	Esophagus, stomach, small intestine, and large intestine
2. Segmentation	Small intestine and large intestine

Peristalsis

Peristalsis is a rhythmic wave of smooth muscle contraction that results in the propulsion of **chyme** through the digestive tract. More specifically, it is the alternating, coordinated stimulation of both the circular and the longitudinal layers of smooth muscle in the muscularis layer.

The following is the series of events illustrated on the facing page:

① The circular layer of smooth muscle contracts behind the mass of materials in the small intestine. The result is that the small intestine is pinched tighter, like rings becoming smaller in size.

② A section of the longitudinal layer in front of the pinched region contracts, which shortens part of the small intestine.

③ This coordinated process of "pinching" and "shortening" continues to alternate at regular intervals along the length of the wall of the small intestine. Viewed in real time, it would look like a gradual, smooth undulating wave of muscle contraction.

Analogy

The action of peristalsis is like squeezing a tube of toothpaste so the toothpaste moves through the tube.

Segmentation

Segmentation is the pinching of the intestine into compartments and subsequent mixing of **chyme** with intestinal secretions. This ensures that the processes of chemical digestion and absorption are both completed. The circular layer of the muscularis is more active than the longitudinal layer. No net movement of materials results in this process as it did in peristalsis.

The following is the series of events illustrated on the facing page:

① The circular layer pinches off the small intestine into similar-sized compartments, or "segments."

② It further pinches to even smaller compartments as the undigested materials are moved back and forth by muscle contractions like an agitator in a washing machine.

③ The undigested materials are mixed thoroughly with intestinal secretions.

Analogy

The action of segmentation is like the movement of a hand mixer combining all the ingredients in cake batter.

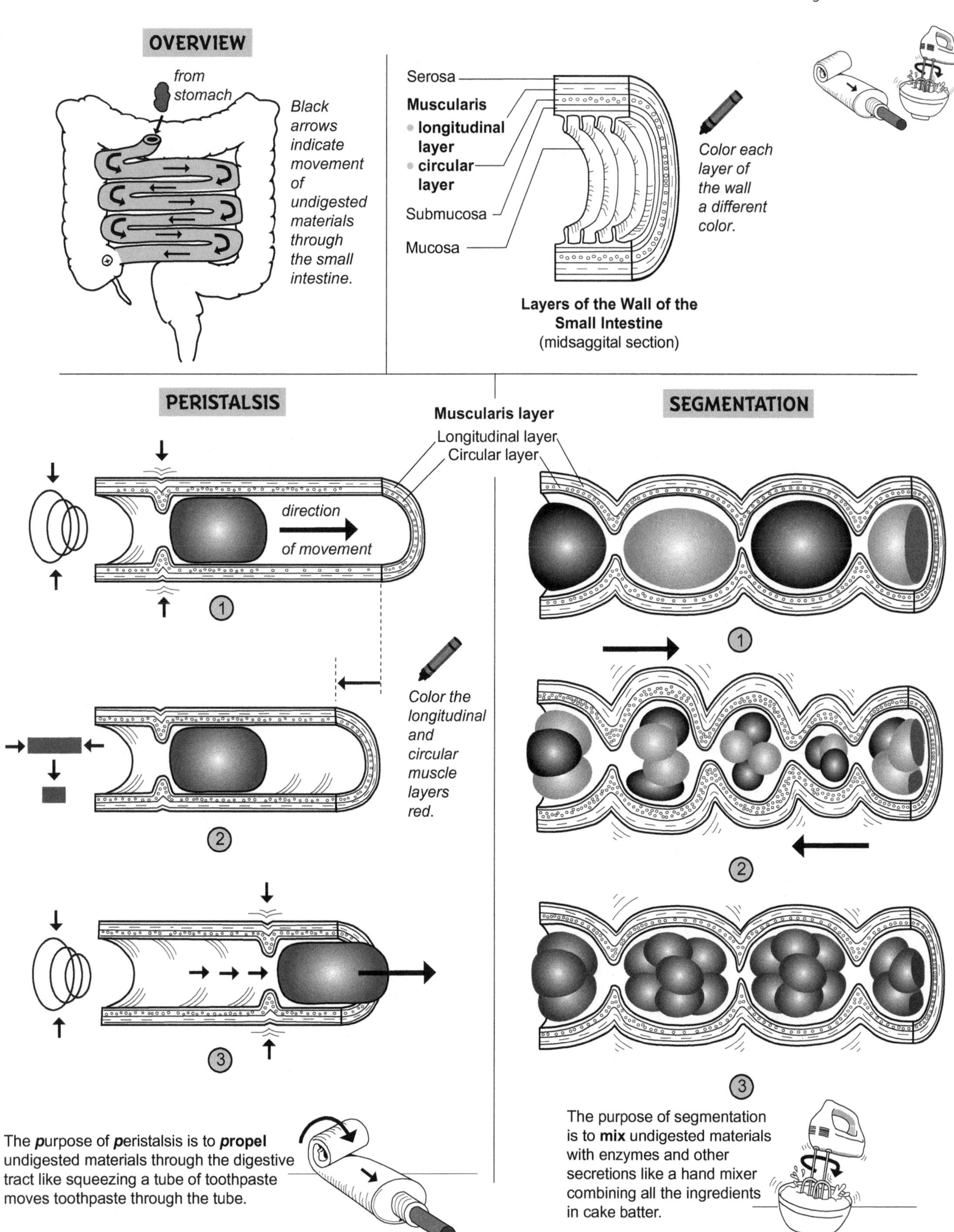

The ***p***urpose of ***p***eristalsis is to ***propel*** undigested materials through the digestive tract like squeezing a tube of toothpaste moves toothpaste through the tube.

The purpose of segmentation is to **mix** undigested materials with enzymes and other secretions like a hand mixer combining all the ingredients in cake batter.

Description

Absorption is the movement of small, digested substances from the digestive tract into either the blood or the lymph. Almost all absorption in the digestive system occurs in the small intestine. Folds are used in organs and structures throughout the body to increase surface area for different purposes. The small intestine uses its many folds to maximize surface area for absorption of nutrients.

The three significant folded structures in the small intestine are the **plicae circulares** (sing. *plica circularis*), **villi** (sing. *villus*), and **microvilli**. The plicae circulares are folds of the inner mucosal layer of the small intestine. Extending out from the plica are numerous fingerlike projections called villi. Each villus is covered by a layer of **simple columnar epithelium**. The **goblet cells** scattered within this tissue secrete **mucus**. Each simple columnar epithelial cell has folds in its plasma membrane called microvilli.

Each cell also has two surfaces through which nutrients must pass: **apical** and **basal**. Nutrients first pass through the apical surface that contains the microvilli, then move through the cytosol, then cross the basal surface that lines the inside of the villi. In summary: nutrients follow this absorption pathway:

Lumen → Epithelial cell: apical surface → Cytosol → Basal surface → Blood or lymph (small intestine)

Process

The purpose of chemical digestion is to break down macromolecules into their simplest building blocks. These building blocks include **monosaccharides** from complex carbohydrates; **amino acids, dipeptides,** and **tripeptides** from **proteins**; and monoglycerides and **fatty acids** from **triglycerides** (lipids). These final products of digestion are small enough to be absorbed into the blood or the lymph and then pass directly into body cells where they can be used.

Some ingested substances already in their simplest form—such as the fructose in fruit juices—are partly absorbed through the wall of the stomach before they even reach the small intestine.

About 90% of all absorption occurs in the small intestine for the following reasons: (1) villi and other structures make it anatomically specialized for the process, (2) the final products of macromolecule catabolism are not even formed until they reach this site, and (3) the 20-plus-foot tube maximizes the time for absorption to take place.

Monosaccharides

Complex carbohydrates must be broken down into **monosaccharides** before they can be absorbed. The three types of monosaccharides are **glucose, galactose,** and **fructose**. Fructose, found in fruit and fruit juices, is transported across both the apical and basolateral surfaces by **facilitated diffusion**. Glucose and galactose are transported by a different method—secondary active transport with sodium ($Na+$). The sodium ion gradient is the driving force that helps move these two sugars across the apical surface and into the cell. Once inside the epithelial cell, both glucose and galactose cross the basolateral surface by facilitated diffusion. Fructose, glucose, and galactose all are absorbed directly into the blood capillary.

Amino acids   Dipeptides   Tripeptides

Most **amino acids** use **active transport** to cross the apical surface. Once inside the epithelial cell, the amino acids are actively transported again across the basolateral surface and then absorbed directly into the blood capillary.

Monoglycerides   Fatty acids

Triglycerides are the most common type of lipids, so they will be considered here. When triglycerides are digested, two of their three fatty acid chains are broken off from their glycerol backbone, resulting in the formation of **monoglycerides** and **fatty acids**. In the chemical rule of "like goes into like," these products of triglyceride digestion easily diffuse through the plasma membrane of the apical surface because they are chemically similar to the phospholipids in the plasma membrane. Some of the fatty acid chains are short-chain fatty acids, and others are long-chain fatty acids. The **short-chain fatty acids** diffuse directly across both the apical and basolateral surfaces and are absorbed directly into the blood capillary.

The pathway for monoglycerides and **long-chain fatty acids** is a bit more complex. After diffusing through the apical surface, the monoglycerides are digested further into glycerol and a single fatty acid chain. The glycerol then is reassembled back into triglycerides by linking with free fatty acid chains. The newly formed triglycerides are packaged in a protein coat to form a structure called a **chylomicron**. Chylomicrons leave the cell by exocytosis (see p. 76). They are too large to enter the blood capillary, so they enter the more permeable lacteal instead. As they travel in the lymph of the **lacteal**, they eventually are deposited into the blood via the left subclavian vein.

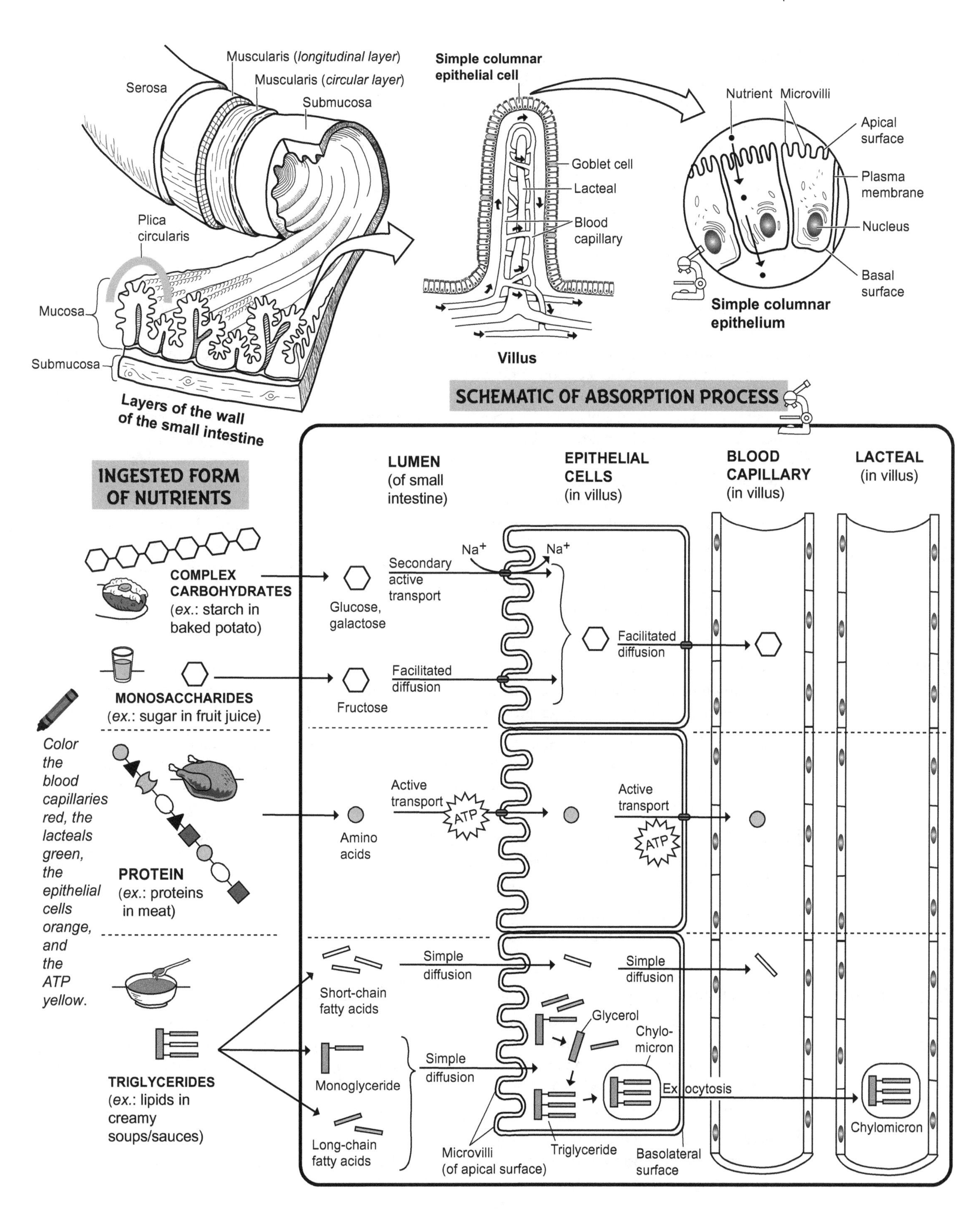
Serosa
Muscularis (longitudinal layer)
Muscularis (circular layer)
Submucosa
Plica circularis
Mucosa
Submucosa
Layers of the wall of the small intestine
Simple columnar epithelial cell
Goblet cell
Lacteal
Blood capillary
Villus
Nutrient
Microvilli
Apical surface
Plasma membrane
Nucleus
Basal surface
Simple columnar epithelium
SCHEMATIC OF ABSORPTION PROCESS
INGESTED FORM OF NUTRIENTS
LUMEN (of small intestine)
EPITHELIAL CELLS (in villus)
BLOOD CAPILLARY (in villus)
LACTEAL (in villus)
COMPLEX CARBOHYDRATES (ex.: starch in baked potato)
Na+
Na+
Secondary active transport
Glucose, galactose
Facilitated diffusion
Facilitated diffusion
Fructose
MONOSACCHARIDES (ex.: sugar in fruit juice)
Color the blood capillaries red, the lacteals green, the epithelial cells orange, and the ATP yellow.
PROTEIN (ex.: proteins in meat)
Active transport
ATP
Amino acids
Active transport
ATP
Simple diffusion
Short-chain fatty acids
Simple diffusion
Glycerol
Chylo-micron
Simple diffusion
Monoglyceride
Exocytosis
Chylomicron
TRIGLYCERIDES (ex.: lipids in creamy soups/sauces)
Long-chain fatty acids
Microvilli (of apical surface)
Triglyceride
Basolateral surface

Description

The **large intestine** (*large bowel*) is a hollow, muscular tube about 5 feet long that connects the end of the small intestine (*ileum*) to the **anus.** It is subdivided into three different parts—the **cecum, colon,** and **rectum.** The cecum is a pouch that marks the beginning of the large intestine. On the posterior surface of this structure, the **appendix** is located. This is a long, slender, hollow tube that opens into the cecum.

The next portion is the colon, the longest part of the intestine. It is subdivided into four segments—the *ascending, transverse, descending,* and *sigmoid* colon. Along the length of the colon is a band of smooth muscle called the **taenia coli.** It bulges the colon into pouches called **haustra** that run along its length and permit expansion and elongation of the colon. At regular intervals along the taenia coli are flaps of fatty tissue called **fatty appendices.** The last segment of the large intestine and the digestive tract is the rectum, which allows for temporary storage of fecal waste.

Process

Most nutrients have been digested and absorbed into the blood by the time they reach the end of the small intestine. The remaining undigested chyme passes through the **ileocecal sphincter** and into the large intestine. Two opposing processes—absorption and secretion—occur. **Absorption** is the movement of nutrients from the large intestine into the blood and then into cells. **Secretion** is the movement of substances from the blood to the large intestine for eventual excretion. Remaining to be absorbed in the large intestine are some electrolytes, water, and vitamins. Absorption and secretion take place in the ascending and transverse colon. The remaining undigestible material is compacted into a semisolid mixture called **feces,** stored temporarily in the descending and sigmoid colon, and finally ejected from the body. The process of eliminating the feces is controlled by a defecation reflex.

We share a mutually beneficial relationship with the various **bacteria**—such as *Escherichia coli* (*E. coli*)—that live inside the large intestine. We provide them a warm environment and food to live and reproduce, and they make vitamins for us that we can't make on our own, such as vitamin K and B-complex vitamins. But bacteria also contribute to *flatus*, or gas. As they metabolize any remaining carbohydrates, they produce hydrogen, carbon dioxide, and methane gases. Another gas produced by bacteria—hydrogen sulfide—gives off a "rotten egg" odor. But the bacteria are not entirely to blame because the major source of intestinal gas is air swallowed during the rapid ingestion of food.

Motility

The **muscularis layer** of the wall of the large intestine contains two layers of **smooth muscle**—an inner, circular layer and an outer, longitudinal layer. Contraction of this smooth muscle in different ways allows for movement of undigested materials inside the large intestine. The three types of motility through the large intestine are:

1. **Segmentation** (*haustral churning*). As undigested material enters the haustra, the expansion triggers muscularis layer contractions that pinch the haustra into separate compartments and turn the contents inside. This serves to enhance absorption and secretion of water and electrolytes as well as to propel the contents to the next haustrum.

2. **Peristalsis.** Stimulation of the muscularis layer results in a weak, rhythmic wave of contraction, a combination of pinching the circular layer and shortening the longitudinal layer. This contraction combination results in the gradual propulsion of undigested material through the tract.

3. **Mass movement.** This strong muscle contraction forces fecal material into the rectum. It is regulated by a reflex and occurs two to three times a day on average.

Absorption and Secretion

Absorption follows a predictable sequence, illustrated on the facing page, as follows:

1. Levels of **sodium** (Na^+) and **potassium** (K^+) are controlled by the **sodium-potassium pump** (see p. 80), which transports these ions in opposite directions. Sodium is actively transported out of the lumen of the large intestine and into the blood, and potassium is transported into the large intestine. The net result is that sodium is reabsorbed into the blood, and potassium is secreted into the large intestine to be removed in the feces. The result is that it causes a higher solute concentration in the blood relative to the lumen of the large intestine.

2. Because **water** (H_2O) always moves passively across a plasma membrane toward the higher solute concentration via osmosis (see p. 70), this is what occurs next: The body forces osmosis by actively transporting sodium out first.

GROSS ANATOMY

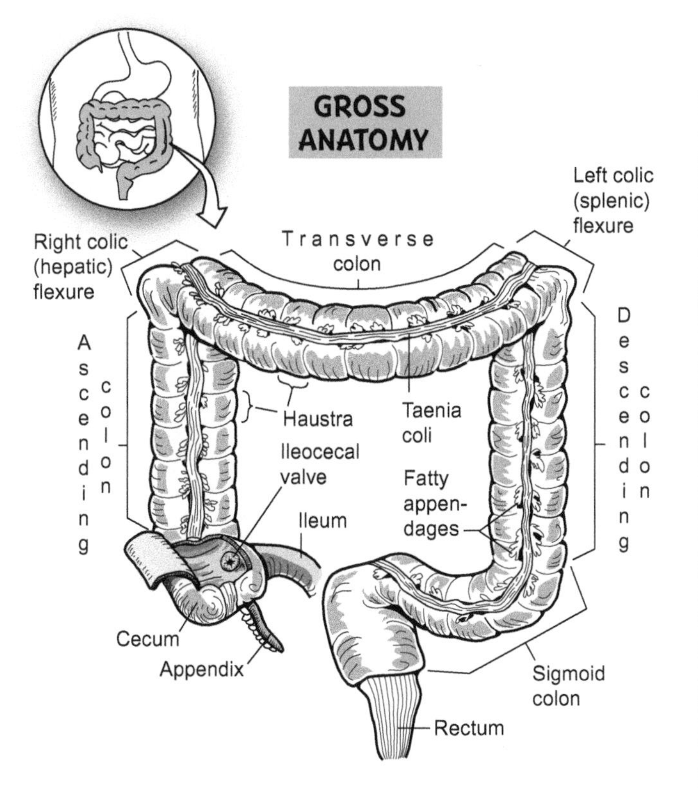

FUNCTIONAL OVERVIEW

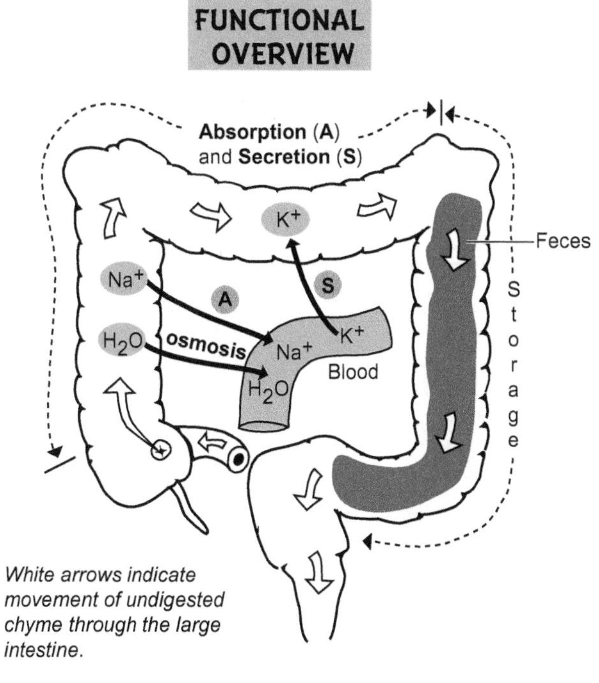

White arrows indicate movement of undigested chyme through the large intestine.

MOTILITY

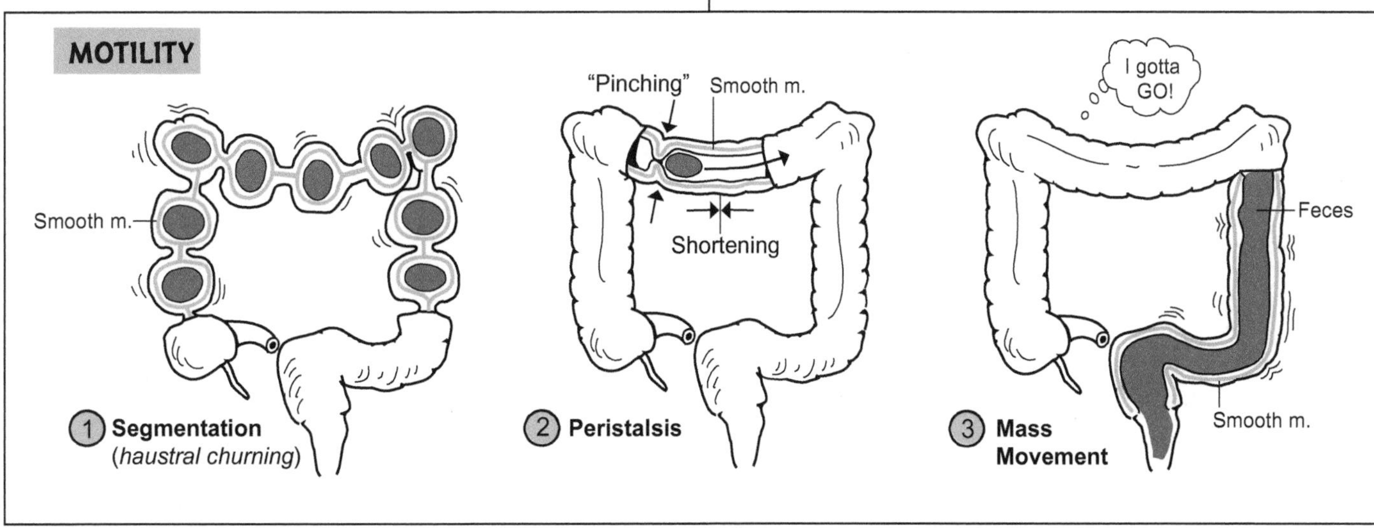

ABSORPTION (A) AND SECRETION (S)

Bacteria are normal inhabitants of the large intestine and manufacture vitamins for the body.

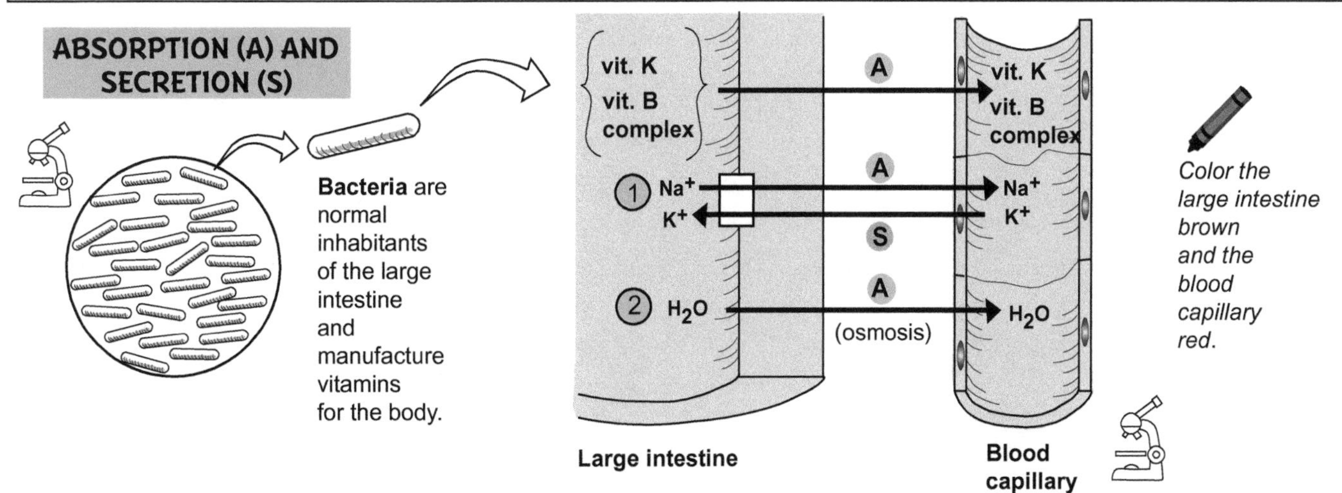

Color the large intestine brown and the blood capillary red.

Notes

METABOLIC PHYSIOLOGY

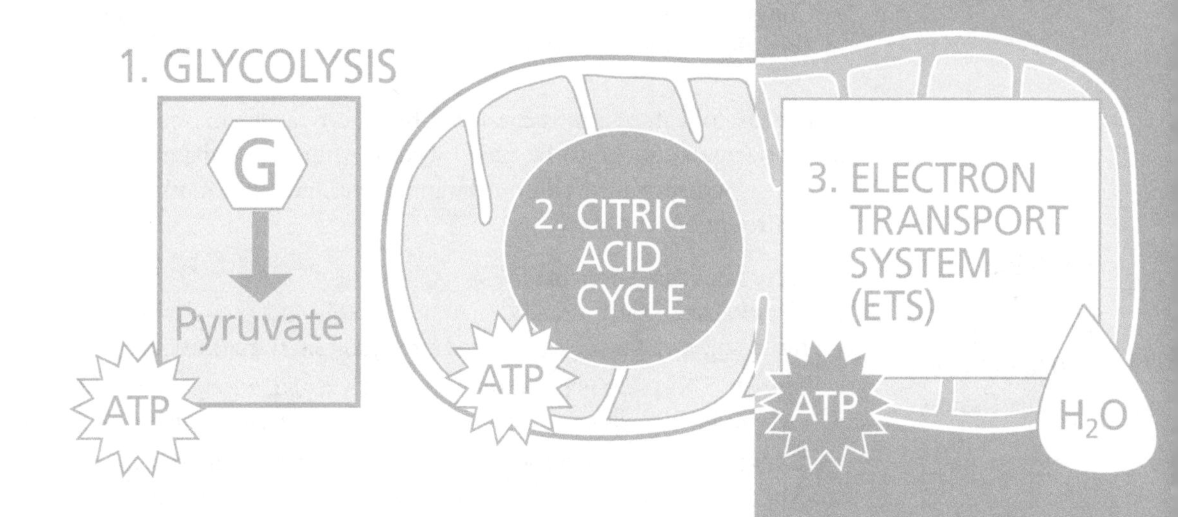

Description

Metabolism is the sum total of all the chemical reactions that occur in the body. It consists of two different parts—catabolic reactions and anabolic reactions. **Catabolic reactions** break down biomolecules into simpler building blocks (e.g., triacylglycerols into fatty acids and glycerol), and **anabolic reactions** use these building blocks to synthesize more complex biomolecules (e.g., amino acids into proteins). Energy metabolism focuses on the elements of metabolism that generate ATP. As a general overview of energy metabolism, let's examine its three main phases:

PHASE 1: The Release of Nutrients in the Digestive Tract Lumen

The digestion of three major nutrients—triacylglycerols, complex carbohydrates, and proteins—is shown below. As these macromolecules are ingested, they enter the digestive tract and gradually are broken down into their fundamental building blocks:

- Proteins → → → → → → **amino acids**
- Complex carbohydrates → → **glucose** (and other simple sugars)
- Triacylglycerols → → → → **monoglycerides** and **fatty acids**

This phase of catabolism is catalyzed by digestive enzymes in the digestive tract. These building blocks now are small enough to be used by cells. After being absorbed into the bloodstream through microscopic structures called **villi** on the inner lining of the small intestine, they are transported to various cells throughout the body.

PHASE 2: The Fate of Nutrients within Cells of Various Tissues

Glucose has three possible fates:
- **Catabolized:** When the body requires energy, glucose is catabolized to produce ATP molecules. The three stages of glucose catabolism are glycolysis, the citric acid cycle, and the electron transport system.
- **Stored:** This takes place in the form of **glycogen**, a long chain of glucose molecules. Skeletal muscle cells and liver cells store lots of glycogen when excess glucose is present in the blood.
- **Converted:** Glucose can be converted into several **amino acids** that may be used to make proteins. Alternatively, when storage of glycogen is maximized, liver cells can convert glucose into **glycerol** and **fatty acids** to form **triglycerides.** The triglycerides then are deposited in fatty tissue.

Fatty acids and **glycerol** have one of two general fates:
- **Converted/catabolized:**

  Glycerol can be converted into glyceraldehyde-3-phosphate, one of the compounds formed during the process of glycolysis. Then, depending on the ATP levels in the cell, it is either converted into **pyruvate** (via glycolysis) or **glucose** (via gluconeogenesis).

  Fatty acids follow a different pathway than glycerol. They can be converted into **acetyl CoA**, which then enters the citric acid cycle to produce ATP.
- **Stored:** Glycerol and fatty acids recombine to form triacylglycerols within liver cells and fat cells.

Amino acids have three different fates:
- **Anabolized:** Amino acids may be used to produce **proteins.**
- **Converted:** Some amino acids can be converted into **acetyl CoA** and enter the citric acid cycle to produce ATP. When there are excess amino acids, they can be converted into either **glucose** (via gluconeogenesis) or **fatty acids.**
- **Deaminated:** Liver cells can remove the amino group from some amino acids to produce carboxylic acids and ammonia (NH_3) in a process called deamination. Then the carboxylic acids can enter different stages of the citric acid cycle.

PHASE 3: Aerobic Catabolism of Nutrients in the Mitochondria of Cells

Aerobic respiration refers to the two parts of cellular respiration that occur inside the mitochondrion: the citric acid cycle (see p. 194) and the electron transport system (see p. 196). **Acetyl CoA,** a key chemical intermediate in metabolism, is generated from fatty acids, pyruvate, and some amino acids within the mitochondrion. Once acetyl CoA is produced, it may enter the citric acid cycle as part of normal catabolism.

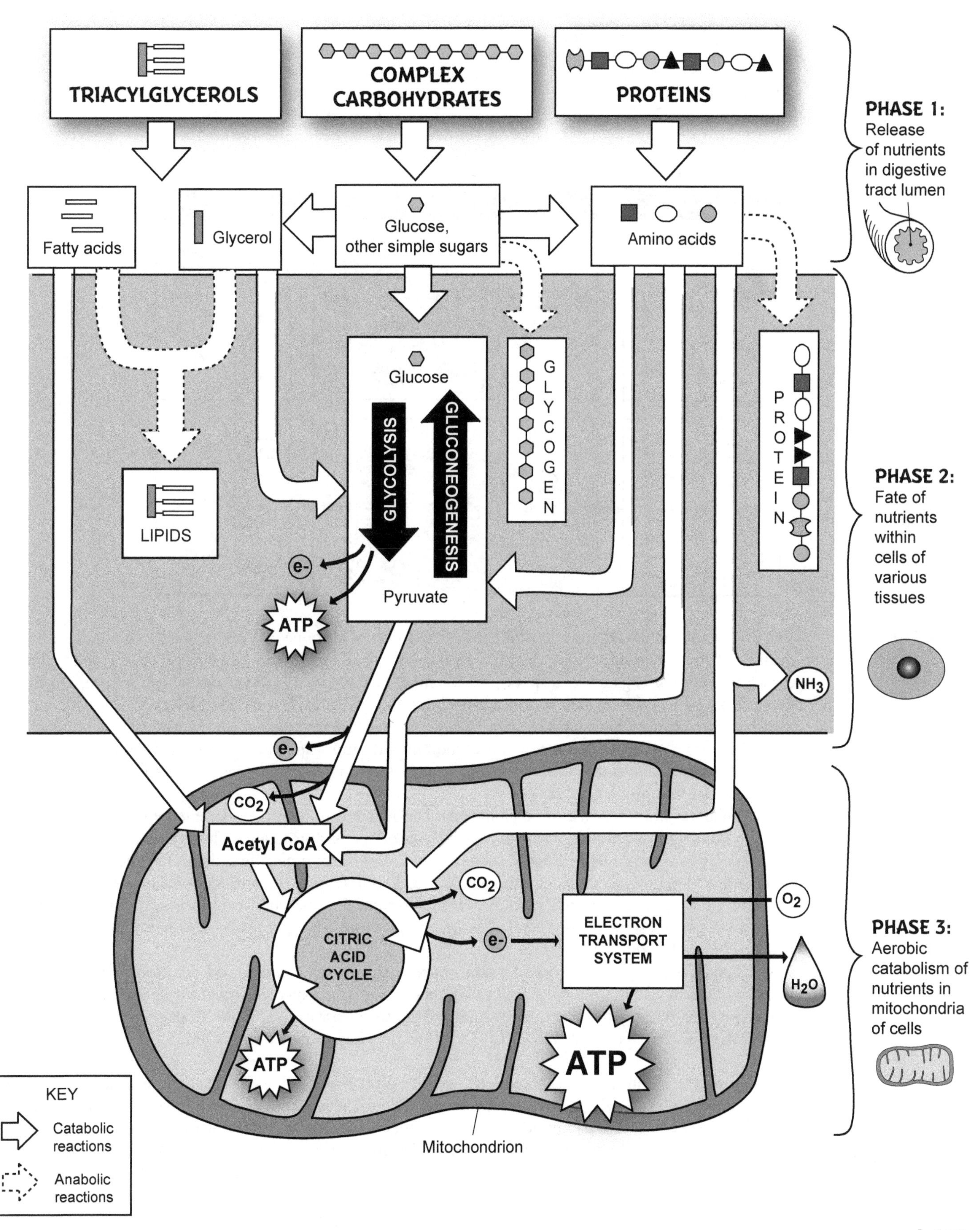
TRIACYLGLYCEROLS
COMPLEX CARBOHYDRATES
PROTEINS
PHASE 1:
Release of nutrients in digestive tract lumen
Fatty acids
Glycerol
Glucose, other simple sugars
Amino acids
Glucose
GLYCOLYSIS
GLUCONEOGENESIS
GLYCOGEN
PROTEIN
LIPIDS
e-
ATP
Pyruvate
PHASE 2:
Fate of nutrients within cells of various tissues
NH_3
e-
CO_2
Acetyl CoA
CO_2
O_2
CITRIC ACID CYCLE
e-
ELECTRON TRANSPORT SYSTEM
H_2O
PHASE 3:
Aerobic catabolism of nutrients in mitochondria of cells
ATP
ATP
Mitochondrion
KEY
Catabolic reactions
Anabolic reactions

Overall Equation for Oxidation of Glucose

In the process of cellular respiration, 1 molecule of glucose reacts with 6 oxygen molecules through a series of chemical reactions to produce 6 carbon dioxide molecules, 6 water molecules, 36 molecules of ATP, and heat energy lost to the environment.

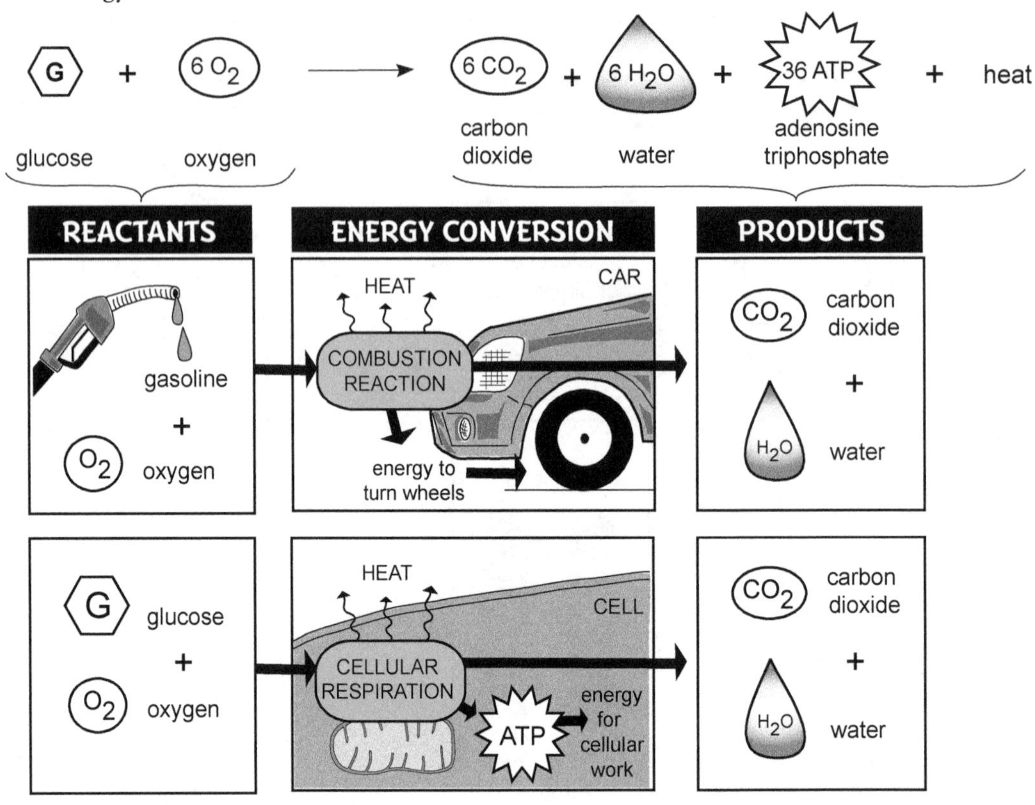

Description

The process of cellular respiration is like the combustion of gasoline in a car engine. Both require a fuel that must be chemically reacted with oxygen. Glucose is the fuel for the cell, whereas gasoline is the fuel for the car engine. The car engine uses a spark plug to ignite the combustible oxygen/gasoline mixture, causing it to produce a small explosion that moves a piston up and down inside a metal cylinder. This kinetic energy is transferred to an axle, causing it to spin. Because the wheels are attached to the axle, they spin as the axle spins. The direct products of this reaction, carbon dioxide and water vapor, are released in the exhaust gases, and heat is lost to the external environment. In summary, the potential energy in gasoline is converted into kinetic energy that propels the car.

Cellular respiration works in a similar manner. Instead of having all the energy released at once, glucose is gradually broken down through a long series of chemical reactions. The overall purpose is to trap as much energy as possible from glucose in the form of an energy molecule called ATP. Then the ATP can be used to do cellular work such as contracting a muscle cell. During the process, some heat is lost to the environment. As in the combustion of gasoline, the final products of this chemical process are carbon dioxide and water. In summary, some of the chemical-bond energy in glucose is transferred slowly into ATP molecules that can be used to do cellular work.

Cellular respiration is a series of oxidation-reduction reactions, or redox reactions. These involve the transfer of electrons from one substance to another. In this case, glucose is oxidized because it loses electrons to oxygen. Oxygen is reduced because it gains the electrons lost from glucose. To distinguish oxidation from reduction remember the words OIL RIG = Oxidation Is Loss, Reduction Is Gain.

Efficiency Rating of Cellular Respiration

Cellular respiration incorporates three processes: (1) glycolysis, (2) the citric acid cycle, and (3) the electron transport system (ETS). Each of these processes will be discussed in more detail in separate modules.

The process of cellular respiration is more efficient than a car engine. During the oxidation of glucose, 686 kilocalories of energy are released. Of this total, 278 are captured in the bonds of ATP molecules, giving it an efficiency rating of 41%. The remaining 59% is lost as heat. Compared to a typical car engine, which is between 10–30% efficient, this rating is looking pretty good!

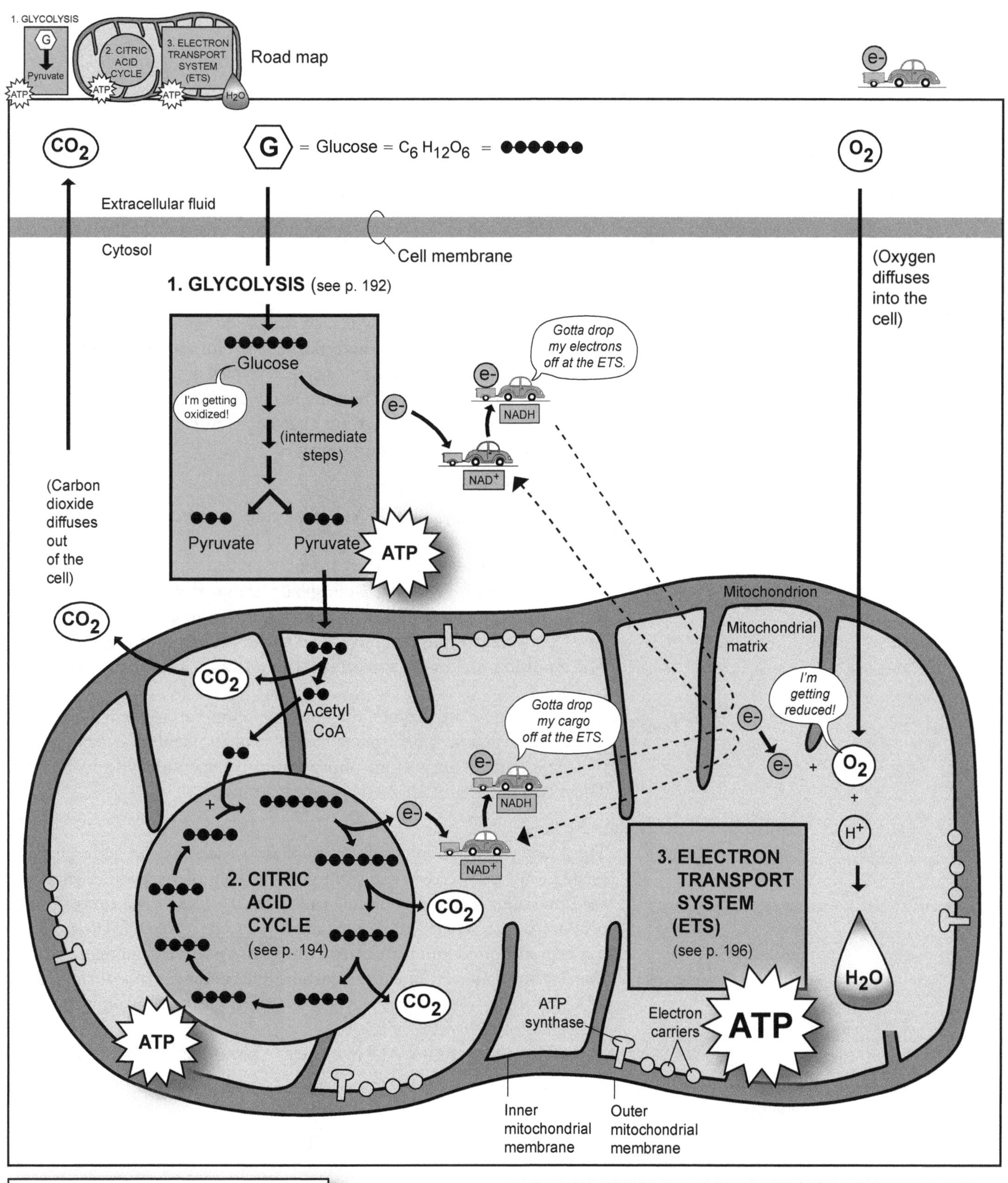

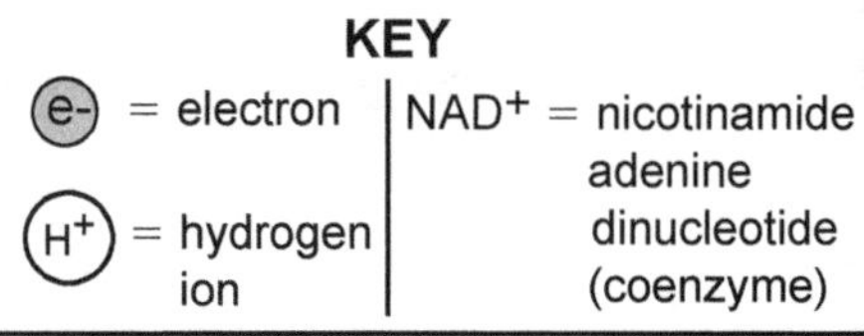

FUN FACT: The carbon atom in each molecule of carbon dioxide gas that you exhale was originally one of the carbon atoms in glucose.

Overview

Location within cell:	cytosol
Aerobic or anaerobic?	anaerobic (doesn't require O_2)
Initial reactant:	glucose
Final product(s):	• 2 pyruvates
Side products:	• 2 NADH (reduced coenzymes)
Net yield of energy:	• 2 ATP molecules (4 created − 2 used)

Description

The term **glycolysis** means the "splitting of glucose," which accurately describes the process. It begins with one glucose molecule containing 6 carbon atoms and ends with the formation of 2 new 3-carbon molecules called **pyruvate.** This process is catalyzed by various enzymes in the **cytosol** of cells and yields a net gain of 2 ATP molecules and 2 NADH. There are numerous intermediate steps.

Instead of presenting the various chemical reactions that occur in each of the intermediate steps, the process will be presented conceptually in two phases:

Phase 1: **Energy Input Phase** (splitting of glucose)

Key idea: **Free energy from the hydrolysis of 2 ATP molecules is invested to help split the original glucose molecule.**

- Glucose enters the cell and remains in the **cytosol.**
- **Step** (1) With the help of an enzyme, 2 phosphate groups from 2 ATP molecules are transferred onto glucose, thereby changing it into **fructose-1,6-bisphosphate.** These phosphate molecules have a net negative charge and repel each other. The repulsion makes the molecule more unstable in preparation for splitting it.
- **Step** (2) With the help of an enzyme, the covalent bond between carbons number 3 and 4 in fructose-1,6-bisphosphate is broken, yielding 2 new 3-carbon molecules: **dihydroxyacetone phosphate** and **glyceraldehyde-3-phosphate.** These two molecules are isomers of each other.

Phase 2: **Energy Capture Stage**

- **Step** (3) The 2 new 3–carbon molecules from phase 1 each have an additional phosphate group transferred onto them with the help of yet another enzyme. At the same time, a carrier molecule called **NAD^+** picks up 2 electrons and 1 proton from each of the molecules and is reduced to **NADH.** These electrons can be transported into the electron transport system for later use.
- **Step** (4) The 2 phosphates from each 3-carbon intermediate are transferred onto ADP to form a total of 4 new ATP molecules. The final result is that 2 new molecules of pyruvate are created.
- The net gain in ATP for glycolysis is 2 ATP (**4 created** in phase 2 minus **2 used** in phase 1).

Analogy

NAD^+ is an ion called **n**icotinamide **a**denine **d**inucleotide. It is derived from the vitamin niacin and functions as a carrier ion that transports electrons to the electron transport system (ETS). When NAD^+ is reduced (gains electrons), it becomes **NADH.** The NAD^+ ion is like a **car pulling a trailer with no cargo,** and NADH is like a **car filled with its cargo of 2 electrons (and 1 proton).**

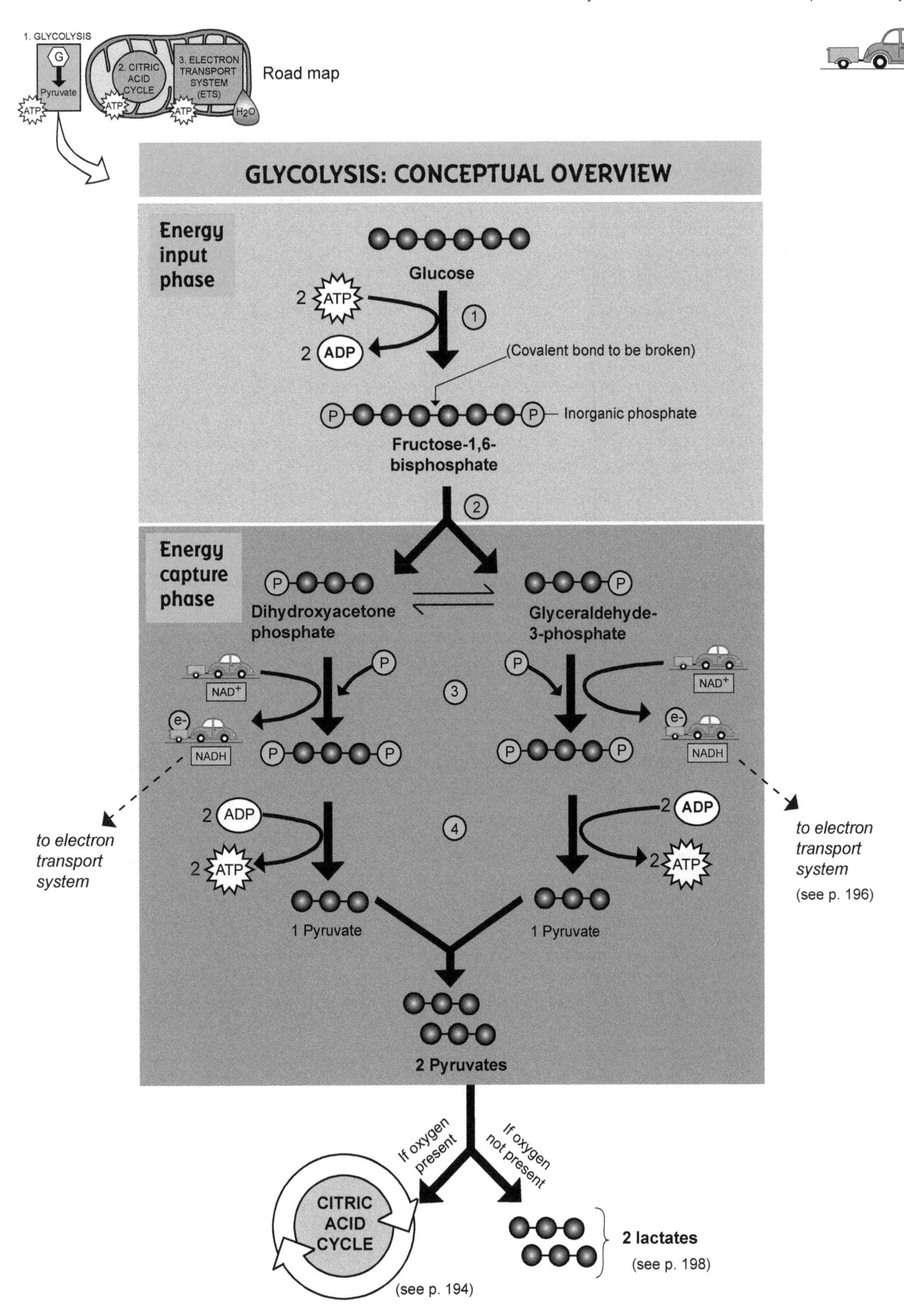
1. GLYCOLYSIS
G
Pyruvate
ATP
2. CITRIC ACID CYCLE
ATP
3. ELECTRON TRANSPORT SYSTEM (ETS)
ATP
H_2O
Road map
GLYCOLYSIS: CONCEPTUAL OVERVIEW
Energy input phase
Glucose
2 ATP
1
2 ADP
(Covalent bond to be broken)
P
P
Inorganic phosphate
Fructose-1,6-bisphosphate
2
Energy capture phase
P
Dihydroxyacetone phosphate
P
Glyceraldehyde-3-phosphate
P
NAD+
3
P
NAD+
e-
NADH
P P
P P
e-
NADH
to electron transport system
2 ADP
4
2 ADP
2 ATP
2 ATP
to electron transport system
(see p. 196)
1 Pyruvate
1 Pyruvate
2 Pyruvates
If oxygen present
If oxygen not present
CITRIC ACID CYCLE
(see p. 194)
2 lactates
(see p. 198)

Overview

Location within cell:	mitochondrion
Aerobic or anaerobic?	aerobic (indirectly)
Initial reactants/substrates:	oxaloacetate and acetyl CoA
Final product:	oxaloacetate (which condenses with acetyl CoA to form citrate)
Important side products: (for **both** pyruvates from glycolysis or two rounds of the citric acid cycle)	• 6 NADH, 2 $FADH_2$ • 6 CO_2 molecules (4 during the cycle; 2 in formation of acetyl CoA)
Net yield of energy:	• 2 ATP molecules

Three Key Events:

1. Conversion of pyruvate (from glycolysis) into **acetyl CoA**. This is what links glycolysis to the citric acid cycle.
2. Formation of CO_2 as a side product. The carbon in carbon dioxide can be traced back to the carbon in glucose. Think about that the next time you exhale some CO_2!
3. Formation of many **reduced coenzymes** (**NADH, $FADH_2$**). These reduced coenzymes link the citric acid cycle to the electron transport system.

Before Citric Acid: Cycle Begins

Pyruvate is first converted into **acetic acid** and then into **acetyl coenzyme A** (CoA).
—**CO_2** is released in the process.
—**CoA** is used to carry the 2-carbon remnant of pyruvate oxidation.
—**NAD^+** is reduced to **NADH** by picking up 2 electrons.

Citric Acid Cycle: Step by Step

NOTE: Each of the steps in the citric acid cycle requires the help of a different enzyme.

Step 1: **Oxaloacetate** condenses with **acetyl coenzyme A** (CoA) to form **citrate.**

Step 2: **Citrate** is converted into **isocitrate.**

Step 3: **Isocitrate** is converted into **alpha** **(α)-ketoglutarate.**
- NAD^+ is reduced to NADH
- CO_2 is released in the process

Step 4: **Alpha** (α)**-ketoglutarate** is converted into **succinyl CoA.**
- CoA enters this step to carry the succinyl group
- NAD^+ is reduced to NADH
- CO_2 is released in the process

Step 5: **Succinyl CoA** is converted into **succinate.**
- Phosphate group is transferred to GDP to create GTP
- Hydrolysis of GTP transfers phosphate group to ADP to form ATP

Step 6: **Succinate** is converted into **fumarate.**
- FAD is reduced to $FADH_2$

Step 7: **Fumarate** is converted into **malate.**

Step 8: **Malate** is converted into **oxaloacetate.**
- NAD^+ is reduced to NADH

Analogy:

NAD^+ and **FAD** are compounds that act like **shuttle buses.** NAD^+ is called **n**icotinamide **a**denine **d**inucleotide, and **FAD** is called **f**lavin **a**denine **d**inucleotide. FAD is derived from vitamin B_2 (riboflavin). Both are classified as carrier molecules, represented as different types of vehicles carrying a cargo. By picking up 2 electrons, **NAD^+** gets reduced to **NADH,** and **FAD** gets reduced to **$FADH_2$**. The **NAD^+** or **FAD** molecule is like a **car pulling a trailer with no cargo,** and **NADH** or **$FADH_2$** is like a **car filled with its cargo of electrons (and protons).**

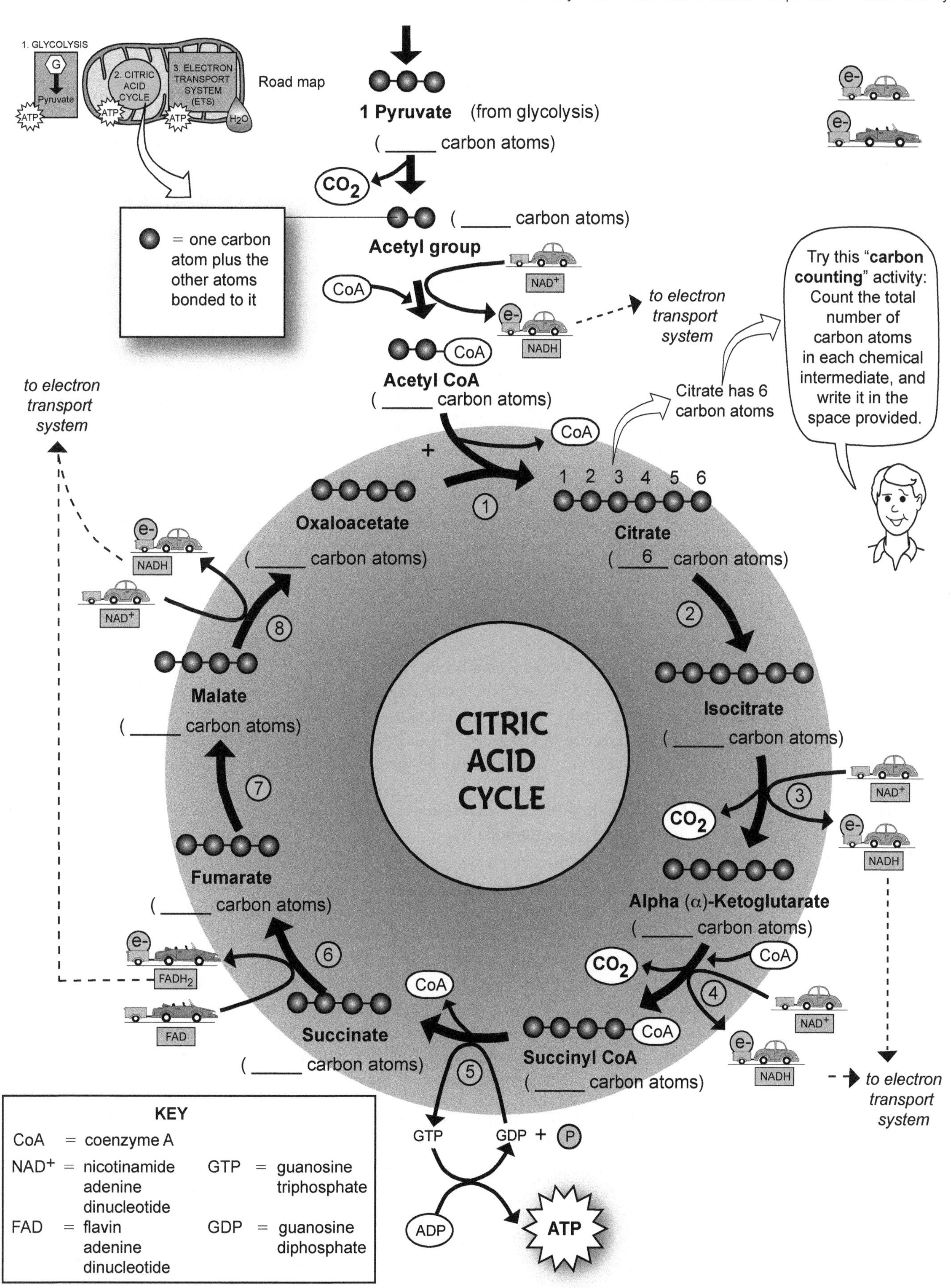
1. GLYCOLYSIS
G
Pyruvate
ATP
2. CITRIC ACID CYCLE
ATP
3. ELECTRON TRANSPORT SYSTEM (ETS)
ATP
H_2O
Road map
1 Pyruvate (from glycolysis)
(_____ carbon atoms)
CO_2
= one carbon atom plus the other atoms bonded to it
(_____ carbon atoms)
Acetyl group
NAD^+
CoA
e-
NADH
to electron transport system
CoA
Acetyl CoA
(_____ carbon atoms)
Try this "carbon counting" activity: Count the total number of carbon atoms in each chemical intermediate, and write it in the space provided.
Citrate has 6 carbon atoms
to electron transport system
CoA
+
1 2 3 4 5 6
1
Oxaloacetate
(_____ carbon atoms)
Citrate
(6 carbon atoms)
e-
NADH
NAD^+
2
8
Malate
(_____ carbon atoms)
Isocitrate
(_____ carbon atoms)
CITRIC ACID CYCLE
NAD^+
3
CO_2
e-
NADH
7
Fumarate
(_____ carbon atoms)
Alpha (α)-Ketoglutarate
(_____ carbon atoms)
e-
$FADH_2$
6
CoA
CO_2
CoA
4
NAD^+
FAD
Succinate
(_____ carbon atoms)
CoA
Succinyl CoA
(_____ carbon atoms)
e-
NADH
5
to electron transport system
KEY
CoA = coenzyme A
NAD^+ = nicotinamide adenine dinucleotide
FAD = flavin adenine dinucleotide
GTP = guanosine triphosphate
GDP = guanosine diphosphate
GTP
GDP + P
ADP
ATP

Overview

Location within cell:	mitochondrion
Aerobic or anaerobic?	aerobic (O_2 used **directly**)
Important side products:	• 10 NADH and 2 $FADH_2$ (total of 12 reduced coenzymes)
Final electron acceptor:	O_2 (oxygen)
Final product:	H_2O (water)
Net yield of energy:	**32 ATP** molecules (maximum)

Description

The **electron transport system** (ETS) is the last step in cellular respiration, yet it produces far more ATP—32—than either of the first two steps. ETS accomplishes three major things: (1) uses the energy from stored electrons (e–) to create a potential energy gradient in the form of hydrogen ions (protons), (2) converts this potential energy into kinetic energy to produce lots of ATP, and (3) regenerates NAD^+ and FAD to allow glycolysis and the citric acid cycle to continue.

The reduced coenzymes (**NADH** and **$FADH_2$**) directly link the citric acid cycle to the ETS. They are carrying electrons from the original glucose molecule that started the whole process. These electrons are delivered to membrane proteins and other electron carriers in the inner mitochondrial membrane where the ETS process occurs. The electrons are shuttled through a series of electron carriers, where they go from a high-energy state to a low-energy state. Oxygen serves as the final electron acceptor. As oxygen bonds with the electron, it also bonds to hydrogen ions (H^+) to create a water molecule in the matrix. Where do the hydrogen ions come from? They are in the solution of the matrix.

The electron carriers use the energy from the shuttled electrons to pump hydrogen ions from the mitochondrial matrix into the intermembrane space. The result is that a gradient of hydrogen ions is created that serves as a source of potential energy. The primary way for the hydrogen ions to move quickly out of the intermembrane space is to flow through a pore in a membrane protein called an **ATP synthase.** As the hydrogen ions diffuse down their gradient by flowing through this pore, the potential energy is converted into kinetic energy. This kinetic energy is used to covalently bond a phosphate group (P_i) to an ADP molecule to produce an ATP molecule. Note that both the phosphate group and the ADP molecule are present already in the solution of the matrix.

Two Key Concepts and Their Analogies:

- Concept 1: **The high-energy electron provides energy to the electron carriers to create a hydrogen ion gradient.**
- Analogy 1: The electron being transported between the proteins in the inner mitochondrial membrane is like a **hot baked potato.** The hot potato has **thermal energy** (heat) just like the electron has **energy.** The **electron carriers in the inner mitochondrial membrane** are like a **row of people** standing next to each other. Imagine that the first person tosses the baked potato to the next in line, who then tosses it to the next, and so on. The result is that each person's hands absorb a little bit of the heat from the potato. Similarly, the electron transfers some energy to the electron carriers that is then used to transport a hydrogen ion (or proton) from the mitochondrial matrix to the inner membrane space. The net result is that a hydrogen ion gradient is formed across the inner mitochondrial membrane.
- Concept 2: **The potential energy in the hydrogen ion gradient is converted into kinetic energy and used to create ATP.**
- Analogy 2: The **potential energy** in the hydrogen ion gradient is like **water behind a dam.** The **dam** is the **inner mitochondrial membrane,** and the **water behind the dam** is like the **gradient of hydrogen ions** in the intermembrane space. The **floodgate** in the dam is like the **ATP synthase,** and the kinetic energy from the flow of water through the dam is like the **flow of hydrogen ions** through the pore in the ATP synthase protein.

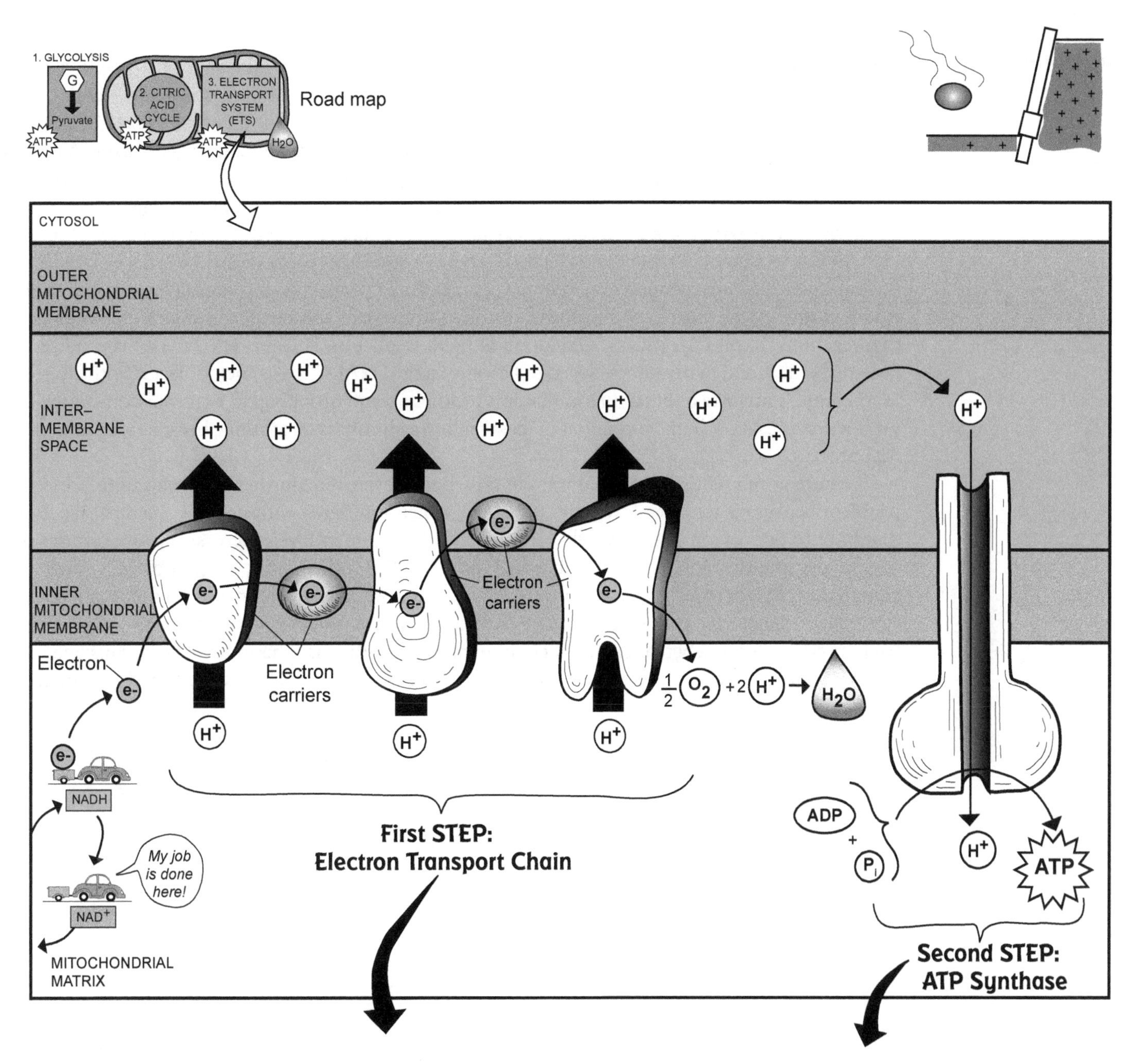

Electron Transport Chain: HOT POTATO ANALOGY

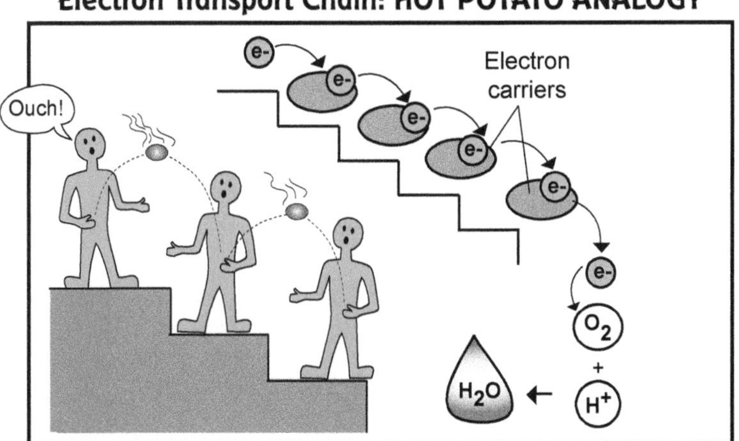

ATP Synthase: DAM ANALOGY

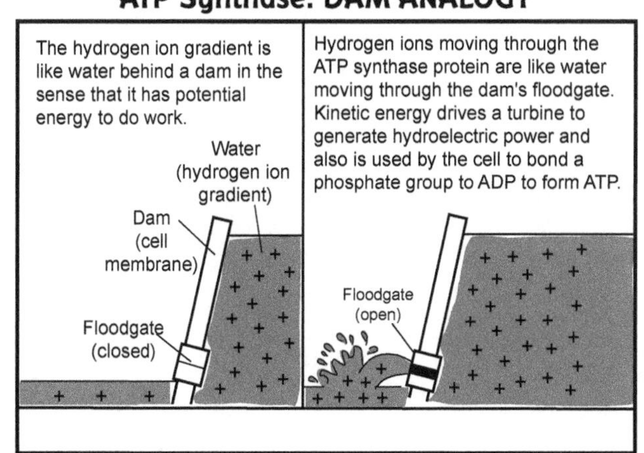

Description

Fermentation is the production of 2 **lactate** molecules from the breakdown of a single **glucose** molecule when no oxygen is present in the cell. This process occurs in the cytosol and yields 2 ATP. Recall that in glycolysis (see p. 192), glucose is broken down into 2 pyruvate molecules. When oxygen is present—**aerobic respiration**—the pyruvic acid is converted into acetyl coenzyme A, enters the citric acid cycle, and then enters the electron transport chain. In this case, the number of ATP produced is about 36. But when no oxygen is present, such as in skeletal muscle cells during vigorous exercise, the pyruvate is converted into lactate. This reduces ATP production to a total of only 2.

The conversion of pyruvate into lactate is a type of chemical reaction called a reduction reaction. Oxidation-reduction reactions are common in metabolic pathways and are always paired. These reactions involve the transfer of electrons from one substance to another. To distinguish between these two reactions, recall OIL RIG = Oxidation Is Loss, Reduction Is Gain. The NADH is oxidized to become NAD^+, and pyruvate is reduced to become lactate. NAD^+ is a carrier molecule with electrons as its cargo. It acts like a shuttle bus in that it is picking up and dropping off electrons constantly. The production of NAD^+ in the formation of lactate allows glycolysis to continue because it is needed for that process.

Lactate accumulates in skeletal muscle cells but doesn't remain there. This accumulation is a problem because it tends to lower the local pH, which interferes with normal chemical reactions that have to occur. Consequently, cells have to get rid of it. As its levels increase, lactate diffuses out of cells and into the blood, from which it is taken up by liver cells. Unlike most body cells, liver cells contain special enzymes that can convert lactate back into pyruvate in the presence of oxygen. After this occurs, liver cells can convert pyruvate back into glucose. The glucose then can diffuse back into the blood, where it can be used as an energy source for other muscle cells. This chemical cycling of lactate back into glucose between muscle cells and liver cells is called the **Cori cycle**.

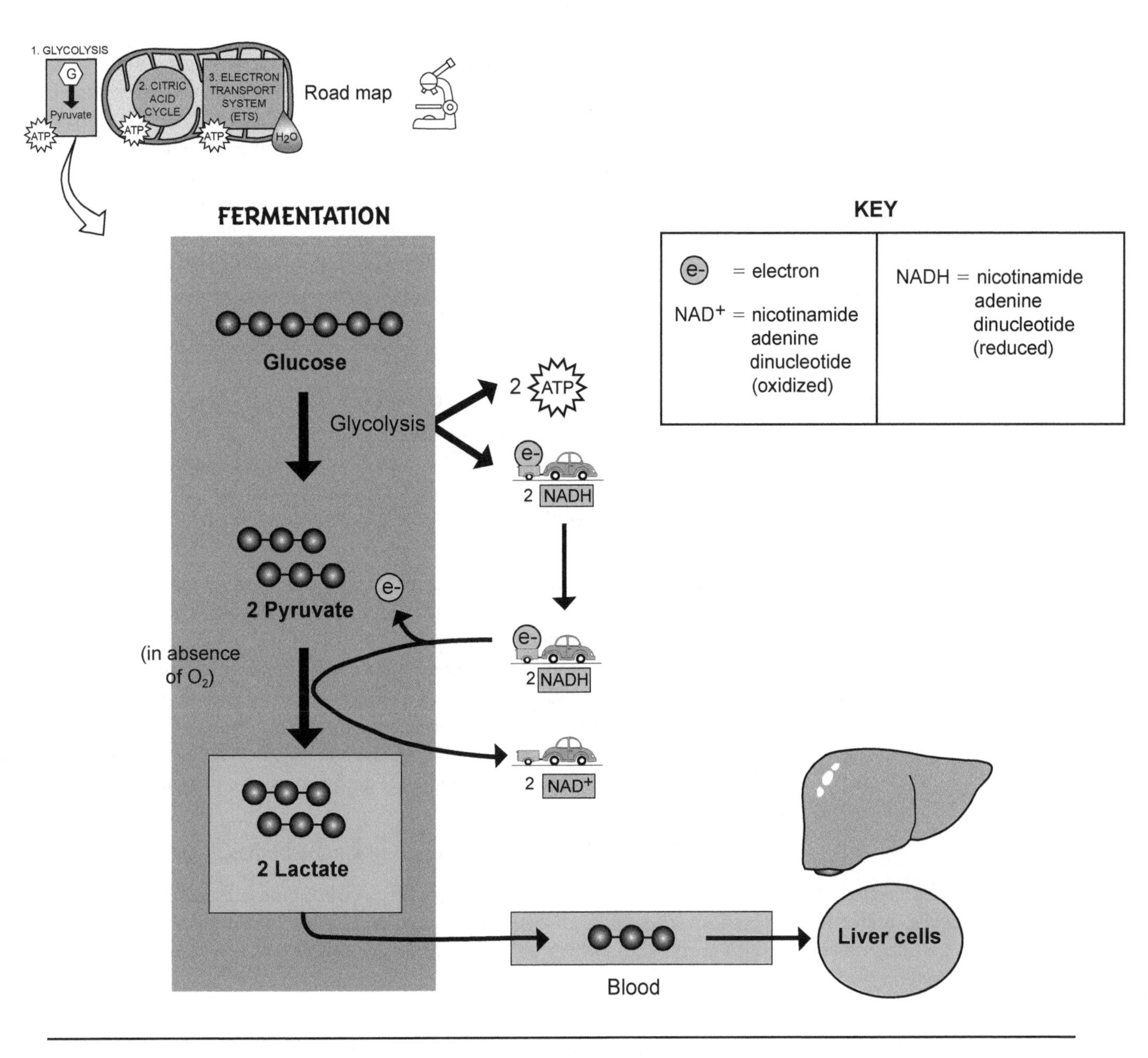
1. GLYCOLYSIS
G
Pyruvate
ATP
2. CITRIC ACID CYCLE
ATP
3. ELECTRON TRANSPORT SYSTEM (ETS)
ATP
H_2O
Road map
FERMENTATION
KEY
e- = electron
NAD^+ = nicotinamide adenine dinucleotide (oxidized)
NADH = nicotinamide adenine dinucleotide (reduced)
Glucose
Glycolysis
2 ATP
e-
2 NADH
2 Pyruvate
e-
(in absence of O_2)
e-
2 NADH
2 NAD^+
2 Lactate
Blood
Liver cells

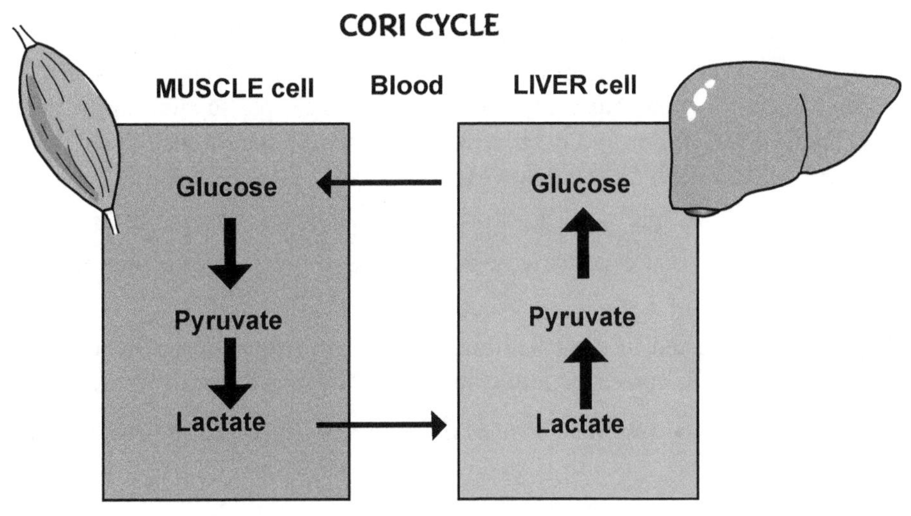
CORI CYCLE
MUSCLE cell
Blood
LIVER cell
Glucose
Pyruvate
Lactate
Glucose
Pyruvate
Lactate

Description

Many different types of lipids exist in the body, including **triglycerides**, **glycerophospholipids**, and **cholesterol.** Some of the stored lipids can be removed and either oxidized to be used as fuel or redeposited in other fat cells. Although many cells in your body prefer to burn a carbohydrate-based fuel such as glucose, they also can burn other fuels such as triglycerides or amino acids when necessary. Alternatively, new lipids also may be synthesized from other compounds. Lipids do not dissolve in water, which makes it difficult to transport them in the blood. To solve this problem, lipids are combined with proteins to form spheres called **lipoproteins** that are more easily transported through the blood and delivered to body cells. Triglycerides are the most common form of lipid in the body, so we will refer to this type when discussing how lipids are metabolized.

Lipolysis

(*lipo* = lipid; *–lysis* = breaking down)

Lipolysis is the process of breaking down lipids to produce ATP.

- **Triglycerides** are broken down via hydrolysis in the small intestine into their component parts: **glycerol** and **fatty acids.** Once inside cells, glycerol can be converted into a glycolysis intermediate, **glyceraldehyde-3-phosphate**, and then into **pyruvate.** Pyruvate is oxidized through the citric acid cycle, producing reduced coenzymes (NADH and $FADH_2$). As these coenzymes are oxidized by the electron transport system, ATP is produced.
- **Fatty acids** follow a different pathway through a process called **beta oxidation.** In this series of chemical reactions, which occur in mitochondria, fatty acids are broken down into 2-carbon fragments. Each of these fragments then is converted into **acetyl CoA**, which is oxidized through the citric acid cycle, producing reduced coenzymes. As these coenzymes are oxidized by the electron transport system, ATP is produced. In total, this accounts for a large production of ATP. For example, some long-chained fatty acids produce about four times more ATP compared to the complete catabolism of a single glucose molecule. In liver cells, fatty acids are catabolized to produce a group of substances called **ketone bodies.** This process is called **ketogenesis**, and it follows this pathway:

  Fatty acids ⟶ acetyl CoA ⟶ **ketone bodies**

These ketone bodies enter the bloodstream and are delivered to cells. Some cells, such as cardiac muscle cells, prefer ketone bodies rather than glucose to produce ATP. They accomplish this by converting ketone bodies back into acetyl CoA, which enters the citric acid cycle and then the electron transport system.

Lipogenesis

(*lipo* = lipid; *–genesis* = generation, birth)

The process of synthesizing lipids is called **lipogenesis.** This occurs in liver cells and adipose cells, which can convert carbohydrate, proteins, and fats into triglycerides. For example, drinking lots of soda pop on a regular basis gives the body more simple sugars than it needs, so they are converted into triglycerides and stored in fat cells. Excess simple sugars such as glucose lead to the production of glycerol or fatty acids by the following pathways:

Glucose ⟶ glyceraldehyde-3-phosphate ⟶ **glycerol** ...OR...
Glucose ⟶ pyruvate ⟶ acetyl CoA ⟶ **fatty acids**

The excess **glycerol** + **fatty acids** ⟶ **triglycerides.**
Proteins are digested in the small intestine to form their component amino acids. Certain types of amino acids can be converted into triglycerides. Here is the pathway for proteins:

Protein ⟶ amino acids (*certain types*) ⟶ acetyl coA ⟶ fatty acids ⟶ **triglycerides**

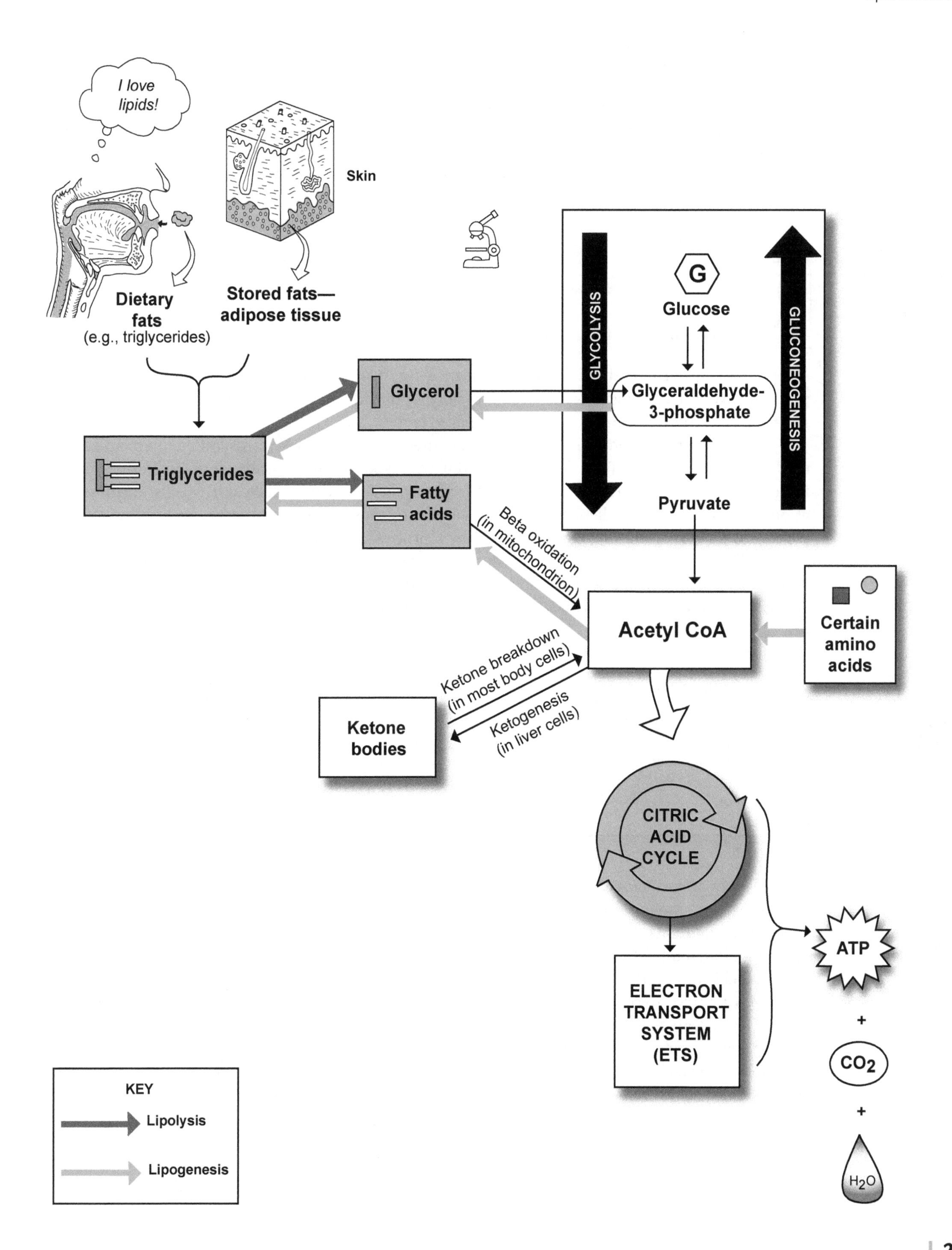
I love lipids!
Skin
Dietary fats
(e.g., triglycerides)
Stored fats—adipose tissue
Glycerol
Triglycerides
Fatty acids
GLYCOLYSIS
G
Glucose
Glyceraldehyde-3-phosphate
Pyruvate
GLUCONEOGENESIS
Beta oxidation (in mitochondrion)
Acetyl CoA
Certain amino acids
Ketone breakdown (in most body cells)
Ketogenesis (in liver cells)
Ketone bodies
CITRIC ACID CYCLE
ELECTRON TRANSPORT SYSTEM (ETS)
ATP
+
CO_2
+
H_2O
KEY
Lipolysis
Lipogenesis

Description

Your body is made primarily of water, fat, and **protein.** A healthy adult body may contain as much as 18% protein. If you eat a chicken sandwich, the protein in the chicken will be broken down gradually into **amino acids** in the digestive tract. The amino acids then are transported into the bloodstream and delivered to body cells. All body cells need amino acids for two purposes: (1) **protein synthesis**—to make new proteins within the cells, or (2) to use as a fuel source to provide energy (to make ATP). If you ingest more protein than your body needs daily, the excess amino acids are converted into either glucose ⟨G⟩ or triglycerides.

Amino Acids for Protein Synthesis

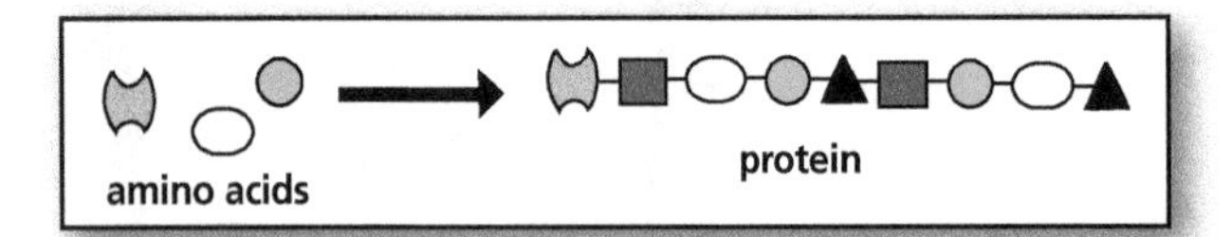

The process of protein synthesis occurs at a ribosome (see p. 62). During human growth and development, proteins have to be produced regularly and rapidly. In the adult, protein synthesis serves the purpose of replacing worn-out proteins and repairing damaged tissues.

Just as the different letters in the alphabet are used to form various words, different amino acids are used to make various proteins. There are 20 different types of amino acids. Your body can synthesize roughly half of them, so the others must come from our diet. Some cells, such as liver cells, are more active at protein synthesis than others. Sometimes a cell has to convert one amino acid into another to aid the process of protein synthesis. This is achieved by a process called **transamination**, in which an α-**amino group** (NH_3^+) from an amino acid is transferred to an α-**keto acid** with the help of the enzyme transaminase. The result is that the α-keto acid is converted into a new amino acid that can be used in protein synthesis. Examples of other tissues active in protein synthesis are cells in the brain, skeletal muscle, and heart.

Amino Acids Used as Fuel Source

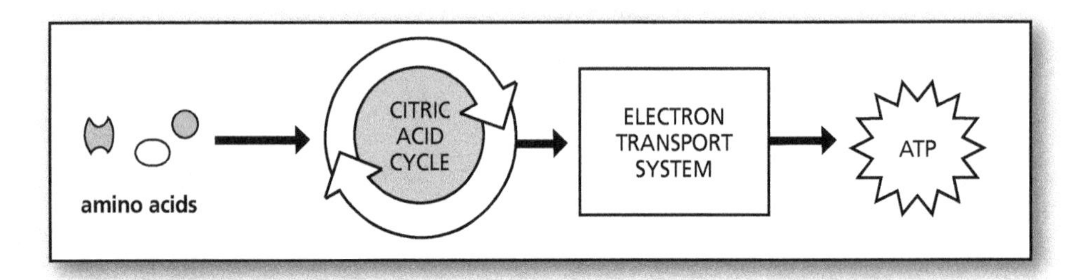

Alternatively, amino acids can be used by body cells to produce ATP by entering the citric acid cycle. For this to occur, amino acids must go through a **deamination** process whereby they lose their α-**amino group** (NH_3^+) and a hydrogen atom with the help of the enzyme deaminase. The result is that an **ammonium ion** (NH_4^+) is produced along with an α-**keto acid.** The keto acid can enter the citric acid cycle and electron transport system to produce ATP for the cell. Because ammonium ions (NH_4^+) in high quantities are toxic to cells, they are further converted into a waste product called urea, a harmless substance. This occurs in liver cells when ammonium ions and carbon dioxide enter a biochemical pathway called the urea cycle, in which urea is the primary side product. The liver is the major site for these deaminations because the liver cells have the proper enzymes to do the job. Urea normally travels through the blood and into the kidneys, where it is filtered out and excreted from the body as part of the urine.

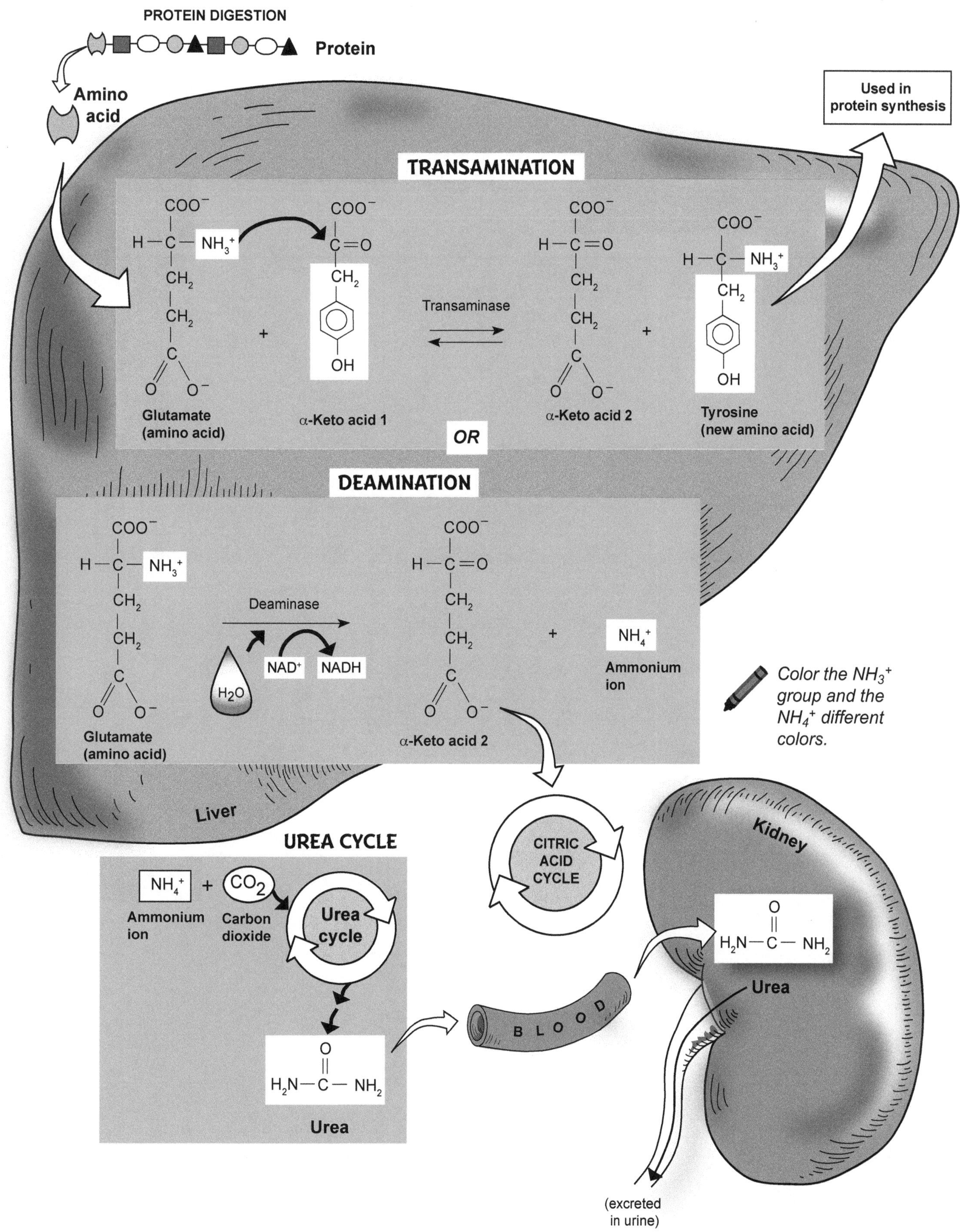

Color the NH_3^+ group and the NH_4^+ different colors.

Notes

LYMPHATIC SYSTEM

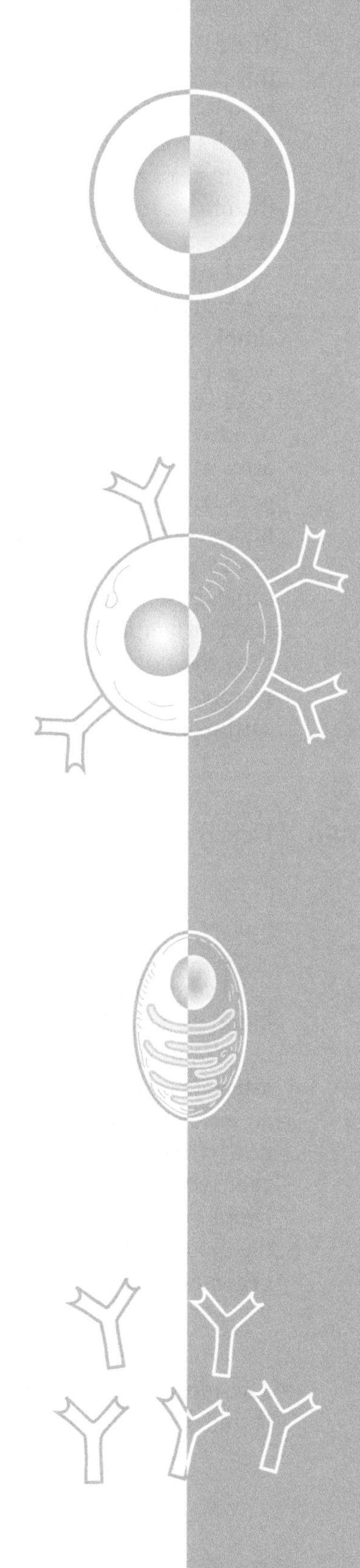

Description

The cardiovascular system has a close relationship with the lymphatic system. Like the veins running through the body, the lymphatic system consists of a network of thin-walled vessels called **lymphatic vessels.** Like veins, they contain one-way valves (semilunar valves) that assist in circulating the lymph, which is under very low pressure.

Instead of carrying blood, the lymphatic vessels carry **lymph**—tissue fluid originally filtered from the blood. In the illustration, the gray areas indicate this filtration process. The composition of lymph is similar to plasma—mostly water along with some solutes such as salts and small proteins.

Lymphatic vessels are connected to lymphatic capillaries and lymph nodes. **Lymphatic capillaries** are structurally similar to blood capillaries. Both are microscopic networks made of a single layer of simple squamous epithelium, but lymphatic capillaries contain flap-like structures that make them more permeable than blood capillaries.

Lymph nodes are pea-sized structures that act as tiny filters to clean the lymph. There are hundreds of nodes scattered within the body, but they tend to be most concentrated in the neck, armpits, and groin. Like an oil filter cleans the motor oil in your car's engine, lymph nodes filter pathogens such as bacteria and viruses out of your lymph before returning it to the plasma. These nodes contain macrophages that ingest and destroy pathogens. Lymphocytes are also present.

Lymph moves through lymph nodes in a one-way direction just as blood flows through blood capillaries. Lymph enters a node through multiple **afferent lymphatic vessels** and leaves through **efferent lymphatic vessels** (**note:** only a single vessel is shown in the illustration). Multiple afferents slow the flow of lymph through the node and allow time for an immune response. In short, the big picture is "unclean lymph in, clean lymph out." The cleansed lymph is finally returned to the cardiovascular system via the subclavian veins.

Flow of Lymph

Here is a summary of the flow of lymph through the lymphatic system:

Lymphatic capillaries ⟶ afferent lymphatic vessels ⟶ lymph nodes ⟶ efferent lymphatic vessels ⟶ subclavian veins

Lymph is really nothing more than filtered blood plasma. All blood capillaries constantly filter the blood as a result of the force of blood pressure (see p. 128). This fluid, called **interstitial fluid**, fills the interstitial spaces between body cells, bathing them in fluid. Filtration occurs at the higher-pressure arteriole end of the capillary. Although some of this fluid is reabsorbed at the venule end, there is still an excess amount. As interstitial fluid pressure builds in the interstitial spaces, it is shunted into the nearby lymphatic capillaries, which act as a drain for the excess fluid. Although this fluid has not changed its chemical composition in any way, once inside the lymphatic capillary, it now is called **lymph.** Think of this **fluid cycle** as recycling of our plasma. This helps maintain normal fluid levels in the blood and tissues. As shown in the illustration: **Plasma** is filtered to become **interstitial fluid**, which becomes **lymph**, which becomes plasma once again. The cycle is complete!

Study Tip

The terms "afferent" and "efferent" apply to multiple organ systems. Here is a way to distinguish them: Afferent as in Approach; Efferent as in Exit.

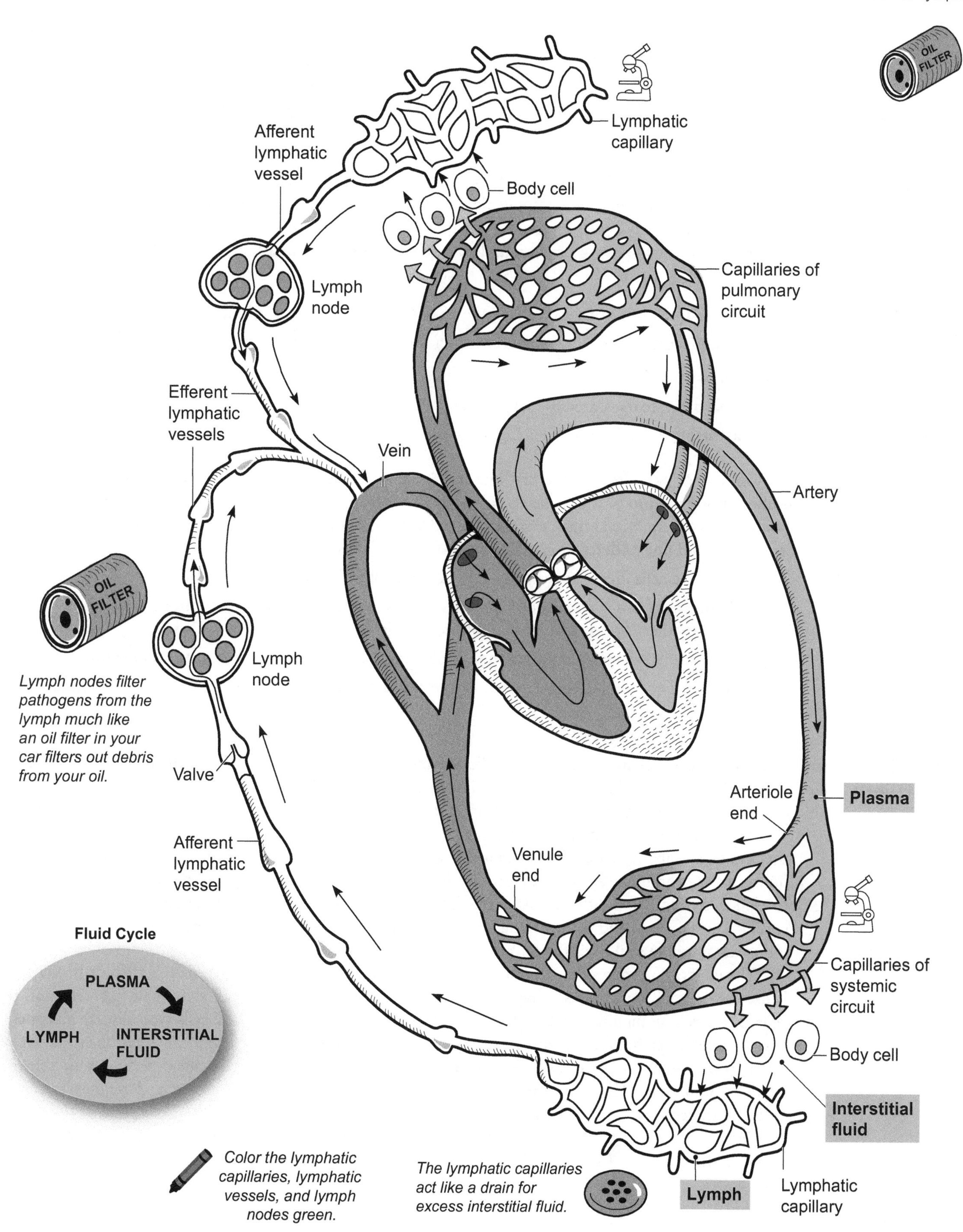
OIL FILTER
Lymphatic capillary
Afferent lymphatic vessel
Body cell
Lymph node
Capillaries of pulmonary circuit
Efferent lymphatic vessels
Vein
Artery
OIL FILTER
Lymph node
Lymph nodes filter pathogens from the lymph much like an oil filter in your car filters out debris from your oil.
Valve
Arteriole end
Plasma
Afferent lymphatic vessel
Venule end
Fluid Cycle
PLASMA
LYMPH
INTERSTITIAL FLUID
Capillaries of systemic circuit
Body cell
Interstitial fluid
Color the lymphatic capillaries, lymphatic vessels, and lymph nodes green.
The lymphatic capillaries act like a drain for excess interstitial fluid.
Lymph
Lymphatic capillary

Description

The **immune system** protects your body against foreign pathogens such as bacteria and viruses using several different lines of defense. The methods it uses can be grouped into two broad categories: (1) **Nonspecific resistance**, and (2) **Specific resistance.** Let's give an overview of each.

Nonspecific Resistance

Nonspecific resistance uses general defenses to prevent pathogenic invasion without targeting specific kinds of pathogens. Think of a pathogenic invasion as one army attacking another army inside a medieval castle. Your body is like the castle. The first line of defense is the wall around the castle. Similarly, your body has the following physical and chemical barriers as its **first line of defense:**

Physical barriers

- **Skin**—a thick layer of dead cells in the epidermis provides protection.
- **Mucous membranes**—the mucus covering these membrane surfaces traps microbes.

Chemical barriers

- **Lysozyme** in tears is an antibacterial agent.
- **Gastric juice** in the stomach is highly acidic (pH 2–3), which destroys bacteria.

Pathogens that pass through the first line of defense must deal with a second line of defense. Imagine archers along the top of the castle shooting arrows into the invading army or soldiers pouring vats of boiling oil on them. Both attack methods would kill clusters of enemy soldiers. Similarly, your body has its own **second line of defense,** and here are a few examples:

- **Phagocytic cells** ingest and destroy all microbes that pass into body tissues.
- **Inflammation** is a normal body response to tissue damage and other stimuli that brings more white blood cells to the site of pathogenic invasion.
- **Fever** inhibits bacterial growth and increases the rate of tissue repair during an infection.
- **Natural killer** (NK) **cells** destroy virus-infected cells and some tumor cells cells in body tissues.

Specific Resistance

The **third line of defense** deals with **specific resistance,** illustrated on the facing page. In modern warfare, these defenses would be like guided missiles that detect and destroy a specific target. In the medieval castle analogy, they might be specially trained soldiers who act as assassins to kill the enemy's general. In short, they have a specific mission with a very specific target. Your immune system also has "assassin" cells that attack specific microbes. Unlike in a war in which soldiers in different armies are wearing different uniforms, your immune system has a more difficult task in distinguishing its own tissues ("self") from foreign microbes ("nonself"). To make this important distinction, it relies on detecting foreign **antigens,** specific substances found on the surface of pathogens. Most are proteins that serve as the stimulus to induce an immune response. The term "antigen" is coined from *anti*body *gen*erator.

The illustration shows the immune response to an antigen. Once the antigen on a virus or bacterium is detected, a dual response is activated by two groups of specialized lymphocytes called **T cells** and **B cells.** These cells communicate with each other through chemical signaling. T cells typically are activated first. After activation, they can either directly destroy the microbes or use chemical secretions to destroy them. At the same time, T cells stimulate B cells to divide, forming other cells, called plasma cells, able to produce **antibodies.** These Y-shaped proteins circulate though the bloodstream and bind to specific antigens, thereby attacking microbes.

Details of the mechanisms T cells and B cells use to attack microbes are examined in subsequent modules. Together, the T cells and B cells provide specific resistance to specific antigens.

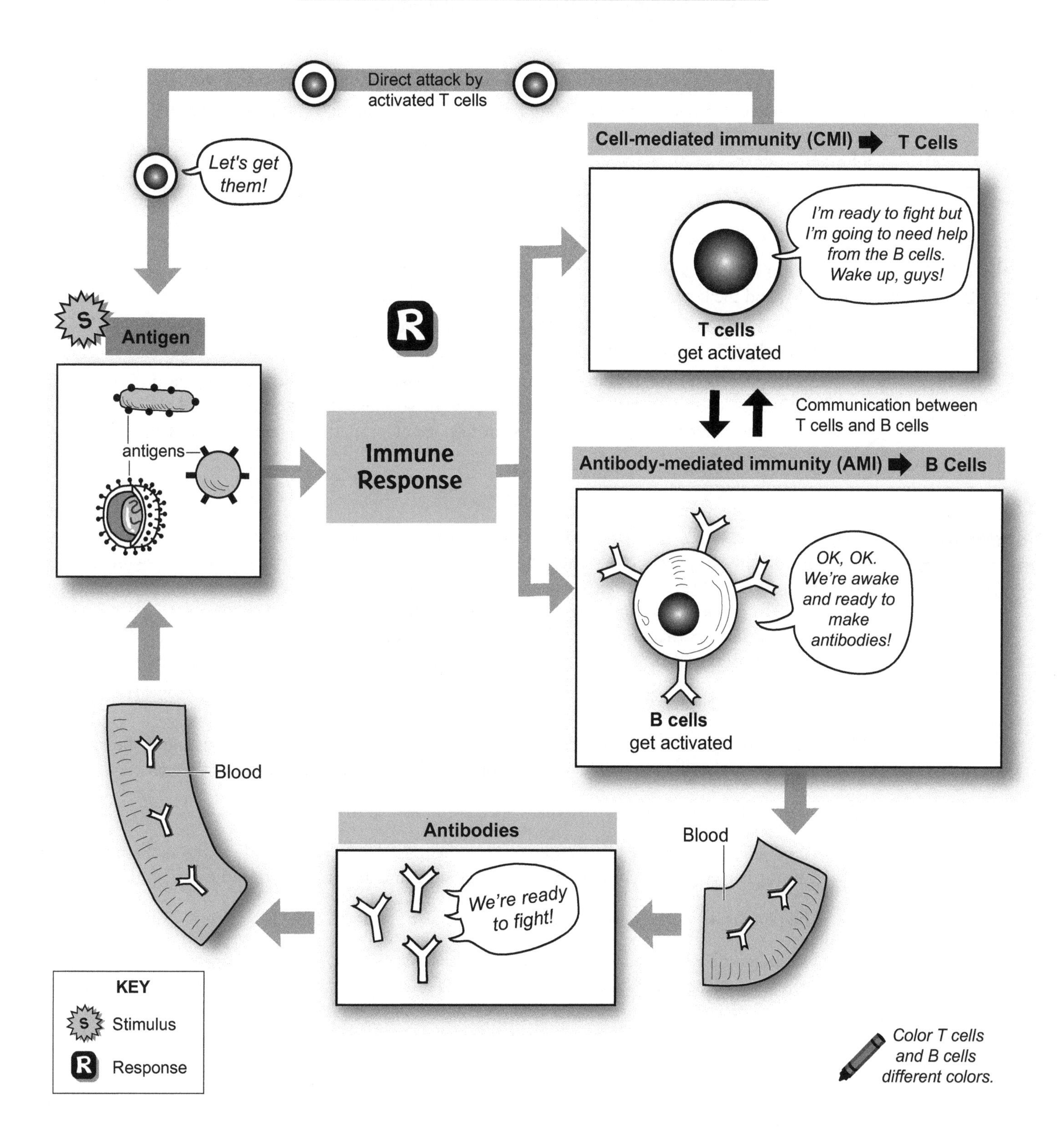
Immune System: Overview of Specific Resistance
Direct attack by activated T cells
Let's get them!
Cell-mediated immunity (CMI) ➡ T Cells
I'm ready to fight but I'm going to need help from the B cells. Wake up, guys!
T cells get activated
S Antigen
antigens
R
Immune Response
Communication between T cells and B cells
Antibody-mediated immunity (AMI) ➡ B Cells
OK, OK. We're awake and ready to make antibodies!
B cells get activated
Blood
Antibodies
We're ready to fight!
Blood
KEY
Stimulus
Response
Color T cells and B cells different colors.

Description

Cell-mediated immunity (CMI) involves the activation of **T cells** (T lymphocytes) by a specific antigen. In total, the body contains millions of different T cells—each able to respond to one specific antigen.

Development of T cells

T cells are a special type of lymphocyte. **Immature lymphocytes** are produced from stem cells in the **red bone marrow.** Some of these cells are processed within the **thymus gland**—hence, **T cell**—during embryological development, then released into the blood. These mature T cells are located in the blood, lymph, and lymphoid organs such as the lymph nodes and spleen.

Common T cells and Their Functions

The three major types of T cells are as follows:

- **Cytotoxic T cells** secrete **lymphotoxin** and **perforin.** The former trigger destruction of the pathogen's DNA, and the latter create holes in the pathogen's plasma membrane, resulting in a **lysed cell.**
- **Helper T cells** secrete **interleukin 2** (I-2), which stimulates cell division of T cells and B cells. This can be thought of as recruiting more soldiers for the fight.
- **Memory T cells** remain dormant after the initial exposure to an antigen. If the same antigen presents itself again—even years later—the memory cells are stimulated to convert themselves into cytotoxic T cells and enter the fight. Memory is an important property of immunity and is the basis for vaccinations.

Phagocytosis and Antigen Presentation

Phagocytes ("eater cells") use the process of **phagocytosis** ("cell eating") to ingest foreign pathogens. An example is a macrophage ("big eater"), derived from the largest white blood cells—monocytes. Macrophages leave the bloodstream and enter body tissues to patrol for pathogens. As some phagocytic cells engage in phagocytosis, they present antigenic fragments on their plasma membrane. This is a very important mechanism that stimulates the activation of more T cells. Consequently, phagocytes are an important part of the CMI. A summary of the phagocytosis process shown in the illustration is:

1. **Microbe** attaches to **phagocyte.**
2. Phagocyte's **plasma membrane** forms armlike extensions that surround and engulf the microbe. The encapsulated microbe pinches off from the plasma membrane to form a **vesicle.**
3. The vesicle merges with a **lysosome,** which contains **digestive enzymes.**
4. The digestive enzymes begin to break down the microbe. The phagocyte extracts the nutrients it can use, leaving the **indigestible material** and **antigenic fragments** within the vesicle.
5. The phagocyte makes **protein markers,** and they enter the vesicle.
6. The indigestible material is removed by exocytosis (see p. 76). The antigenic fragments bind to the protein marker and are displayed on the plasma membrane surface. This serves to activate T cells.

Development of T Cells

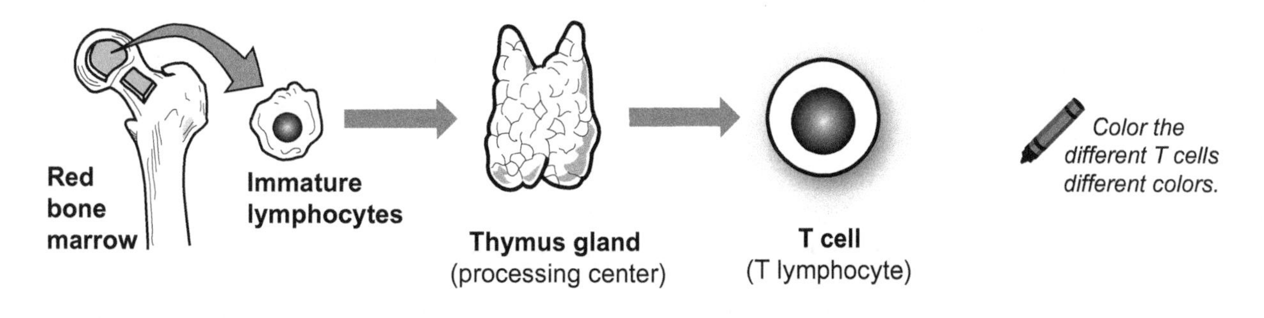

Common Types of T Cells and Their Functions

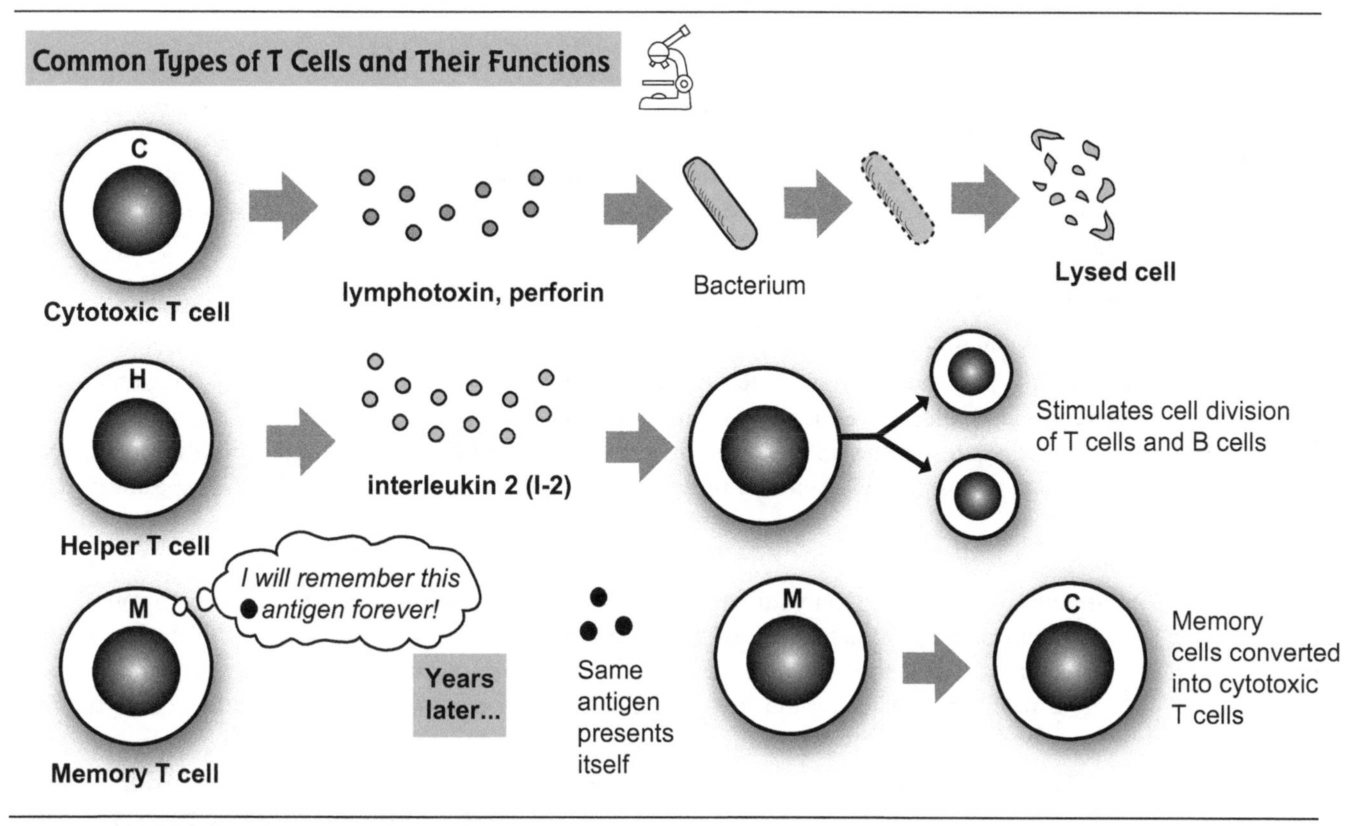

Phagocytosis and Antigen Presentation

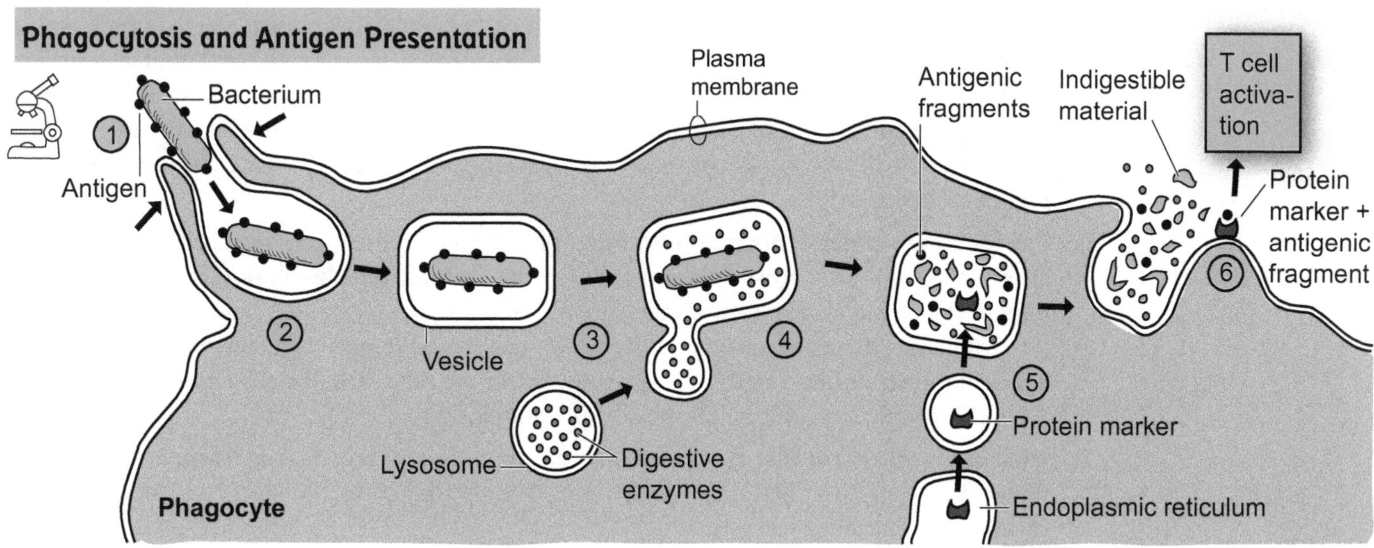

Description

Antibody-mediated (*humoral*) **immunity** (**AMI**) involves the activation of **B cells** (B lymphocytes) by a specific antigen. This triggers the B cells to transform into **plasma cells** able to secrete special proteins called **antibodies.** The antibodies are transported through the blood and the lymph to the pathogenic invasion site. In total, the body contains millions of different B cells—each able to respond to one specific **antigen** and produce one specific antibody. Amazing!

Development of B Cells

B cells are a special type of lymphocyte. **Immature B cells** are produced from stem cells in the **red bone marrow**. These immature cells are later processed within the red bone marrow—hence, B cell—during embryological development to become mature B cells, then are released into the blood. The mature B cells are located in the blood, lymph, and lymphoid organs such as the lymph nodes and spleen.

Antibody Production

B cells can be stimulated to divide, forming two types of numerous cells: (1) **plasma cells,** which secrete **antibodies,** and (2) **B memory cells,** which exist in the body for many years, ensuring a quick response to the same antigen. Antibodies (immunoglobulin, or Ig) are Y-shaped proteins subdivided into five classes: IgG, IgM, IgA, IgE, IgD. These are listed in order from the *most* common to the *least* common.

Study Tip: Mnemonic: *Get Me Another Excellent Donut!*

The basic structure on an antibody consists of four polypeptide chains—two **heavy chains** and two **light chains.** Both heavy chains and both light chains are identical to the other, and each contains a **constant region** and a **variable region.** The constant region forms the "trunk" of the molecule, and the variable region forms the **antigen-binding site** on the antibody. Note that each antibody has two of these antigen-binding sites. Think of these like claws on a lobster used to "grab" its specific antigen.

How Do Antibodies Work?

Antibodies work through many different mechanisms, of which the following are major ones:

1. **Neutralizing antigen:** The **antibody** can bind to an **antigen**, forming an **antigen–antibody complex.** This forms a shield around the antigen, preventing its normal function. In this way, a toxin from a bacterium may be neutralized or a viral antigen may not be able to bind to a body cell, thereby preventing infection.
2. **Activating complement:** "Complement" refers to a group of plasma proteins made by the liver that normally are inactive in the blood. An **antigen-antibody complex** triggers a cascade reaction that activates these proteins to induce beneficial responses. For example, some of these activated proteins can cluster together to form a pore or channel that inserts into a microbial plasma membrane. This results in a **lysed cell.** Other responses include **chemotaxis** and **inflammation.** Both of these mechanisms serve to increase the number of white blood cells at the site of invasion. (*Please note this mechanism is much more complex than what is presented here*).
3. **Precipitating antigens:** Numerous antibodies can bind to the same free antigens in solution to cross-link them. This cross-linked mass then precipitates out of solution, making it easier for phagocytic cells to ingest it by phagocytosis (see p. 76).

   Similarly, microbes (such as bacteria) can be clumped together by a process called *agglutination* (not illustrated). The antigens within the cell walls of the bacteria are cross-linked. As with precipitation, this is followed by phagocytosis.
4. **Facilitating phagocytosis:** An **antigen-antibody complex** acts like a warning sign to signal phagocytic cells to attack. In fact, the complex also binds to the surface of **macrophages** to further facilitate phagocytosis.

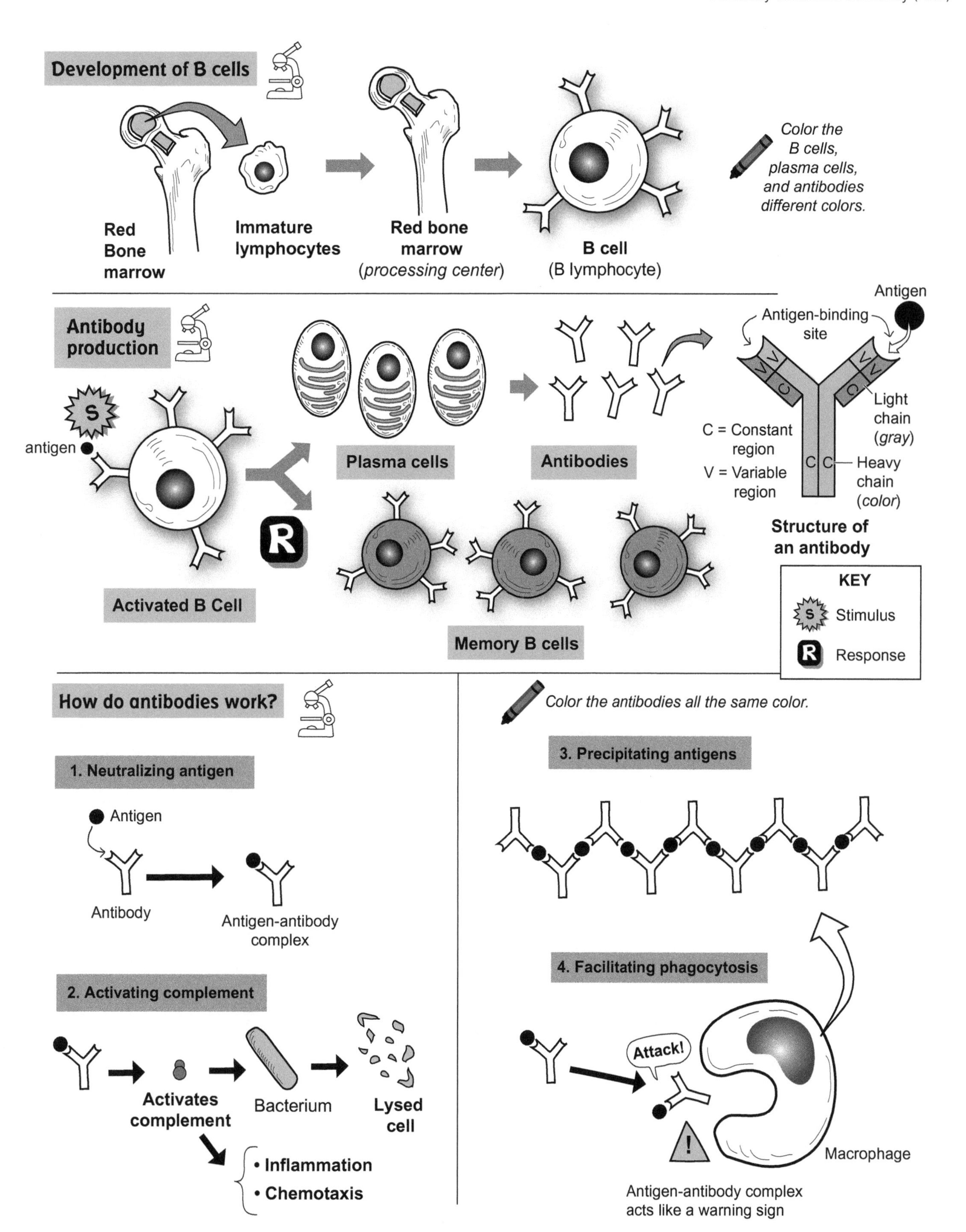
Development of B cells
Red Bone marrow
Immature lymphocytes
Red bone marrow (processing center)
B cell (B lymphocyte)
Color the B cells, plasma cells, and antibodies different colors.
Antibody production
S
antigen
R
Activated B Cell
Plasma cells
Antibodies
Memory B cells
Antigen
Antigen-binding site
V
C
Light chain (gray)
C = Constant region
V = Variable region
Heavy chain (color)
Structure of an antibody
KEY
Stimulus
Response
How do antibodies work?
Color the antibodies all the same color.
1. Neutralizing antigen
Antigen
Antibody
Antigen-antibody complex
2. Activating complement
Activates complement
Bacterium
Lysed cell
• Inflammation
• Chemotaxis
3. Precipitating antigens
4. Facilitating phagocytosis
Attack!
Macrophage
Antigen-antibody complex acts like a warning sign

Notes

URINARY SYSTEM

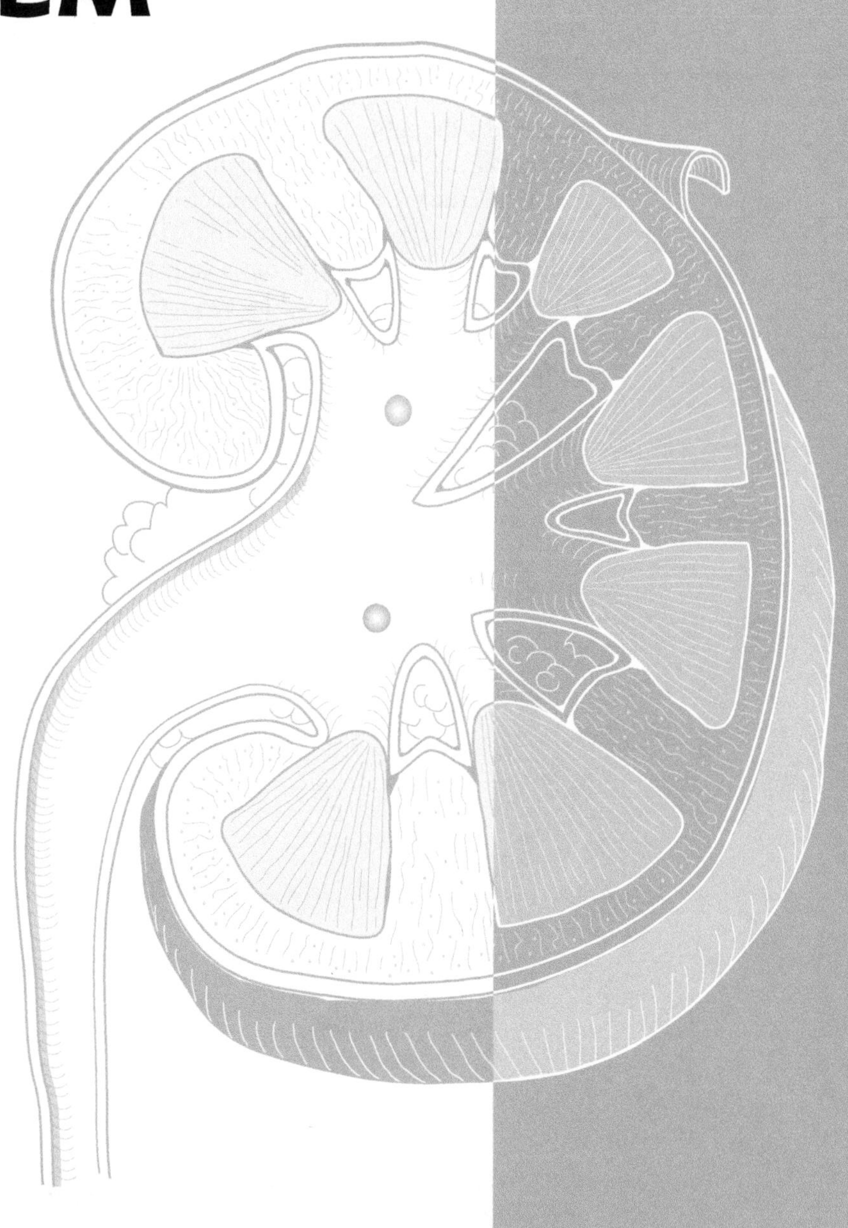

Description

The urinary system is composed of the **kidneys, ureters, urinary bladder**, and **urethra.** The kidneys are located on either side of the upper lumbar region of the vertebral column. They lie between the dorsal body wall and the **parietal peritoneum** in a retroperitoneal position. They are tightly covered by the parietal peritoneum on their anterior surface and further protected by layers of fatty tissue. Major abdominal organs such as the small and large intestines are all positioned anterior to the kidneys.

The **renal arteries** branch off the **abdominal aorta** and supply oxygenated blood to the kidneys, and the **renal veins** drain blood from the kidneys and empty into the **inferior vena cava.** The blood that enters the kidney via the renal artery contains various waste products that must be removed from the blood. The functional units within the kidney are microscopic structures called **nephrons.** They filter waste products and nutrients out of the blood and transport them through a tubular system. The fluid that enters the tubular system is called the filtrate. The filtrate is processed in a manner that transports nutrients back to the bloodstream and retains waste products in the tubular system.

When the processing is completed, the liquid inside the tube is called **urine.** In reality, urine is nothing more than processed blood plasma. The urine then moves out of the nephrons and into the pelvis of the kidneys. Slender, muscular tubes called **ureters** transport urine from the kidney to the urinary bladder. The muscular **urinary bladder** can expand to hold as much as 600–800 mL of urine. Finally, urine is removed from the body by the process of **micturition** (*urination*), which uses forceful muscular contractions in the wall of the urinary bladder to transport urine into the urethra and out of the body.

Key to Illustration

Major Organs/Structures

1. Kidney
2. Ureter
3. Urinary bladder

Blood Vessels (B)

B1. Inferior vena cava
B2. Abdominal aorta
B3. Renal a.
B4. Renal v.
B5. Common iliac a.
B6. Internal iliac a.
B7. External iliac a.
B8. Gonadal v. (*testicular v. in male; ovarian v. in female*)
B9. Gonadal a. (*testicular a. in male; ovarian a. in female*)

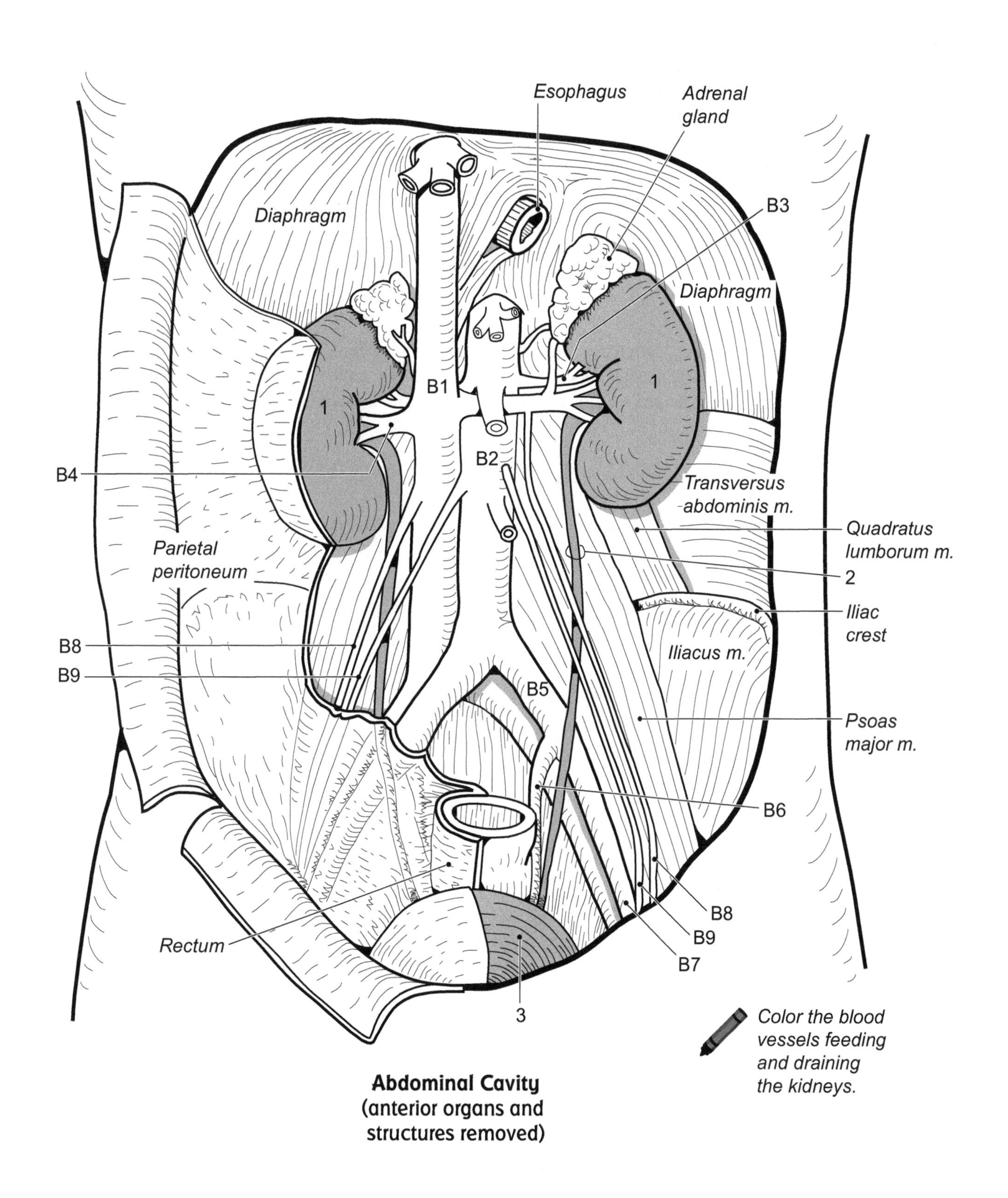

Abdominal Cavity
(anterior organs and structures removed)

Description

The kidneys are covered by a fibrous tissue called the **renal capsule.** The **renal hilus,** a cleft or indentation on the medial surface of the kidneys, is a handy landmark for locating the **ureters,** blood vessels (*renal a., renal v.),* and nerves that serve this organ. Each kidney is divided into three regional areas: cortex, medulla, and pelvis. The **cortex** is the outermost region, like the bark around a tree, and appears granular. The darker colored **medulla** is the middle region, containing numerous funnel-shaped structural units called **renal pyramids,** each with a **papilla** at its tapered end.

Separating the pyramids are projections of cortical tissue called **renal columns.** The central region is the **pelvis**—a pouch that narrows and extends directly into the **ureter.**

Near the medulla, the pelvis branches off into structures called **calyces** (sing., *calyx*). Each cup-shaped **minor calyx** surrounds each papillus of a renal pyramid to receive urine from it. When several of these minor calyces join, they form a larger chamber called a **major calyx.** As urine is collected in the pelvis, it moves down the **ureter** and into the urinary bladder.

Analogy

The **calyces** (sing., *calyx*) are functionally like a **plumbing system** of smaller-diameter pipes leading to larger ones. Each **smaller pipe** is a **minor calyx,** and each **larger pipe** is a **major calyx.** The flow of water through the pipes follows a path similar to the flow of urine through the calyces.

Key to Illustration

Regional areas

1. Cortex
2. Medulla
3. Pelvis

Structures

4. Renal papillus
5. Renal pyramids
6. Renal capsule
7. Renal column
8. Minor calyx
9. Major calyx
10. Adipose tissue in renal sinus
11. Renal hilus
12. Ureter

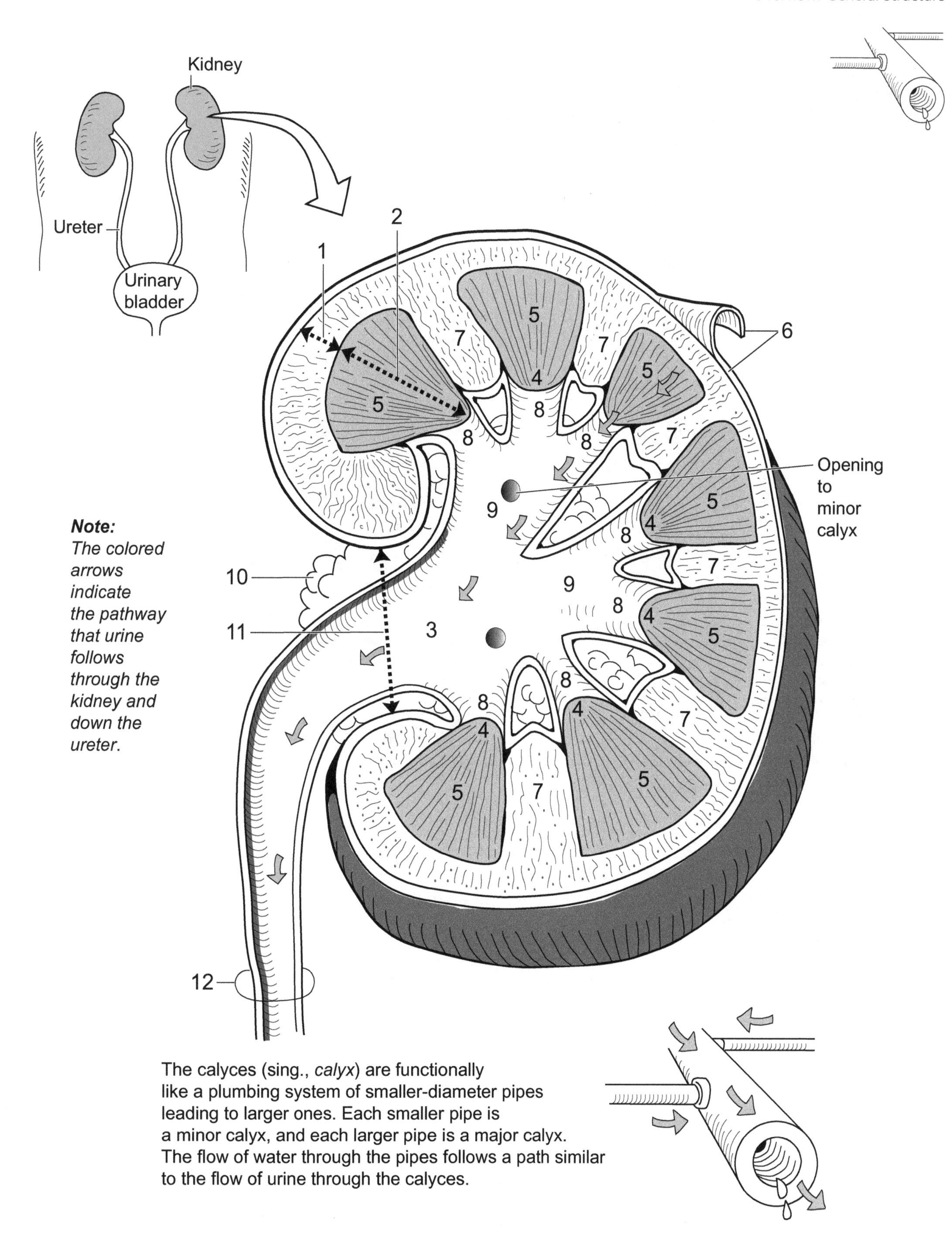

Note:
The colored arrows indicate the pathway that urine follows through the kidney and down the ureter.

The calyces (sing., *calyx*) are functionally like a plumbing system of smaller-diameter pipes leading to larger ones. Each smaller pipe is a minor calyx, and each larger pipe is a major calyx. The flow of water through the pipes follows a path similar to the flow of urine through the calyces.

Blood Flow through the Kidneys

The renal artery feeds oxygenated blood to the kidneys, and the **renal vein** drains deoxygenated blood from them.

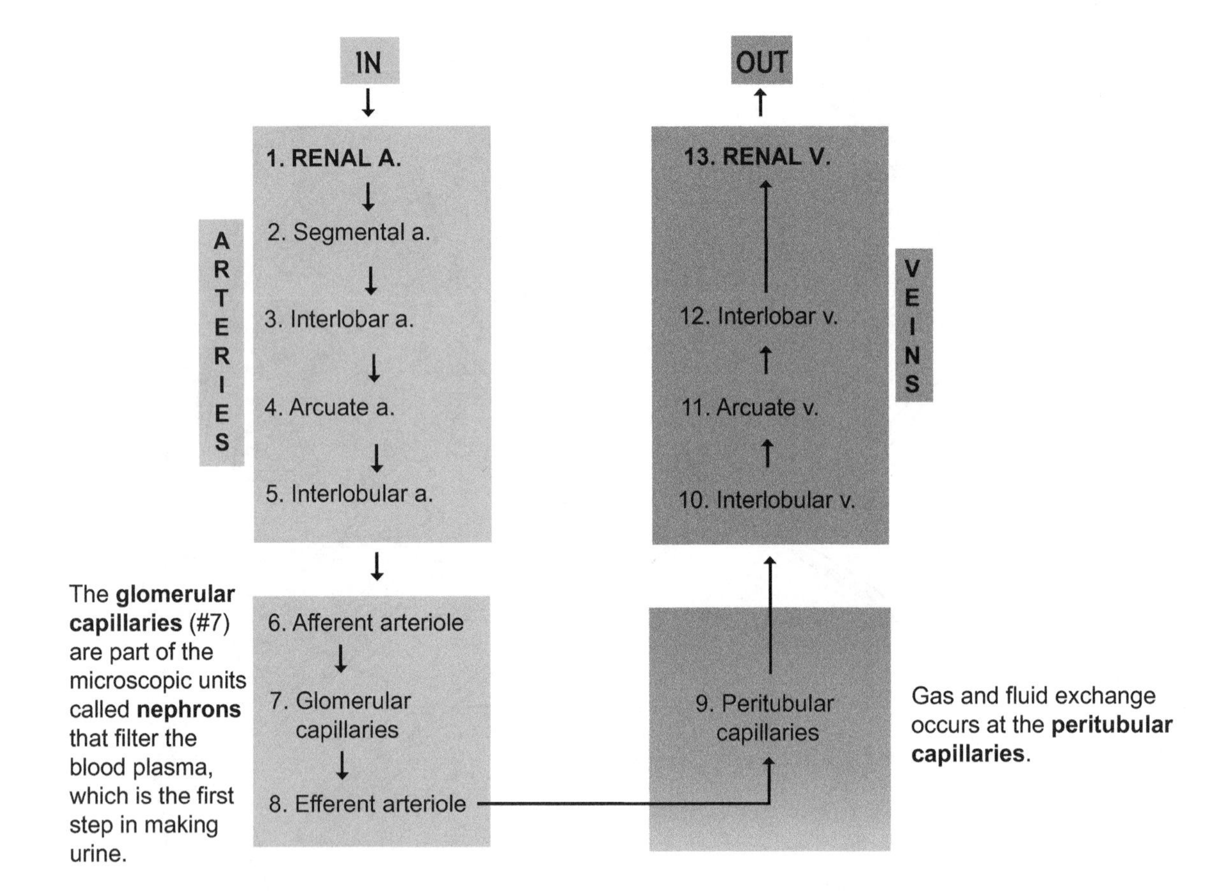

The **glomerular capillaries** (#7) are part of the microscopic units called **nephrons** that filter the blood plasma, which is the first step in making urine.

Gas and fluid exchange occurs at the **peritubular capillaries**.

Blood Flow through the Kidneys

Trace the pathway of blood flowing into and out of the kidneys.

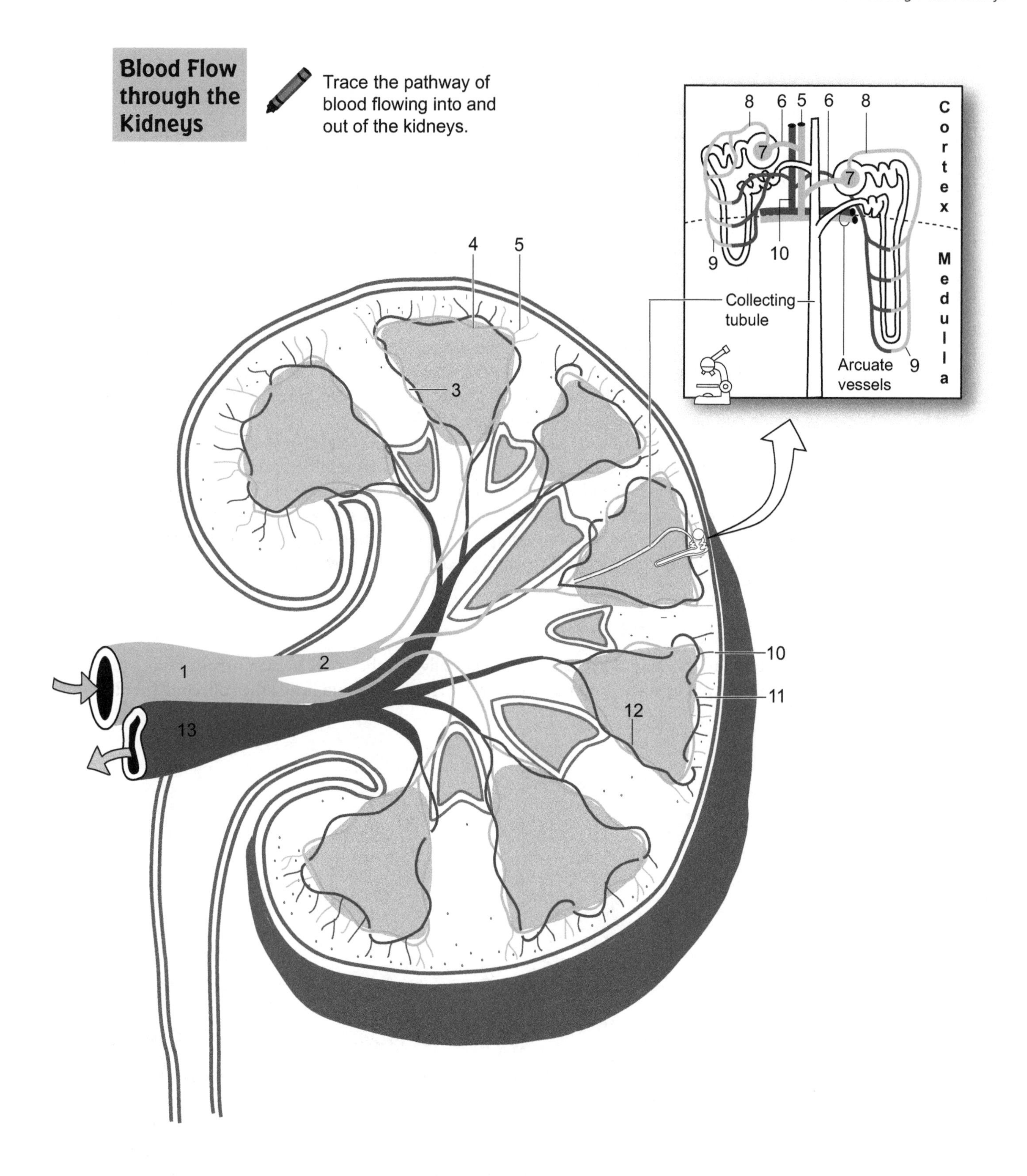

Description

The **nephron** is the microscopic functional unit of the kidney. Each kidney has more than a million of these units. Each nephron is divided into six distinct parts: *glomerulus, glomerular capsule, proximal convoluted tubule, nephron loop, distal convoluted tubule,* and *collecting tubule.* As blood enters the kidney via the **renal artery**, it branches into smaller vessels until it becomes the interlobular artery in the **renal cortex.** The afferent arteriole branches off the interlobular artery and connects to a special coiled ball of capillaries called the **glomerulus.** Completely surrounding the glomerulus is a cuplike structure called the **glomerular capsule.**

Connected to the capsule is a long tubular system. The glomerulus is highly permeable because of pores in its walls called **fenestrae.** Wrapped around the glomerular capillaries are cells called **podocytes**—modified simple squamous epithelial cells. The blood in the glomerulus is under high pressure so the plasma is passively filtered into the glomerular capsule. The fluid inside the glomerular capsule is referred to as *filtrate.* This filtrate flows from the glomerular capsule into the first part of the tubular system called the **proximal convoluted** *(coiled)* **tubule.** Next it flows into the **nephron loop**, which consists of a descending limb where the filtrate flows downward and an ascending limb where the filtrate moves upward. Then the filtrate enters the **distal convoluted tubule** and finally moves into the **collecting tubule.**

The general function of the nephron is to collect the filtrate (*filtered blood plasma*) in a separate tubular system and process it. The filtrate in the glomerular capsule is a solution that contains both nutrients and waste products mixed together. The nephron processes the filtrate by returning the nutrients to the bloodstream and retaining the waste products in the tubular system. By the time this processing job is completed, the resulting fluid is referred to as **urine**. The urine then follows this pathway: *minor calyx*, *major calyx*, *ureter*, *urinary bladder*, *urethra*. Then it exits the body.

Analogy

The tubular system in the nephron is **structurally** like the sewer lines in private homes that connect to a larger public sewer drain under a street. The **sewer drain** is like the **collecting tubule.**

Key to Illustration

Blood Vessels (B)

B1. Interlobular artery

B2. Afferent arteriole

B3. Efferent arteriole

Nephron Structures

1. Glomerular (*Bowman's*) capsule
2. Glomerulus
3. Renal corpuscle
4. Proximal convoluted tubule
5. Descending limb
6. Ascending limb
7. Nephron loop (*loop of Henle*)
8. Distal convoluted tubule
9. Collecting tubule (*duct*)

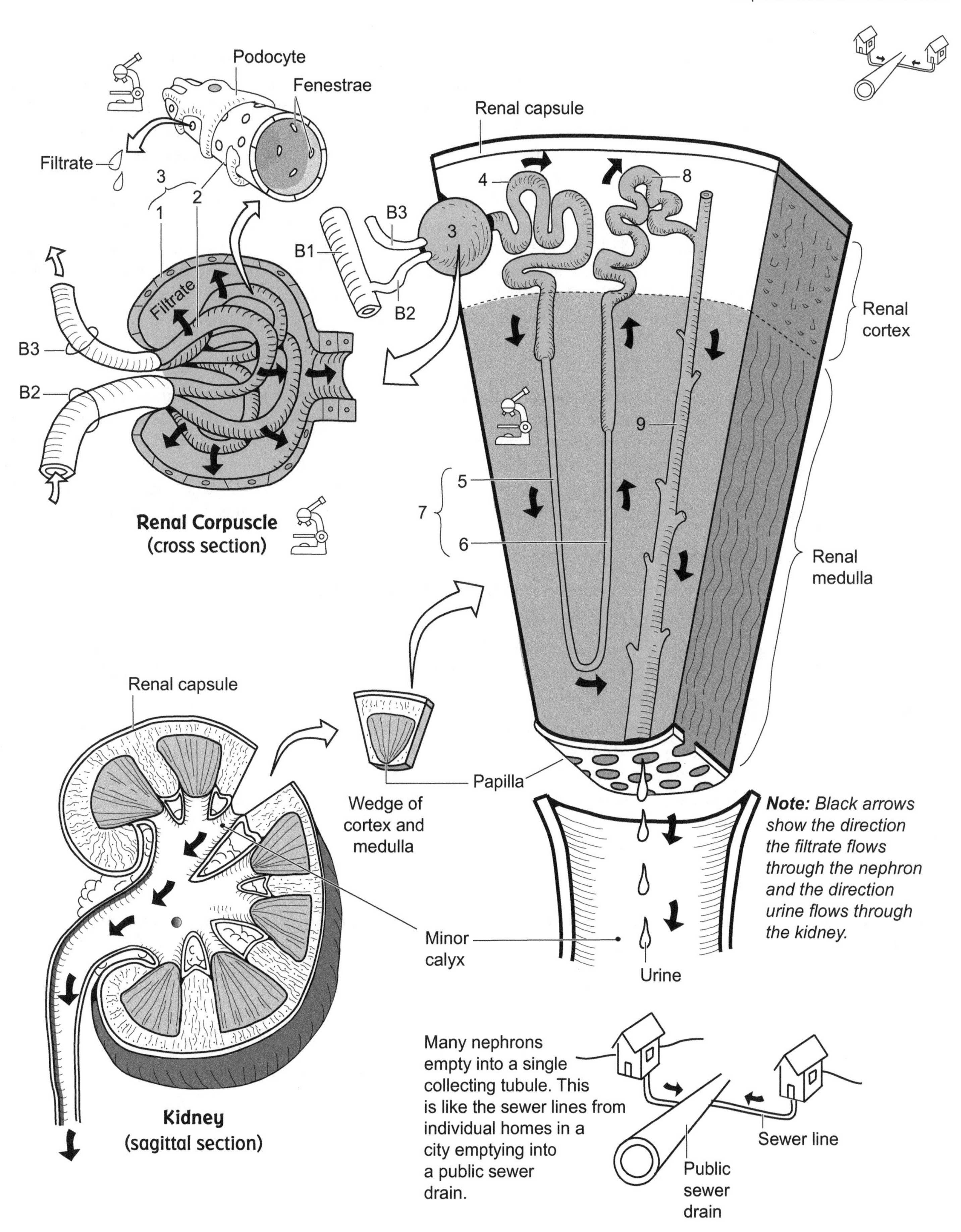
Podocyte
Fenestrae
Filtrate
3
2
1
Filtrate
B3
B2
B1
B3
B2
3
Renal capsule
4
8
Renal cortex
9
5
7
6
Renal medulla
Renal Corpuscle
(cross section)
Renal capsule
Papilla
Wedge of cortex and medulla
Minor calyx
Urine
Note: Black arrows show the direction the filtrate flows through the nephron and the direction urine flows through the kidney.
Kidney
(sagittal section)
Many nephrons empty into a single collecting tubule. This is like the sewer lines from individual homes in a city emptying into a public sewer drain.
Sewer line
Public sewer drain

Description

The major function of the **kidneys** is to filter and process the blood plasma to produce **urine.** This occurs within the kidney's millions of microscopic filter units called **nephrons.** The first step in this process is **filtration** of the blood (see pp. 132–133), which occurs in the nephron's coiled ball of capillaries called the **glomerulus.** During this process, the blood pressure in the glomerulus forces a liquid mixture of nutrients and waste products to move into the cuplike structure surrounding the glomerulus, called the glomerular capsule.

This mixture is referred to as the filtrate. It contains substances such as water, amino acids, glucose, sodium, chloride, urea, uric acid, and many others. Working together, both kidneys filter about 48 gallons (182 liters) of plasma per day. This is equivalent to 91 two-liter bottles of soda pop! In total, about 99% of this large volume with all its nutrients will be reabsorbed back into the blood, and the remaining 1% will be released from the body as urine. To better understand the process of glomerular filtration, we will examine the forces and counterforces involved.

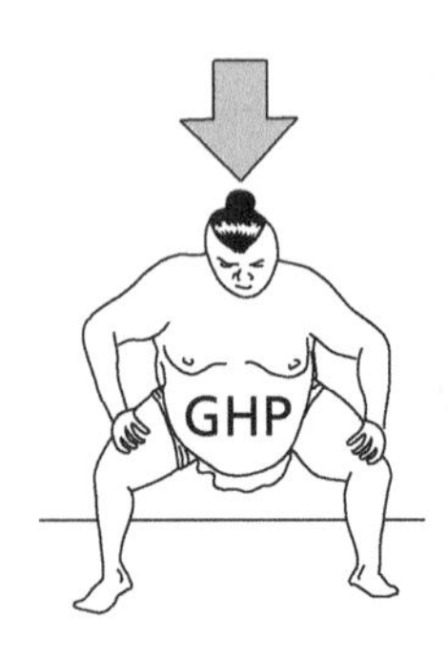

Force

Glomerular hydrostatic pressure (GHP): This is the force of the blood pressure against the walls of the glomerulus. Filtration is completely dependent on pressure. If the blood pressure drops too low, filtration stops. The blood pressure in the glomerulus averages about 50 mm Hg and cannot be measured directly, so it must be determined mathematically. As a comparison, recall that the typical pressure on the ateriole end of a capillary bed is 35 mm Hg (see pp. 132–133). Think of the GHP like the force exerted by a sumo wrestler as he thrusts himself forward against his opponent(s).

Counterforces

Two forces work against the GHP: the CHP and the BCOP. Each is like a boy sumo wrestler pushing against the adult sumo wrestler.

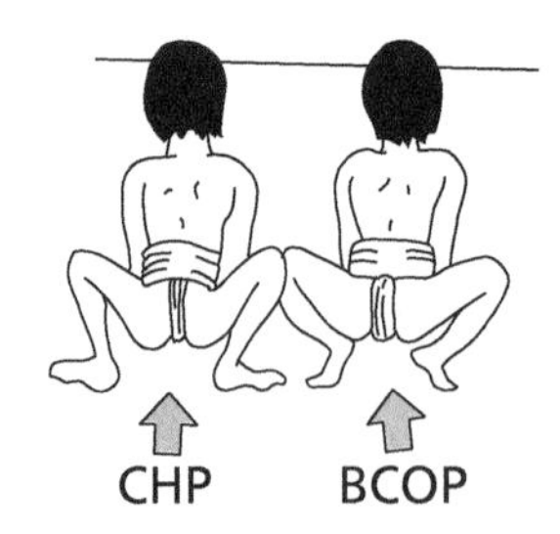

1. **Capsular hydrostatic pressure (CHP):** Like the force of water in a water balloon, this is the force of the filtrate fluid against the wall of the glomerular capsule. It averages about 15 mm Hg and works against filtration.
2. **Blood colloidal osmotic pressure (BCOP):** This is the only force that is not a hydrostatic pressure of fluid against a wall. Instead, it is an osmotic pressure, which refers to the tendency of water to move toward the solution with the higher concentration of nonpenetrating solutes. Because plasma proteins like albumin remain in the blood in the glomerulus, the water in the filtrate tends to move back into the glomerulus. This osmotic pressure is a force working against filtration, and it averages about 25 mm Hg.

Net Filtration Pressure Formula

Net filtration pressure = GHP − (CHP + BCOP) or "forces = sum of all counterforces"

$= 50 - (15 + 25)$
$= 50 - (40)$
$=$ **10 mm Hg**

In other words, a net positive pressure is maintained so the process of filtration can continue. If this were not the case, renal failure could result.

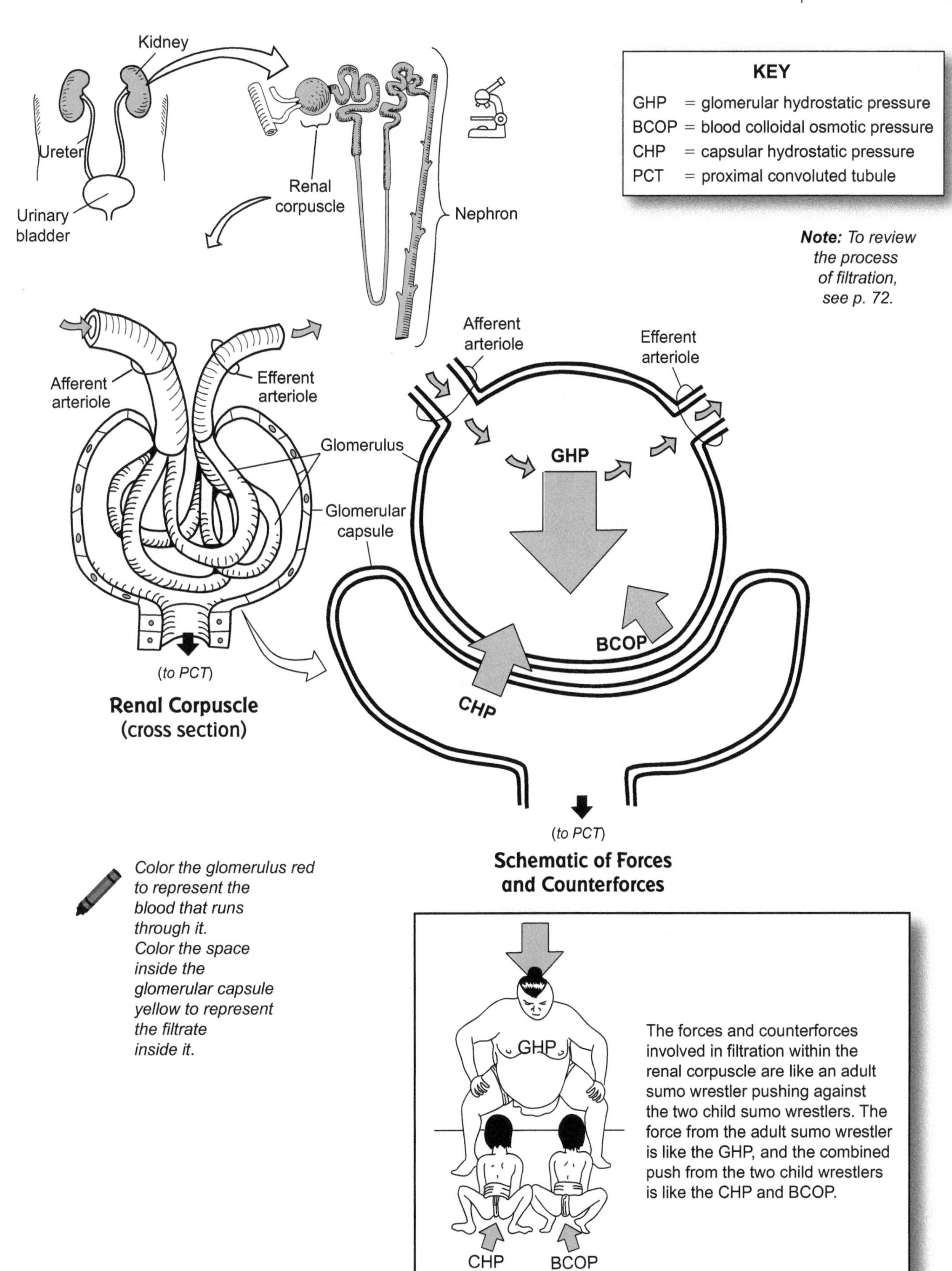

Note: *To review the process of filtration,* *see p. 72.*

Renal Corpuscle (cross section)

Schematic of Forces and Counterforces

Color the glomerulus red to represent the blood that runs through it.
Color the space inside the glomerular capsule yellow to represent the filtrate inside it.

The forces and counterforces involved in filtration within the renal corpuscle are like an adult sumo wrestler pushing against the two child sumo wrestlers. The force from the adult sumo wrestler is like the GHP, and the combined push from the two child wrestlers is like the CHP and BCOP.

Description

The **juxtaglomerular apparatus** is vital to the **glomerular filtration rate** (GFR) and consists of the following two structures:

① **Juxtaglomerular (JG) cells:** a group of modified smooth muscle cells around the afferent arteriole. They secrete the enzyme **renin** (see pp. 242–243).

② **Macula densa:** a group of modified epithelial cells within the wall of the distal convoluted tubule adjacent to the afferent arteriole. They act as *chemoreceptors* that sense changes in solute concentration of the filtrate.

Each glomerulus in every nephron filters the blood to produce filtrate. The GFR is the volume of filtrate produced every minute by all the nephrons in both kidneys and is equal to about 125 mL filtrate/min. The blood pressure (BP) in each glomerulus must be kept relatively constant to ensure that filtration continues. If the GFR rises or falls too far from this normal level, **nephron** function is impaired.

As shown in the illustration, a general scheme is used to either increase or decrease the GFR as needed. This involves controlling the blood flow through the glomerulus.

Analogy

Let's compare the glomerulus to a loop of garden hose that has some holes for filtration. If we expand the incoming end of the hose, this allows more water in and increases the water pressure. If we squeeze the outgoing end, this causes a backup of fluid in the loop and increases water pressure. Similarly, the glomerulus uses vasodilation and vasoconstriction of its afferent and efferent arterioles to change the BP in the glomerulus. By doing so, this directly changes the GFR.

Mechanisms

To control the GFR, three major mechanisms are used: (1) autoregulation, (2) neural regulation, and (3) hormonal regulation. Let's summarize each, in turn.

- *Autoregulation* is the dominant control mechanism at rest. It is subdivided into two mechanisms: (1) *smooth muscle mechanism*, and (2) *tubular mechanism*. The first uses stretching of the wall of the afferent arteriole as a stimulus to induce vasoconstriction of itself. This leads to a decrease in blood flow to the glomerulus and decreases GFR back to normal levels. The second mechanism involves the **macula densa.** For example, when BP increases, the GFR also increases. This results in an increase in filtrate flow in the renal tubules. Because this also increases the level of solute (such as Na^+ and Cl^-) in the filtrate, the macula densa detects this change and responds by constricting the afferent arteriole. This reduces blood flow to the glomerulus and decreases the GFR back to normal levels.
- *Neural regulation* occurs during periods of physical activity and stressful situations. The idea here is you don't need to make urine if you are running or fighting for your life. Because it is the sympathetic division of the autonomic nervous system (ANS) that helps us respond to these situations, it also controls the GFR. Neurons are linked to the afferent arteriole through a reflex pathway. In an effort to deliver more blood to vital organs, less blood should be going to the kidneys. To accomplish this, nerve impulses stimulate the smooth muscle around the afferent arteriole to constrict, which decreases the blood flow to the glomerulus and decreases GFR.
- *Hormonal regulation* involves the hormone angiotensin II and is triggered by a decrease in blood volume or BP. This is part of the normal renin-angiotensin aldosterone system (see p. 242). Juxtaglomerular cells release the enzyme renin into the bloodstream, which leads to the production of angiotensin II, a powerful vasoconstrictor. By causing peripheral blood vessels to constrict, it increases systemic BP, which increases GFR.

Juxtaglomerular apparatus

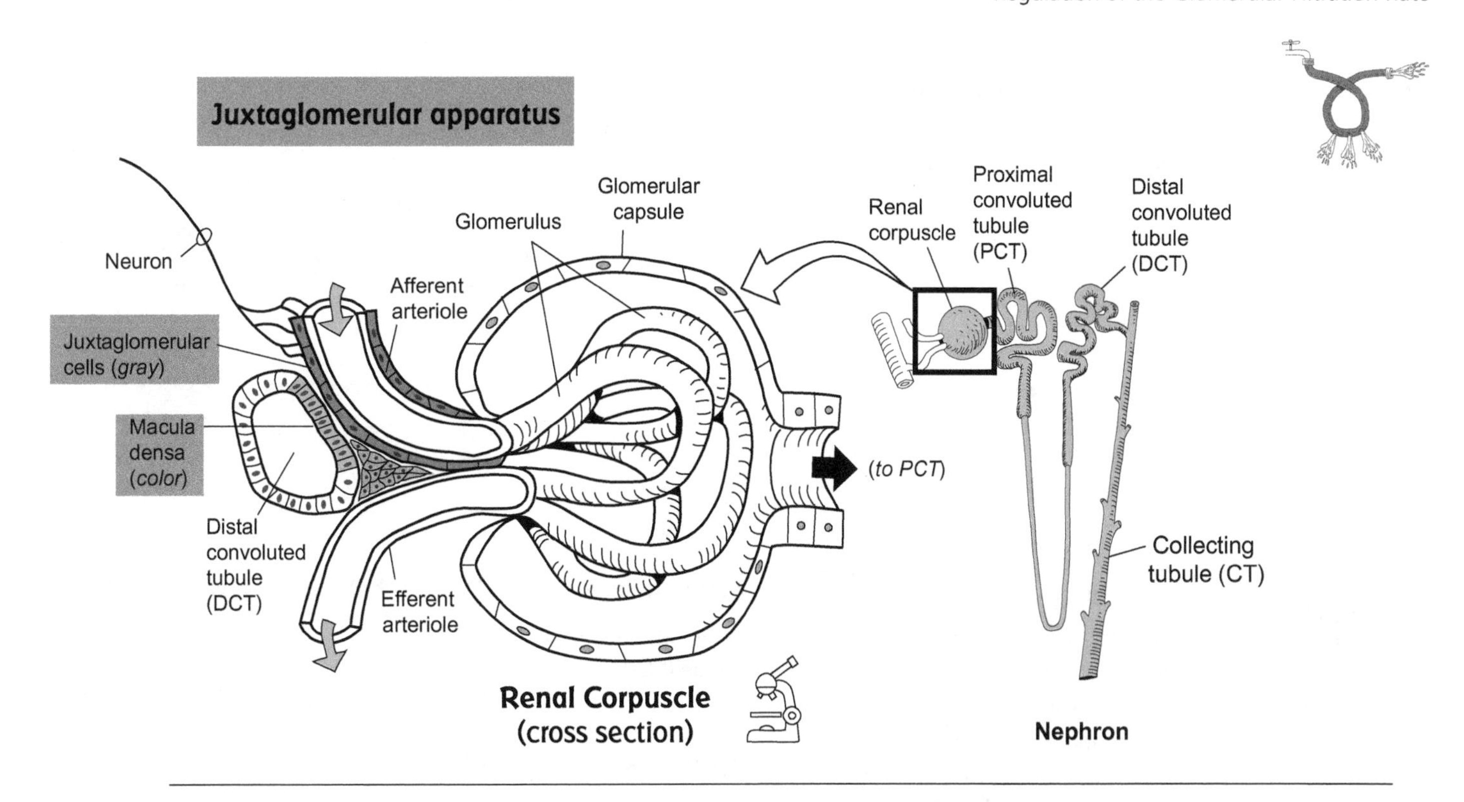

Renal Corpuscle
(cross section)

General Scheme for Controlling the GFR

By increasing or decreasing the BP in the glomerulus

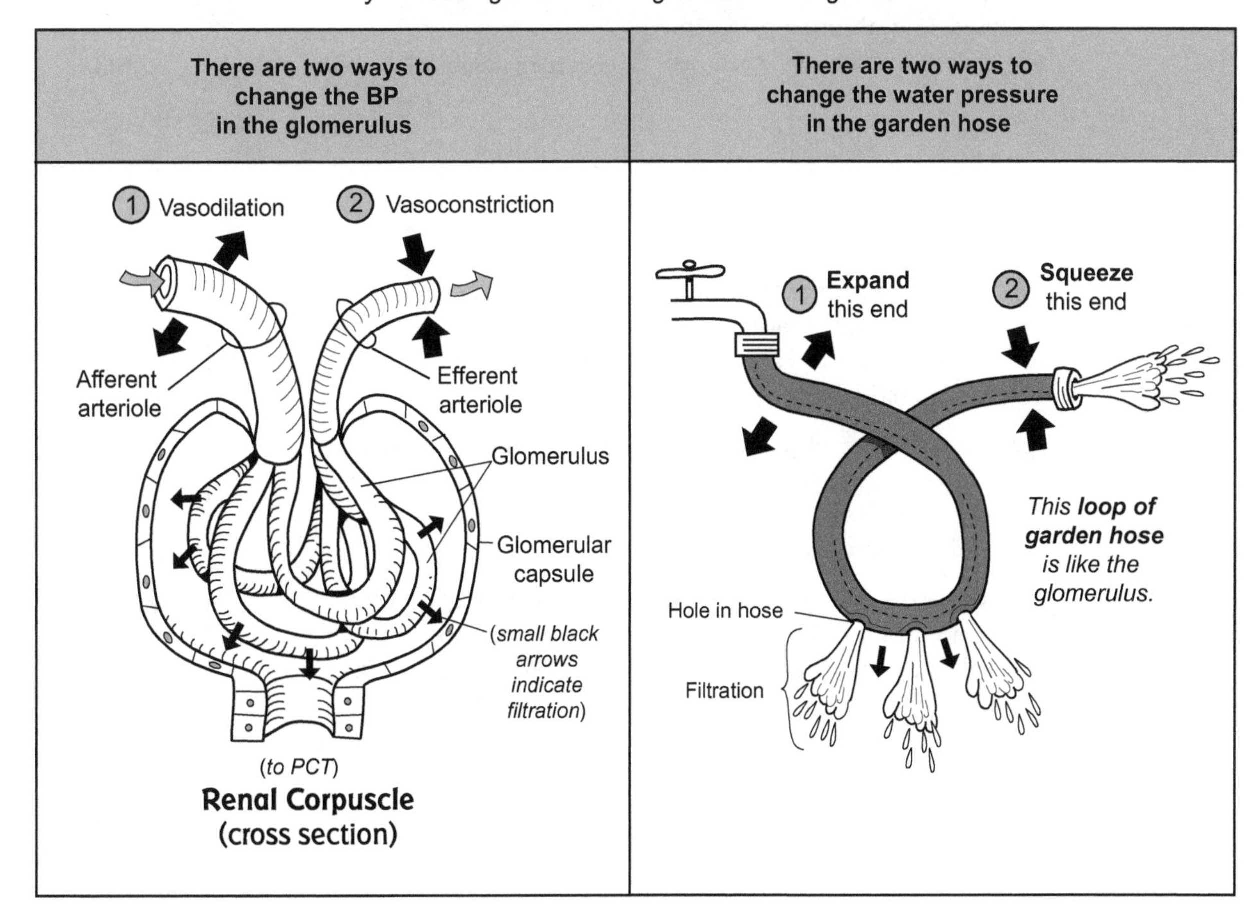

Renal Corpuscle
(cross section)

Description

Reabsorption is the transport of nutrients such as sodium ions or water from the renal tubule into the blood of the peritubular capillaries.

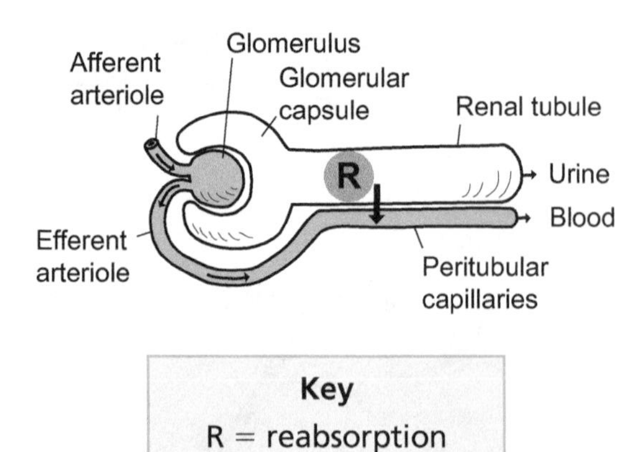

Key
R = reabsorption

Reabsorption by Nephron Region

Some of the major substances reabsorbed include ions such as sodium (Na^+), chloride (Cl^-), potassium (K^+), calcium (Ca^{+2}), magnesium (Mg^{+2}), and bicarbonate (HCO^{3-}), along with water, glucose, and amino acids. The table shows what substances are reabsorbed in each part of the nephron. Both active and passive transport processes are used in reabsorption (see p. 66). Some substances, such as glucose, are cotransported.

Nephron Region	Major Substances Reabsorbed			
Proximal convoluted tubule (PCT)	• Na^+ • Cl^- • K^+	H_2O	• G • amino acids	• HCO_3^- • Ca^{+2} • MG^{+2}
Nephron loop	• Na^+ • Cl^- • K^+	H_2O	• HCO_3^- • Ca^{+2} • MG^{+2}	
Distal convoluted tubule (DCT)	• Na^+ • Cl^-	H_2O	• Ca^{+2}	
Collecting tubule (CT)	• Na^+	H_2O	• HCO_3^-	

General Scheme for Reabsorption

① Sodium ions (Na^+) typically are transported actively into the **interstitial space.** These ions then diffuse into the blood. This creates a gradient of positive (+) charge in the interstitial space.

② The gradient of *positive* charge in the interstitial space draws *negatively* charged anions such as chloride ions (Cl^-) passively out of the **tubule** into the interstitial space and then into the **blood.**

③ Water (H_2O) follows the salt. Because of the higher solute concentration in the interstitial space relative to the tubule, water leaves the tubule via osmosis (see p. 70). Therefore, by creating this gradient of sodium and chloride ions, the nephron has really forced osmosis to occur.

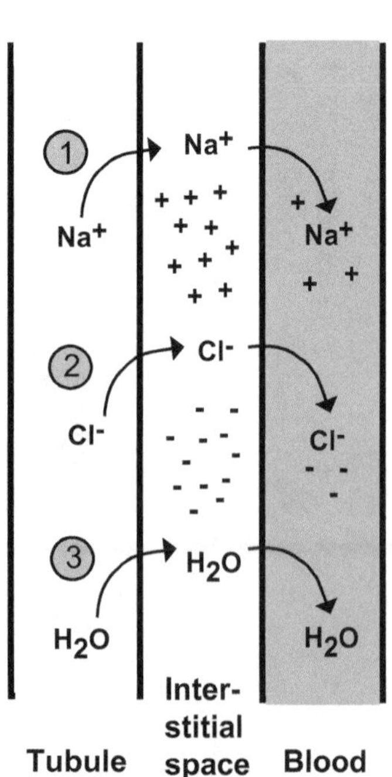

Reabsorption in the Nephron

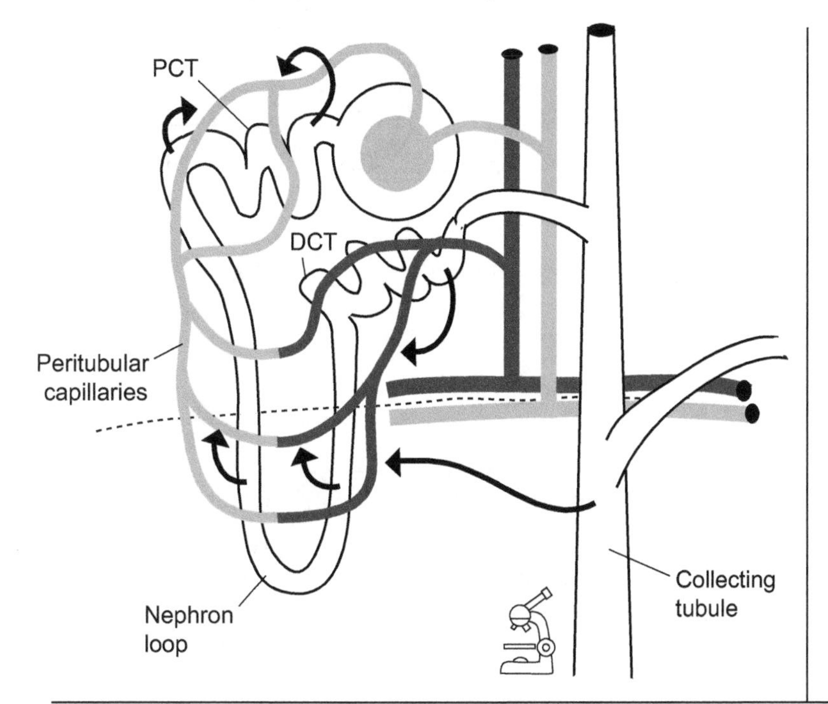

Amount of Filtrate Reabsorbed by Nephron Region

NEPHRON REGION	AMOUNT REABSORBED (out of 125 ml filtrate)
Proximal convoluted tubule (PCT)	100 ml
Nephron loop	7 ml
Distal convoluted tubule (DCT)	12 ml
Collecting tubule (CT)	5 ml

TOTAL REABSORBED: 124 ml

This is based on a rate of filtrate being formed at 125 ml/min. by both kidneys. Therefore, 124 of the 125 ml is reabsorbed into the blood. About 80% of this total occurs in the PCT.

Functional Analogy

The reabsorption process in the nephron is like workers sorting different products on a conveyor belt. The substances filtered from the blood in the glomerulus contain a mixture of nutrients (*ex.:* water, glucose, amino acids) and waste products (*ex.:* urea, uric acid). The waste products remain in the tubules (PCT, nephron loop, DCT, and collecting tubule), and the nutrients must be reabsorbed back into the blood. The workers are like the membrane proteins in the tubules that transport specific substances into the blood. The tubule conveyor represents the filtrate moving through the renal tubules, and the blood conveyor represents the blood in the peritubular capillaries that surround the tubules. In the end, the waste products are collected in the urine, stored in the urinary bladder, and excreted from the body, and the nutrients travel through the blood to finally be used by body cells.

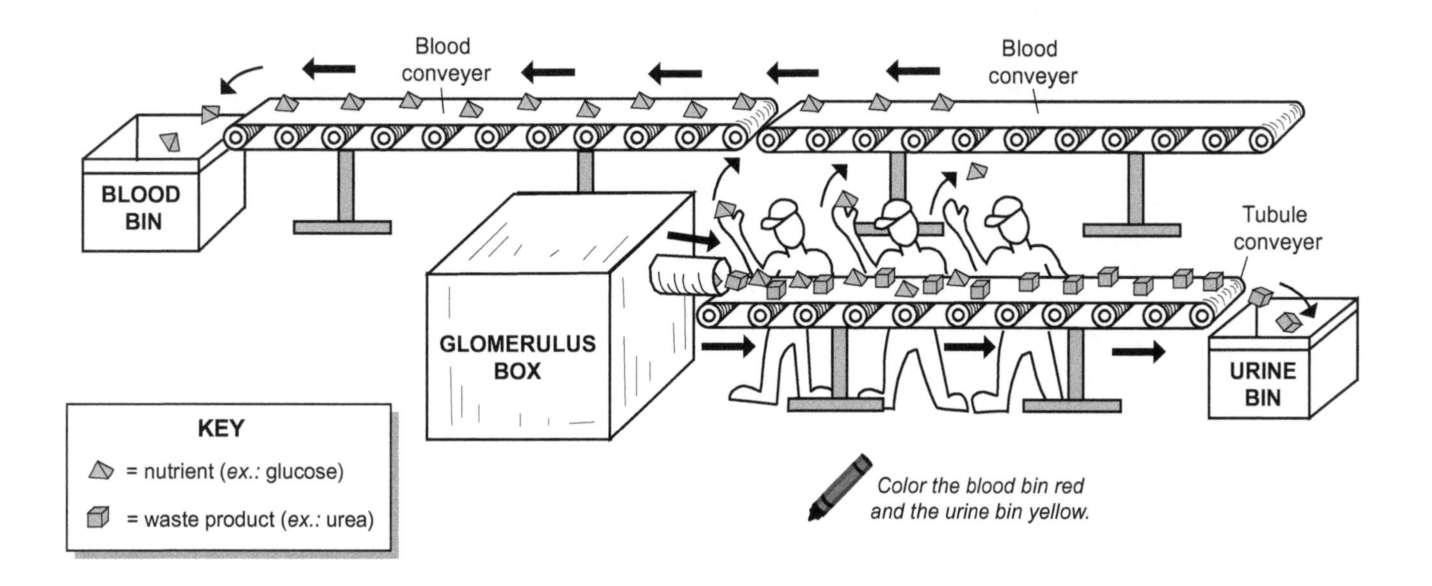

Description

Secretion is the transport of substances such as hydrogen ions from the blood of the **peritubular capillaries** into the **renal tubules**. The purpose is to remove these substances from the body in the urine.

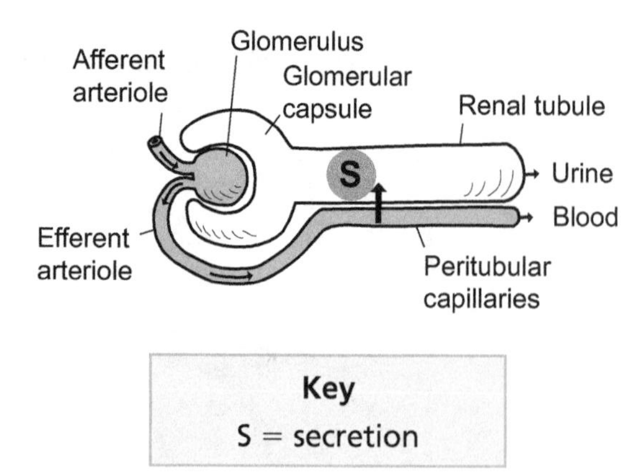

Key
S = secretion

Secretion by Nephron Region

The table shows the substances secreted in each part of the nephron.

Key

H_3O^+ = hydronium ions*
K^+ = potassium ions
NH_4^+ = ammonium ions

* **Note:** In solution, hydrogen ions (H^+) bind to water molecules to form hydronium ions (H_3O^+).

Nephron Region	Substance(s) Secreted
Proximal convoluted tubule (PCT)	H_3O^+ NH_4^+ Urea Creatine
Nephron loop	Urea
Distal convoluted tubule (DCT)	H_3O^+ K^+
Collecting tubule (CT)	H_3O^+ K^+

Outcomes of Secretion

Tubular secretion has three important outcomes:

- to control blood pH: achieved by the secretion of hydronium ions;
- to rid the body of nitrogen wastes; substances such as urea, creatinine, and ammonium ions are all products of metabolic processes that must be released in the urine; and
- to rid the body of unwanted drugs; drugs such as penicillin also are secreted and released in the urine.

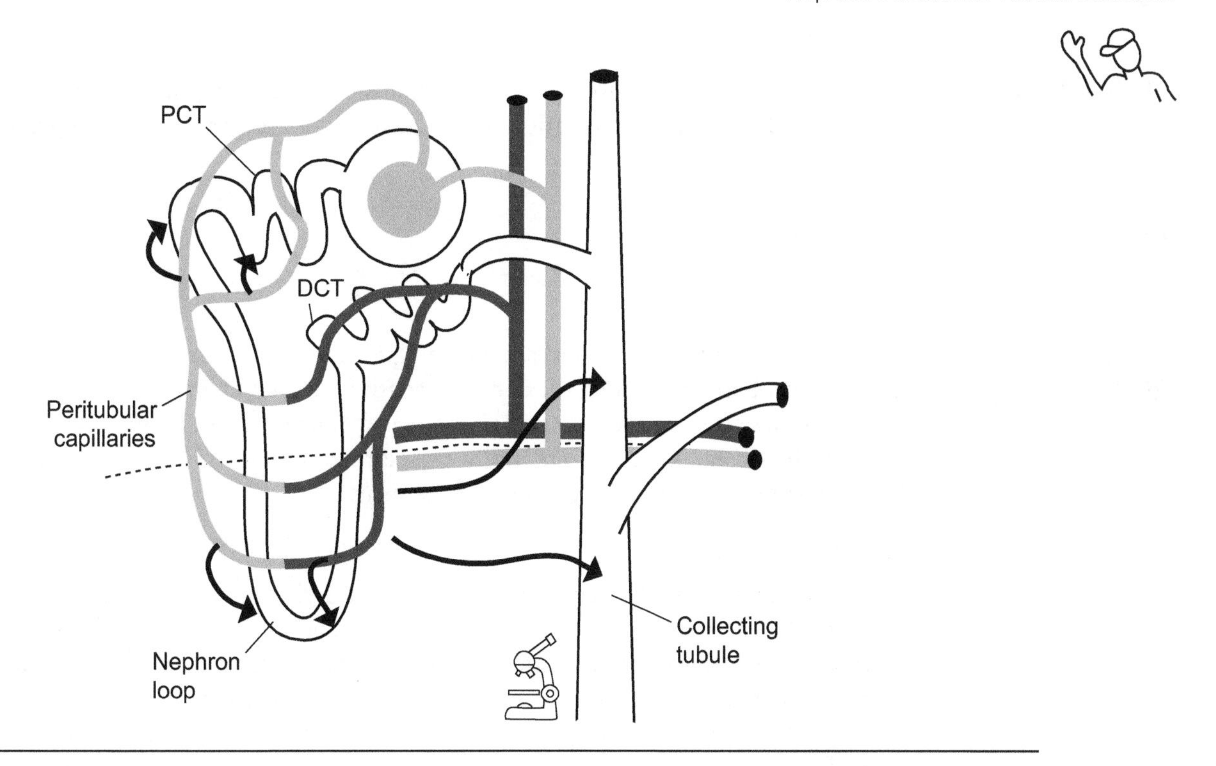

Functional Analogy

The secretion process in the nephron is like workers moving products from one conveyor belt to another. When certain substances such as potassium ions and hydrogen ions reach high levels in the blood, they are transported into the renal tubules and eventually released from the body in the urine. The workers are like the membrane proteins in the blood capillaries that transport substances into the renal tubules. The tubule conveyor represents the filtrate moving through the renal tubules, and the blood conveyor represents the blood in the peritubular capillaries that surround the tubules.

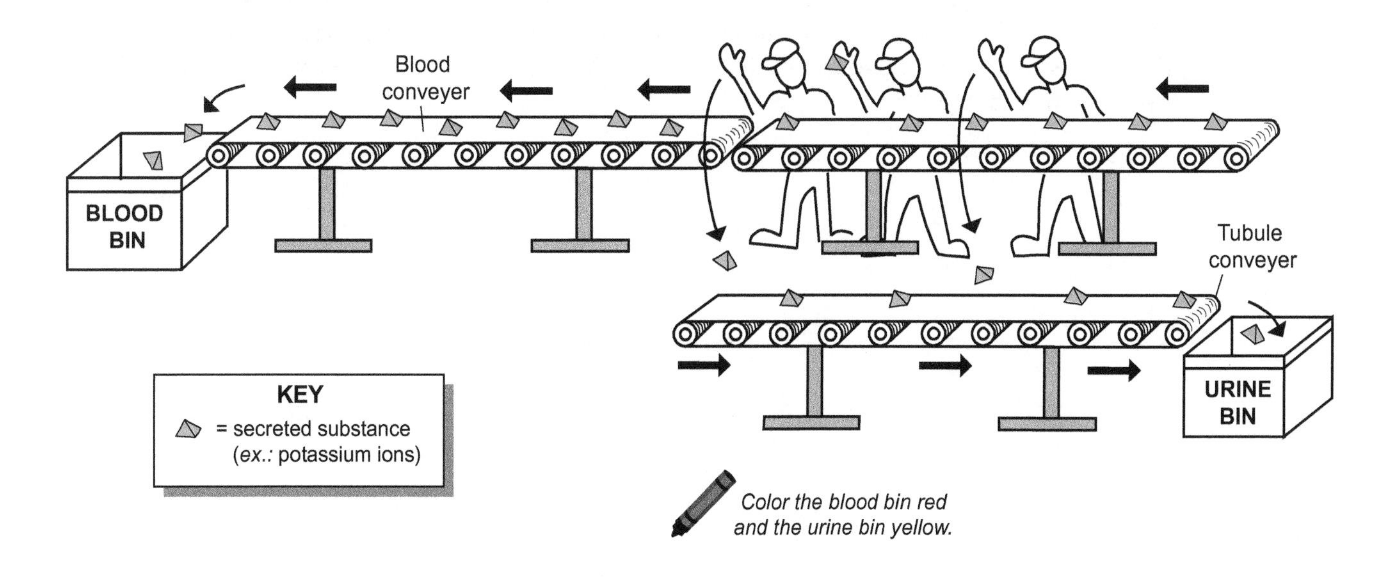

Description

This module covers the three general mechanisms used to regulate acid-base balance in the body. These are buffer systems, exhalation of CO_2, and kidney excretion of hydrogen ions (H^+).

pH Scale

See pages 28–29.

Major Mechanisms

It is vital that body fluids remain in a normal pH range. The **normal blood pH** is 7.35–7.45. Enzymes and other proteins in the blood are highly sensitive to pH changes and can lose their function if the blood becomes either too acidic (<7.35) or too alkaline (>7.45). By-products from normal metabolism make the blood more acidic. The body uses different mechanisms to raise the pH back to normal. Here is a summary of the three major mechanisms:

① **Buffer systems**

Buffers allow a pH to maintain a stable pH at a given pH value. It's like a vise that clamps down on a pH value and prevents it from shifting. The most important buffering system is called the **carbonic acid-bicarbonate buffering system.** When carbon dioxide (CO_2) diffuses into the blood it reacts with water (H_2O) in the plasma to produce carbonic acid (H_2CO_3). This weak acid then dissociates into hydronium ions (H_3O^+) and bicarbonate ions (HCO_3^-). When CO_2 levels are high, as during exercise, more carbonic acid is produced, and the blood becomes more acidic. But when CO_2 levels are low, the reaction can run in reverse. In short, the HCO_3^- binds the free H_3O^+ to produce carbonic acid and convert it into CO_2 and H_2O. This makes the blood pH more alkaline.

② **Exhalation of CO_2**

Respiration centers in the brainstem control the rate and depth of breathing. Increasing these factors results in a faster, more forceful exhalation, which removes CO_2 from the body. As explained about the **carbonic acid-bicarbonate buffering system** above, less CO_2 in the blood means less carbonic acid being formed, making the blood pH more alkaline.

③ **Kidney excretion of hydronium ions** (H_3O^+)

Unlike buffering systems, the kidneys actually *remove* the excess H_3O^+ from the body and excrete them in the urine. This makes the urine *more* acidic and the blood *less* acidic. For the details of this process, see p. 236. (**Note:** In solution, hydrogen ions [H^+] bind to water molecules to form hydronium ions [H_3O^+].)

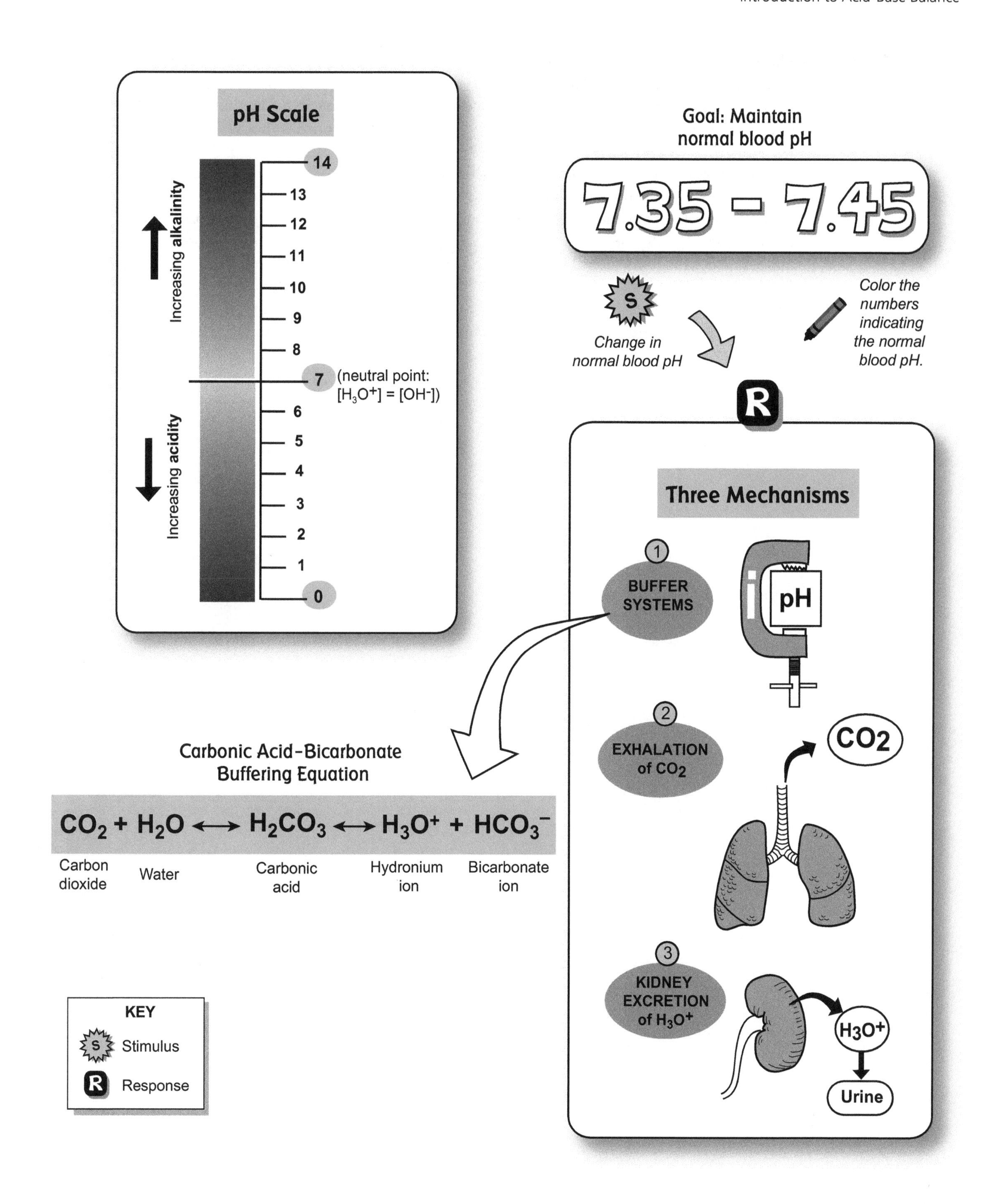
pH Scale
14
13
12
11
10
9
8
7 (neutral point: $[H_3O^+] = [OH^-]$)
6
5
4
3
2
1
0
Increasing alkalinity
Increasing acidity
Goal: Maintain normal blood pH
7.35 - 7.45
S
Change in normal blood pH
Color the numbers indicating the normal blood pH.
R
Three Mechanisms
1
BUFFER SYSTEMS
pH
2
EXHALATION of CO_2
CO_2
3
KIDNEY EXCRETION of H_3O^+
H_3O^+
Urine
Carbonic Acid-Bicarbonate Buffering Equation
$CO_2 + H_2O \longleftrightarrow H_2CO_3 \longleftrightarrow H_3O^+ + HCO_3^-$
Carbon dioxide
Water
Carbonic acid
Hydronium ion
Bicarbonate ion
KEY
S Stimulus
R Response

Description

The kidneys regulate acid-base balance in the body. Normal metabolic reactions in body cells produce acids, such as sulfuric acid, deposited in the blood. As with any acid, this increases the amount of free hydronium ions (H_3O^+) in solution and makes the blood pH more **acidic** than normal. Because these acids are produced day after day, they must be eliminated from the body. This is accomplished by excreting the H_3O^+ in the urine.

Let's explain this idea a bit further. Within the kidneys are millions of microscopic functional units called **nephrons** that constantly filter the blood plasma and process it into urine. Two specific parts of the nephron—the **proximal convoluted tubule** (PCT) and the **collecting tubule** (CT)—are involved in excreting H_3O^+ into the urine. Any time a substance is moved out of the blood and into the renal tubules, it is called *secretion*. Consequently, it is more accurate to say that H_3O^+ is secreted into the **renal tubules** (PCT and CT). At the same time H_3O^+ is being secreted, bicarbonate ions (HCO_3^-) are being reabsorbed back into the blood so they are not lost in the urine. This constant removal of metabolic acids ensures that the blood pH remains in its normal range of 7.35–7.45.

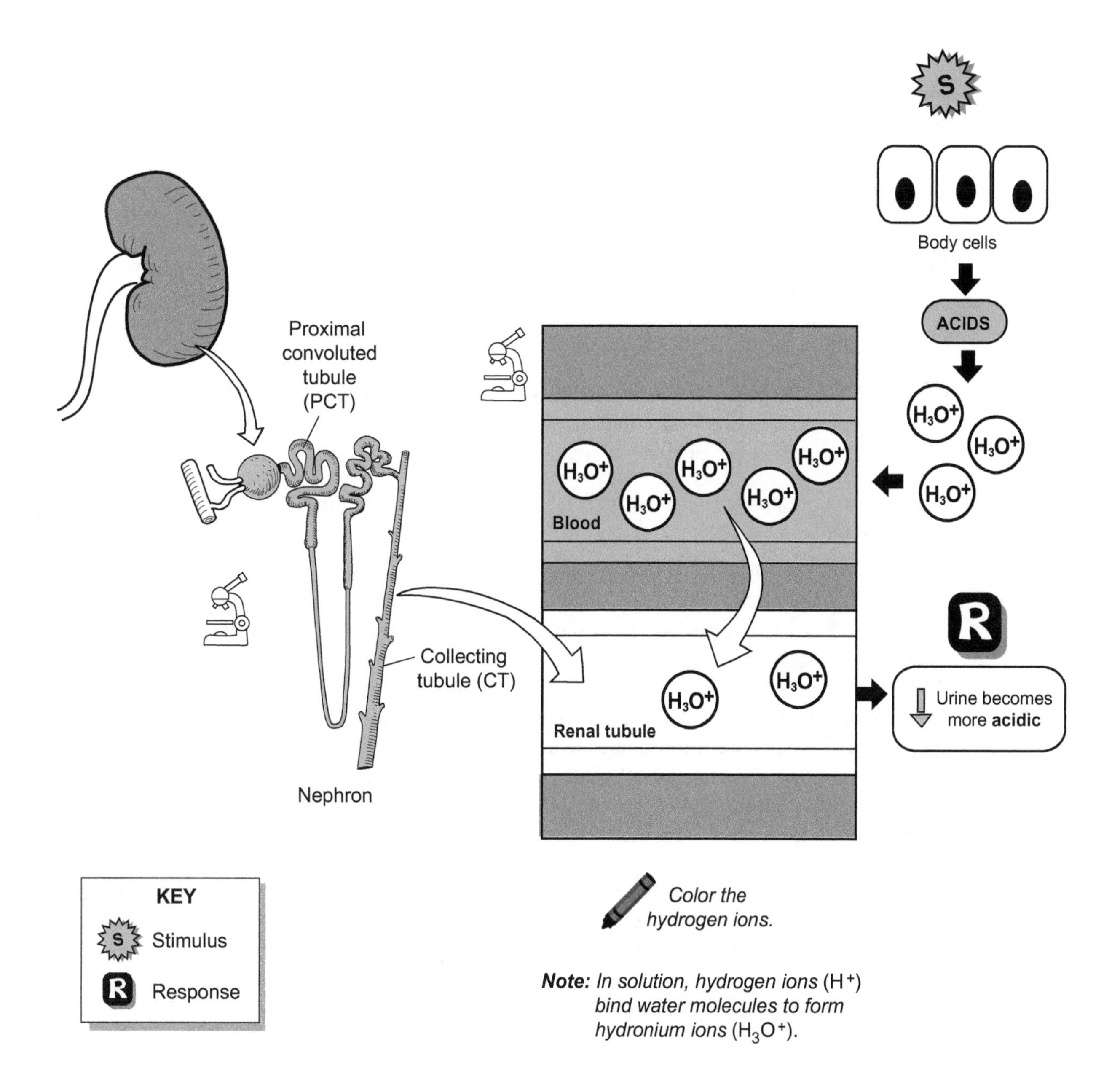

Color the hydrogen ions.

Note: *In solution, hydrogen ions* (H^+) *bind water molecules to form hydronium ions* (H_3O^+).

Description

Antidiuretic hormone (ADH) helps conserve water. Imagine a hot day where you are working outdoors, sweating, and getting dehydrated. This triggers the following series of events:

1. Sweating increases the **blood osmotic pressure.** This serves as a stimulus (S).
2. This increase in osmotic pressure is detected by **osmoreceptors** within the **hypothalamus** that constantly monitor the osmolarity ("saltiness") of the blood.
3. Osmoreceptors stimulate groups of neurons within the **hypothalamus** to release **ADH** from the **posterior lobe** of the **pituitary gland.**
4. Like all hormones, ADH travels through the bloodstream to its various target organs.*
5. One of its targets, the **renal tubules** in the **kidney**, contains receptors for ADH. Most of these receptors are in the **collecting tubules** (CT). When ADH binds to its receptor, it makes the membrane more permeable to water.
6. The final response (R) is an increase in water **reabsorption**, which leads to a decrease in **urine output.**

Now imagine that you drink lots of water to get rehydrated after working outdoors. What happens? The excess water decreases the blood osmotic pressure. This inhibits osmoreceptors, which, in turn, stops secretion of ADH. If no ADH is released, the renal tubules remain impermeable to water. Therefore, the excess water is retained by the renal tubules, and the result is an increase in urine output. In short, although ADH helps the body conserve water, it also controls whether the urine is diluted or concentrated. That's one important little hormone!

Have you ever wondered why alcohol consumption makes you urinate more frequently? It's because alcohol inhibits the release of ADH, thus increasing urine production. Contrary to what many people think, drinking alcohol actually *dehydrates* you.

* Other targets that have receptors for ADH are sweat glands and arterioles. In sweat glands, ADH decreases perspiration to conserve water. In arterioles, it causes the smooth muscle in the wall to constrict, narrowing the vessel diameter and increasing blood pressure.

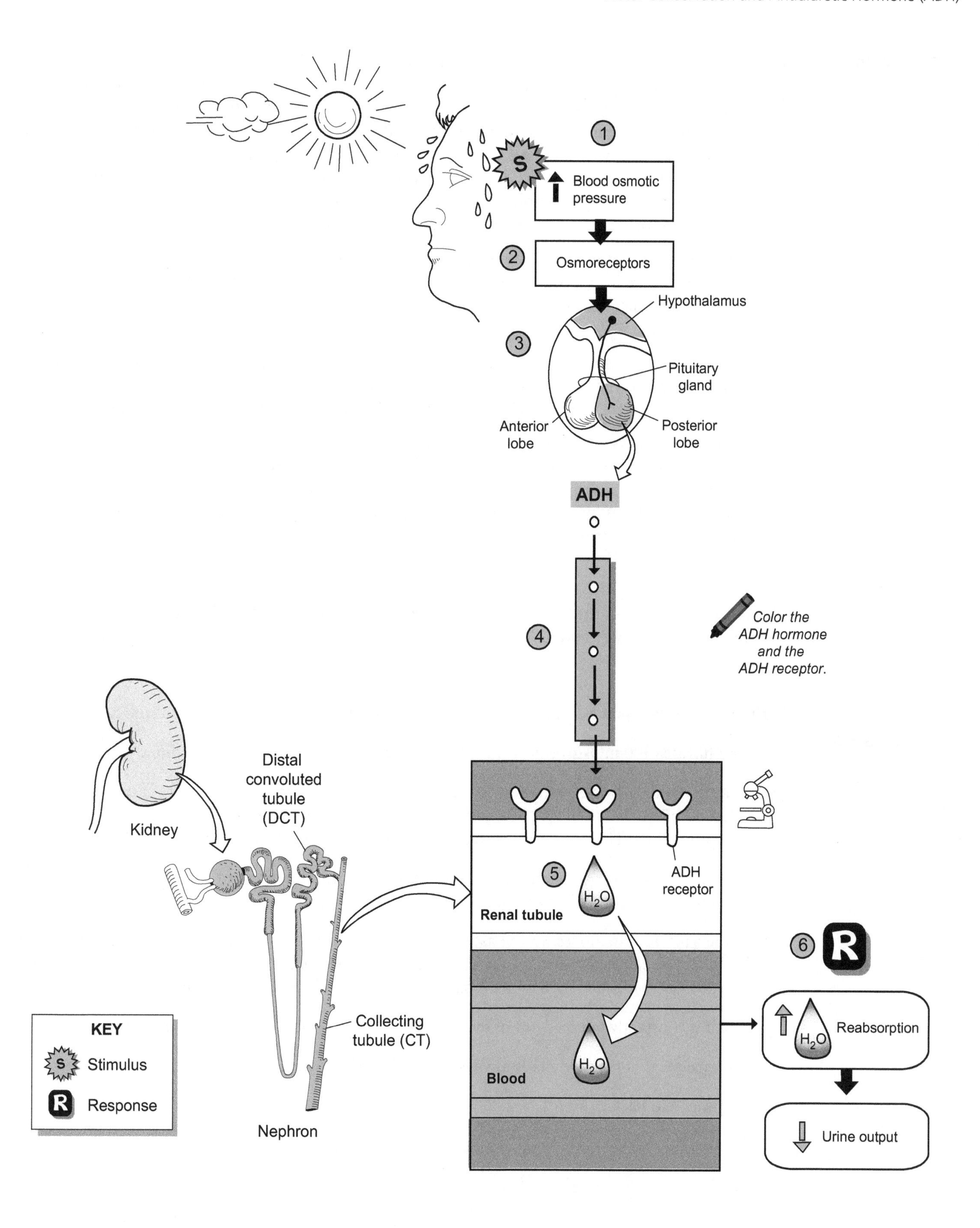
1
S
Blood osmotic pressure
2
Osmoreceptors
Hypothalamus
3
Pituitary gland
Anterior lobe
Posterior lobe
ADH
4
Color the ADH hormone and the ADH receptor.
Kidney
Distal convoluted tubule (DCT)
5
H_2O
ADH receptor
Renal tubule
Blood
H_2O
6 R
H_2O
Reabsorption
Urine output
Collecting tubule (CT)
Nephron
KEY
S Stimulus
R Response

Description

This module covers the **countercurrent multiplier** mechanism in the **nephron loop** of the **nephrons**. First, though, let's summarize a few key things about the nephron loop. It is composed of a **descending limb** (DL) and an **ascending limb** (AL). Filtrate from the glomerular capsule flows down the DL and up the AL before entering the **distal convoluted tubule** (DCT), followed by the **collecting tubule** (CT). The DL has a thinner wall permeable to water but not salt, sodium ions (Na^+) and chloride ions (Cl^-). In contrast, the wall of the AL becomes thicker (as indicated by a heavier line in the illustration). This thicker portion is impermeable to water and contains protein pumps used to actively transport Na^+ and Cl^- out of the AL and into the **extracellular fluid** (ECF).

- Name? Why is it called the **countercurrent multiplier?** The term "countercurrent" refers to the flow of filtrate in opposite directions in the nephron loop—down the DL and up the AL. The term "multiplier" refers to the fact that it increases (or multiplies) the salt concentration in the renal medulla.
- **Osmolarity?** This is a measure of the amount of dissolved particles in 1 liter of solution. In living systems, smaller units called **milliosmoles** (**mOsm**) are used. For our purposes, this is a measure of the saltiness of the solution. The higher the number, the saltier the solution, and vice versa.
- Purpose? Notice from the illustration that the ECF deep in the medulla is four times saltier than the ECF near the cortex (1,200 mOsm versus 300 mOsm). The general purpose of the countercurrent multiplier mechanism is to maintain a constant gradient of salt deep within the renal medulla. Why? *This allows the collecting tubule (CT) to reabsorb water via osmosis, thereby concentrating the urine.*

Analogy

The salt in your ECF is the same as that in table salt—sodium chloride (NaCl). In solution, sodium chloride ionizes into Na^+ and Cl^-. A salt shaker will remind you of this connection to everyday life.

Mechanism Step by Step

For the step-by-step process of this cyclical mechanism, refer to 1–5 in the illustration:

① Salty filtrate (300 mOsm) enters the nephron loop from the **proximal convoluted tubule** (PCT).

② Because the ECF is saltier than the filtrate, water leaves the descending limb via osmosis. Then, it is reabsorbed into the blood. The saltier the ECF becomes, the more water is removed from the descending limb. This loss of water decreases the volume and makes the filtrate saltier.

③ As water leaves the descending limb, the filtrate becomes saltier until it reaches a peak of about 1,200 mOsm at the bottom of the nephron loop.

④ As the salty filtrate moves up the ascending limb, salt (Na^+ and Cl^-) is actively transported into the ECF. The saltier the filtrate, the more salt is pumped out. As this occurs, the filtrate becomes less salty but the volume remains the same, moving into the DCT at 100 mOsm.

⑤ As salt is constantly pumped out of the ascending limb, the ECF becomes saltier in the renal medulla.

This ensures that the medulla is always saltier than the cortex.

Cycle repeats. Go back to step 1.

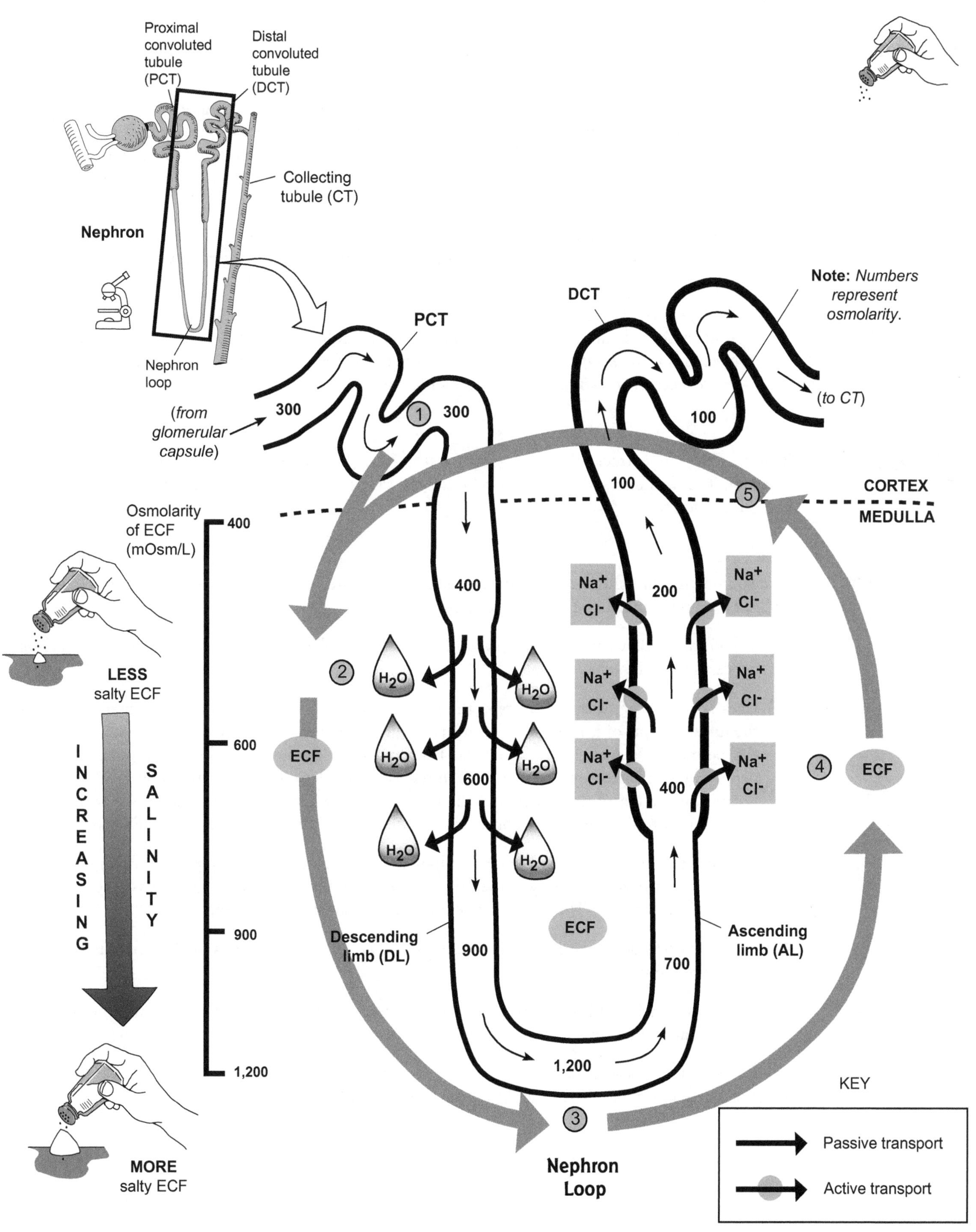
Proximal convoluted tubule (PCT)
Distal convoluted tubule (DCT)
Collecting tubule (CT)
Nephron
Nephron loop
PCT
DCT
Note: Numbers represent osmolarity.
(from glomerular capsule)
300
1
300
100
(to CT)
100
5
CORTEX
MEDULLA
Osmolarity of ECF (mOsm/L)
400
600
900
1,200
LESS salty ECF
INCREASING SALINITY
MORE salty ECF
400
200
Na+ Cl-
2
H2O
ECF
600
400
4
ECF
Descending limb (DL)
900
ECF
700
Ascending limb (AL)
1,200
3
Nephron Loop
KEY
Passive transport
Active transport

Description

This module covers the **countercurrent exchanger** mechanism in the **nephron loop** of the nephrons. First, let's go over a few fundamentals:

- Name? Why is it called the countercurrent exchanger? The term "countercurrent" refers to the flow of blood in opposite directions in the **vasa recta**—down one side and up the other. The vasa recta is blood capillary branches located next to the nephron loop in the same looping pattern. The term "exchanger" refers to the fact that water is exchanged for salt along the length of the vasa recta.
- **Osmolarity**? This is a measure of the amount of dissolved particles in 1 liter of solution. In living systems, smaller units called **milliosmoles (mOsm)** are used. For our purposes, this is a measure of the saltiness of the solution. The higher the number, the saltier the solution, and vice versa.
- Purpose? The general purpose of the countercurrent exchanger mechanism is to maintain the gradient of salt deep within the renal medulla established by the countercurrent multiplier mechanism. *Without the countercurrent exchanger, the salt would be carried away by the blood in the vasa recta, and the gradient would be lost.*

Analogy

The salt in your extracellular fluid (ECF) is the same as that in table salt—sodium chloride. In solution, it ionizes into sodium ions (Na^+) and chloride ions (Cl^-). A salt shaker will remind you of this connection to everyday life.

Mechanism

The **countercurrent multiplier mechanism** (see p. 238) created a salt gradient in the extracellular fluid (**ECF**) of the renal medulla. Here are the steps:

① The gradient allows water to be reabsorbed from the **collecting tubule** (**CT**) by osmosis, thereby concentrating the urine.

② **Urea** is one of the waste products in the nephron tubules. Although most of the urea remains in the tubules, some of it is recycled by diffusing out of the permeable lower end of the CT and into the permeable descending limb of the nephron loop. Because the thickened portions of the nephron (as indicated by a heavier line) in the ascending limb and DCT are impermeable to urea, it can't diffuse out of that portion of the tubule. The net result of this recycling is that urea adds substantially to the total solute concentration in the ECF, which increases the ECF's osmolarity.

③ The **vasa recta** is adjacent to the nephron loop and follows the same looping pattern with opposite flow. As mentioned previously, if the salt were to diffuse into the blood in the vasa recta, the salt gradient could be lost. The countercurrent exchanger stops this problem.

④ As blood flows downward in the vasa recta, salt (Na^+ and Cl^-) diffuses into the blood, and water leaves the blood by osmosis.

⑤ In contrast, as blood flows upward in the vasa recta, water in the ECF moves into the blood by osmosis, and salt diffuses out of the blood.

In short, the blood carries away more water than salt. The salt is recycled in the vasa recta, so most of it remains in the ECF, which stabilizes the salt gradient.

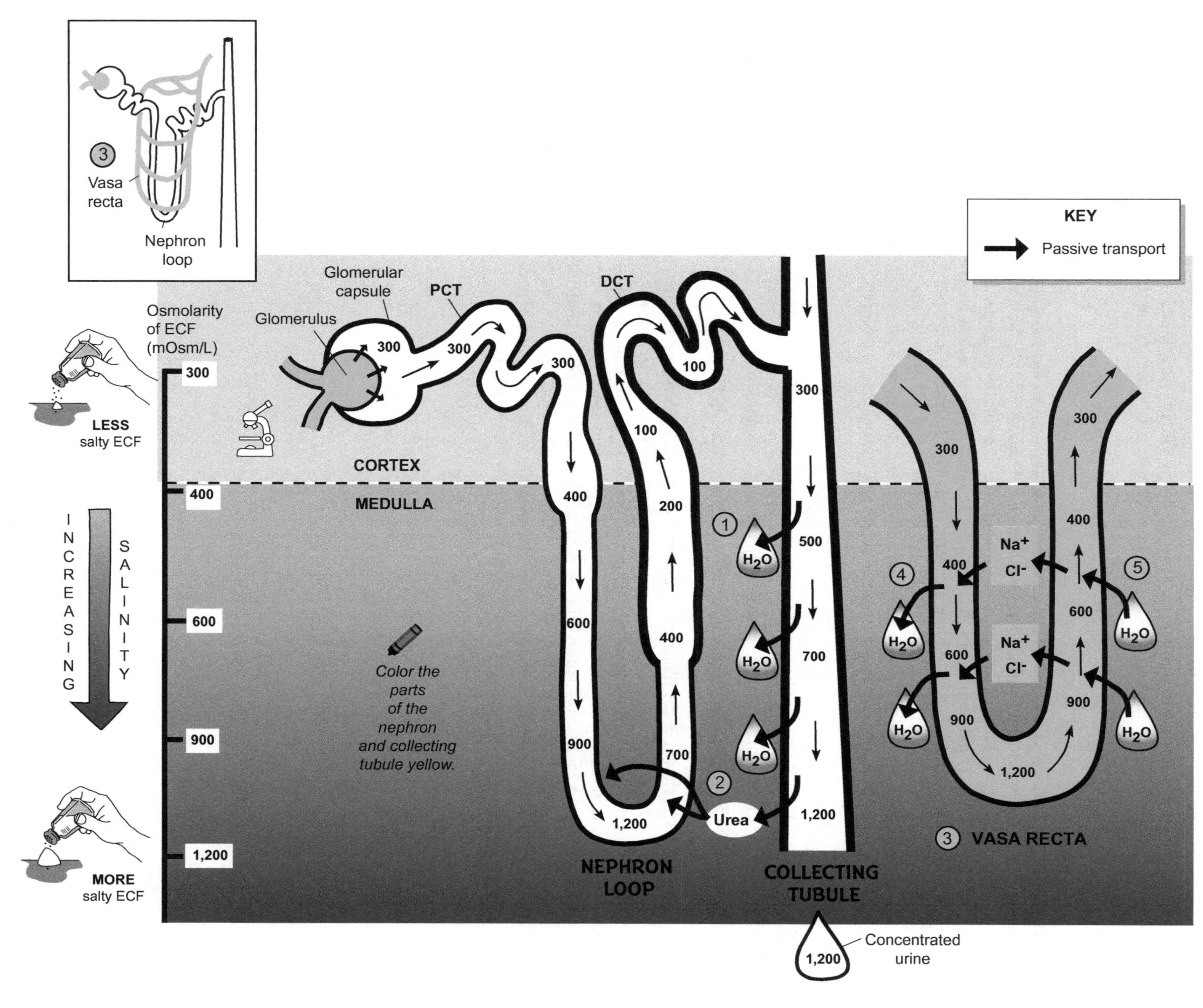
Vasa recta
Nephron loop
KEY
Passive transport
Osmolarity of ECF (mOsm/L)
300
400
600
900
1,200
LESS salty ECF
MORE salty ECF
INCREASING SALINITY
Glomerulus
Glomerular capsule
PCT
DCT
CORTEX
MEDULLA
Color the parts of the nephron and collecting tubule yellow.
H_2O
Urea
Na^+
Cl^-
NEPHRON LOOP
COLLECTING TUBULE
VASA RECTA
Concentrated urine

Description

The **renin-angiotensin-aldosterone (RAA)** system is one of the body's mechanisms to detect falling blood pressure (BP) and bring it back to normal. There are serious consequences for allowing BP to drop below normal levels. For example, the kidneys depend on a constant, normal BP to filter the blood and remove waste products. If the BP drops too low, shock and even renal failure could result. Not surprisingly, the body has numerous mechanisms to detect falling blood pressure and bring it back to normal.

RAA System: Overview

Let's say you were in a car accident and suffered a bad laceration that caused you to lose some blood. This reduction in blood volume causes a decrease in BP, which, in turn, sets a cascade of events in motion. First, this decrease in BP would be detected by juxtaglomerular cells within the kidney, and they would respond by secreting the enzyme renin, which through a chain of events leads to the activation of the hormone angiotensin II. This hormone functions to increase BP in two important ways: (1) vasoconstriction of blood vessels, and (2) stimulation of the adrenal cortex to secrete **aldosterone**, which increases blood volume. Aldosterone does this by increasing reabsorption of sodium ions. Because the general rule is that "water follows the salt," water also is reabsorbed.

Formation of Angiotensin II: Detail

How does renin lead to the formation of angiotensin II? This process actually involves two steps:

① **Renin** helps convert **angiotensinogen** into **angiotensin I**

② ACE helps convert **angiotensin I** into **angiotensin II.**

As shown in the illustration, angiotensinogen is a plasma protein made by the liver. It normally travels through the bloodstream without doing much of anything. But when the enzyme renin is released into the blood from the kidney, it specifically catalyzes the conversion of angiotensinogen into a slightly different protein called angiotensin I. This conversion involves the cutting off of one part of the protein. As angiotensin I travels through the capillaries of the lung and other tissues, it gets exposed to **angiotensin-converting enzyme** (ACE), which catalyzes the conversion of angiotensin I into **angiotensin II.** This conversion also involves cutting off one part of the protein and completes the process. Now angiotensin II is ready to do its important work.

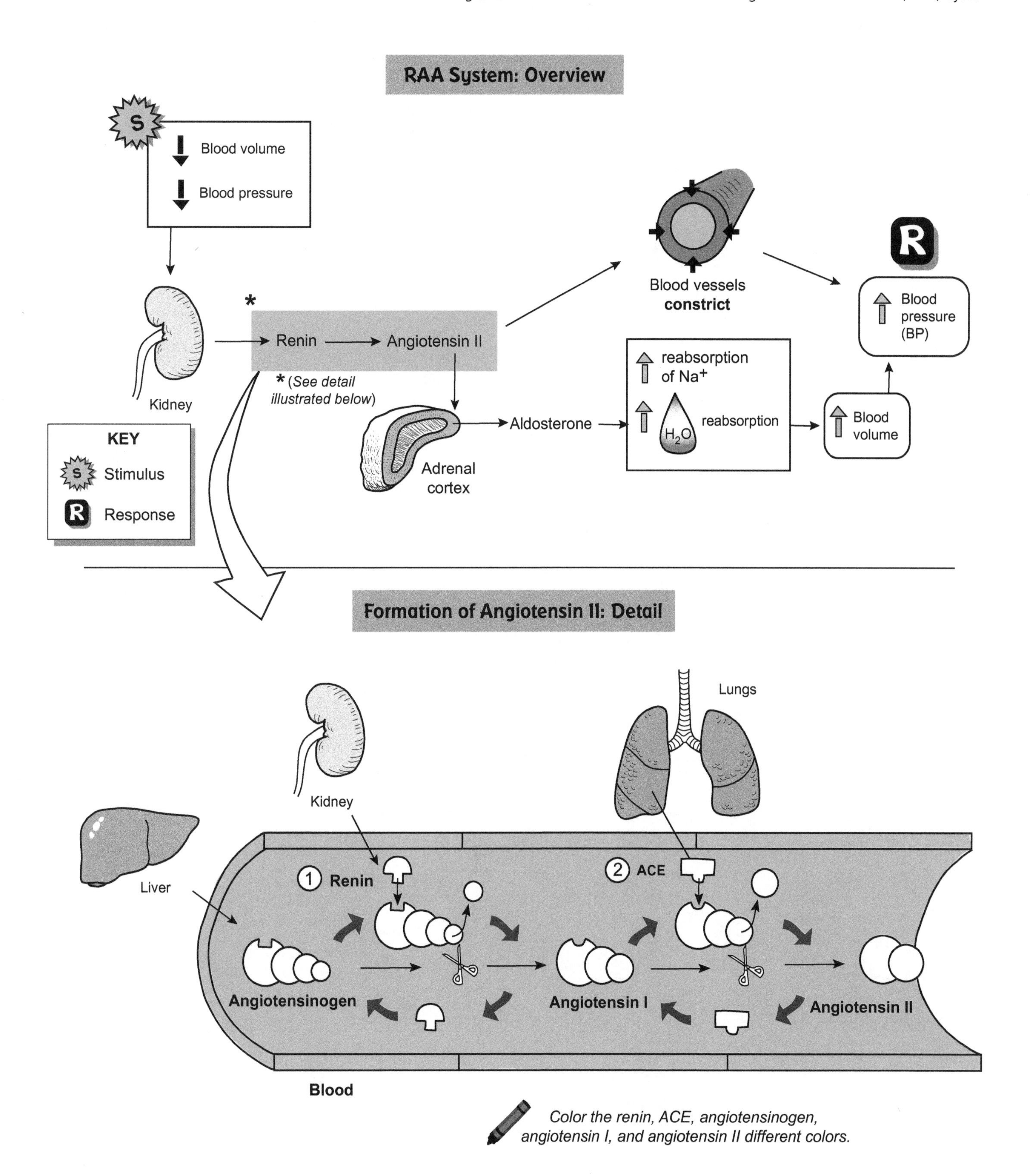

Color the renin, ACE, angiotensinogen, angiotensin I, and angiotensin II different colors.

Notes

NERVOUS SYSTEM

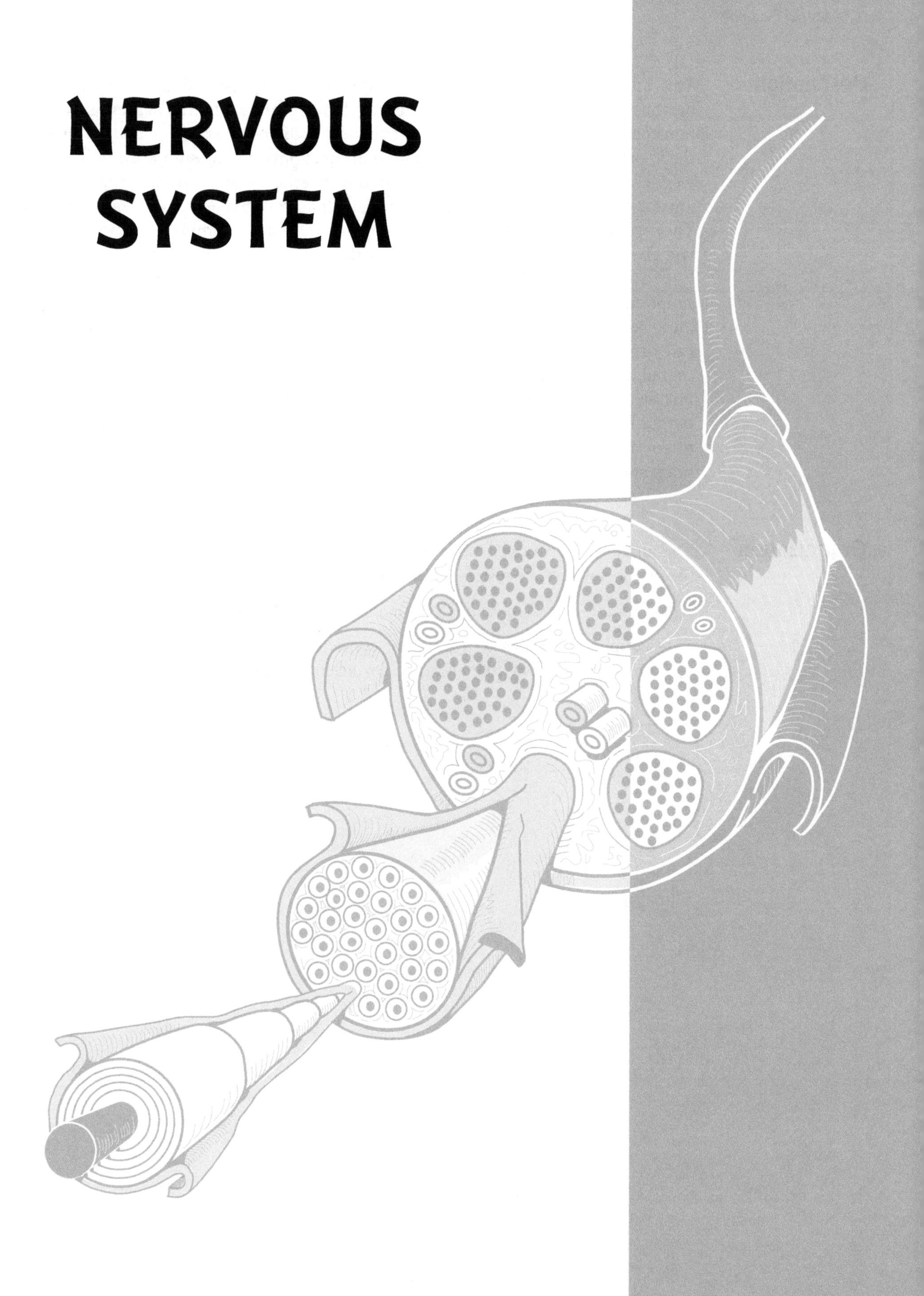

Description

This module provides an overview of the nervous system. Along with the endocrine system, the **nervous system** is one of the important regulators of the body. The nervous system helps maintain our homeostasis. For example, by sensing changes in blood pressure or body temperature, it can trigger the appropriate responses to keep our vital functions running smoothly. Without the nervous system, we literally would be disconnected from our inner and outer worlds. Imagine not being able to sense anything—no sights, no sounds, no smells! What if all 600+ muscles in your body were paralyzed? They would be unable to move because you would have nothing to stimulate them to contract. This will give you an appreciation of the importance of the nervous system.

To illustrate further, consider how much nervous system activity is involved in the seemingly simple act of getting your morning cup of coffee. All the sights, sounds, smells, and tastes of the coffee must be detected by various **sensory receptors.** These receptors are connected to **sensory nerves** that carry the information to the **brain** like data along a computer cable. All this information must be routed to the correct interpretation center in the cerebrum of the brain. When you decide to take that first sip, you need to contract your biceps brachii to bring the cup to your mouth. This is motor output that begins as a nerve impulse in the brain, runs down the spinal cord, and is carried by a **motor nerve** to the muscle. In summary, without the nervous system, no sensory input would go to the brain and no motor output to muscles or glands.

Organization

The nervous system can be divided into two major divisions: the **central nervous system** (CNS) and the **peripheral nervous system** (PNS). Here is a brief summary of each:

- Central nervous system (CNS): consists of the **brain** and **spinal cord.**
- Peripheral nervous system (PNS): consists of all the **sensory nerves** and **motor nerves** that travel through the body like electrical wiring in a house; more broadly, any nerve tissue outside the CNS.

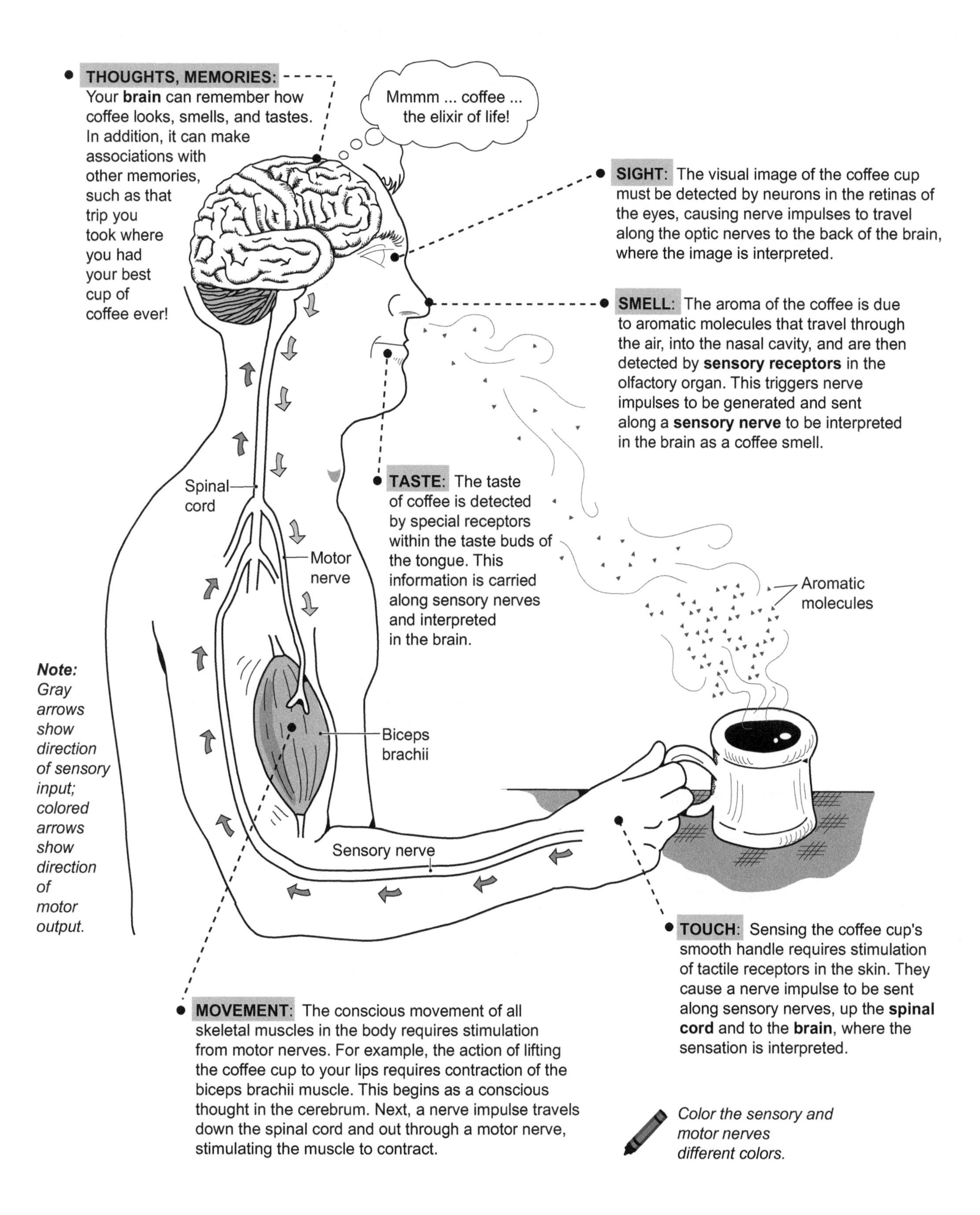

Color the sensory and motor nerves different colors.

Description

Peripheral **nerves** spread throughout the body like electrical wiring through a house. They are a collection of bundles of microscopic axons. Examples include the sciatic and femoral nerves in the thigh and the brachial nerve in the arm. Nerves are visible to the naked eye and structured as a series of tubes within tubes. Each tubelike structure is wrapped in a protective connective tissue. Each of the labeled structures in the illustration is explained in the tables below.

Tubes within Tubes

Structure	Description
1. Nerve	long, macroscopic, cable-like structure containing bundles of axons
2. Fascicle	a single bundle of axons
3. Myelinated axon	an axon wrapped in a protective myelin sheath
4. Axon	long, thin part of a neuron that carries nerve impulses along its length

Connective Tissue Organization

Structure	Description
5. Epineurium	the thick layer of dense irregular connective tissue that wraps around the outside of a nerve
6. Perineurium	the cellular connective tissue layer that wraps around each fascicle
7. Endoneurium	the thin layer of areolar connective tissue that wraps around each axon and binds one myelinated axon to another within a fascicle

Neuronal Structures

Structure	Description
8. Neurolemmocyte (*Schwann cell*)	a cell that wraps itself around the axon of a neuron numerous times during development of the nervous system; the end result is that a myelin sheath is created
9. Myelin sheath	the layers of the neurolemmocyte's cell membrane that form a protective, insulating coating of lipoprotein around the axon

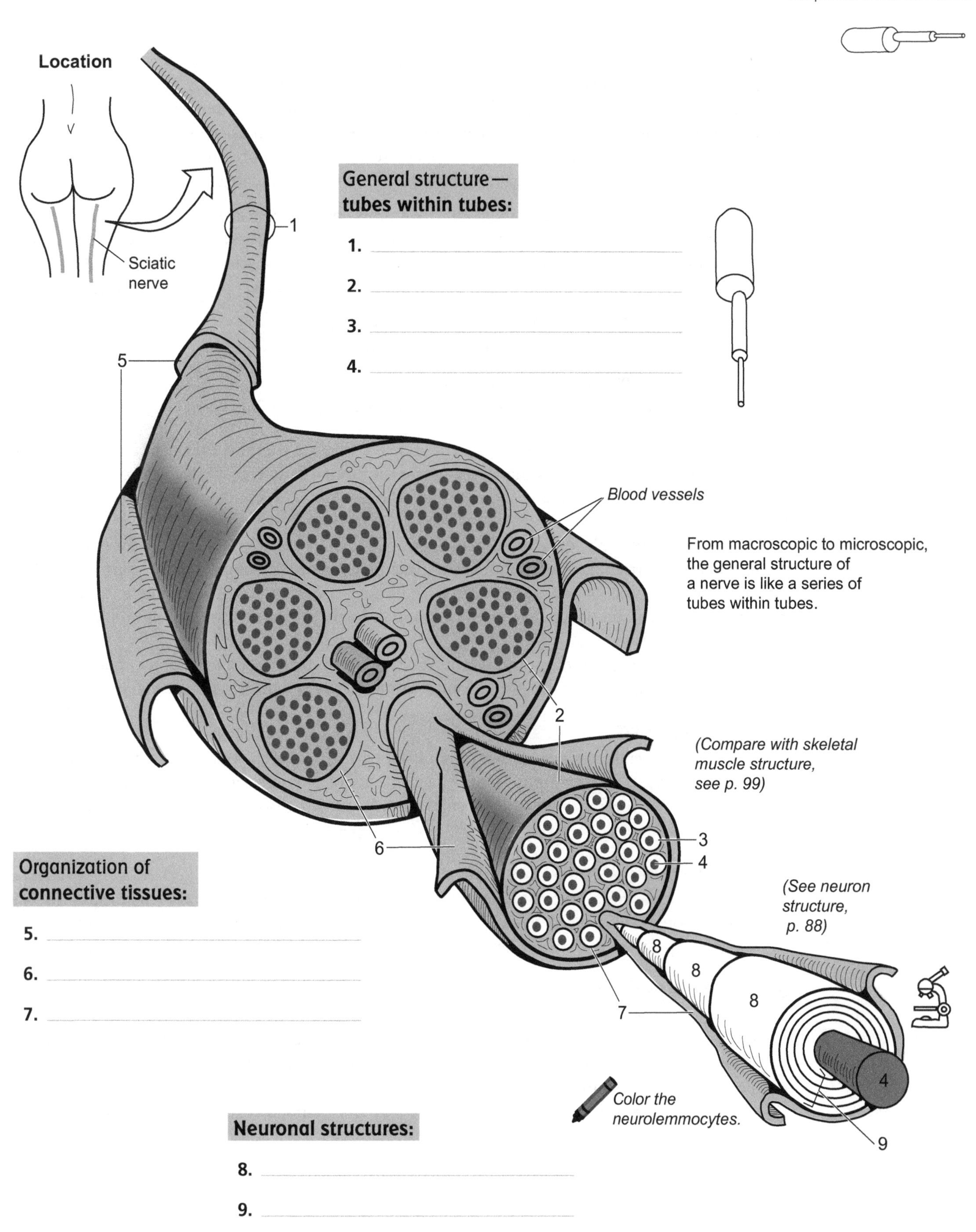
Location
Sciatic nerve
1
2
3
4
5
6
7
8
9
Blood vessels
General structure—tubes within tubes:
1.
2.
3.
4.
From macroscopic to microscopic, the general structure of a nerve is like a series of tubes within tubes.
(Compare with skeletal muscle structure, see p. 99)
(See neuron structure, p. 88)
Organization of connective tissues:
5.
6.
7.
Color the neurolemmocytes.
Neuronal structures:
8.
9.

Description

This module gives an overview of how sensory receptors detect a stimulus, convert it into an electrical impulse, and send it along a **sensory neuron** to be interpreted in the brain. A **sensory receptor** is a specialized structure to detect a specific stimulus. Many are simply the dendrites of sensory neurons encapsulated by some other tissue. They are found in locations such as the skin, joints, tendons, and blood vessel walls. Each type of receptor responds to a single type of stimulus—touch, temperature, pressure, pain. Receptors can be organized into the following major categories:

- **Mechanoreceptors:** detect stimuli that compress, bend, or stretch cells; allow for sensations such as touch, pressure, and vibration. *Ex.:* Lamellated corpuscles: mostly detect deep pressure; locations include lower dermis of the skin, joints, tendons, muscles, periosteum, and pancreas.
- **Chemoreceptors:** detect changes in chemical concentrations (H^+, O_2, CO_2) in a solution
- **Nociceptors:** detect pain resulting from tissue damage
- **Photoreceptors:** sensitive to light that strikes the retina of the eye
- **Thermoreceptors:** detect changes in temperature
- **Osmoreceptors:** detect changes in osmotic pressure of body fluids

Sensory transduction refers to the process of converting the stimulus energy into a nerve impulse. This conversion is essential, allowing all the sensory information to travel through the nervous system along sensory nerves to eventually be interpreted in the **brain**. Without it, no sensations would be perceived.

Let's use the example of a **lamellated corpuscle** as our sensory receptor. It primarily detects pressure but also stretching and vibration. Its structure consists of a single dendrite surrounded by an oblong-shaped **capsule** of up to 60 layers of fibrous connective tissue, giving it a sliced-onion appearance. Between the layers is a viscous material. In the capsule's core is a fluid-filled space that surrounds the dendrite.

Imagine pressing your finger against a flat surface and feeling the pressure that results. The illustration shows the process of sensory transduction. Here is a summary of each step:

① **Stimulus: pressure**

The normal membrane potential is symbolized by the " + " signs outside the membrane and the " – " signs inside the membrane (see p. 78). The act of pressing serves as a stimulus to compress the concentric layers within the capsule. This, in turn, deforms the plasma membrane of the dendrite.

② **Graded potential generated**

In response to this membrane deformation, the membrane becomes more permeable to Na+ (sodium ions). This triggers protein channels to open, and Na+ rapidly diffuses into the dendrite, thereby depolarizing the membrane. The small current generated is called a **graded potential.**

③ **Action potential generated**

A graded potential has different orders of magnitude. If it reaches a threshold or minimum strength, it can induce an **action potential**, or nerve impulse, in the first **node of Ranvier.** Because this sensory neuron is myelinated, it results in saltatory conduction (see p. 94). This is where the nerve impulse is shuttled from the first node to the second, then third, and so on. Finally, the nerve impulse travels from the sensory nerve and up the **spinal cord**, where it is interpreted in the cerebral cortex.

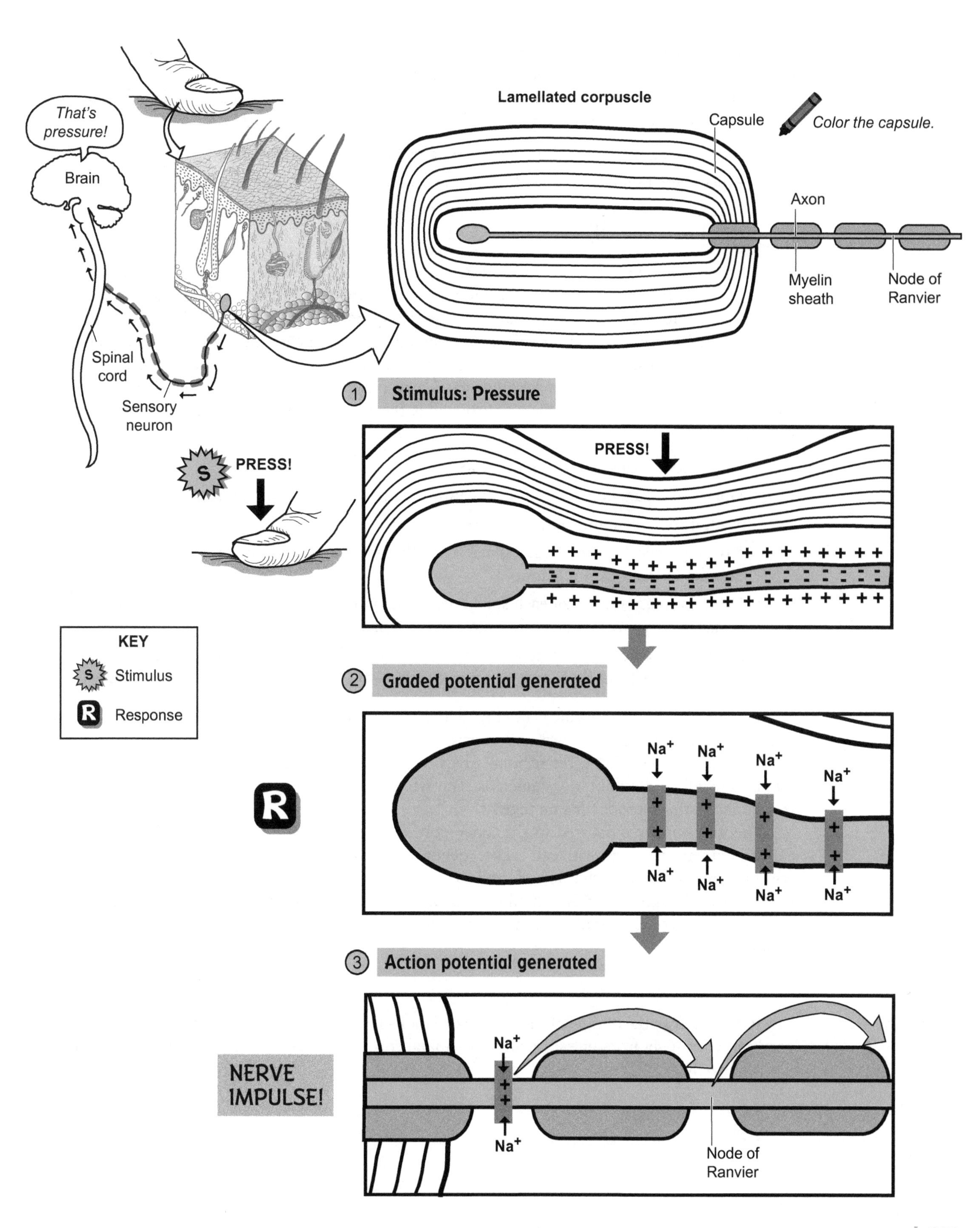
That's pressure!
Brain
Spinal cord
Sensory neuron
Lamellated corpuscle
Capsule
Color the capsule.
Axon
Myelin sheath
Node of Ranvier
1 Stimulus: Pressure
PRESS!
PRESS!
KEY
Stimulus
Response
2 Graded potential generated
Na+
3 Action potential generated
NERVE IMPULSE!
Na+
Node of Ranvier

Description

The **spinal cord** is a long, slender structure that links the body with the brain. Most of the cord is protected by the bony vertebrae because it runs through the vertebral canal of the vertebral column. Three layers of protective membranes called **meninges** surround the spinal cord and brain. From outermost to innermost, these are as follows:

- **Dura mater:** thickest and strongest; contains fibrous connective tissue
- **Arachnoid:** thin layer made of simple squamous epithelium; lacks blood vessels
- **Pia mater:** tightly adheres to the spinal cord and follows every surface feature; supplies many blood vessels directly to the spinal cord

Below the arachnoid is a potential space called the **subarachnoid space** that is filled with **cerebrospinal fluid.** This serves as a cushion to protect the spinal cord and functions as a medium through which to deliver nutrients and remove wastes. Extending laterally off the spinal cord are 31 pairs of **spinal nerves.** Each short spinal nerve immediately branches into different parts, such as the smaller dorsal ramus, the larger ventral ramus, and the rami communicantes. Each spinal nerve, along with these branches, is *mixed*, meaning it carries both sensory and motor axons. These become the various peripheral nerves that spread throughout the body. The spinal nerves and their associated structures are:

- **Dorsal root:** contains only *sensory* axons
- **Dorsal root ganglion:** contains the neuron cell bodies of sensory neurons
- **Ventral root:** contains only motor axons
- **Dorsal ramus:** branch of each spinal nerve that innervates certain parts of the skin and muscles of the neck and back
- **Ventral ramus:** branch of each spinal nerve that innervates the ventral and lateral body surfaces, structures in the body wall, and the limbs
- **Rami communicantes:** branches of each spinal nerve that contain axons related to the autonomic nervous system (ANS); consist of a pair of rami—one white ramus and one gray ramus

The spinal cord contains areas of **gray matter** and **white matter.** The white matter is located in the outer portion of the spinal cord and consists of myelinated axons that run along its length. It is divided into the following three regional areas called **funiculi** (sing. *funiculus*): a **posterior funiculus, lateral funiculi**, and an **anterior funiculus.** The **white commissure** is a narrow band of white matter that connects the anterior funiculi together.

The **gray matter** is located in the inner portion of the spinal cord and includes short neurons called interneurons along with cell bodies, dendrites, and axon terminals of other neurons. The three regional areas of gray matter are referred to as horns: **posterior horn, lateral horn, anterior horn.** In the center of the spinal cord is a small passageway called the **central canal** that contains cerebrospinal fluid. The horizontal band of gray matter that surrounds the central canal is called the **gray commissure.**

Analogy

The gray matter in the middle of the spinal cord looks like a butterfly. Note that the exact shape of the gray matter changes from one segment of the spinal cord to another, so it does not always look exactly like a butterfly.

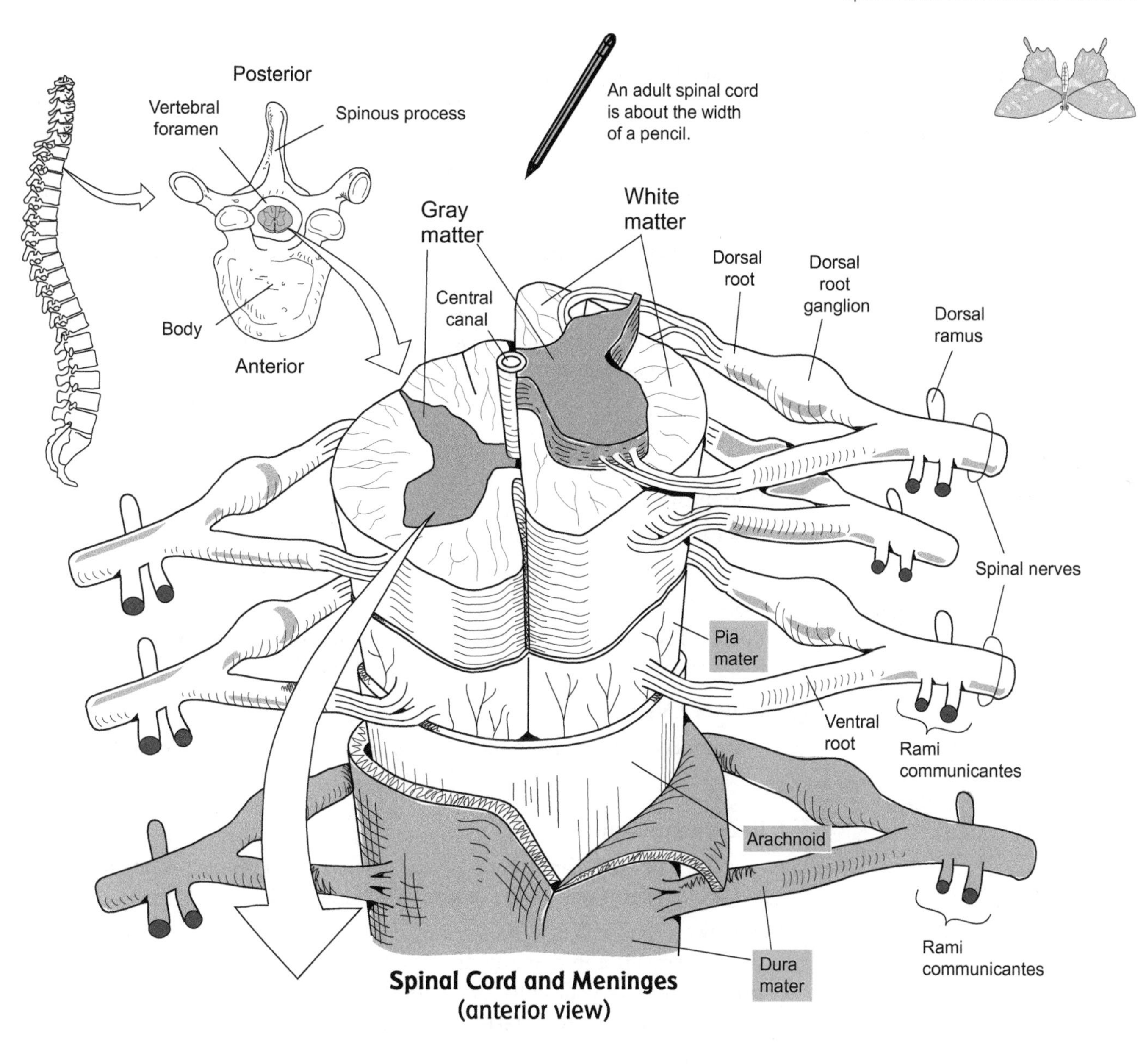

Spinal Cord and Meninges
(anterior view)

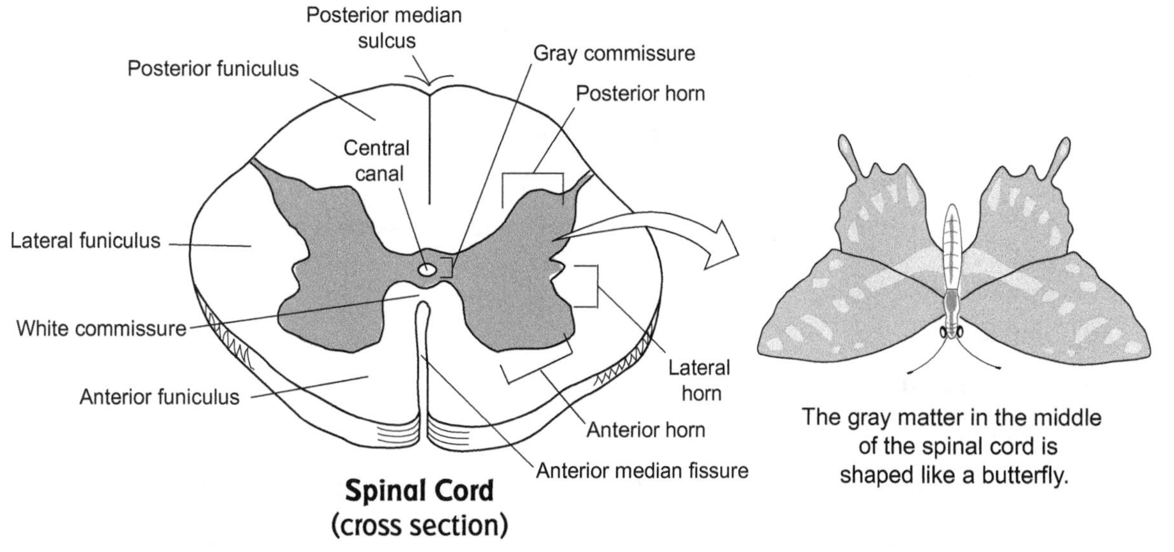

Spinal Cord
(cross section)

Description

A **reflex** is a rapid, involuntary response to a stimulus. A reflex arc refers to the neural pathway involved in a single reflex. The two general reflex categories are somatic reflexes and autonomic reflexes. Only **somatic** (*body*) **reflexes**—which result in the contraction of skeletal muscles—will be illustrated in this module. An example is the withdrawal reflex. If the skin on your finger is poked by a thumbtack, this stimulus is detected by pain receptors that trigger nerve impulses to contract flexor muscles to allow you to pull your hand away. During a physical exam, a physician may strike your patellar tendon with a percussion hammer. This stimulates your quadriceps femoris muscle to contract, resulting in extension of the knee joint.

The autonomic reflexes are so named because they are part of the autonomic nervous system (ANS). These reflex arcs stimulate responses in smooth muscle, cardiac muscle, and glands that allow for control of functions such as digestion, urination, and heart rate.

Somatic Reflex Arc

The parts of a typical somatic reflex arc are:

Input

① **Receptors**

—located at the end of sensory neurons

—respond to a specific stimulus (*ex.:* touch, pressure, or pain stimuli).

—different for different stimuli

In the illustration on the facing page, a thumbtack is shown penetrating the skin; this would be detected by pain receptors in the skin.

② **Sensory Neuron**

—conducts impulses from the receptor to the spinal cord (or brain)

Processing

③ **Integrating Center**

—regions of gray matter in the spinal cord or brain that act as the integrating center; simple reflexes pass only through the spinal cord

A polysynaptic reflex is shown on the facing page; it always includes at least one short **interneuron** in the gray matter of the spinal cord that transmits nerve impulses from the sensory neuron to the motor neuron

—monosynaptic reflexes lack the interneuron and simply directly connect the sensory neuron to the motor neuron

Output

④ **Motor Neuron**

—conducts impulses from the integration center to the effector

⑤ **Effector**

—the last component in a somatic reflex arc; skeletal muscle

Analogy

① **Input**

② **Processing**

③ **Output**

This is like typing the letter k in a word processor on your laptop. Pressing on the keyboard is the input. This sends a message to the microchip—the brain of the computer or processing center. Then the microchip sends a signal for the final output—the appearance of the letter k on the screen.

SOMATIC REFLEX ARC

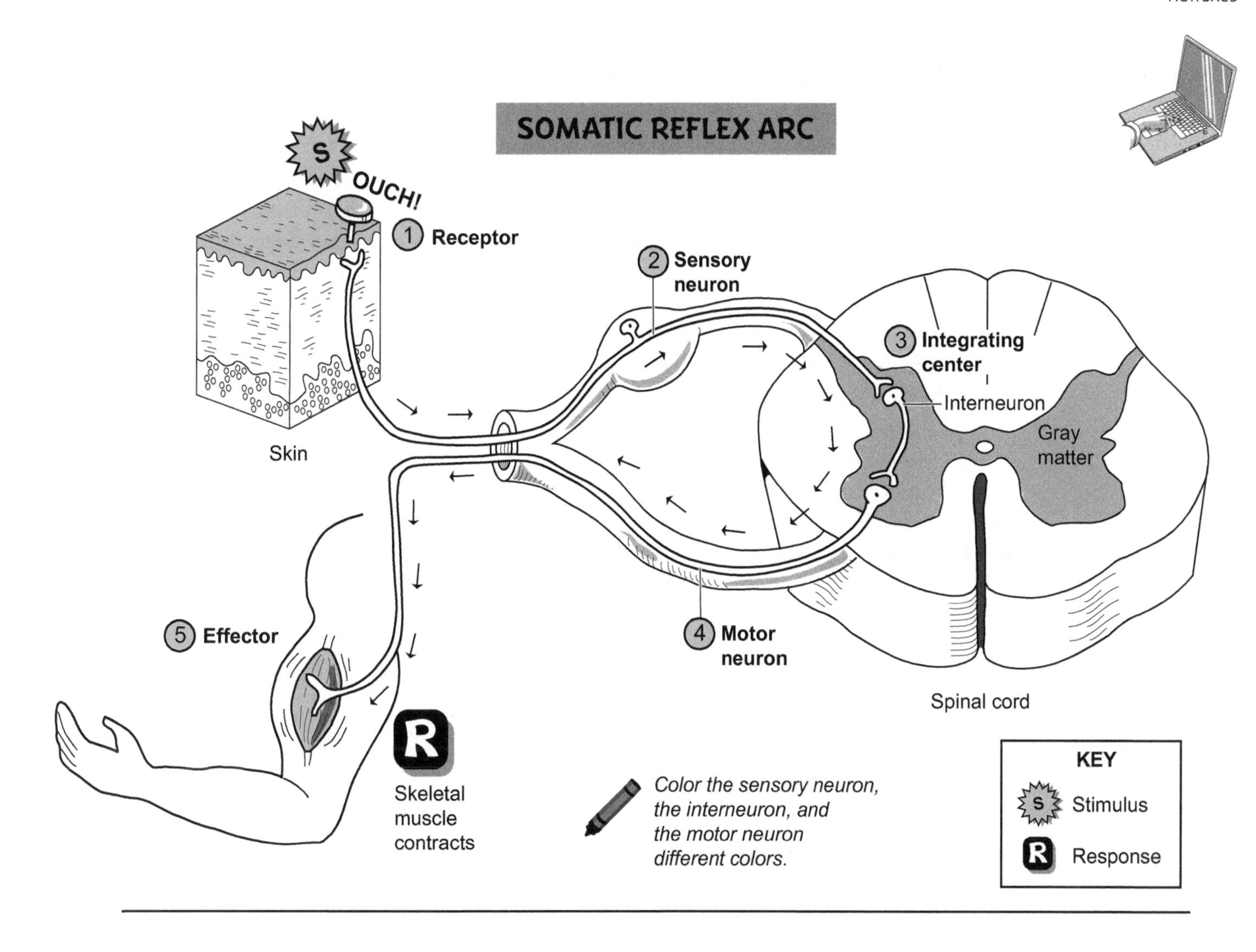

Color the sensory neuron, the interneuron, and the motor neuron different colors.

KEY

S Stimulus

R Response

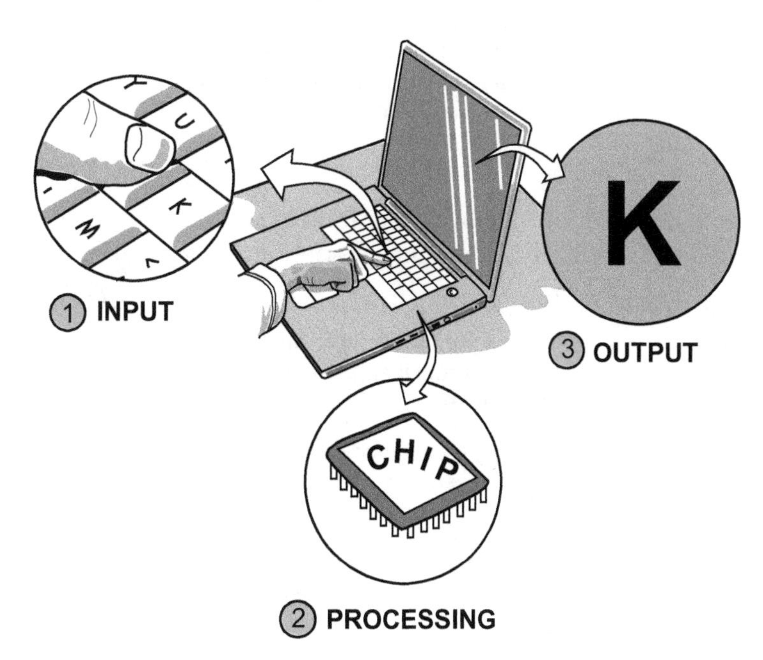

Description

The three *largest regions* of the brain are the **brainstem, cerebellum,** and **cerebrum.**

1. The **brainstem** is located at the base of the brain and contains regulatory centers to control things we take for granted, such as respiration, cardiovascular activities, and digestion.
2. The **cerebellum** is located posterior to the brainstem and inferior to the cerebrum. It is divided into two left or right halves, or **hemispheres,** and is extensively folded to increase surface area. Its general function is to work with the cerebrum to coordinate skeletal muscle movements, and it also allows the body to maintain proper balance and posture.
3. The **cerebrum** is the largest part of the brain and contains billions of neurons. Like the cerebellum, it is divided into two **hemispheres.** The deep division between these two hemispheres is called the **longitudinal fissure.** The term **fissure** indicates a deep groove or depression that separates major sections of the brain.

The surface of the cerebrum is not smooth but is folded into many little hills and gulleys. Each hill is called a **convolution** (or *gyrus*), and each valley is a shallow groove called a **sulcus.**

The cerebrum is the part of the brain associated with higher brain functions including planning, reasoning, analyzing, and storing/accessing memories. Ironically, without it, you would not be able to read and learn about the brain as you are doing now. It also perceives and interprets sensory information and coordinates various motor functions such as those involved in speech. The cerebrum is divided into four major lobes named after the bones that cover them: *frontal*, *parietal*, *temporal*, and *occipital.*

Brainstem

The **brainstem** consists of three parts: *medulla oblongata*, *pons*, and *midbrain.* The table below gives a description and general function of each part.

Brainstem Region	Description	General Functions
Medulla oblongata	Between spinal cord and pons	Respiratory control center; cardiovascular control center
Pons	Between medulla and midbrain; bulges out as widest region in brainstem	Controls respiration along with medulla; relays information from cerebrum to cerebellum
Midbrain	Between diencephalon and pons; includes corpora quadrigemina (*sensory relay station*) and cerebral aqueduct (*connects third and fourth ventricles; contains cerebrospinal fluid*)	Visual and auditory reflex centers; provides pathway between brainstem and cerebrum

Diencephalon

The **diencephalon** is located above the brainstem and contains three parts: *epithalamus, thalamus,*and *hypothalamus.* The table below gives a description and general function of each part.

Brainstem Region	Description	General Functions
Epithalamus	Roof of third ventricle; includes pineal gland; choroid plexus (forms cerebro-spinal fluid)	Pineal gland makes hormone melatonin, which regulates day-night cycles.
Thalamus	Two egg-shaped bodies that surround the third ventricle	Relays sensory information to cerebral cortex; relays information for motor activities; information filter
Hypothalamus	Forms floor of third ventricle; between thalamus and chiasm	Controls autonomic centers for heart rate, blood pressure, respiration, digestion, hunger center, thirst center, regulation of body temperature, production of emotions

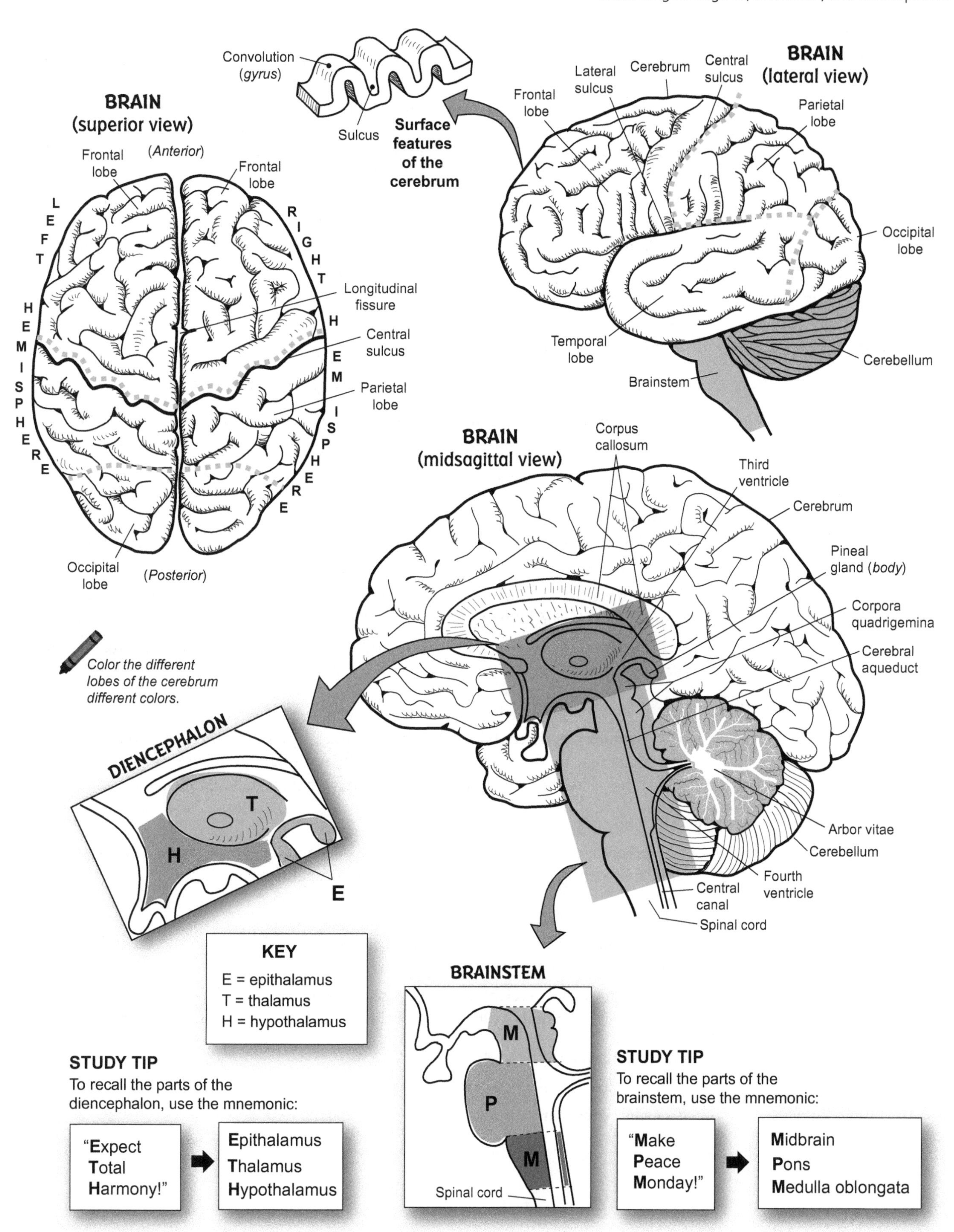
BRAIN
(superior view)
(Anterior)
Frontal lobe
Frontal lobe
LEFT HEMISPHERE
RIGHT HEMISPHERE
Longitudinal fissure
Central sulcus
Parietal lobe
Occipital lobe
(Posterior)
Convolution (gyrus)
Sulcus
Surface features of the cerebrum
BRAIN
(lateral view)
Frontal lobe
Lateral sulcus
Cerebrum
Central sulcus
Parietal lobe
Occipital lobe
Temporal lobe
Brainstem
Cerebellum
BRAIN
(midsagittal view)
Corpus callosum
Third ventricle
Cerebrum
Pineal gland (body)
Corpora quadrigemina
Cerebral aqueduct
Arbor vitae
Cerebellum
Fourth ventricle
Central canal
Spinal cord
Color the different lobes of the cerebrum different colors.
DIENCEPHALON
T
H
E
KEY
E = epithalamus
T = thalamus
H = hypothalamus
STUDY TIP
To recall the parts of the diencephalon, use the mnemonic:
"Expect Total Harmony!"
Epithalamus
Thalamus
Hypothalamus
BRAINSTEM
M
P
M
Spinal cord
STUDY TIP
To recall the parts of the brainstem, use the mnemonic:
"Make Peace Monday!"
Midbrain
Pons
Medulla oblongata

Description

Though small in size, the **hypothalamus** in the **brain** controls many vital body functions. In general, it controls autonomic nervous system (**ANS**) centers for heart rate, blood pressure, respiration, and digestion. It also serves as our hunger center and thirst center, regulator of body temperature, and producer of emotions. It even plays a role in memory. As part of the diencephalon, it is located between the **thalamus** and the **optic chiasm.** The **anterior commissure** and **mamillary body** form the borders on either side. The **infundibulum** is a stalk-like structure that connects the hypothalamus to the **pituitary gland.** Neurons in the hypothalamus are wired directly into the **posterior pituitary** gland to create a vital link between the nervous system and the endocrine system.

The hypothalamus consists of many different nuclei—clusters of cell bodies. These nuclei correlate to the specific functions listed below.

Functions

The hypothalamus has five major functions:

1. Controls the autonomic nervous system (**ANS**) and the **endocrine system.**

   For the **ANS**, which is divided into two divisions—sympathetic and parasympathetic:

   - The **anterior nucleus** controls the parasympathetic division.
   - The **dorsomedial nucleus** controls the sympathetic division.

   For the **endocrine system:**

   - Notice the colored lines in the illustration linking the paraventricular nucleus and the supraoptic nucleus to the posterior pituitary gland.
   - The **paraventricular nucleus** releases **OT** (*oxytocin*), the hormone that stimulates uterine contractions during labor and delivery.
   - The **supraoptic nucleus** releases **ADH** (*antidiuretic hormone*), important for maintaining water balance.

2. Regulates emotional behavior.

   - The **mamillary body** contains relay stations for the sense of smell. In addition, it links the hypothalamus to the limbic system—our "emotional brain" (see p. 264). The limbic system helps us regulate basic emotions such as rage, fear, happiness, and sadness.

3. Regulates body temperature.

   - The **preoptic area** works like a thermostat to maintain our normal body temperature of 98.6°F by regulating shivering and sweating mechanisms.

4. Regulates food and water intake.

   - The **anterior nucleus** is the thirst center, checking dissolved solutes in the blood. As solute levels increase, fluid intake is stimulated.
   - The **ventromedial nucleus** is the hunger center, checking levels of glucose and other nutrients in the blood. As these nutrient levels decrease, food intake is stimulated.

5. **Regulates sleep/wake cycles.**

   - These cycles are called *circadian rhythms*.
   - The **suprachiasmatic nucleus** is the control center.

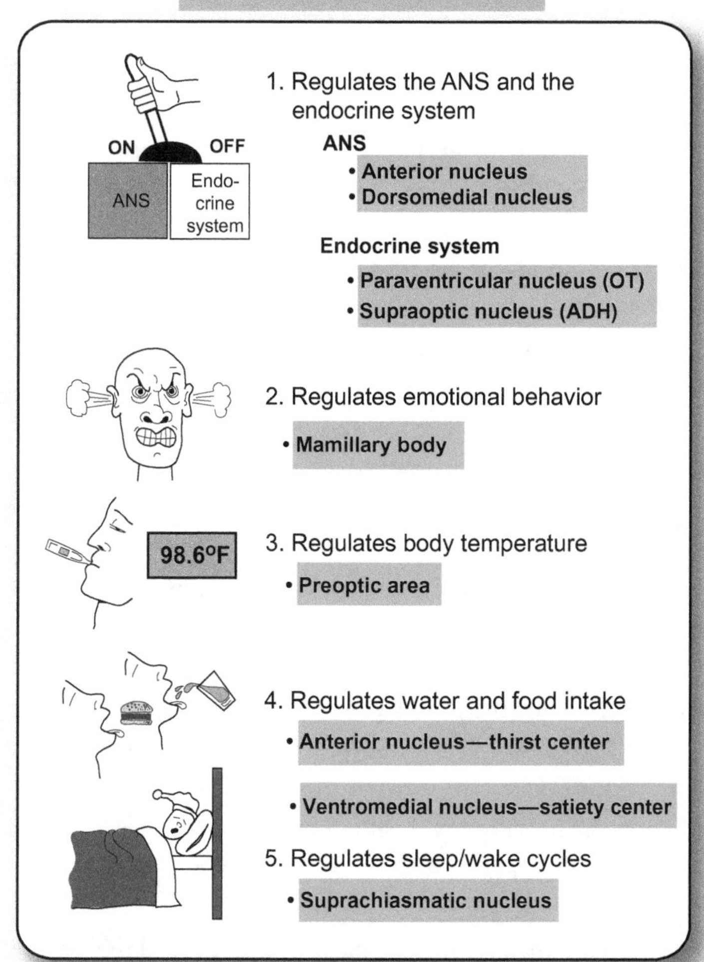

BRAIN
(midsagittal view)

Thalamus
Anterior commissure
Paraventricular nucleus
Dorsomedial nucleus
Posterior nucleus
Preoptic area
Anterior nucleus
Supraoptic nucleus
Suprachiasmatic nucleus
Mamillary body
Ventromedial nucleus
Infundibulum
Optic chiasm
Pons
Anterior pituitary
Posterior pituitary
Pituitary gland

Color each of the different nuclei a different color.

HYPOTHALAMUS
(midsaggittal view)

This module will describe some of the selected functional areas of the cerebral cortex. These areas have been divided into three general groups: **sensory areas, motor areas**, and **association areas.** Note that the words *cortex* and *area* are often used interchangeably.

SENSORY AREAS Control regions where sensations are perceived

1. **Primary somatic sensory cortex**	This important region is shown in dark gray behind the central sulcus. General sensory input (e.g., touch, temperature, pressure, and pain) from all parts of the body is perceived here.
2. **Gustatory cortex**	Located in the parietal lobe; taste sensations are perceived here, such as the flavors of the ice cream shown in the icon.
3. **Auditory cortex**	Located in the temporal lobe; auditory stimuli are processed by the brain here.
4. **Visual cortex**	Located in the occipital lobe; visual images are perceived here (like the star shown in the icon).

MOTOR AREAS Control centers for conscious muscle movements

1. **Primary motor cortex**	This important area is shown in color in front of the central sulcus. It controls voluntary muscle movements throughout the body, including those of the hands and feet, arms and legs, face and tongue.
2. **Premotor cortex**	This area serves as the "choreographer" for the primary motor cortex. It decides which muscle groups will be used and how they will be used prior to stimulating the primary motor cortex.
3. **Motor speech area** (*Broca's area*)	This area controls and coordinates the muscles involved in normal, fluent speech. Damage to this area can result in strained speech with disconnected words. Located in only one hemisphere, usually left.
4. **Frontal eye field**	This area controls muscle movements of the eye, such as those needed to read this page.

ASSOCIATION AREAS Control regions—near sensory areas—involved in recognizing and analyzing incoming information

1. **Prefrontal area**	This area is most highly developed in humans and other primates. It regulates emotional behavior and mood and also is involved in planning, learning, reasoning, motivation, personality, and intellect.
2. **Somatic sensory association area**	This area allows you to *predict* that sandpaper is rough, for example, even without looking at it. It also stores memories about previous sensory experiences, so you can determine when blindfolded, for example, that the object placed in your hand was a pair of scissors.
3. **Sensory speech area** (*Wernicke's area*)	This area seems to be an important part of language development—processing words we hear being spoken. It also appears important for children when they are sounding out new words. Damage to this area may result in deficiencies in recognizing written and spoken words.
4. **Auditory association area**	This area allows you to comprehend, interpret, analyze, and question what you are hearing. For example, it enables you to recognize a familiar song or disregard noise.
5. **Visual association area**	This allows you to associate the perceived image of the star with the letters "S-T-A-R". You connect the word "star" with the image of a star.

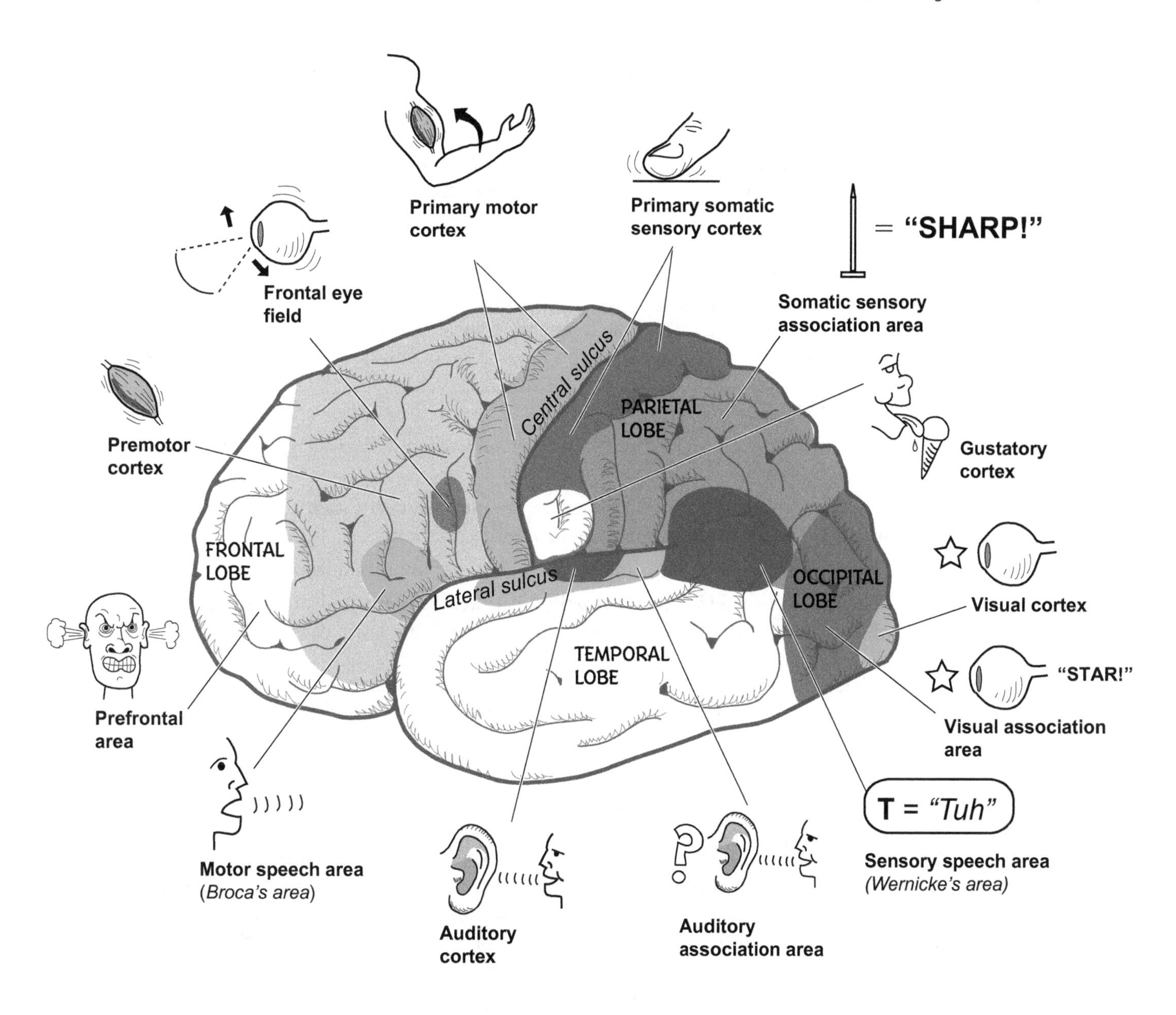

Left Cerebral Hemisphere
(lateral view)

Description

This module explains electroencephalogram (EEG), sleep/wakefulness, and reticular formation.

EEG

An **electroencephalogram** (EEG) is to the nervous system what an **ECG** (or EKG) is to the cardiovascular system (see p. 116). By detecting the electrical activity of the billions of neurons in the brain, an EEG measures normal brain function. **Electrodes** are placed around the scalp to detect the small voltage generated from these nerve impulses. Wires connect the electrodes to an EEG machine, where the voltage is amplified so it can be seen as characteristic waves. Each person has her own unique brain waves, displayed as a graph of **voltage** (millivolts or mV) over **time**. The height of the wave is the **amplitude** (in mV), and the frequency is measured in **hertz** (Hz), or cycles per second. The following is a summary of the four common types of brain waves:

Sleep/Wakefulness

- **Alpha** waves (8–13 Hz) occur when a person is calm and awake with eyes closed. This is like the restful state of the brain.
- **Beta** waves (14–30 Hz) occur when we are mentally alert, such as balancing your checkbook or studying a map to find your way to your destination.
- **Theta** waves (4–7 Hz) are common in children but abnormal in wakeful adults.
- **Delta** waves (< 4 Hz) occur normally in wakeful infants and in adults during deep sleep or while under anesthesia. In wakeful adults, they indicate brain damage.

EEGs also can be used to detect abnormalities in the brain, such as epilepsy, brain tumors, and traumatic injuries. The absence of any brain waves is the legal definition of what is called *brain dead*.

Reticular Formation

The brain never really "turns off." Neurons are sending impulses to each other constantly whether a person is awake or asleep, but the brain needs to have methods to control these states of consciousness. Sleep states are regulated through the hypothalamus. Wakeful states are controlled by the **reticular formation**—a cluster of neurons centrally located and extending like columns through the brainstem—*midbrain*, **pons**, and **medulla oblongata.** One subset of these neurons—the **reticular activating system** (**RAS**) acts like an alarm system that keeps the brain alert. The RAS sends stimulatory impulses through nerve pathways to other key parts of the brain, such as the cerebral cortex, thalamus, hypothalamus, cerebellum, and spinal cord. Sensory input is delivered to the RAS so it can filter out everything that is not important. Otherwise, we would go insane from sensory overload. Only the most important sensory input is routed to the cerebral cortex so we can be aware of it. For example, you probably are unaware of the sensation of your silky, long-sleeved shirt against your arm, but you would notice if someone suddenly were to pull on it.

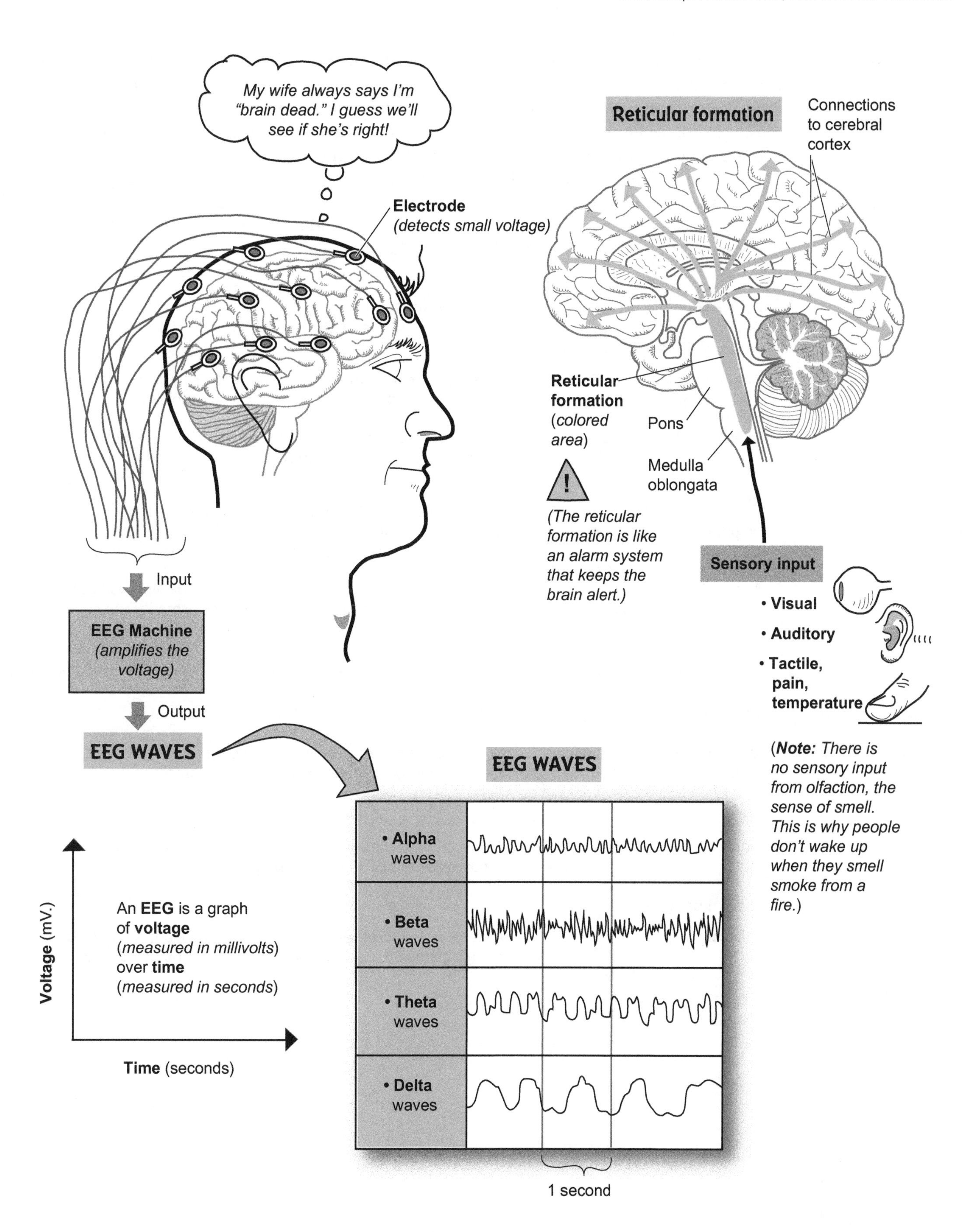
My wife always says I'm "brain dead." I guess we'll see if she's right!
Electrode
(detects small voltage)
Input
EEG Machine
(amplifies the voltage)
Output
EEG WAVES
Reticular formation
Connections to cerebral cortex
Reticular formation
(colored area)
Pons
Medulla oblongata
(The reticular formation is like an alarm system that keeps the brain alert.)
Sensory input
• Visual
• Auditory
• Tactile, pain, temperature
(Note: There is no sensory input from olfaction, the sense of smell. This is why people don't wake up when they smell smoke from a fire.)
EEG WAVES
• Alpha waves
• Beta waves
• Theta waves
• Delta waves
1 second
Voltage (mV.)
An EEG is a graph of voltage (measured in millivolts) over time (measured in seconds)
Time (seconds)

Description

The **limbic** (*limbus* = border) **system** is called our "emotional brain" because it processes our basic emotions such as fear, happiness, and sadness. To achieve this, it must integrate different brain centers, such as sensory systems and conscious control centers. Physically, it is a collection of structures found mostly in the cerebrum and diencephalon (thalamus, hypothalamus, epithalamus). It is so named because these structures form a border around the diencephalon.

Scientists still do not universally agree on the structures that compose the limbic system. The table below lists the most commonly recognized structures, with similar structures grouped together. The term *nuclei* refers to clusters of cell bodies in the CNS, and *tracts* are bundles of axons in the CNS. The structures in the table are illustrated on the facing page.

Limbic System Structure	Description
Cerebral Cortex Structures	
Cingulate (*cingulum* = girdle, to surround) **gyrus**	A mass of cerebral cortex located above the corpus collosum; its name describes the fact that it surrounds the diencephalon
Parahippocampal gyrus	A mass of cerebral cortex in the temporal lobe; works with the hippocampus to store memories
Nuclei	
Hippocampus (*hippocampos* = seahorse)	Anatomists thought this nucleus looked like a seahorse; located above the parahippocampal gyrus; connects to the diencephalon through the fornix; important in learning and storing long-term memories
Amygdaloid (*amygda* = almond shaped) **body**	Nucleus located in front of and connected to the hippocampus; directly involved in emotions such as fear; can store memories and correlate them with specific emotions
Anterior thalamic nucleus	Located in the thalamus; relays visceral sensations to the cingulate gyrus
Septal nucleus	Nucleus located near the anterior commissure
Mamillary (*mammilla* = nipple) **body**	Nipple-like structures in the floor of the hypothalamus that serve as relay stations for reflexes related to sense of smell
Tracts	
Fornix (arch)	A tract of white matter that connects the hippocampus to the hypothalamus; a portion of it is shaped like an arch
Other Structures	
Olfactory bulbs	All serve to interpret various odors in the brain; odors are linked to specific memories, so certain odors can trigger powerful emotions

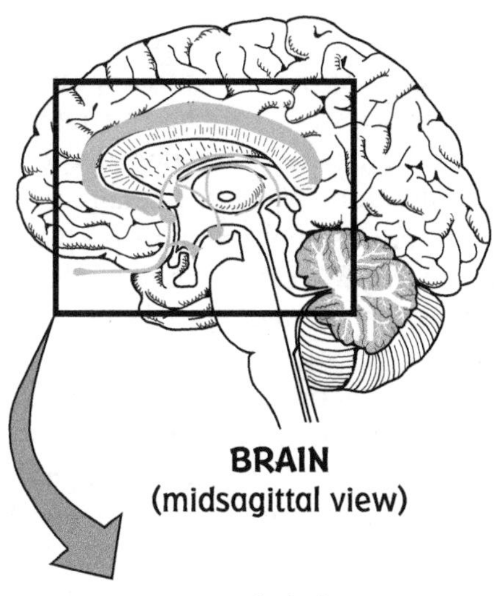

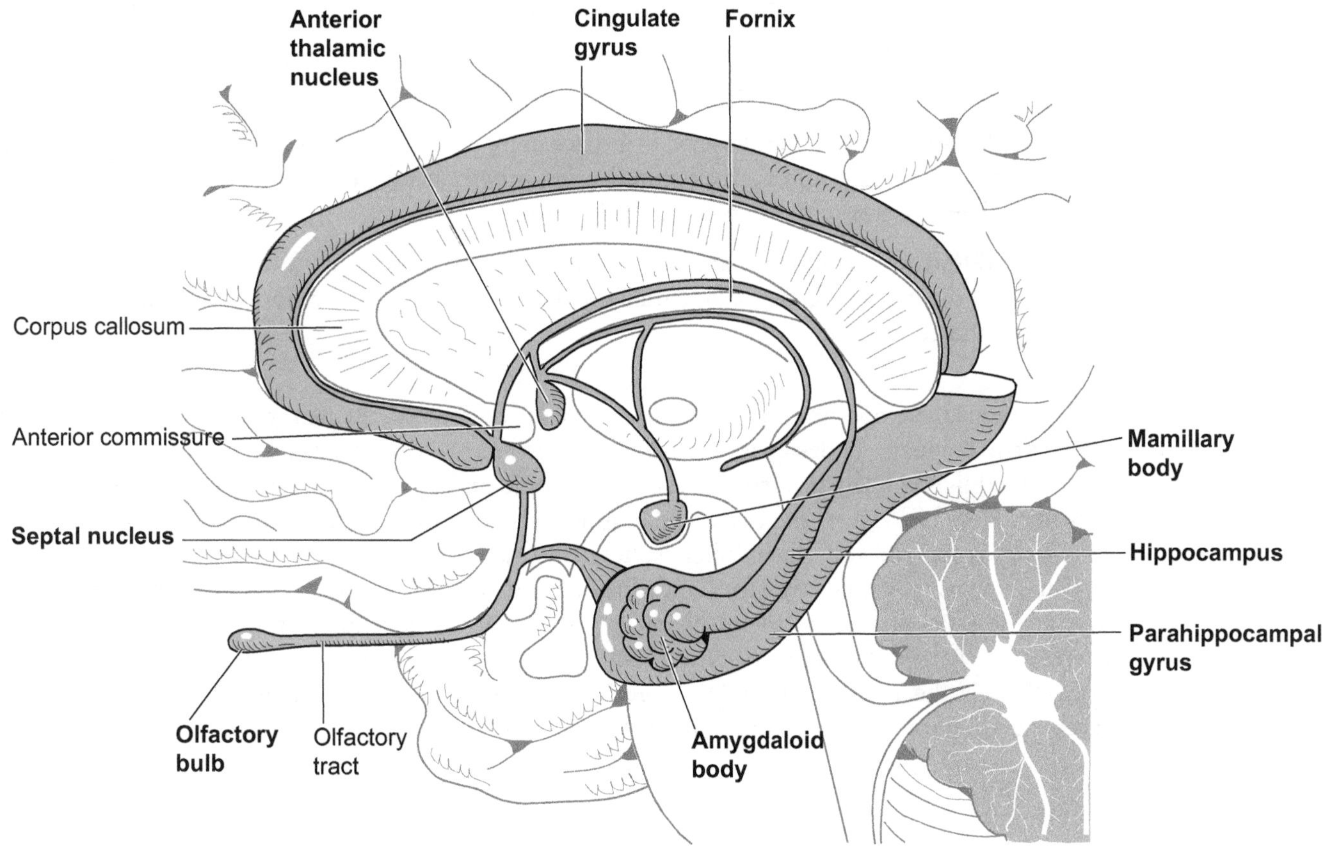

Midsagittal View

Note: *Major structures in the limbic system are shown in color.*

Description

This module explains the concept of lateralization between the left and right cerebral hemispheres. Looking at the illustration of the **brain**, the left and right cerebral hemispheres appear very similar. In fact, they are anatomically similar, and the two hemispheres work together for many functions. This is evidenced by the **corpus callosum**—a thick band of axons connecting the left and right hemispheres. Each hemisphere also has functional specialization. There is lateralization in which certain functions are found in only one of the hemispheres. For example, in most people, the Broca's area for speech production is found only in the left hemisphere. We can make generalizations about the functional differences between the two hemispheres that apply to most people. Consider your left hemisphere your "analytical" hemisphere and your right your "creative" hemisphere.

The illustration lists the functional differences. Your left brain excels at language and logic. It deals with information in an organized, logical way as a scientist would. It helps you work with mathematical equations, write, and follow directions step by step. In contrast, your right hemisphere excels at musical and artistic abilities. It helps you understand shape and pattern relationships useful for facial recognition and drawing. It also is the seat of insight and inspiration.

But these generalizations are not set in stone. Here are some variables that are exceptions to the rule:

- **Individual differences.** Some individuals have one or more control centers in the hemisphere opposite from the one where it normally is found.
- **Gender differences.** Lateralization is greater in males than females. In typical females, a portion of the corpus callosum is thicker, indicating greater hemispheric integration. This means that both hemispheres work together more frequently.
- **Age differences.** Children can "rewire" their brains more easily than adults. For example, if part of the brain is damaged or surgically removed in a child, the opposite hemisphere can take over and compensate.

One hemisphere doesn't actually dominate the other. Even so, the hemisphere that controls spoken and written language is designated as the *categorical* (or "dominant") hemisphere. As mentioned previously, this is the left hemisphere for most people and correlates to handedness. Because nerves cross over from one side of the body to the opposite side in the brain, motor activity on the right side of the body is controlled by the left hemisphere, and vice versa.

The same usually holds true for handedness. About 91% of the population is right-handed, and in most of these people the left hemisphere is the categorical one. Interestingly, the situation is a bit different for "lefties." In the majority of them, the left hemisphere is still their categorical one. In only about 15% is the right hemisphere categorical. In summary, although the two hemispheres work together all the time, they also specialize in specific functions.

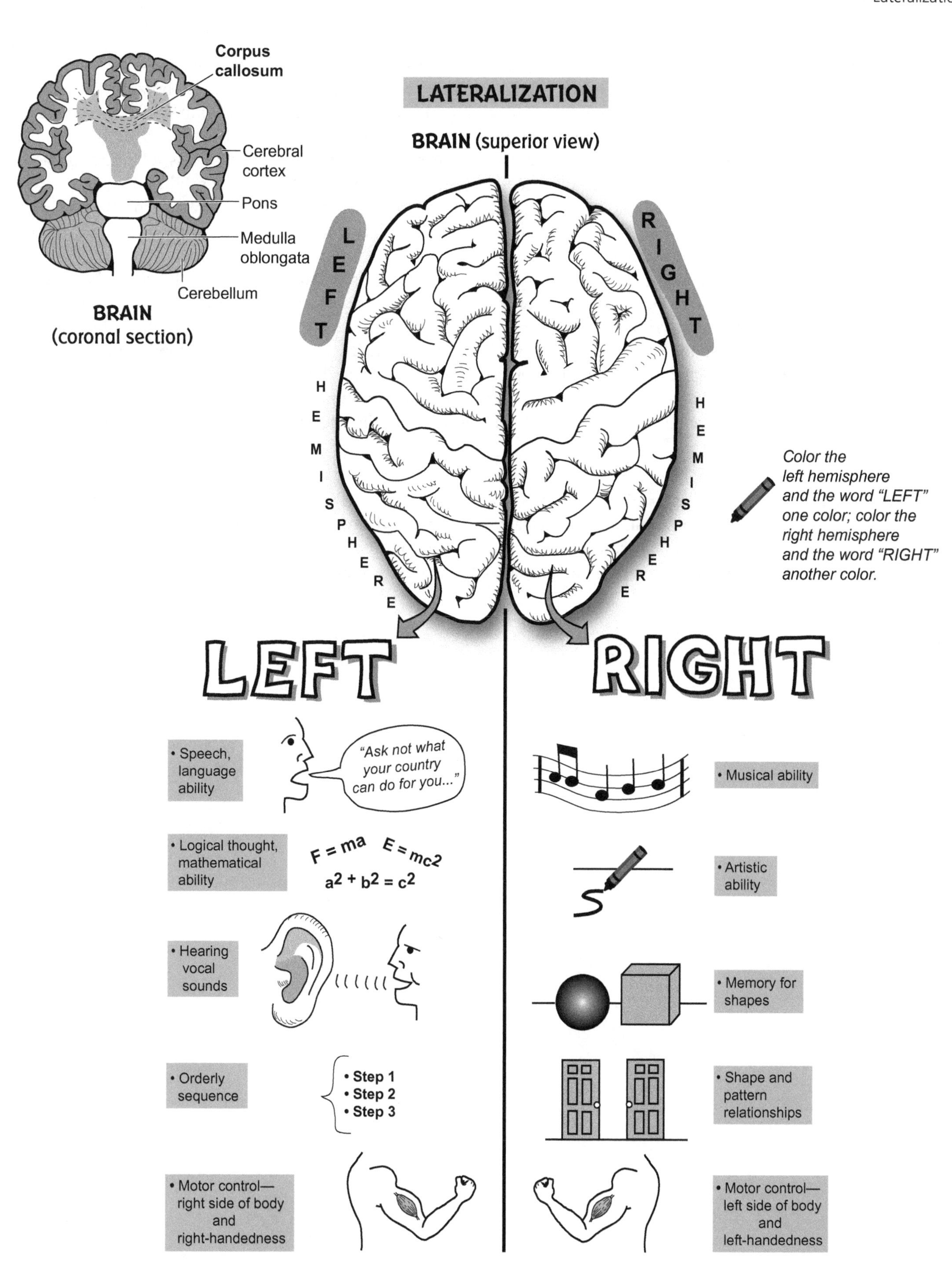
Corpus callosum
Cerebral cortex
Pons
Medulla oblongata
Cerebellum
BRAIN
(coronal section)
LATERALIZATION
BRAIN (superior view)
LEFT
HEMISPHERE
RIGHT
HEMISPHERE
Color the left hemisphere and the word "LEFT" one color; color the right hemisphere and the word "RIGHT" another color.
LEFT
RIGHT
• Speech, language ability
"Ask not what your country can do for you..."
• Logical thought, mathematical ability
F = ma E = mc2
a2 + b2 = c2
• Hearing vocal sounds
• Orderly sequence
• Step 1
• Step 2
• Step 3
• Motor control—right side of body and right-handedness
• Musical ability
• Artistic ability
• Memory for shapes
• Shape and pattern relationships
• Motor control—left side of body and left-handedness

Description

Though the **brain** makes up only 2% of body mass, it uses about 20% of the body's glucose and oxygen. The brain never really shuts off; it's even active during sleep. This is because the billions of **neurons** in the brain are constantly sending nerve impulses along their axons, which requires large amounts of **ATP**. This process occurs in mitochondria within the **cell body** of the neuron. These hungry cells require a constant supply of **glucose** and **oxygen** to produce this ATP through **aerobic respiration**. For each glucose molecule broken down, up to 36 ATP can be produced.

Unlike skeletal muscle cells, which contain a stored form of glucose called **glycogen**, the brain has no glycogen stores. Skeletal muscle also contains a protein called **myoglobin** that temporarily binds and stores large amounts of oxygen.

Unfortunately, the brain has no myoglobin deposits. To compensate, the brain has elaborate networks of blood vessels to directly deliver all the glucose and oxygen it needs through the blood. The **carotid arteries** transport oxygenated blood to the brain. But to protect vital neurons, a brain barrier system restricts what specific substances can be transported from the blood to the brain tissues. This prevents toxins that may enter the bloodstream from damaging neurons. Fortunately, this system is permeable to glucose, oxygen, carbon dioxide, sodium, potassium, and chloride. As a result, it doesn't present a problem for delivering nutrients to the neurons or eliminating wastes from them.

Cardiac Arrest

If a person has cardiac arrest, blood stops flowing to the brain. Of all the different cells in the body, neurons in the brain are most likely to die if blood flow is not restored. This is why **CPR** (cardio-pulmonary resuscitation) is performed. Without it, neurons begin to die in about 3 to 4 minutes. In fact, this occurs any time blood flow to the brain is interrupted.

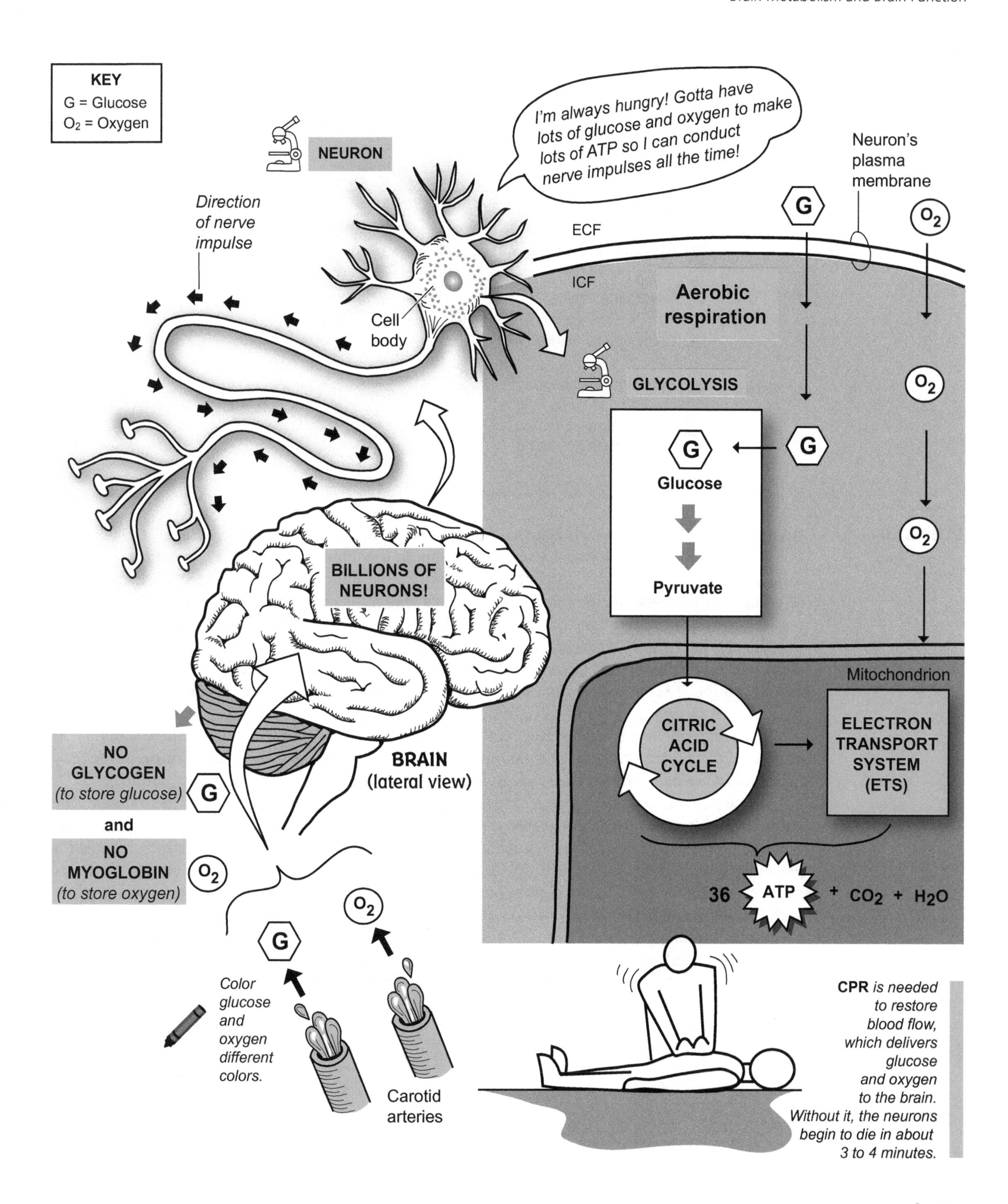
KEY
G = Glucose
O2 = Oxygen
NEURON
I'm always hungry! Gotta have lots of glucose and oxygen to make lots of ATP so I can conduct nerve impulses all the time!
Neuron's plasma membrane
Direction of nerve impulse
ECF
ICF
Aerobic respiration
Cell body
GLYCOLYSIS
G
Glucose
Pyruvate
O2
BILLIONS OF NEURONS!
Mitochondrion
CITRIC ACID CYCLE
ELECTRON TRANSPORT SYSTEM (ETS)
NO GLYCOGEN (to store glucose)
and
NO MYOGLOBIN (to store oxygen)
BRAIN (lateral view)
36 ATP + CO2 + H2O
Color glucose and oxygen different colors.
Carotid arteries
CPR is needed to restore blood flow, which delivers glucose and oxygen to the brain. Without it, the neurons begin to die in about 3 to 4 minutes.

Description

Somatic reflexes were covered previously (see p. 254). This module compares and contrasts somatic reflexes and autonomic reflexes. Somatic reflexes stimulate the contraction of skeletal muscle. **Autonomic reflexes** are used by the autonomic nervous system (ANS) to control many of the involuntary functions we take for granted, such as digestion, urination, and heart rate.

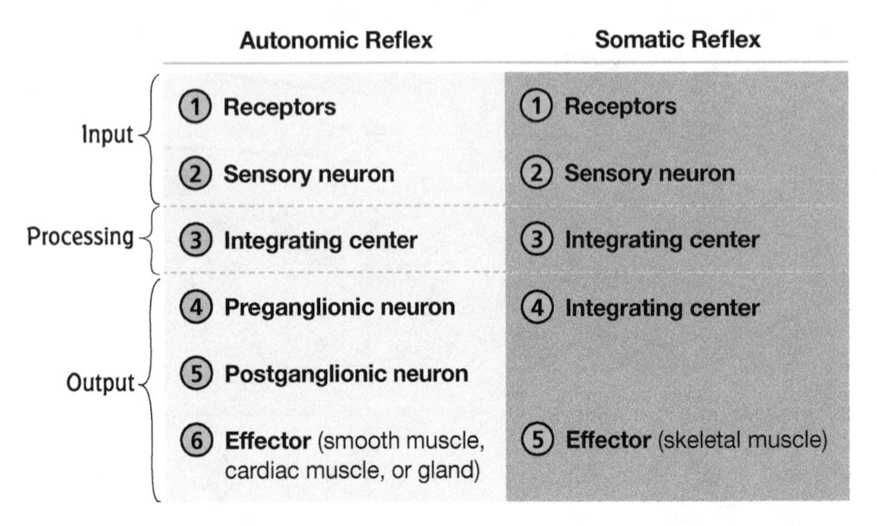

Pathway

All reflexes involve a pathway called an arc that includes input, processing, and output.

Input involves a receptor and a **sensory neuron.** In an autonomic reflex, the receptor typically is located on the organ/gland it regulates. For example, the illustration shows a section of the small intestine. During digestion, the smooth muscle in its wall has to be stimulated to contract and thereby propel digested food through the intestine. This has to occur only when digested food is moving through the tract. In this example, a stretch receptor is located in the wall of the small intestine to detect when food is present. In contrast, most somatic receptors are located in the skin, as shown in the illustration. The sensory neuron has the same function in both reflexes: It conducts the impulse from the receptor to the integrating center.

Processing involves the **integrating center**—the spinal cord and/or the brain—where the input is connected to the output. In the illustration both reflex types contain a short **interneuron** in the gray matter of the spinal cord that transmits impulses from the sensory input to the motor output.

The biggest difference between somatic and autonomic reflexes is seen in the output. The autonomic reflex uses two motor neurons—a **preganglionic neuron** and a **postganglionic neuron**, as opposed to the single neuron in a somatic reflex. This introduces an additional structure called a **ganglion**—a collection of neuronal cell bodies outside the CNS. Notice that the autonomic reflex contains a ganglion in its motor output, but the somatic reflex does not. Now compare that ganglion with the one found in the sensory neuron. Both contain cell bodies, but only the ganglion in the motor output has a new synapse where the preganglionic neuron meets the postganglionic neuron.

Finally, the **effector** is also different. In somatic reflexes skeletal muscle is the only effector. An autonomic reflex has three different types of effectors: smooth muscle, cardiac muscle, or glands.

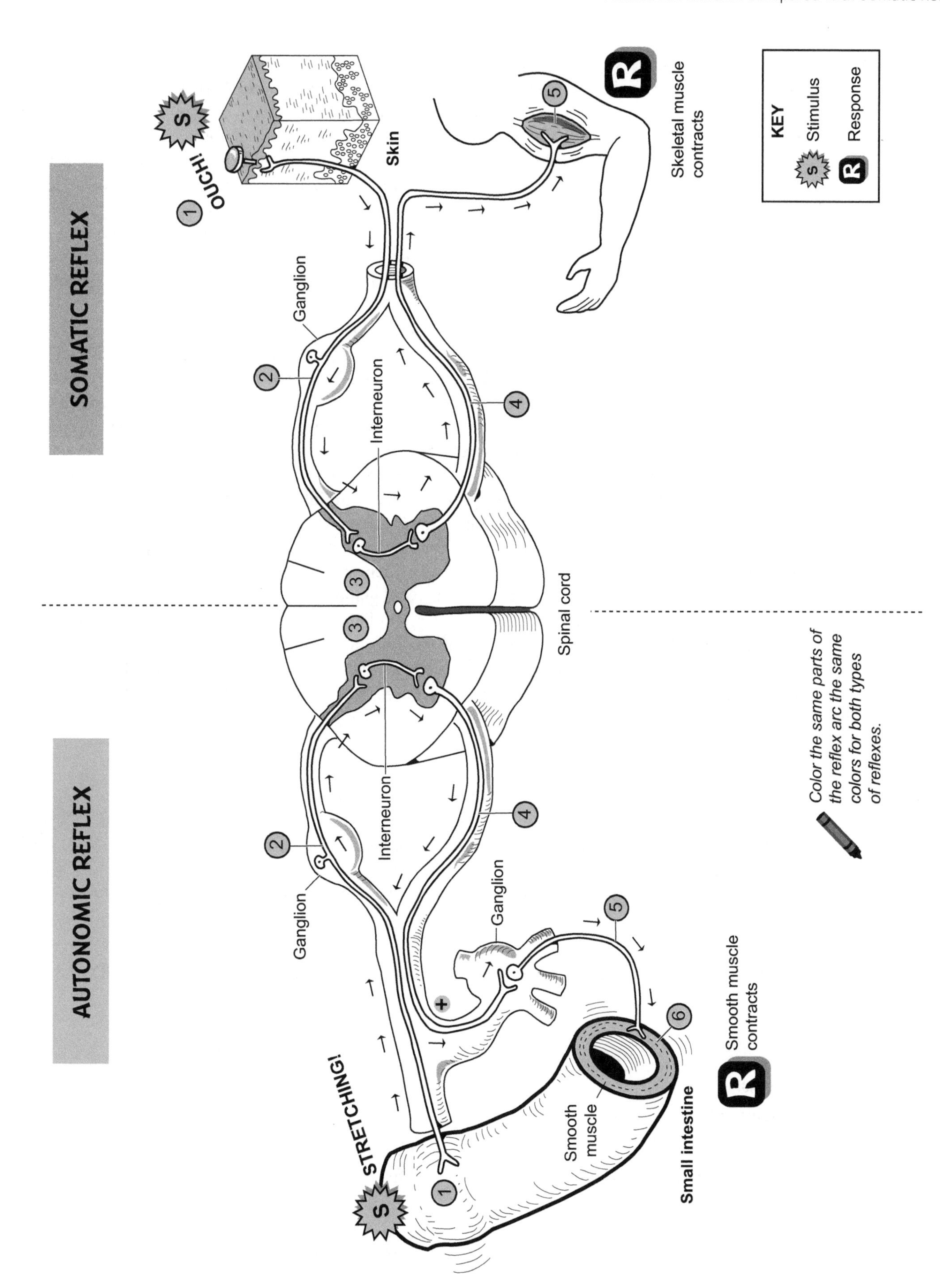
SOMATIC REFLEX
1
OUCH!
S
Skin
R
5
Skeletal muscle contracts
Ganglion
2
Interneuron
4
3
KEY
S Stimulus
R Response
Spinal cord
3
AUTONOMIC REFLEX
2
Interneuron
4
Ganglion
Ganglion
5
STRETCHING!
S
1
Smooth muscle
Small intestine
6
R
Smooth muscle contracts
Color the same parts of the reflex arc the same colors for both types of reflexes.

Description

The **autonomic nervous system** (ANS) controls involuntary activities such as digestion and heart rate. It is subdivided into two divisions: **sympathetic** (**SD**) and **parasympathetic** (**PD**). These two divisions work together to regulate normal functions and maintain **homeostasis.** Each division has its own groups of nerves that connect to the major visceral organs in the body. But there are some exceptions. For example, the adrenal gland is controlled only by the SD because the PD has no nerves connected to it.

This module gives an overview of the SD, also referred to as the *thoracolumbar division*, as motor output arises from these sections of the spinal cord. Because it prepares the body for emergency situations, it is nicknamed the "fight-or-flight" system. Consider what happens when you get nervous before a test you are not prepared to take. Your heart rate increases, your pupils dilate, your breathing rate increases, and your palms perspire. All these are a direct result of your sympathetic division in action.

The SD has a more complex anatomical structure than the PD. The illustration on the facing page shows only the motor output to the effector. The nerves involved are like the electrical wiring in a house. They consist of a **preganglionic neuron** linked to a **postganglionic neuron.** Preganglionic neurons arise from the T1–L2 segments of the spinal cord. Notice that the preganglionic axons are quite short, and the postganglionic axons are relatively long. Recall that ganglia (sing. *ganglion*) are clusters of neuron cell bodies. The **paravertebral ganglia**—two columns of ganglia like strings of beads that run along either side of the spinal cord—are unique to the SD. Included in this chain are the cervical ganglia, which consist of three parts: **superior**, **middle**, and **inferior**. Postganglionic neurons are wired from these hubs to target organs in the head and thoracic cavity.

Three other ganglia, all located near the abdominal aorta and classified as the **prevertebral ganglia**, are: (1) **celiac**, (2) **superior mesenteric**, and (3) **inferior mesenteric.** Postganglionic neurons arising from these hubs connect to all the target organs and glands below the diaphragm. The ANS has three possible effectors: smooth muscle, cardiac muscle, and glands.

To clarify the illustration, when you see the postganglionic neurons connecting to organs, they actually are connecting to muscle in the walls of these target organs. Notice that many of the postganglionic neurons pass through **plexuses** (sing. *plexus*), or neural networks, to link with their target organs.

Analogy

Regulating the visceral organs is like driving a car. Sometimes you have to step on the gas, and other times you have to step on the brake. Most of the time, the SD is like stepping on the gas. For example, sympathetic impulses sent along the accelerator nerve to the heart increase the heart rate. Unfortunately, it does not work this way with all organs/glands. There are exceptions to this rule.

Study Tips

Sympathetic = Stressful. This simple alliteration helps you distinguish the function of the SD from the PD.

Summary

ITEM	DESCRIPTION
Nickname	*Fight-or-flight* system
Location of preganglionic neuron cell bodies	T1–L2 regions of spinal cord (*thoracolumbar*)
Length of preganglionic axon	Short
Length of postganglionic axon	Long
Location of ganglia	Near the spinal cord
Neurotransmitters	Preganglionic neurons secrete acetylcholine (ACh). Postganglionic neurons secrete norepinephrine (NE).

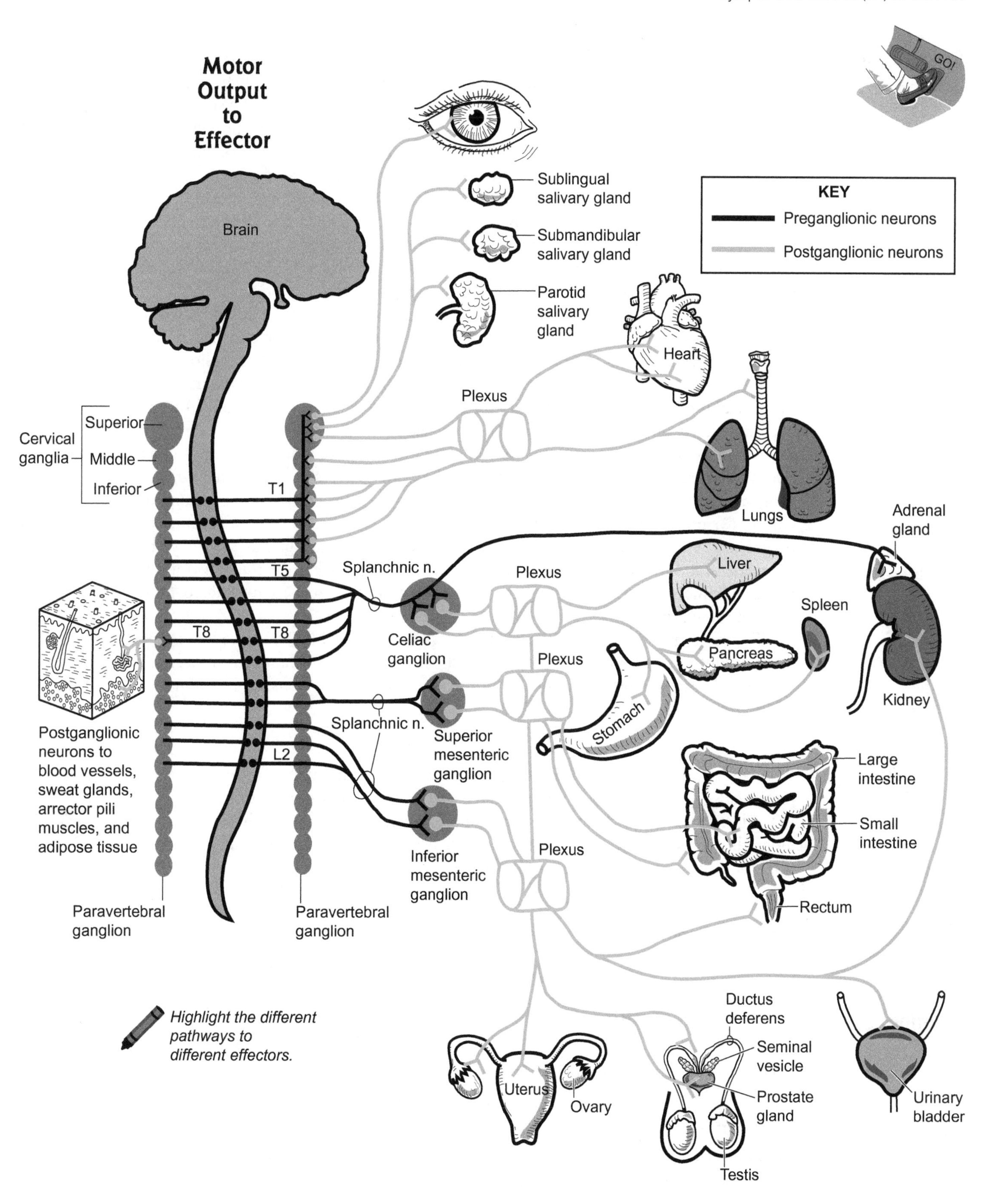

Highlight the different pathways to different effectors.

Description

The autonomic nervous system (ANS) controls involuntary activities such as digestion and heart rate. It is subdivided into two divisions: **sympathetic** (**SD**) and **parasympathetic** (**PD**). These two divisions work together to regulate normal functions and maintain **homeostasis.** Each division has its own groups of nerves that connect to the major visceral organs in the body. But there are some exceptions. For example, the lacrimal gland is controlled only by the PD, because the SD has no nerves that connect to it.

This module gives an overview of the PD—also referred to as the *craniosacral division* because motor output arises from some *cranial* nerves and from the *sacral* region of the spinal cord. Because the PD is most active when the body is at rest, it is nicknamed the "resting and digesting" system. Consider what happens when you lie down on the couch after eating a heavy meal. Your digestive function increases, pupils constrict, heart rate decreases, breathing rate decreases, and blood pressure decreases. All of these illustrate your parasympathetic division in action.

The PD has a simpler anatomical structure than the SD. The illustration on the facing page shows only the motor output to the effector. The nerves involved are like the electrical wiring in a house and consist of a **preganglionic neuron** linked to a **postganglionic neuron.** Preganglionic neurons arise from the following cranial nerves (CN): **III** (oculomotor), **VII** (facial), **IX** (glossopharyngeal), and **X** (vagus). In the sacral region of the spinal cord, preganglionic neurons arise from **S2–S4** to become the **pelvic nerves,** which connect to target organs in the pelvic region.

As a general rule, the preganglionic axons tend to be quite long, and the postganglionic axons are relatively **short.** Recall that ganglia (sing. *ganglion*) are clusters of neuron cell bodies. Four important ganglia are located in the head, where preganglionic neurons link to postganglionic neurons: (1) **ciliary**, (2) **pterygopalatine**, (3) **submandibular**, and (4) **otic.** These major hubs connect to target organs in the head, such as the eye, lacrimal gland, and salivary glands. The ANS has three possible effectors: *smooth muscle*, *cardiac muscle*, and *glands*. To clarify the illustration, when you see postganglionic neurons connecting to organs, they actually are connecting to muscle tissue in the walls of these organs.

The most important nerve in the PD is the **vagus nerve** (CN X) because it accounts for as much as 80% of all the preganglionic axons in the PD. It innervates all the thoracic organs and most of the abdominal organs. Notice that it runs through **plexuses** (sing. *plexus*) or neural networks to link with all its target organs.

Analogy

Regulating the visceral organs is like driving a car. Sometimes you have to step on the gas, and other times you have to step on the brake. Most of the time, the PD is like stepping on the brake. For example, parasympathetic impulses sent along the vagus nerve to the heart decrease the heart rate. Unfortunately, it does not work this way with all organs/glands. There are exceptions to this rule.

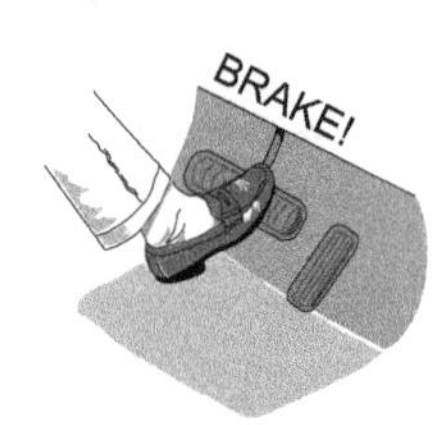

Study Tip

Parasympathetic = Peaceful. This simple alliteration helps you distinguish the function of the SD from the PD.

Summary

ITEM	DESCRIPTION
Nickname	*Resting and digesting* system
Location of preganglionic neuron cell bodies	Brainstem and S2–S4 regions of spinal cord (*craniosacral*)
Length of preganglionic axon	Long
Length of postganglionic axon	Short
Location of ganglia	Usually near or within the wall of the target organ
Neurotransmitters	Preganglionic neurons secrete acetylcholine (ACh). Postganglionic neurons secrete acetylcholine (ACh).

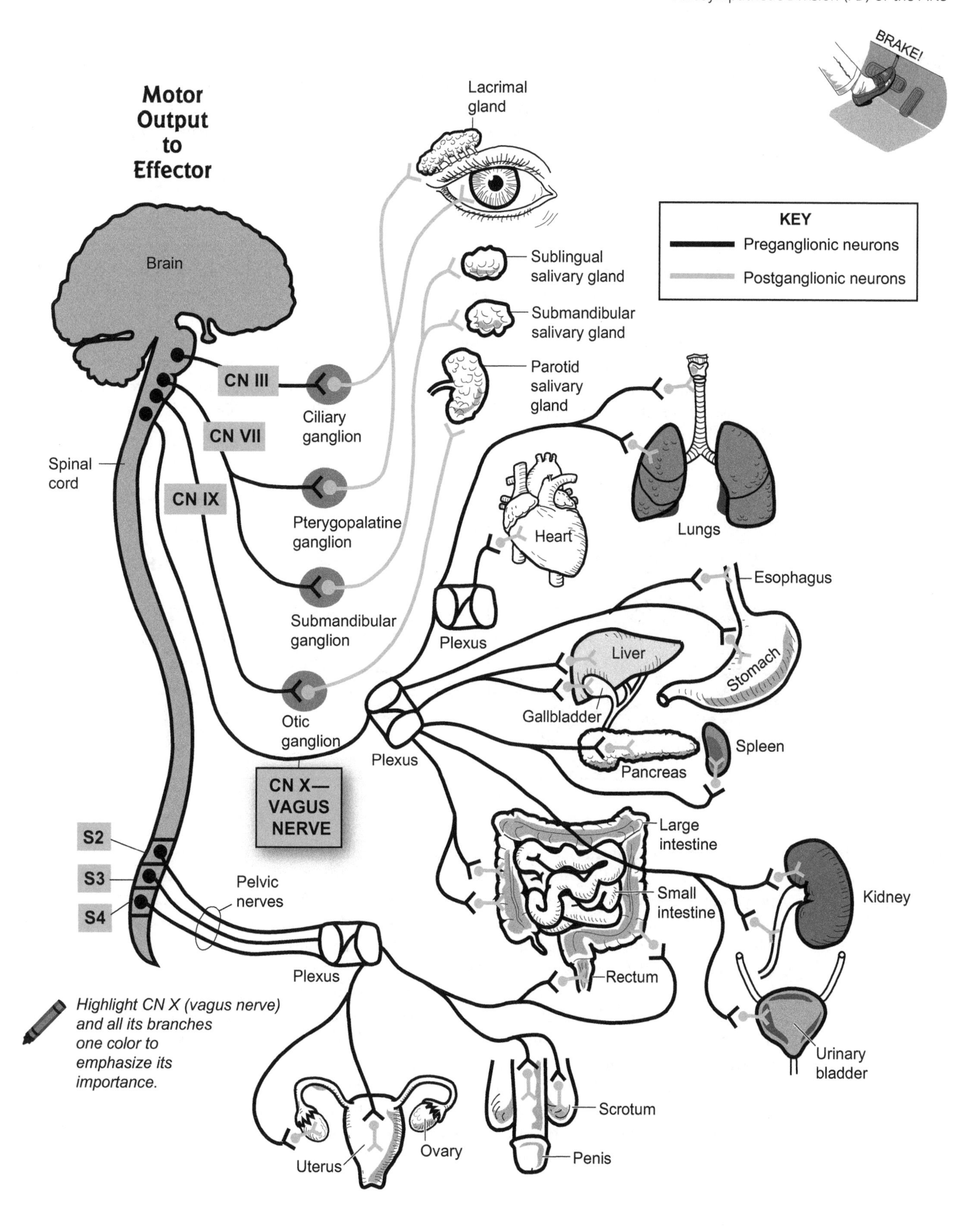
BRAKE!
Motor Output to Effector
Lacrimal gland
KEY
Preganglionic neurons
Postganglionic neurons
Brain
Sublingual salivary gland
Submandibular salivary gland
Parotid salivary gland
CN III
Ciliary ganglion
CN VII
Spinal cord
CN IX
Pterygopalatine ganglion
Submandibular ganglion
Otic ganglion
Heart
Lungs
Esophagus
Plexus
Liver
Stomach
Gallbladder
Plexus
Pancreas
Spleen
CN X— VAGUS NERVE
Large intestine
Small intestine
Kidney
S2
S3
S4
Pelvic nerves
Plexus
Rectum
Urinary bladder
Scrotum
Penis
Uterus
Ovary
Highlight CN X (vagus nerve) and all its branches one color to emphasize its importance.

Description

The autonomic nervous system (ANS) uses reflexes to regulate many of the unconscious activities we take for granted, such as digestive activities, heart rate, and pupillary responses. These responses are induced by specific neurotransmitters combining with specific receptors. The three major neurotransmitters secreted in the ANS are **acetylcholine** (**ACh**), **norepinephrine** (**NE**), and **epinephrine** (**EPI**). Acetylcholine binds to either **muscarinic** (**M**) or **nicotinic** (**N**) **receptors,** and NE and EPI bind to either alpha or beta receptors. There are two types of **alpha** (μ) **receptors** (*alpha 1* and *alpha 2*) and three kinds of **beta** (β) **receptors** (*beta 1*, *beta 2*, and *beta 3*).

When a neurotransmitter binds to a receptor, it induces a response in its **target cells.** The response is determined by which specific neurotransmitter combines with which specific receptor. For example, when **NE** binds to an alpha 1 receptor, it typically elicits a different response than when NE binds to a beta 1 receptor. Although there are exceptions to the rule, we can generalize about the different responses induced by various neurotransmitter/receptor combinations. When ACh binds to muscarinic receptors, for example, it sometimes results in an excitatory response. Other times, it results in an inhibitory response. And when ACh binds to nicotinic receptors, it usually produces an excitatory response. When NE (or EPI) binds to an alpha receptor, it usually induces an excitatory response, such as constriction of smooth muscle within blood vessels walls. When NE binds to a beta receptor, it often induces an inhibitory response, such as dilation of the bronchioles in the lungs.

ACh Receptors

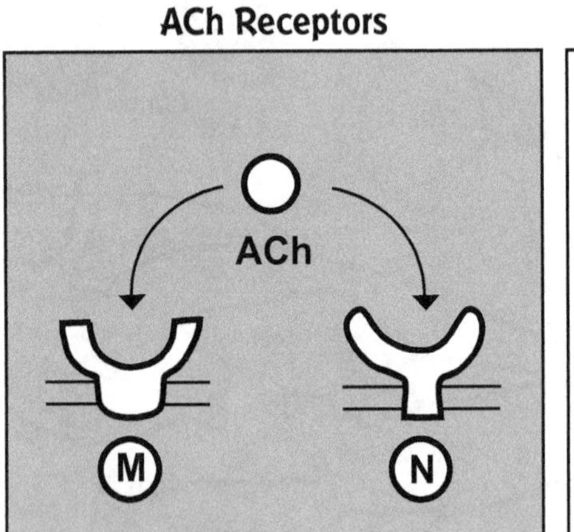

NE and EPI Receptors

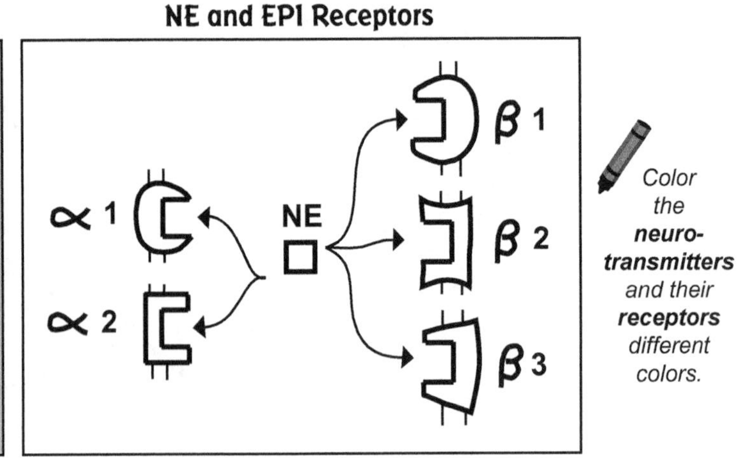

Cholinergic and Adrenergic Neurons

In the ANS, nerve fibers that secrete *only* ACh are called **cholinergic neurons,** and nerve fibers that secrete *only* NE or EPI are called **adrenergic neurons.** In the parasympathetic division, all the **preganglionic** neurons and **postganglionic** neurons are cholinergic. The ACh released from the preganglionic neurons binds to **nicotinic neurons** in the dendrites of the postganglionic neurons. This stimulates the **postganglionic neuron,** in turn, to release ACh, which binds to **muscarinic receptors** in the **effector** (*smooth muscle, cardiac muscle,* or a *gland*).

In the sympathetic division, the preganglionic neurons also are cholinergic, so they release ACh into the **synaptic cleft.** The ACh then binds to the nicotinic receptors in dendrites of the postganglionic neurons. Because all the postganglionic neurons are adrenergic, they release NE, which binds to alpha or beta receptors in the effector. In one special case that does not involve a postganglionic neuron, the sympathetic division also directly innervates hormone-producing cells in the inner medulla of the adrenal gland. The preganglionic neurons secrete ACh, which binds to the nicotinic receptors in the hormone-producing cells. This stimulates the secretion of both NE and EPI into the blood. These hormones travel through the bloodstream to their target cells, which contain either alpha or beta receptors.

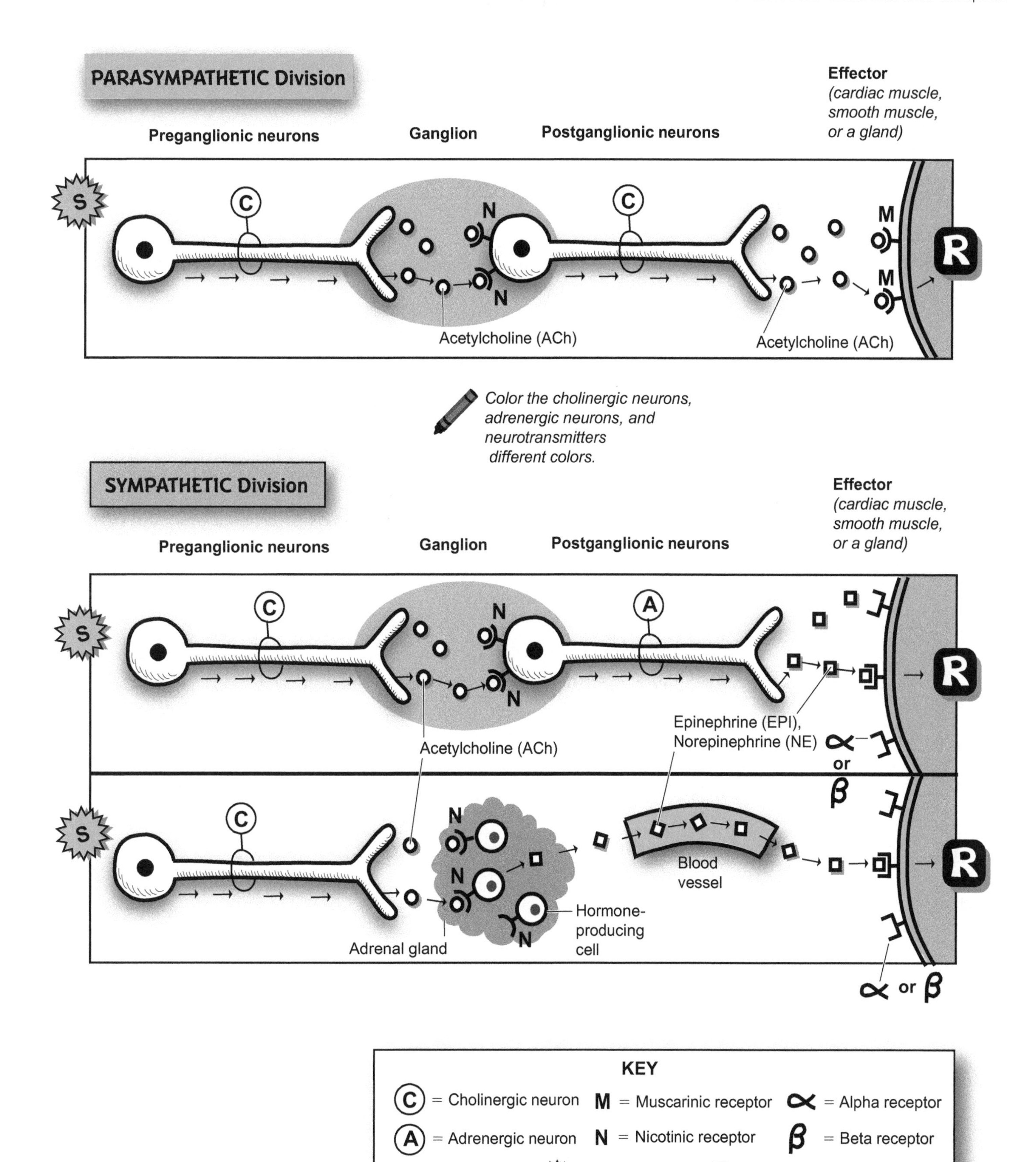

KEY

(C) = Cholinergic neuron	**M** = Muscarinic receptor	α = Alpha receptor
(A) = Adrenergic neuron	**N** = Nicotinic receptor	β = Beta receptor
(S) Stimulus	(R) Response	

Notes

ENDOCRINE SYSTEM

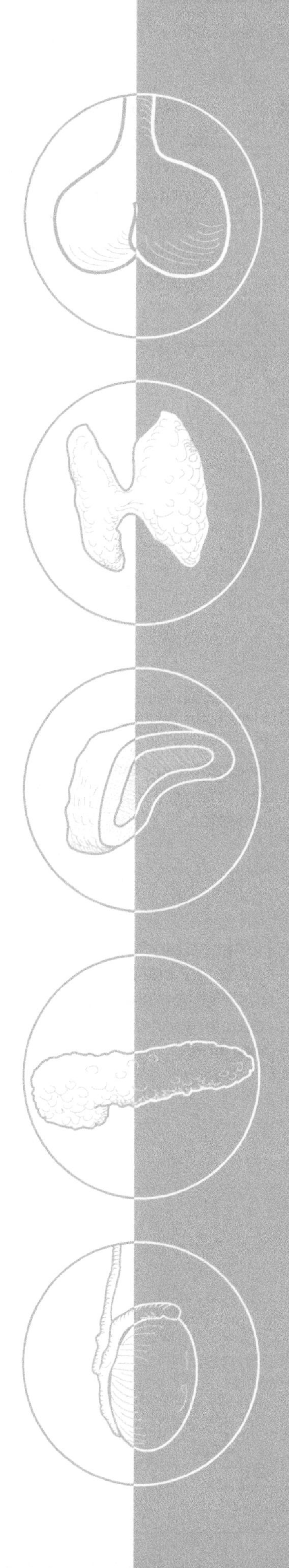

Description

Along with the nervous system, the endocrine system is one of the great regulators of activities in the human body. Its major purpose is to maintain **homeostasis** and to regulate various processes such as growth and human development. It consists of many different glands containing specific cells that synthesize and release chemical messengers called **hormones**. These hormones enter the bloodstream, where they travel to a specific cell called a **target cell**. This target cell has a receptor for that specific hormone. Once the hormone binds to the receptor, it induces a response in that cell. The general concept of how the endocrine system functions is illustrated below:

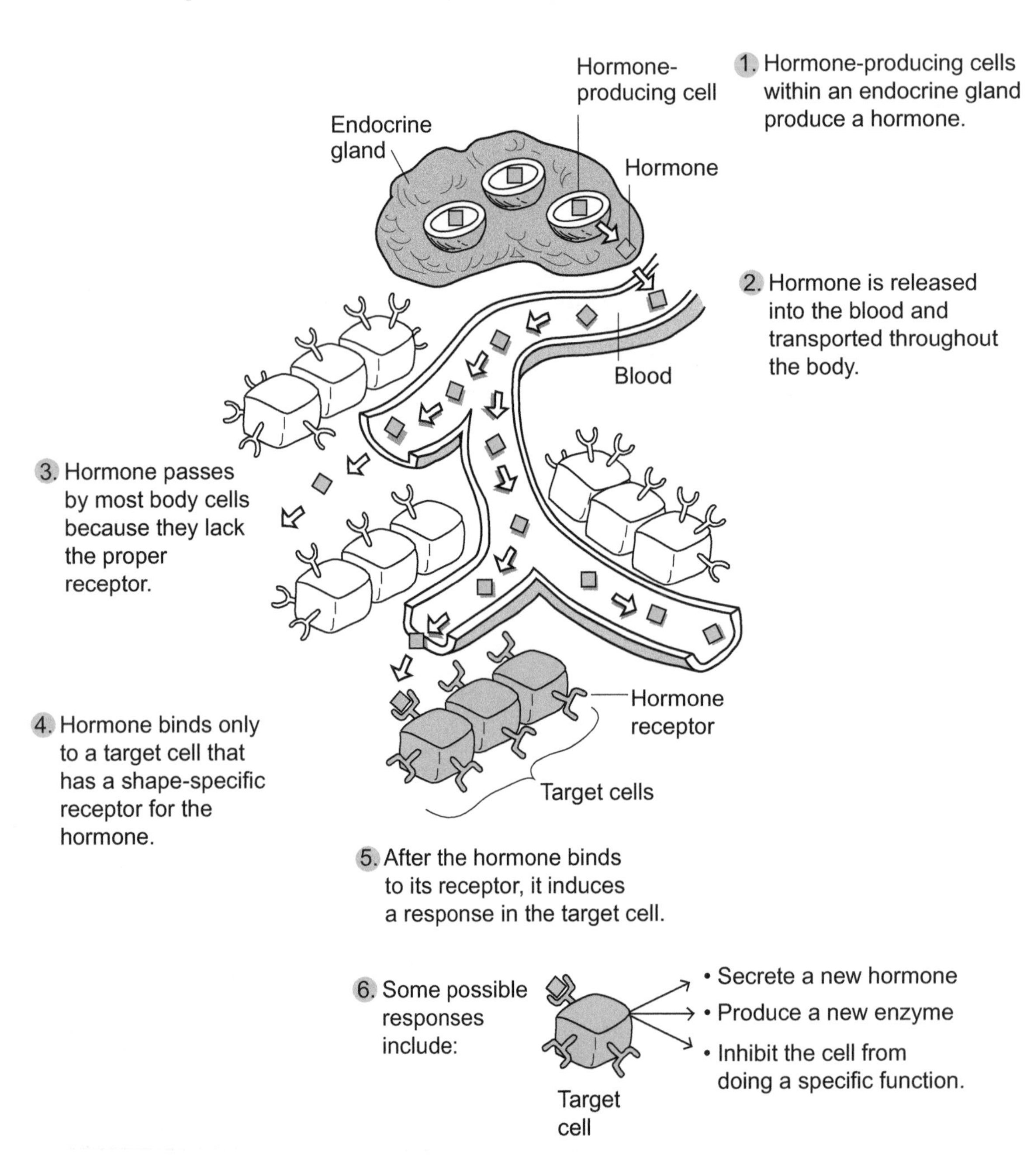

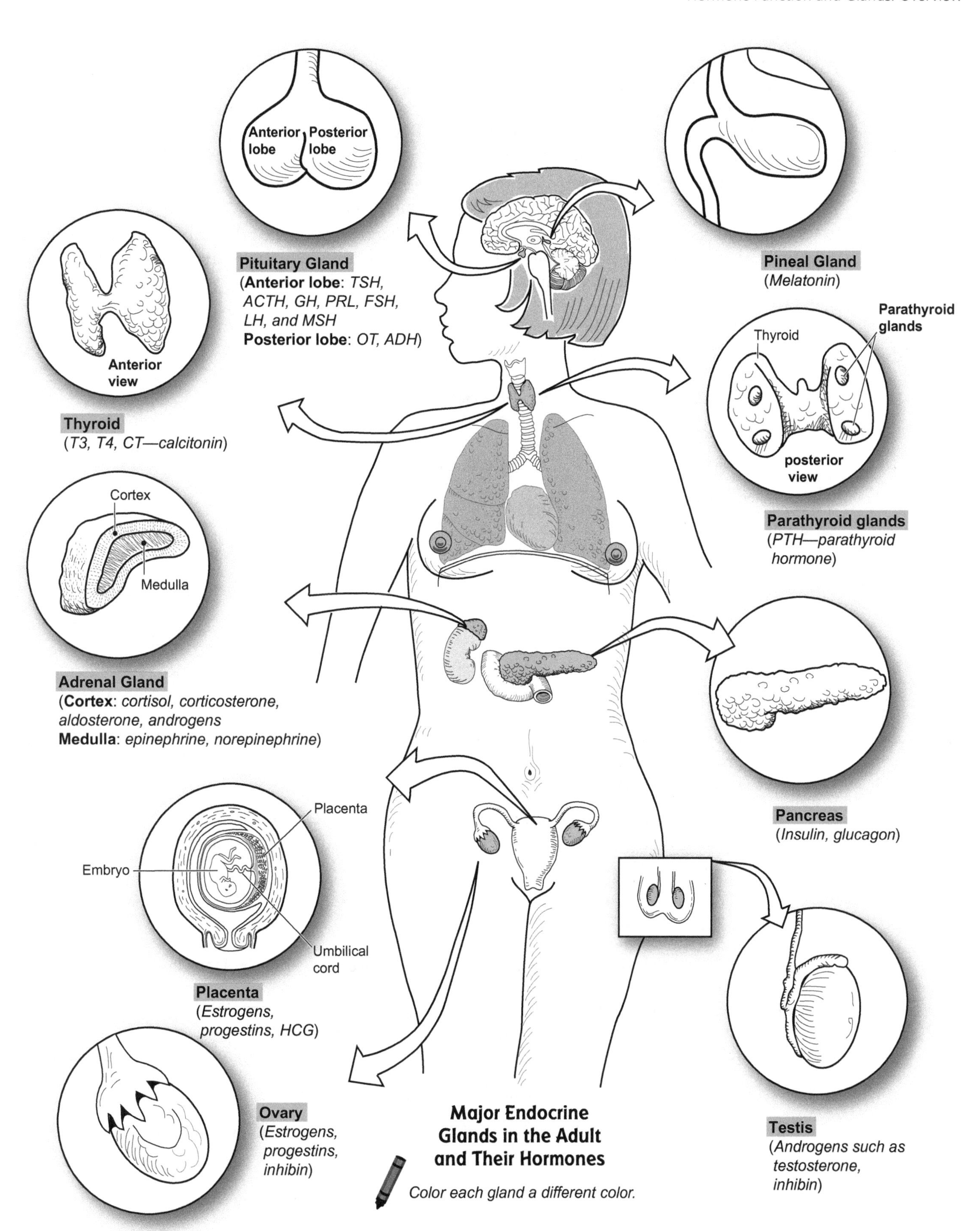

Major Endocrine Glands in the Adult and Their Hormones

Color each gland a different color.

Description

Steroid hormones are lipid-soluble and synthesized from **cholesterol.** Because they must be transported through the watery blood, they need special transporters called **protein carriers.** These carriers also prevent hormones from being degraded by enzymes in the blood. Once the hormones arrive at their target cells, they enter the nucleus to stimulate DNA to begin protein synthesis. The effects of any steroid hormone have the following key features:

- **Slower action:** They produce a response in hours or days after the initial binding of hormone with its receptor—slower than nonsteroid hormone.
- **No amplification:** The effects produced are proportional to the amount of hormone secreted; a greater amount of hormone is needed to induce a response compared with a nonsteroid, in which the effects of a hormone are amplified by a chain reaction.

Steroid

Examples of steroid hormones are given in the table below.

Steroid Hormones	Site of Secretion
• Aldosterone • Cortisol • Androgens (e.g., testosterone)	Adrenal glands (cortex)
• Testosterone	Testes
• Estrogens • Progesterone	Ovaries
• Calcitriol	Kidneys

Sequence

The following sequence of events is a simplified mechanism describing how steroid hormones induce a response in their target cells, illustrated on the facing page. Note that thyroid hormones (T_3 and T_4) use a similar mechanism of action.

① The steroid hormone travels through the blood with the help of a protein carrier. After being released from this carrier, the hormone diffuses through the plasma membrane into the **cytoplasm** and into the **nucleus** through the **nuclear pore.**

② Once inside the nucleus, the hormone binds with its **hormone receptor** to form a **hormone-receptor complex.**

③ The hormone-receptor complex binds to a **shape-specific receptor on DNA** like a key in a lock.

④ This stimulates DNA to begin the process of protein synthesis. First, DNA makes a copy of itself in the form of a mobile, single-stranded molecule called **mRNA** (messenger RNA). The smaller mRNA is able to leave the nucleus through the nuclear pore, and it moves into the cytoplasm.

⑤ The mRNA molecule binds to the **ribosome,** the protein factory of the cell.

⑥ The message in mRNA is read by the ribosome, and it links free **amino acids** to form a new protein.

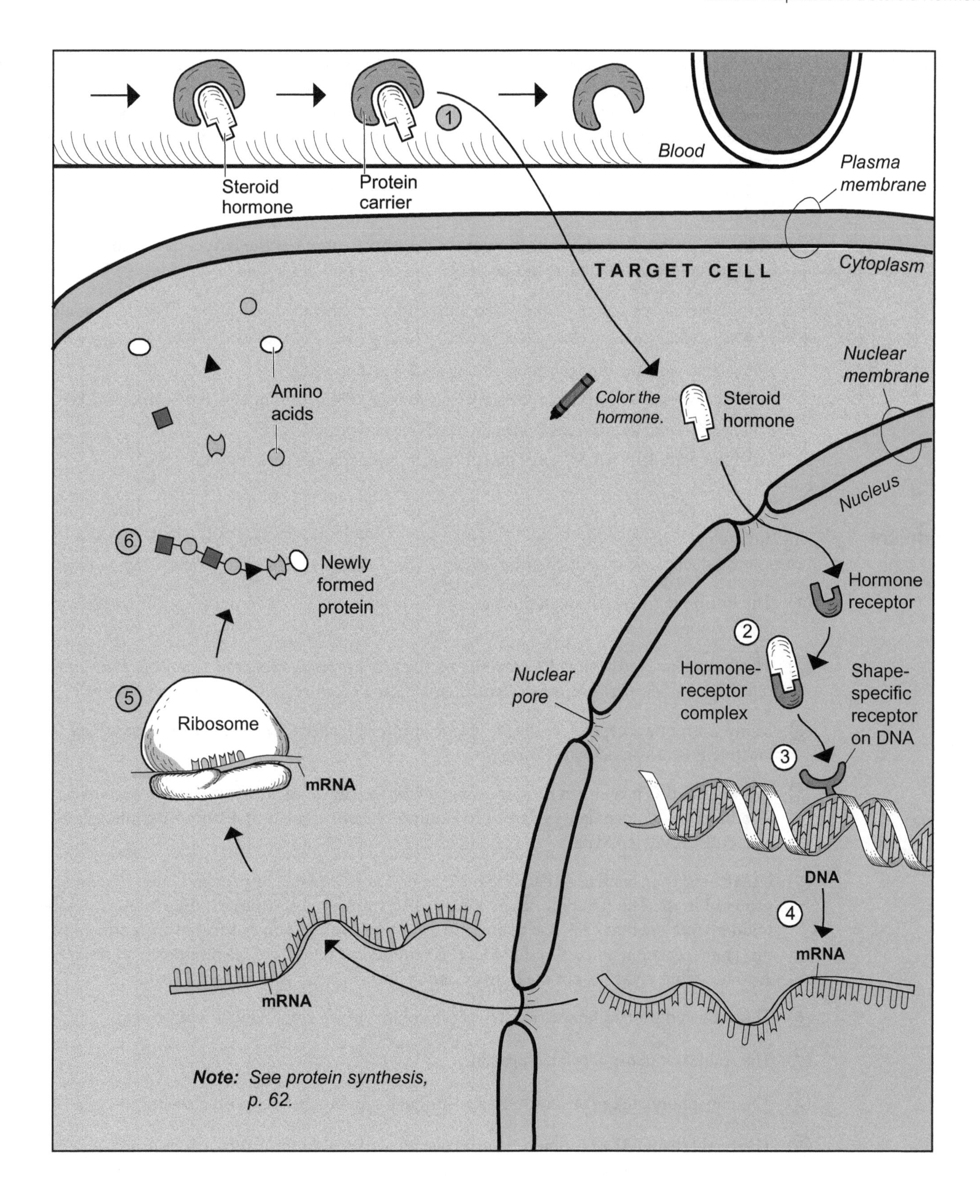

Note: See protein synthesis, p. 62.

Description

Nonsteroid hormones are water soluble, so they are easily transported through the blood. Many are protected and transported by protein carriers that prevent them from being degraded by enzymes in the blood. Once the hormones arrive at their **target cells**, they first bind at receptors located at the plasma membrane. This triggers a cascade inside the cell that eventually leads to cellular changes. The effects of a nonsteroid hormone have the following key features:

- **Rapid action:** They produce a response within *seconds* or *minutes* after the hormone initially binds to its receptor.
- **Amplification:** The chain reaction in the target cell amplifies the effects of the hormone; a little hormone produces a big response.

Nonsteroid hormones can be subdivided into different chemical categories. The four major subgroups are:

- Proteins—*ex.:* **insulin**, **glucagon**, and **growth hormone** (GH)
- Glycoproteins—*ex.:* **follicle-stimulating hormone** (FSH), and **luteinizing hormone** (LH)
- Peptides—*ex.:* **antidiuretic hormone** (ADH) and **oxytocin** (OT)
- Amino acid derivatives—*ex.:* **epinephrine**, **norepinephrine**, and **thyroxine** (T_4)

Mechanism

The following sequence of events describes one of the mechanisms by which *some* nonsteroid hormones induce a response in their target cells. These events are illustrated on the facing page.

1. The **hormone** travels through the blood to its target cell, which contains membrane **receptors** for the hormone.
2. The hormone binds with the receptor to form a **hormone-receptor complex.** This complex is referred to as the *first messenger* because it begins a chain reaction within the target cell.
3. Many hormone-receptor complexes trigger proteins called **G proteins** to be converted from an inactive form to an activated form.
4. The **activated G protein,** in turn, triggers an enzyme called **adenylate cyclase** to become activated. The function of adenylate cyclase is to catalyze the conversion of **adenosine triphosphate** (ATP) into **cyclic AMP** (cAMP).
5. **cAMP** acts as an enzyme to catalyze the conversion of an **inactive protein kinase** into an **activated protein kinase.** The function of any kinase is to transfer a phosphate group from one substance to another. Note that cAMP is referred to as the *second messenger* because it's a critical player in this chain reaction, and a threshold level is required to finally induce a response in the target cell. Also note that cAMP is not the *only* second messenger used within all target cells.
6. The **activated protein kinase** transfers a phosphate group from ATP onto a protein.
7. The result is a **phosphorylated protein.**
8. The phosphorylated proteins eventually produce cellular changes within the target cell.
9. The final fate of **cAMP** is that it is either deactivated within the target cell, or it diffuses out of the cell. This ensures that the chain reaction stops.

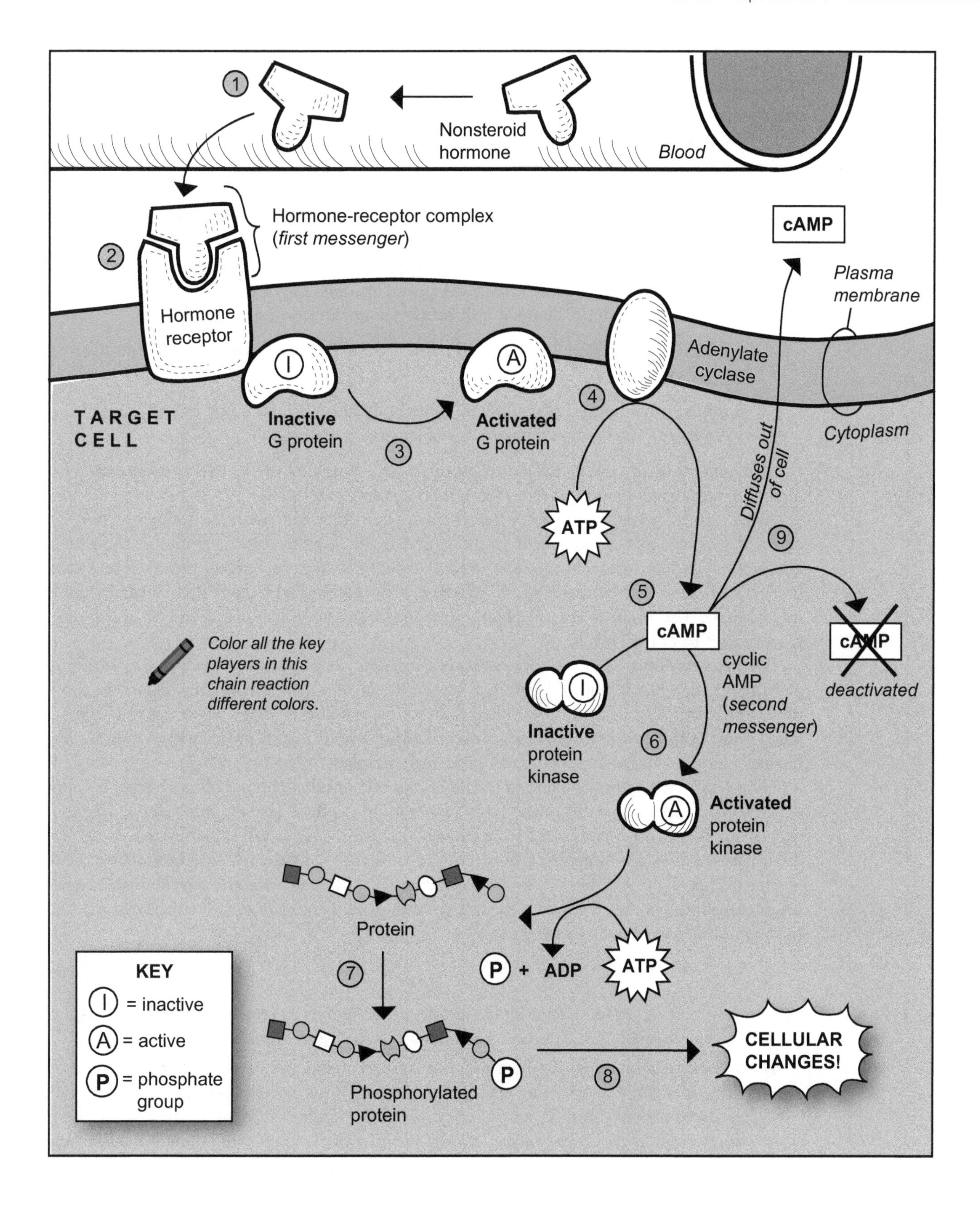
1
Nonsteroid hormone
Blood
2
Hormone-receptor complex (*first messenger*)
Hormone receptor
cAMP
Plasma membrane
Adenylate cyclase
TARGET CELL
Inactive G protein
3
Activated G protein
4
Cytoplasm
Diffuses out of cell
ATP
9
5
cAMP
cAMP deactivated
Color all the key players in this chain reaction different colors.
cyclic AMP (*second messenger*)
Inactive protein kinase
6
Activated protein kinase
Protein
KEY
I = inactive
A = active
P = phosphate group
7
P + ADP
ATP
Phosphorylated protein
8
CELLULAR CHANGES!

Description

Regulation of **blood glucose** levels is controlled by two hormones: **insulin** and **glucagon**. These two hormones are antagonists because they have opposite functions. Insulin causes blood glucose levels to decrease, and glucagon causes blood glucose levels to increase. Both hormones are produced by cells within the **pancreas**.

Mechanism

Let's examine the mechanism of action. The normal level of blood glucose is about 70–100 mg/dL and has to be maintained. Glucose levels typically rise after a meal rich in carbohydrates, such as baked potatoes or corn flakes. These high glucose levels are detected by insulin-secreting cells within the pancreas. This triggers these cells to secrete more insulin—the "feeding hormone"—into the blood. As insulin travels through the blood, it induces two major responses:

1. Stimulates the storage of excess glucose in the form of **glycogen** within the liver and some other tissues.
2. Stimulates the uptake of glucose by most body cells. Insulin binds to receptors in the plasma membrane, which opens channels for glucose to pass through.

The glucose then is metabolized to produce large amounts of the energy molecule ATP. Insulin secretion stops when blood glucose levels fall back to normal levels.

If you haven't eaten anything for a long time, your blood glucose levels fall below normal. This is detected by glucagon-secreting cells in the pancreas. In response to this stimulus, these cells secrete glucagon—the "starvation hormone"—into the blood. Glucagon targets the liver and some other tissues, such as skeletal muscle, where glycogen is stored, and stimulates the conversion of glycogen into **glucose**. The result is that the blood glucose level rises. Glucagon secretion stops when blood glucose rises to normal levels.

Diabetes mellitus is caused by the body's inability to produce or use insulin, which results in abnormally high blood glucose levels. It is one of the most common endocrine disorders and is one of the leading causes of death in the United States. One method of diagnosis is detecting high levels of glucose in the urine. If left untreated, it can result in serious health problems, such as cardiovascular disease, vision problems, kidney damage, and nerve damage.

There are two types of diabetes mellitus, **type 1** (**insulin-dependent**) and **type 2** (**non-insulin-dependent**). Type 1 is a result of the body's inability to produce insulin, and there is no cure. It often occurs in children so it is referred to as *juvenile diabetes*. This is the less common of the two types. Treatment requires self-monitoring of blood glucose levels and daily insulin injections or the use of an insulin pump. Type 2 is far more common than Type 1. Sufferers are typically obese adults who produce insulin, but their body cells become resistant to it. Treatment involves behavior modification in the form of diet, exercise, and weight loss.

Study Tips

- A simple association to distinguish insulin from glucagon is to remember that insulin is nicknamed the "feeding hormone" and glucagon is nicknamed the "starvation hormone."
- The pancreas is included in the endocrine system and digestive system because it is both an endocrine gland and an exocrine gland. Only about 1% of its cells serve the endocrine function to secrete hormones like insulin and glucagon. The remaining 99% of its cells serve the exocrine function to secrete digestive enzymes for the purpose of chemically digesting lipids, carbohydrates, and proteins.

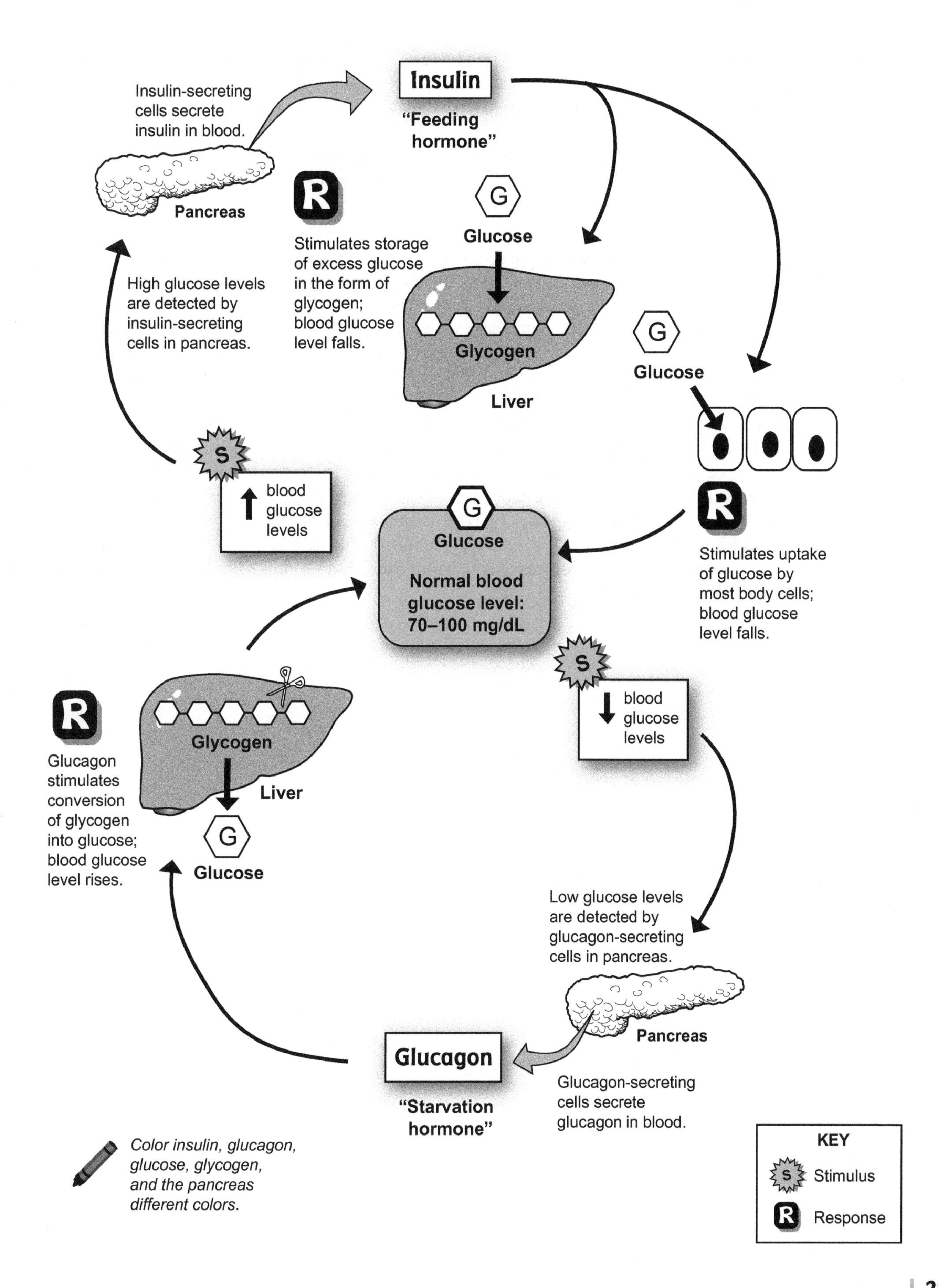
Insulin-secreting cells secrete insulin in blood.
Insulin
"Feeding hormone"
Pancreas
R
Stimulates storage of excess glucose in the form of glycogen; blood glucose level falls.
G
Glucose
High glucose levels are detected by insulin-secreting cells in pancreas.
Glycogen
Liver
G
Glucose
S
blood glucose levels
G
Glucose
Normal blood glucose level: 70–100 mg/dL
R
Stimulates uptake of glucose by most body cells; blood glucose level falls.
S
blood glucose levels
R
Glucagon stimulates conversion of glycogen into glucose; blood glucose level rises.
Glycogen
Liver
G
Glucose
Low glucose levels are detected by glucagon-secreting cells in pancreas.
Pancreas
Glucagon
"Starvation hormone"
Glucagon-secreting cells secrete glucagon in blood.
Color insulin, glucagon, glucose, glycogen, and the pancreas different colors.
KEY
S Stimulus
R Response

Description

Blood calcium levels are regulated by two hormones: **calcitonin** (CT) and **parathyroid hormone** (PTH) or parathormone. CT is produced by the **thyroid gland**, and PTH is produced by the **parathyroid glands**. These two hormones are **antagonists** because they have opposite functions: CT causes blood calcium levels to decrease, and PTH causes blood calcium levels to increase. It's interesting to note that PTH is essential for life, but CT is not (only in pregnant women). Without PTH, calcium levels would fall so low that it would impair normal muscle contraction and blood clotting.

Mechanism

Let's examine the mechanism of action. The normal level of blood calcium is in the range of 9–11 mg/dL, which has to be maintained. As calcium-rich foods such as milk, broccoli, and tofu are ingested, blood calcium levels rise. Calcitonin-producing cells within the thyroid gland sense this, and they secrete CT into the blood. CT targets the bone-degrading cells within the skeletal system called **osteoclasts**. The plasma membranes in osteoclasts contain receptors for CT. Once CT binds to this receptor, it inhibits these cells from doing their normal job of degrading bone. CT does not affect **osteoblasts**—the bone-forming cells—they carry on with their normal function of depositing calcium salts in bone. This causes the blood calcium levels to fall. CT stops being produced when blood calcium levels return to normal.

If you haven't eaten any calcium-rich foods or taken any calcium supplements for a long time, your blood calcium levels fall. This is detected by the two pairs of parathyroid glands located on the posterior surface of the thyroid gland, which are stimulated to secrete PTH into the blood. PTH targets **osteoclasts**—bone-degrading cells within the skeletal system. Osteoclasts contain receptors for PTH.

Once PTH docks at its PTH receptor, it stimulates the osteoclasts to decompose bone and release calcium into the blood. Consequently, the blood calcium level rises. PTH stops being produced when blood calcium levels return to normal.

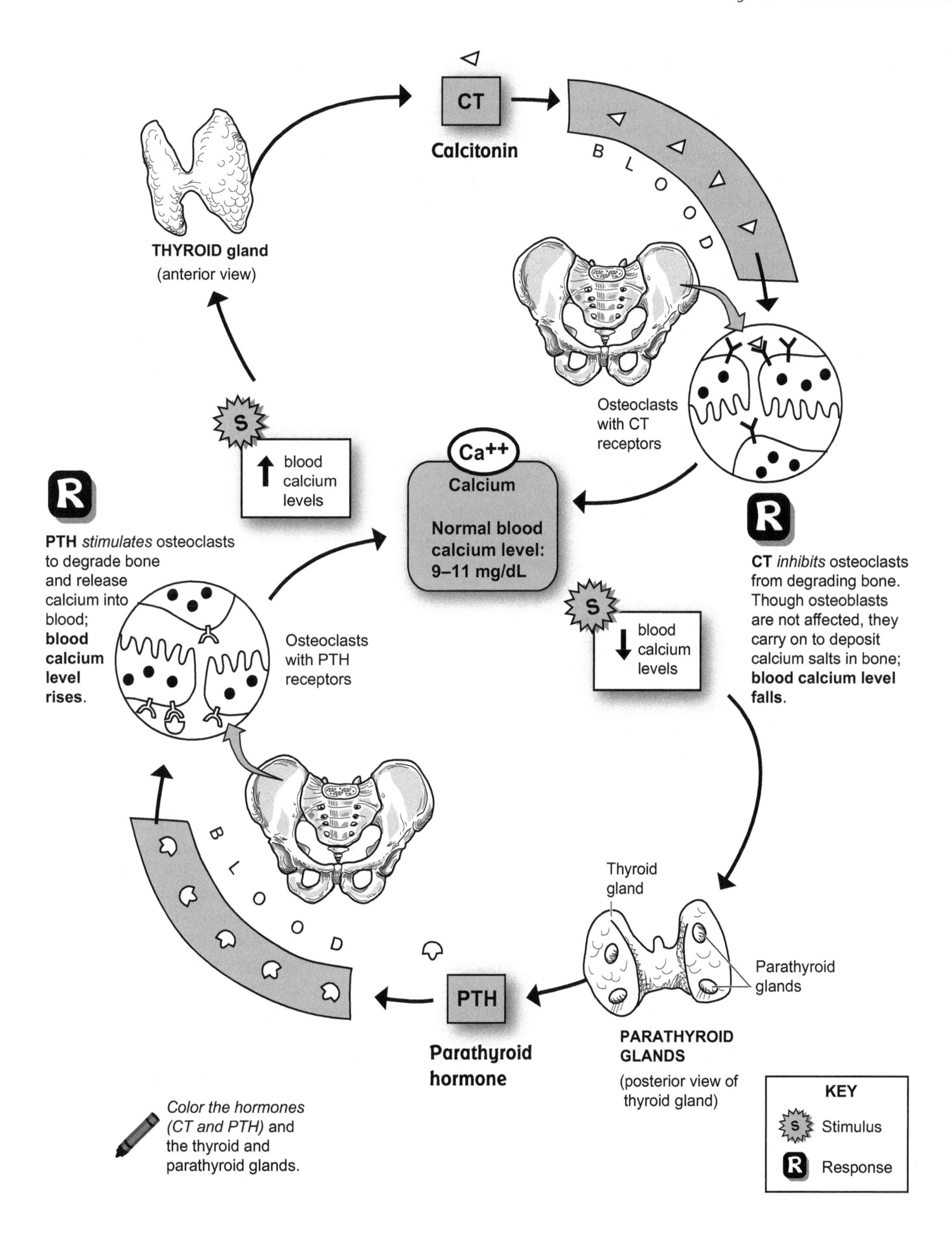

Color the hormones (CT and PTH) and the thyroid and parathyroid glands.

Description

The **pituitary gland** (*hypophysis*) is a small but powerful gland located at the base of the brain. It consists of two distinct lobes: the **anterior lobe** (*adenohypophysis*) and the **posterior lobe** (*neurohypophysis*). Because each lobe functions differently and secretes different hormones, each is presented in a separate module. This one covers the **anterior pituitary.**

Surrounding the pituitary is a protective pocket of bone called the **sella turcica** of the **sphenoid bone.** The pituitary is nicknamed the "master gland" because it makes many different hormones that control the other endocrine glands in the body. But the **hypothalamus** in the brain actually controls the pituitary gland with specific bundles of neurons. This provides a *vital link* between the nervous system and the endocrine system. The pituitary is connected to the hypothalamus by a slender stem-like structure called the **infundibulum** (*pituitary stalk*).

Hormones Produced

Five different types of **hormone-producing cells** within the anterior pituitary directly secrete the following seven major hormones:

- Adrenocorticotropic hormone (**ACTH**)
- Follicle-stimulating hormone (**FSH**)
- Growth hormone (**GH**)
- Luteinizing hormone (**LH**)
- Prolactin (**PRL**)
- Thyroid-stimulating hormone (**TSH**)
- Melanocyte-stimulating hormone (**MSH**)

Sequence

The anterior pituitary relies on a specialized group of blood vessels collectively called a portal system. The body has two important portal systems: the hepatic portal system (*in the liver*) and the **hypophyseal portal system** (*in the anterior pituitary*). Portal systems link two capillary networks to more efficiently deliver chemical messengers to a specific location before they are released to the general circulation:

Superior hypophyseal artery → Primary capillary network → Hypophyseal portal veins → Secondary capillary network → Anterior hypophyseal veins

Mechanism

The anterior pituitary contains a mass of various **hormone-producing cells** (**HPCs**) controlled by a group of neurons in the **ventral hypothalamus.** These neurons can either stimulate or inhibit hormone production by secreting either **releasing factors** (**RFs**) or inhibiting factors, respectively. Let's examine the stimulation process. **RFs** stimulate the **HPCs** to secrete hormones. Different RFs are needed to make the seven major hormones produced by the anterior pituitary. These RFs are produced in the neuron's **cell body** and released from the **axon terminal.** After a nerve impulse triggers them to release, they diffuse rapidly into the **primary capillary network** of the portal system, and then the hypophyseal portal veins quickly shuttle them into the secondary capillary network surrounding the HPCs. As the RFs diffuse out of the blood, they immediately bind to shape-specific receptors in the HPC's plasma membrane, which induces the production of specific hormones. These hormones then enter the secondary capillary network and exit the anterior pituitary via the anterior hypophyseal vein. As the hormones travel through the blood, they eventually bind to receptors in their target cells and induce a response.

For example, a thyroid-stimulating releasing factor is needed to stimulate the secretion of thyroid-stimulating hormone (TSH). After TSH leaves the anterior lobe via the **anterior hypophyseal vein,** it travels through the blood to all the different organs but binds only at those targets that have receptors for TSH. As its name implies, TSH targets the thyroid gland to induce the secretion of thyroid hormones such as triiodothyronine (T_3) and thyroxine (T_4).

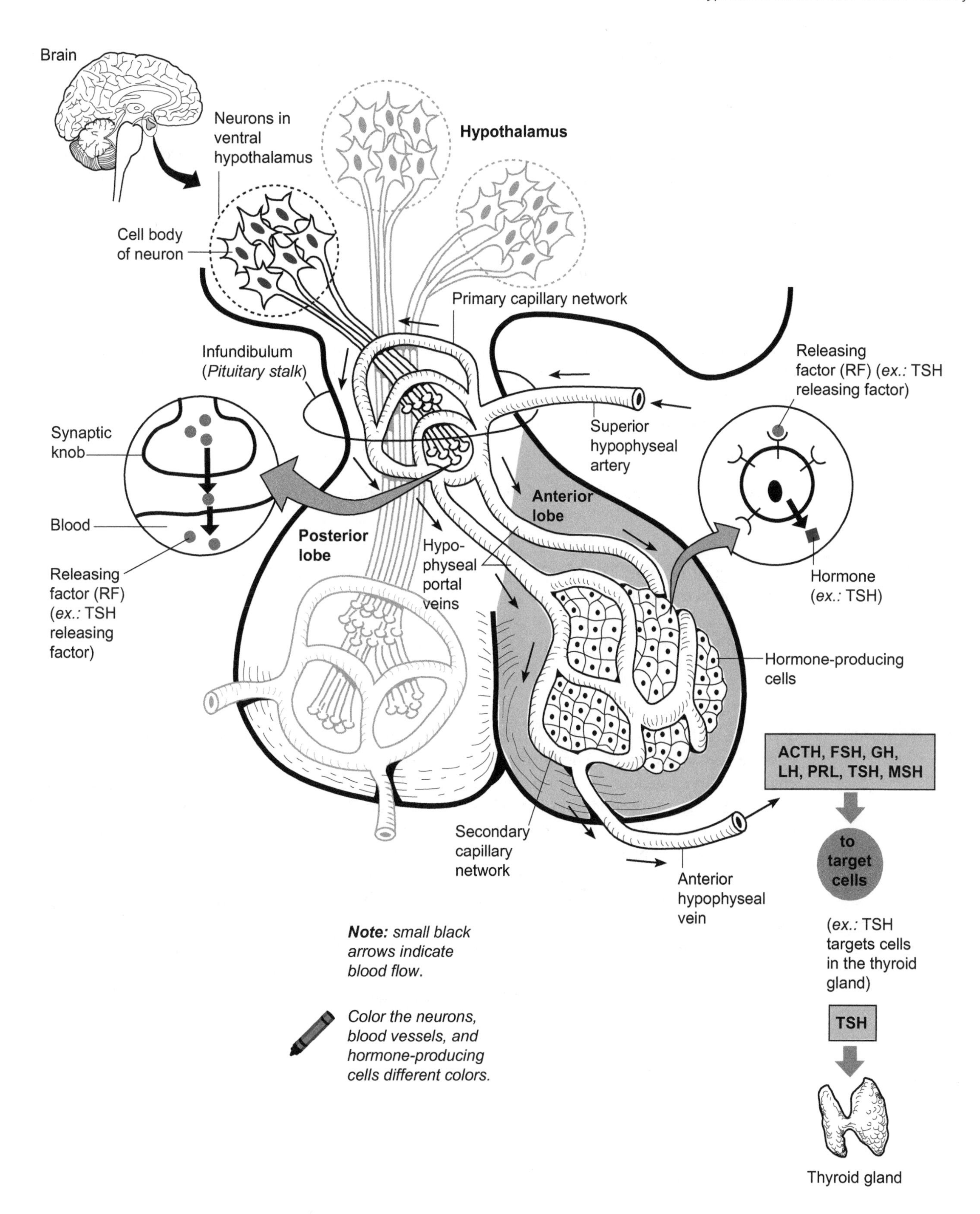

Note: *small black arrows indicate blood flow.*

Color the neurons, blood vessels, and hormone-producing cells different colors.

Description

The **pituitary gland** (*hypophysis*) is a small but powerful gland located at the base of the brain. It consists of two distinct lobes: the **anterior lobe** (*adenohypophysis*) and the **posterior lobe** (*neurohypophysis).* Because each lobe functions differently and secretes different hormones, each is presented in a separate module. This one highlights the **posterior pituitary.**

Surrounding the pituitary is a protective pocket of bone called the **sella turcica** of the **sphenoid bone.** The pituitary is nicknamed the "master gland" because it makes many different hormones that control the other endocrine glands in the body. But the **hypothalamus** in the brain actually controls the pituitary gland with specific bundles of neurons. This provides a *vital link* between the nervous system and the endocrine system. The pituitary is connected to the hypothalamus by a slender, stem-like structure called the **infundibulum** (*pituitary stalk*).

Hormones Produced

The cells in the posterior pituitary do *not* produce any hormones. Instead, neurons originating in the hypothalamus that extend into the posterior pituitary produce, store, and release the following two hormones:

- **Antidiuretic hormone** (ADH)
- **Oxytocin** (OT)

In short, both ADH and OT are released by **neurosecretion.**

Sequence

The posterior pituitary contains a simple **capillary network.** The following flowchart indicates blood flow through the posterior pituitary:

Inferior hypophyseal a. → capillary network → Posterior hypophyseal v.

Mechanism

Two different clusters of neurons in the hypothalamus produce hormones: the **paraventricular nuclei** and the **supraoptic nuclei.** The former produces OT, and the latter produces ADH. Each neuron in the cluster is a long, slender cell that has a kind of head and a tail, the **cell body** and the **axon terminal.**

Hormones are produced in the cell body and then stored in membrane-bound chambers called vesicles. These vesicles travel through the neurons and into the axon terminals located in the posterior pituitary. Nerve impulses trigger the release of these hormones from the axon terminals and into the blood. ADH and OT diffuse into the capillary network and exit the pituitary gland via the posterior hypophyseal vein. As they travel through the blood, they bind to receptors in their target cells and induce a response.

For example, ADH targets cells in the kidney to reabsorb more water into the blood, thereby decreasing urine output. OT has multiple targets, one of which is the smooth muscle in the wall of the uterus. During childbirth, OT is released to trigger the muscular contractions needed for labor and delivery.

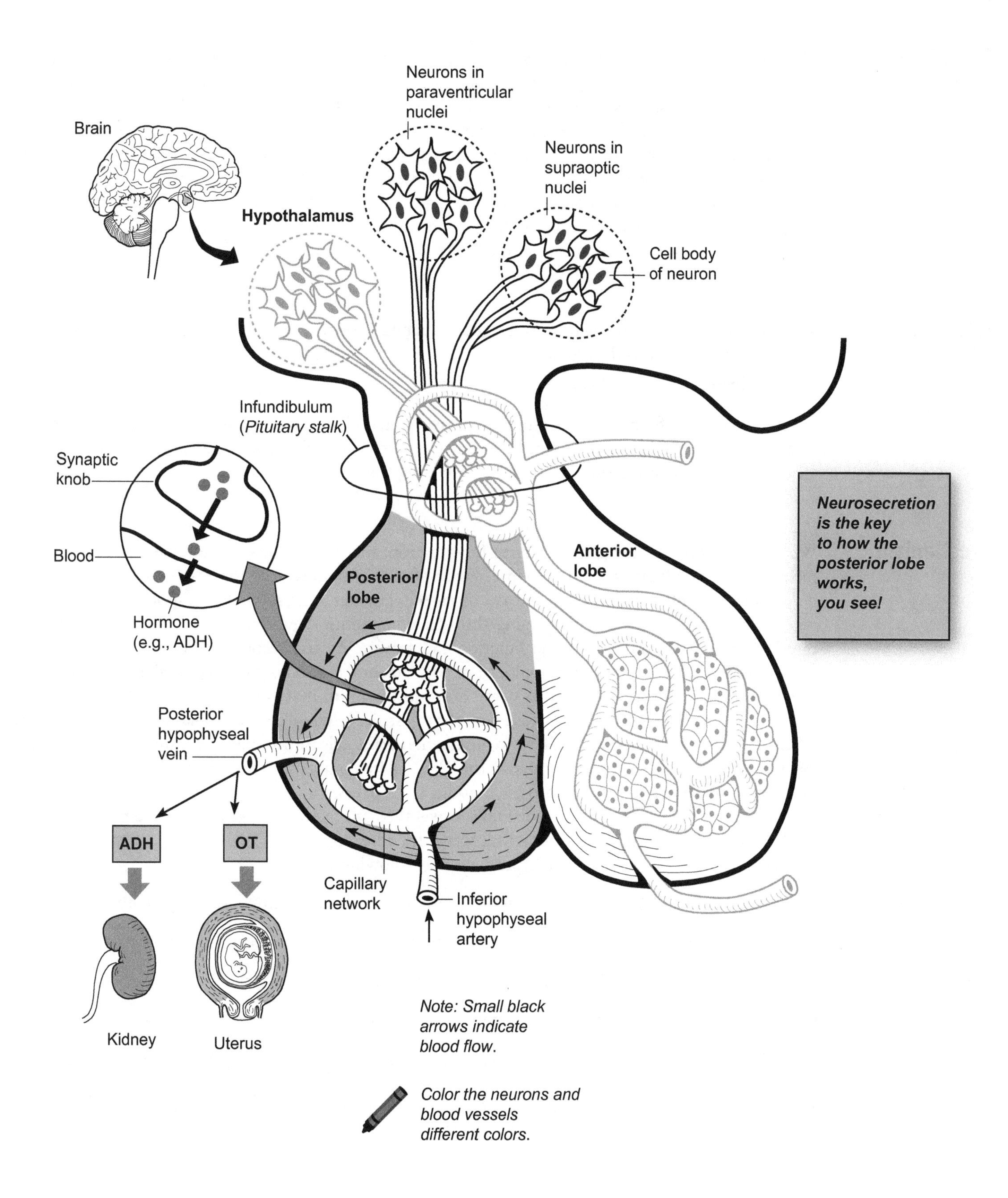

Note: Small black arrows indicate blood flow.

Color the neurons and blood vessels different colors.

Description

The highly vascular **adrenal** (*suprarenal*) **glands** are small structures located on top of both kidneys. They are divided into two major regions: the *outer* **adrenal cortex** and the *inner* **adrenal medulla.** The adrenal cortex constitutes most of the mass of the adrenal glands. Each of these regions contains different types of cells that produce different hormones. This module focuses only on the adrenal medulla. **Catecholamines** are a group of chemicals derived from the amino acid **tyrosine. Catecholamine-producing cells** (*chromaffin cells*), or **CPCs**, are located within the medulla and are directly stimulated by nerves from the **sympathetic division** of the autonomic nervous system (see p. 272). Because the sympathetic division helps the body respond to stressful situations on its own, the secretion of catecholamines simply boosts this response.

Hormones Produced

The catecholamine-producing cells in the adrenal medulla secrete the following two hormones:

- **Norepinephrine** (NE) (*noradrenaline*)
- **Epinephrine** (*adrenaline*)

As you can see from the illustration, they have similar chemical structures. In fact, the body is able to convert norepinephrine into epinephrine with the help of a converting enzyme. About 80% of the CPCs secrete epinephrine, and 20% secrete norepinephrine.

Mechanism

When faced with an emergency or stressful situation, the body reacts in multiple ways collectively referred to as the **fight-or-flight** response. Most of these responses are the result of the sympathetic nervous system at work, but the release of catecholamines intensifies this response. For example, imagine that you discover a fire raging through your home. Your body needs to prepare you for physical activity so you can flee the situation and survive. As you recognize the fire as a danger, this conscious thought sends a nerve impulse from your **cerebral cortex** to your **hypothalamus** in the brain. Then an impulse is stimulated along nerve fibers in the brain that connect the hypothalamus to the spinal cord. The nerve impulse carried along the last nerve connects the spinal cord directly to the CPCs in your adrenal medulla. These sympathetic nerves release the neurotransmitter **acetylcholine** (**ACh**), and it binds to ACh receptors in the CPCs. After binding, it stimulates the CPCs to secrete norepinephrine and epinephrine into the blood.

Like all hormones, these catecholamines travel through the blood like cars on the highway. Their final destination—**target cells**—contain receptors for catecholamines. The binding of the hormone to its receptor in the various target cells produces the following responses:

① Heart rate and force of contraction increase (this speeds blood circulation throughout the body)

② Blood vessels feeding kidneys and digestive tract constrict (this reduces blood flow to those tissues)

③ Bronchioles in lungs dilate (this moves air in and out of lungs faster)

④ Liver converts glycogen to glucose (this increases blood glucose levels)

All these responses are for the purpose of helping you flee the burning house. To run away from the fire, more blood will have to flow to your skeletal muscles to deliver more glucose and oxygen to the muscle cells. These nutrients will provide energy for muscle contraction as you flee. The following responses also occur (not included in the illustration):

- Metabolic rate increases.
- Urine output decreases.
- Blood is redirected from the digestive system toward the brain, skeletal muscles, and heart.

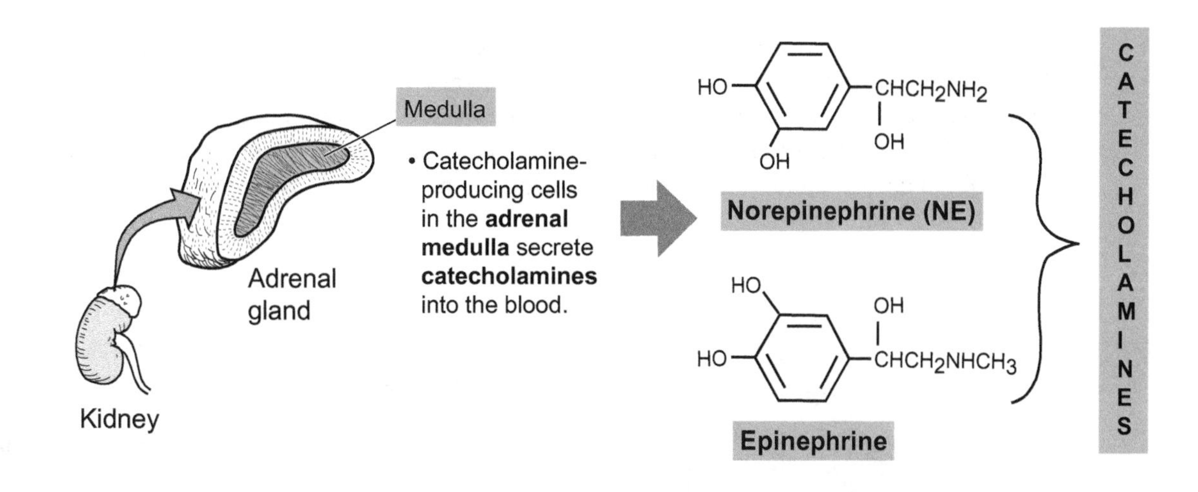
Medulla
• Catecholamine-producing cells in the **adrenal medulla** secrete **catecholamines** into the blood.
Adrenal gland
Kidney
HO
CHCH2NH2
OH
OH
Norepinephrine (NE)
HO
OH
HO
CHCH2NHCH3
Epinephrine
CATECHOLAMINES

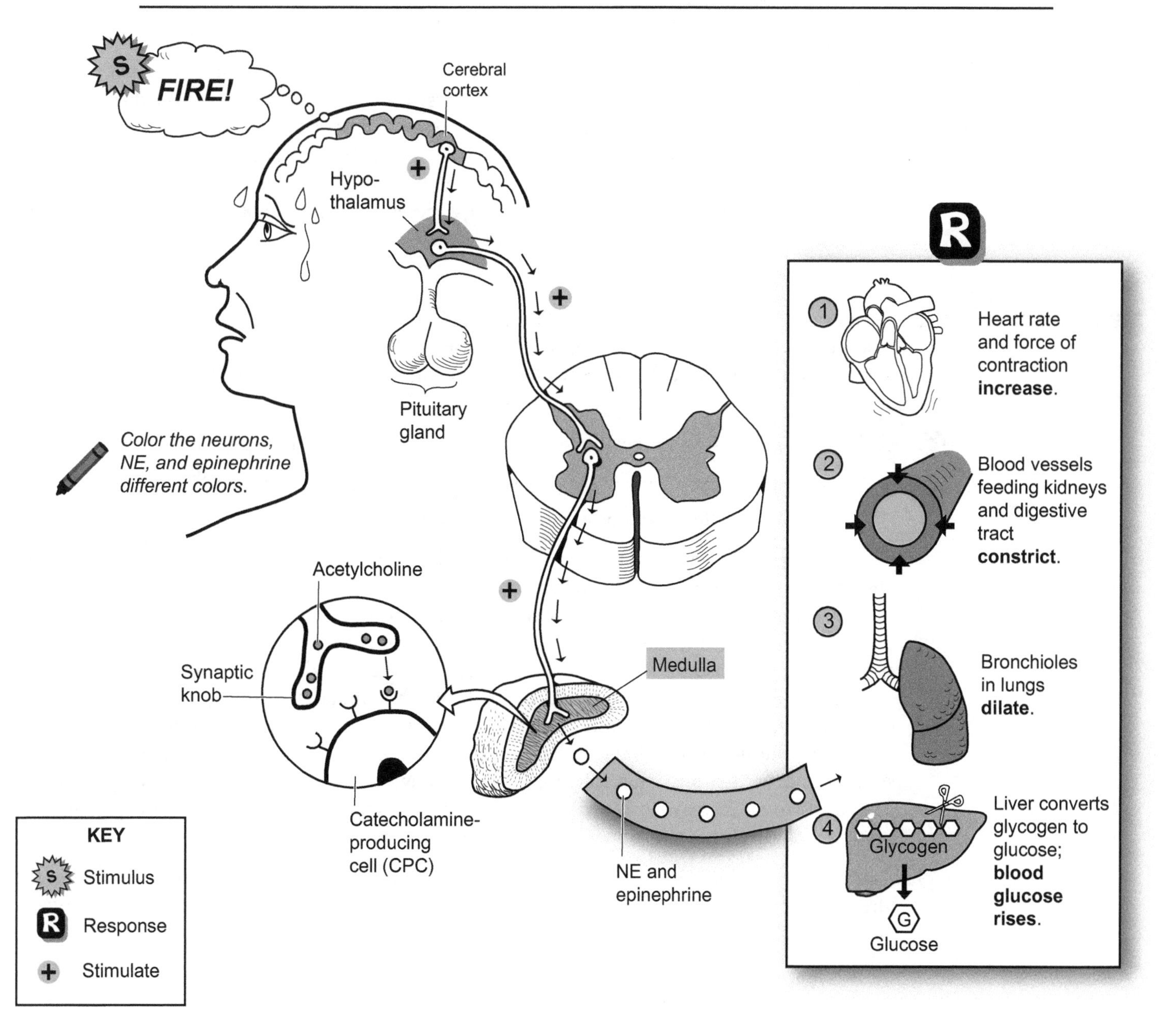
S
FIRE!
Cerebral cortex
Hypo-thalamus
Pituitary gland
Color the neurons, NE, and epinephrine different colors.
Acetylcholine
Synaptic knob
Catecholamine-producing cell (CPC)
Medulla
NE and epinephrine
R
1
Heart rate and force of contraction **increase**.
2
Blood vessels feeding kidneys and digestive tract **constrict**.
3
Bronchioles in lungs **dilate**.
4
Glycogen
G
Glucose
Liver converts glycogen to glucose; **blood glucose rises**.
KEY
S Stimulus
R Response
+ Stimulate

Description

The highly vascularized **adrenal** (*suprarenal*) **glands** are small structures located on top of both kidneys. They are divided into two major regions: the *outer* **adrenal cortex** and the *inner* **adrenal medulla.** The adrenal cortex constitutes most of the mass of the adrenal glands. Each of these regions contains different types of cells that produce different hormones. This module focuses on one aspect of the adrenal cortex, namely, its role in producing the hormone **aldosterone.**

Hormones Produced

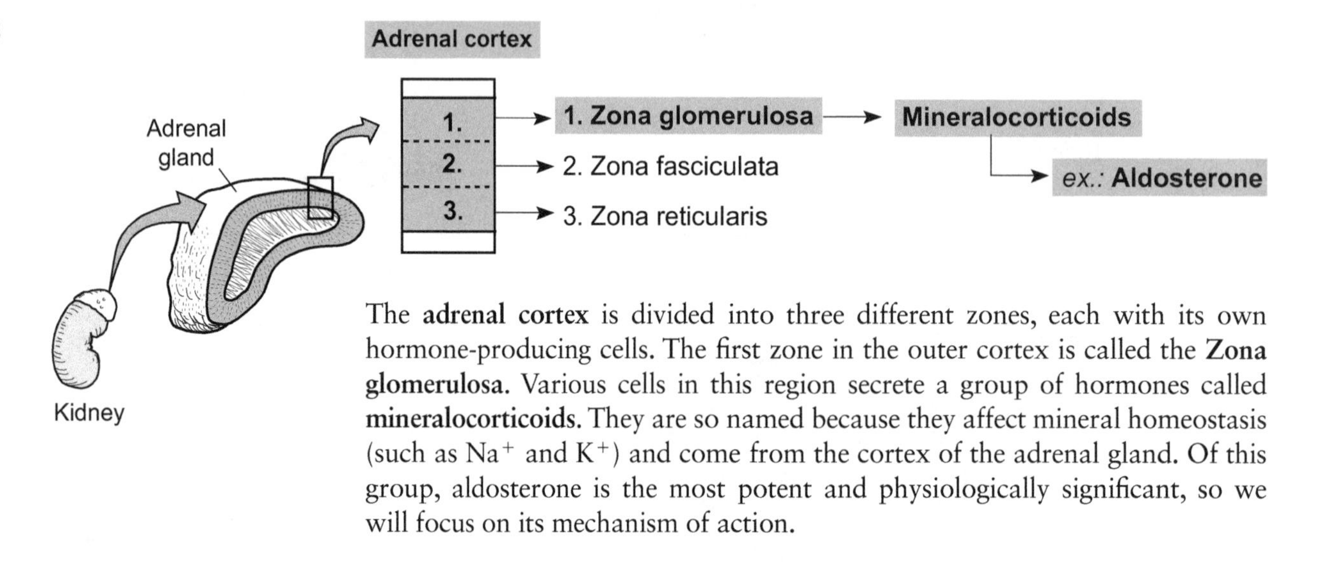

The **adrenal cortex** is divided into three different zones, each with its own hormone-producing cells. The first zone in the outer cortex is called the **Zona glomerulosa.** Various cells in this region secrete a group of hormones called **mineralocorticoids.** They are so named because they affect mineral homeostasis (such as Na^+ and K^+) and come from the cortex of the adrenal gland. Of this group, aldosterone is the most potent and physiologically significant, so we will focus on its mechanism of action.

Mechanism

Aldosterone controls the levels of $\mathbf{Na^+}$ (sodium) and $\mathbf{K^+}$ (potassium) ions in extracellular fluids (such as the blood). The net result of its action is to reabsorb Na^+ ions into the blood and simultaneously excrete K^+ ions into the urine. Because "water follows the ions," as Na^+ is reabsorbed, water also is reabsorbed.

① As shown in the illustration, two types of stimuli stimulate aldosterone release:

1. a decrease in Na^+ levels in the plasma or an increase in K^+ levels, and
2. a decrease in **blood volume** and/or **blood pressure.**

The kidneys detect these stimuli and cause a chemical chain reaction referred to as the **renin-angiotensin-aldosterone (RAA) mechanism.**

This is the most important pathway to trigger aldosterone release, the details of which are discussed in another module (see p. 242). In short, certain cells in the nephrons of the kidney are stimulated to release an enzyme called renin into the blood. This eventually leads to the activation of a hormone called **angiotensin II,** which stimulates secretory cells in the adrenal cortex to release aldosterone into the blood. The primary targets for aldosterone are the renal tubules in the **nephrons** of the kidneys. The response is that protein transporters reabsorb sodium and excrete potassium. Because water follows the sodium ions, it is also reabsorbed. This leads to an increase in blood volume and a corresponding increase in blood pressure. As blood pressure returns to normal, it shuts off the RAA mechanism.

② Changes in plasma levels of either Na^+ or K^+ also can directly stimulate the secretory cells in the adrenal cortex to release aldosterone. Of these two ions, the secretory cells are *more sensitive* to changes in K^+.

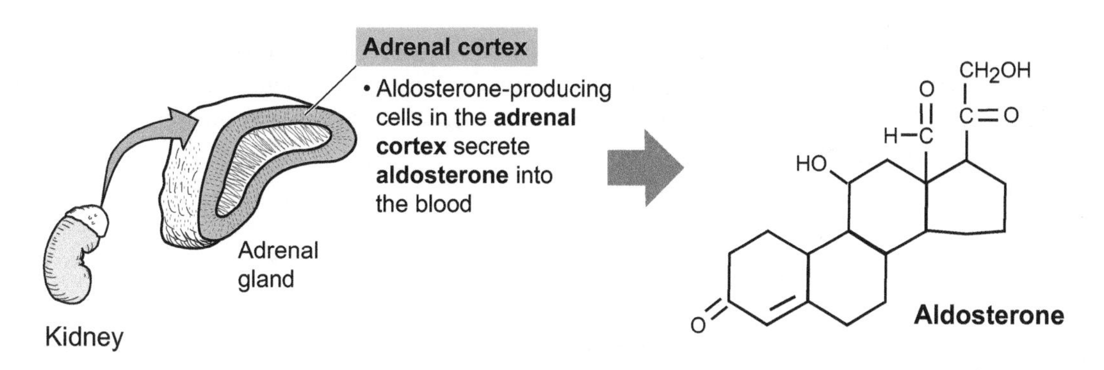
Adrenal cortex
• Aldosterone-producing cells in the adrenal cortex secrete aldosterone into the blood
Adrenal gland
Kidney
CH2OH
O
H–C
C=O
HO
O
Aldosterone

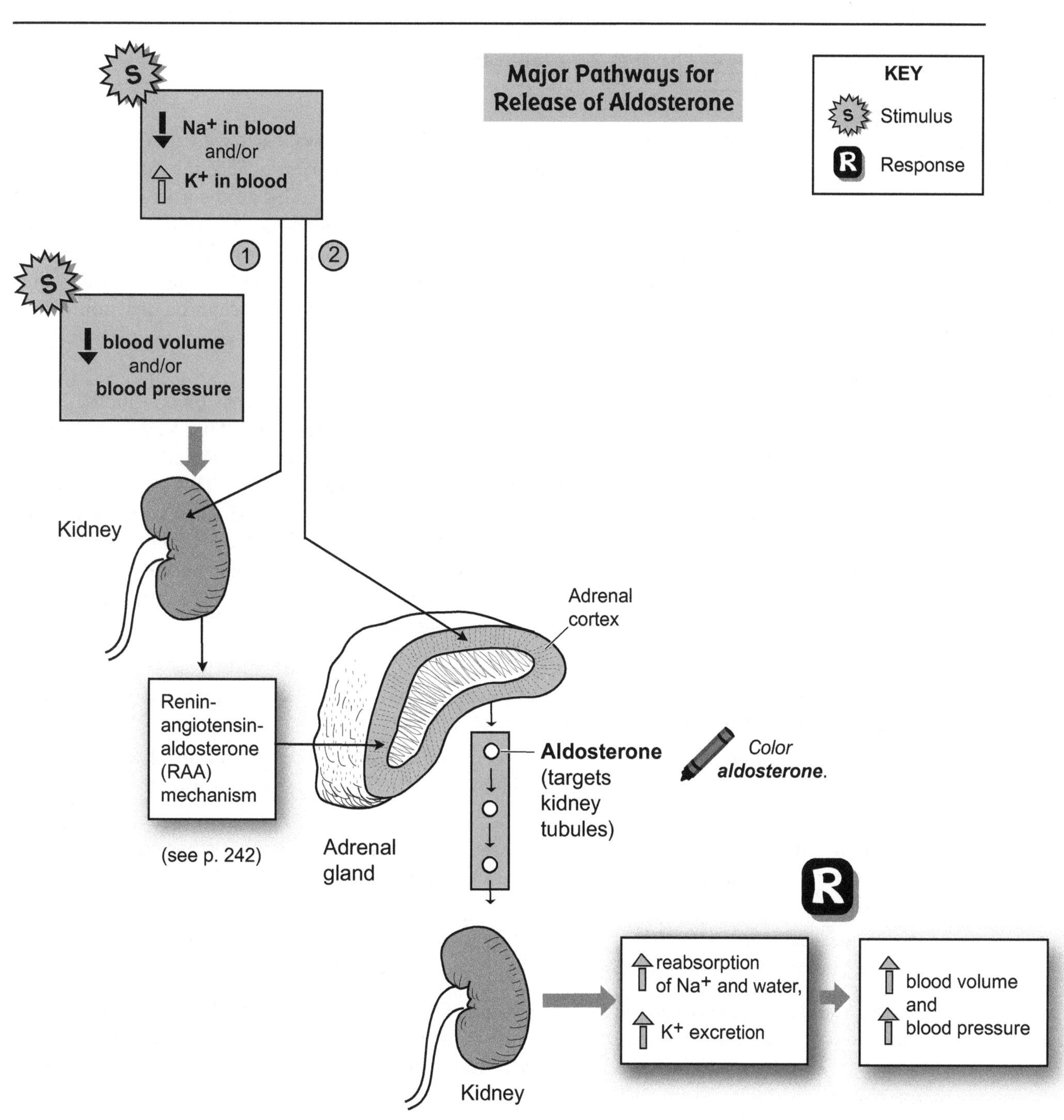
Major Pathways for Release of Aldosterone
KEY
S Stimulus
R Response
S
Na+ in blood
and/or
K+ in blood
1
2
S
blood volume
and/or
blood pressure
Kidney
Renin-angiotensin-aldosterone (RAA) mechanism
(see p. 242)
Adrenal cortex
Adrenal gland
Aldosterone (targets kidney tubules)
Color aldosterone.
R
Kidney
reabsorption of Na+ and water,
K+ excretion
blood volume and blood pressure

Description

The **adrenal** (*suprarenal*) **glands**, located on top of both kidneys, are divided into two major regions: the *outer* **adrenal cortex** and the *inner* **adrenal medulla.** Each of these regions contains different types of cells that produce different hormones. This module focuses on the role of the adrenal cortex, which produces the hormone **cortisol.**

Hormones Produced

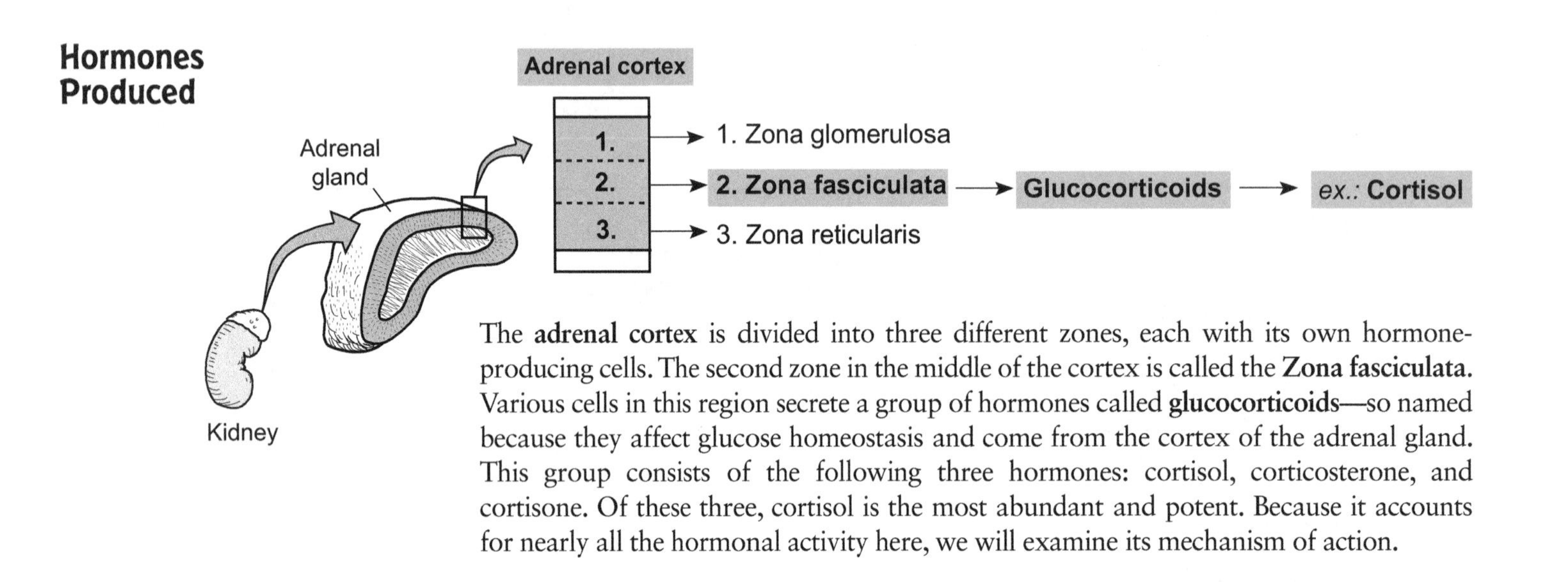

The **adrenal cortex** is divided into three different zones, each with its own hormone-producing cells. The second zone in the middle of the cortex is called the **Zona fasciculata.** Various cells in this region secrete a group of hormones called **glucocorticoids**—so named because they affect glucose homeostasis and come from the cortex of the adrenal gland. This group consists of the following three hormones: cortisol, corticosterone, and cortisone. Of these three, cortisol is the most abundant and potent. Because it accounts for nearly all the hormonal activity here, we will examine its mechanism of action.

Mechanism

We can't live without glucocorticoids, because they help us maintain normal blood glucose levels, normal blood volume, and resistance to stress—to name just a few functions. First let's examine how cortisol is secreted, then the responses it produces. Cortisol follows a predictable daily cycle in which its levels ***peak** after we awake* and ***dip to their lowest point** after we fall asleep*. This cycle is regulated by a negative feedback loop. As blood levels of glucocorticoids fall, this stimulates neurons in the **hypothalamus** to secrete **CRH** (corticotropin-releasing hormone) into the blood. CRH targets cells in the **anterior pituitary** and stimulates them to secrete **ACTH** (adrenocorticotropic hormone). As ACTH travels through the blood, it targets cells in the adrenal cortex to secrete cortisol. As cortisol travels to its target cells in the body, it induces the following responses:

1. **Protein breakdown:** Proteins are broken down into **amino acids** (mostly in skeletal muscle). These amino acids enter the blood and travel to body cells, where they can be used to either make new proteins or be used to produce ATP.
2. **Gluconeogenesis:** This is a process that occurs in liver cells, where a noncarbohydrate such as an amino acid or lactic acid is converted into glucose. The glucose then is used by body cells to produce ATP.
3. **Lipolysis:** This is the process of taking triglycerides stored in adipose tissue and converting them into fatty acids and glycerol. Cells can then use the glycerol and fatty acids to produce ATP.
4. **Stress resistance:** The glucose produced by liver cells in gluconeogenesis allows body cells to produce ATP. This supply of ATP enables the body to battle many different stresses.

Stress also triggers the release of cortisol. Chronic stress can lead to elevated levels of cortisol, and this can affect your health negatively. This leads to depression of the immune system. As a result, wounds may take longer to heal, and you may be more susceptible to contracting a virus such as the common cold.

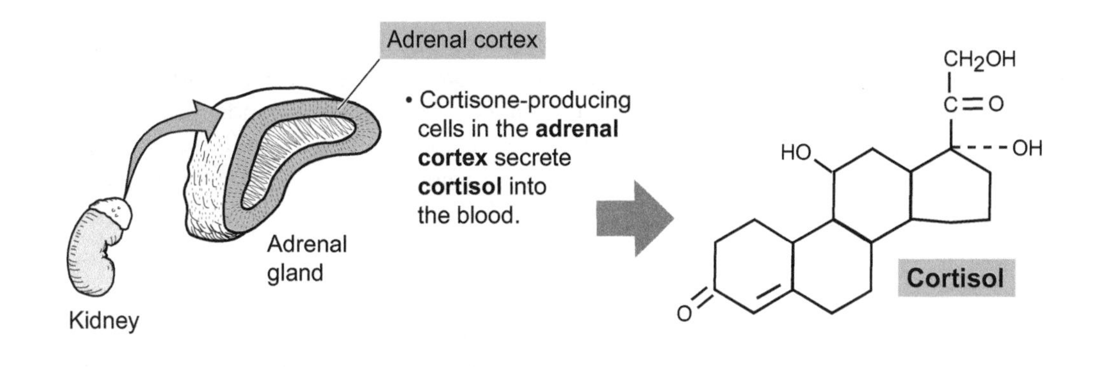
Adrenal cortex
• Cortisone-producing cells in the **adrenal cortex** secrete **cortisol** into the blood.
Adrenal gland
Kidney
CH2OH
C=O
HO
OH
O
Cortisol

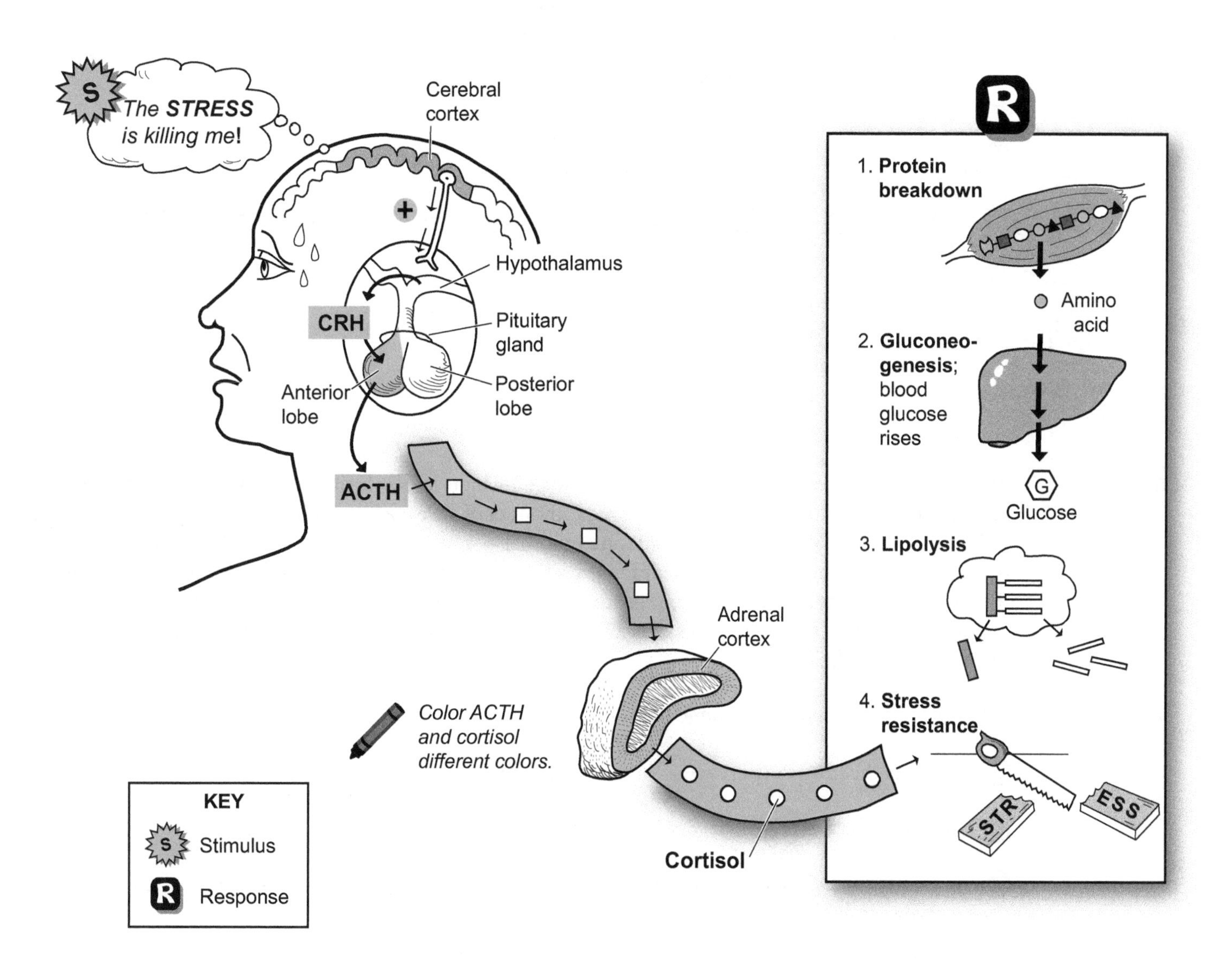
S
The STRESS is killing me!
Cerebral cortex
+
Hypothalamus
CRH
Pituitary gland
Anterior lobe
Posterior lobe
ACTH
Adrenal cortex
Color ACTH and cortisol different colors.
Cortisol
R
1. Protein breakdown
Amino acid
2. Gluconeogenesis; blood glucose rises
G
Glucose
3. Lipolysis
4. Stress resistance
STR
ESS
KEY
S Stimulus
R Response

Description

The highly vascularized **adrenal** (*suprarenal*) **glands** are small structures located on top of both kidneys. They are divided into two major regions: the *outer* **adrenal cortex** and the *inner* **adrenal medulla.** The adrenal cortex constitutes most of the mass of the adrenal glands. Each of these regions contains different types of cells that produce different hormones. This module focuses on one aspect of the adrenal cortex—its role in producing the masculinizing hormones called **androgens** (*andros* = male human being).

Hormones Produced

- The **adrenal cortex** is divided into three different zones, each with its own hormone-producing cells. The innermost zone is called the **Zona reticularis.** Various cells in this region secrete a group of male sex steroids called androgens. The most potent type of androgen is **testosterone.** Like all steroidal hormones, it is derived from **cholesterol.**

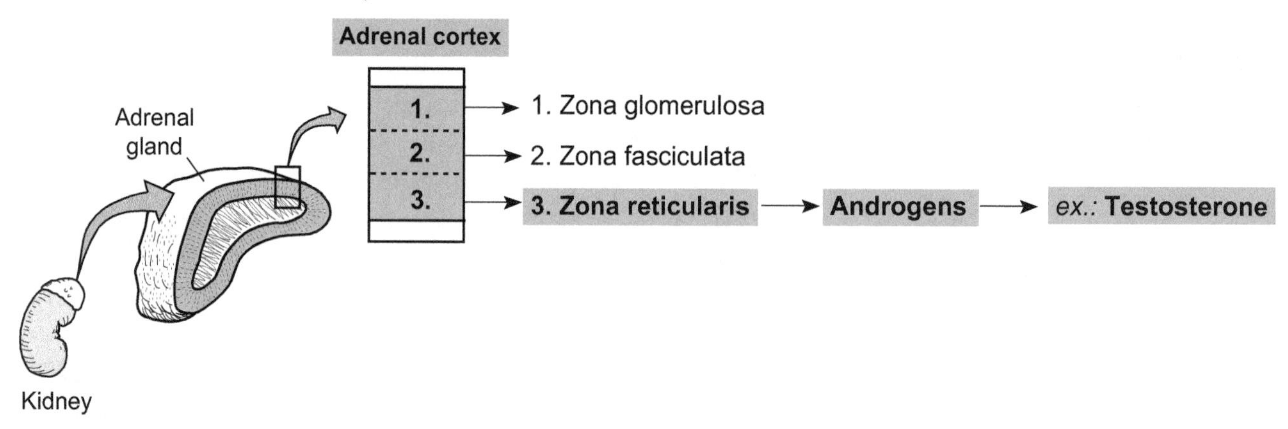

Mechanism

Testosterone is produced by the testes in the male at high levels during puberty and adulthood. The adrenal cortexes of both sexes produce small amounts of an androgen called **dehydroepiandrosterone** (DHEA). This is converted into **testosterone** by other tissue cells. At the onset of puberty, testosterone triggers similar responses in both sexes, such as an increase in underarm hair and pubic hair. But in the male, the amount of testosterone produced by the adrenal glands is insignificant compared with the quantity produced by the testes. So let's focus on the responses testosterone produces in the female throughout her lifetime:

1. **Develops underarm and pubic hair**—occurs before puberty

2. **Triggers the growth spurt**—also occurs before puberty

3. **Increases the sex drive**—occurs in the adult female

4. **Converts testosterone into estrogens**—postmenopausal women convert testosterone into estrogens (female sex hormones).

Details of the mechanism by which the adrenal glands produce androgens is not well understood. Production seems to be stimulated by **adrenocorticotropic hormone** (ACTH) secreted by the **anterior pituitary.**

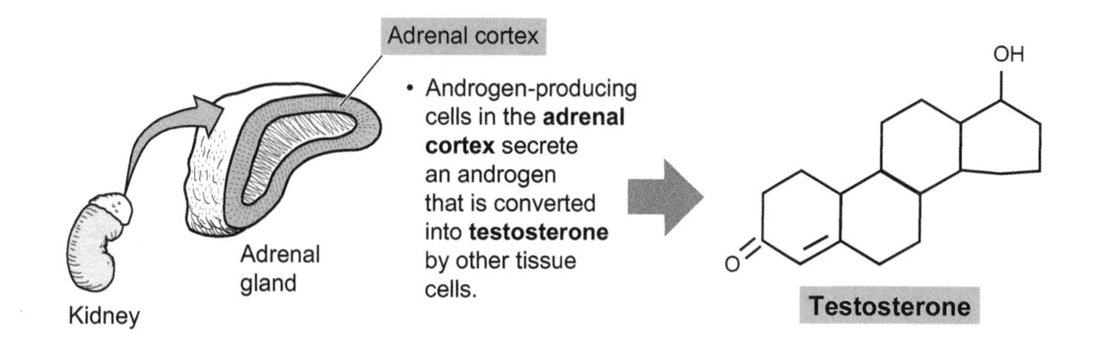
Adrenal cortex
• Androgen-producing cells in the **adrenal cortex** secrete an androgen that is converted into **testosterone** by other tissue cells.
Adrenal gland
Kidney
OH
O
Testosterone

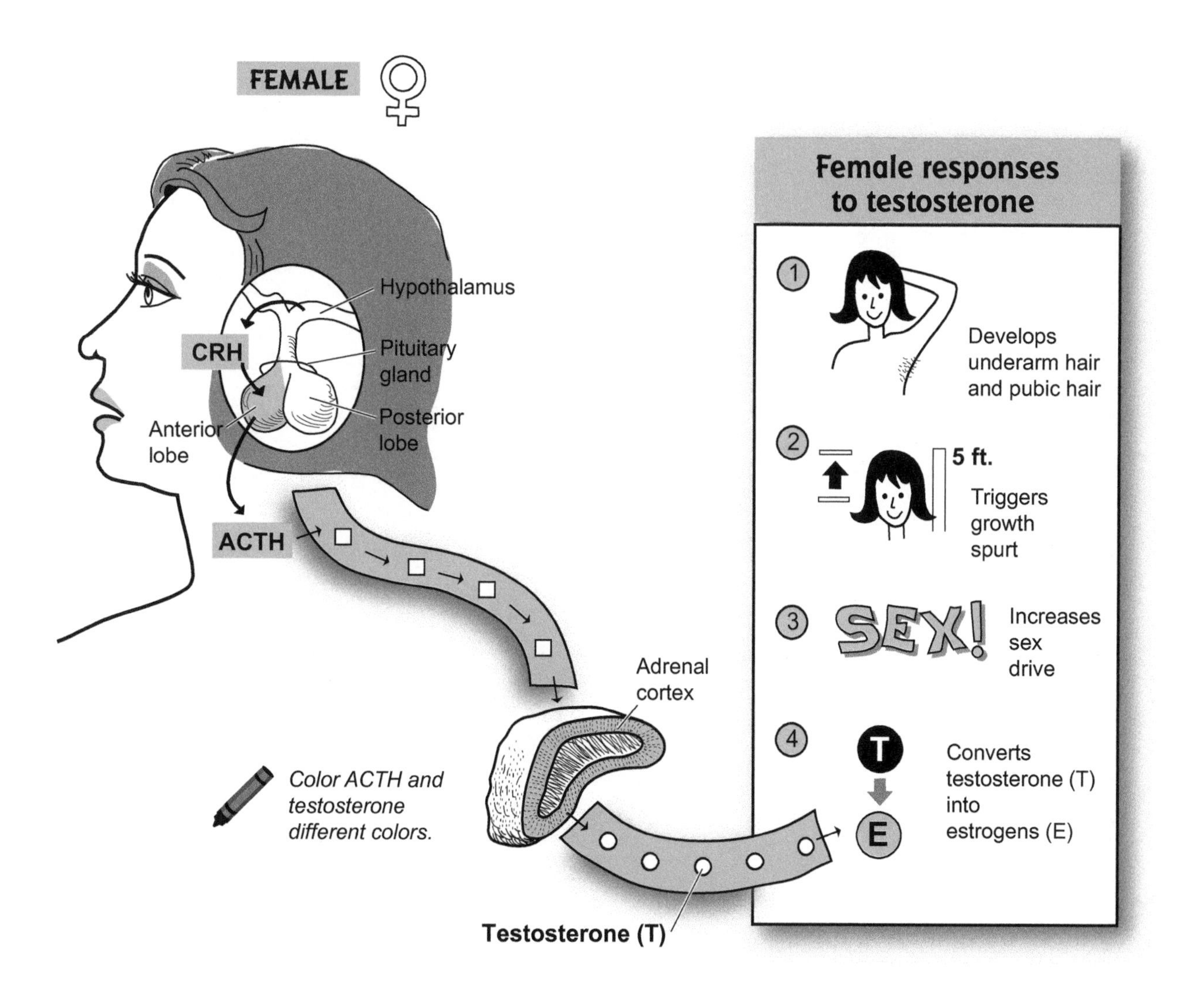
FEMALE
Hypothalamus
CRH
Pituitary gland
Anterior lobe
Posterior lobe
ACTH
Adrenal cortex
Color ACTH and testosterone different colors.
Testosterone (T)
Female responses to testosterone
1 Develops underarm hair and pubic hair
2 5 ft. Triggers growth spurt
3 SEX! Increases sex drive
4 T E Converts testosterone (T) into estrogens (E)

Notes

REPRODUCTIVE SYSTEMS

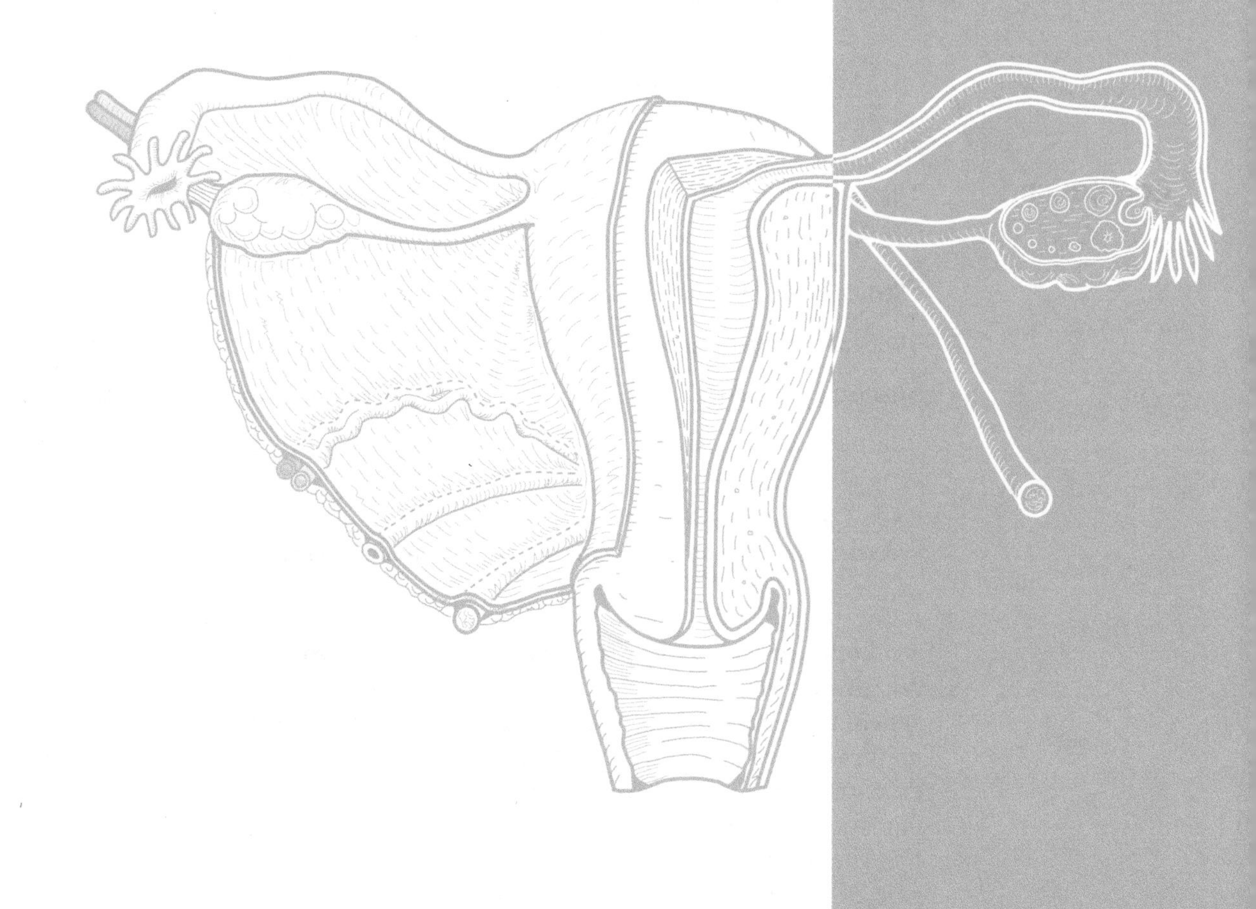

Description

The male and female reproductive systems are introduced and compared here. Both reproductive systems are derived from the same general tissues and are composed of the following three general structures:

- **Gonads:** produce the gametes (*sex cells*)
- **Ducts:** transport the gametes to the site of fertilization or outside the body (the "plumbing" of the reproductive system)
- **Accessory glands:** add liquid secretions to the reproductive tract to act as a lubricant or a medium in which the gametes are transported

For the external genitalia, similar structures between the sexes are:

- **Penis** and **clitoris:** both contain erectile tissues
- **Scrotal sac** and **labial folds**

Comparison

The table below gives a simplified, not comprehensive, structural comparison of male and female reproductive systems.

Structure/Product	Male	Female
Gonads (*glands that produce gametes*)	Testes	Ovaries
Gametes (*sex cells*)	Spermatozoa (*sperm cells*)	Ova (*egg cells*)
Ducts (*transport gametes*)	Epididymis, ductus deferens, and urethra	Uterine tube
Accessory glands (*add liquid secretions to the reproductive tract*)	Seminal vesicles, prostate gland, bulbourethral glands	Greater vestibular glands

Both sexes have a pair of gonads: in the male, two **testes** within the scrotal sac and, in the female, two ovaries in the pelvic cavity. The testes produce **sperm cells**, and the **ovaries** produce **ova**. The duct system in the male is longer and more complex than in the female. It connects the testes to the external urethral orifice, where sperm cells are released from the body. The female has only one type of main duct, the uterine tubes, which connect each ovary to the uterus. If fertilization occurs, the embryo implants in the **endometrium**, the innermost lining of the uterus.

The three accessory glands in the male produce a collective secretion called the **seminal fluid**. A pair of **seminal vesicles** secrete most of this, the **prostate gland** adds a milky secretion, and the **bulbourethral glands** contribute the smallest amount. In the female, the **greater vestibular glands** secrete a mucus that lines the inside of the **vagina**.

Both sexes use many of the same hormones to regulate reproductive processes. Both use the anterior pituitary hormones—**follicle-stimulating hormone** (FSH) and **luteinizing hormone** (LH) to regulate gonadal activity. The details are explained in other modules. In the testes, hormone-producing cells called **interstitial cells** secrete the "masculinizing" hormone **testosterone**. During puberty, it stimulates the development of all the secondary sexual characteristics in the male, such as deepening of the voice, enlargement of the genitals, and distribution of body hair. Similarly, the ovaries produce the "feminizing" hormone **estrogen**, responsible for the secondary sexual characteristics in the female, such as enlargement of the breasts, widening of the pelvis, and development of body hair. Although it is easy to focus on gender differences, the similarities are numerous.

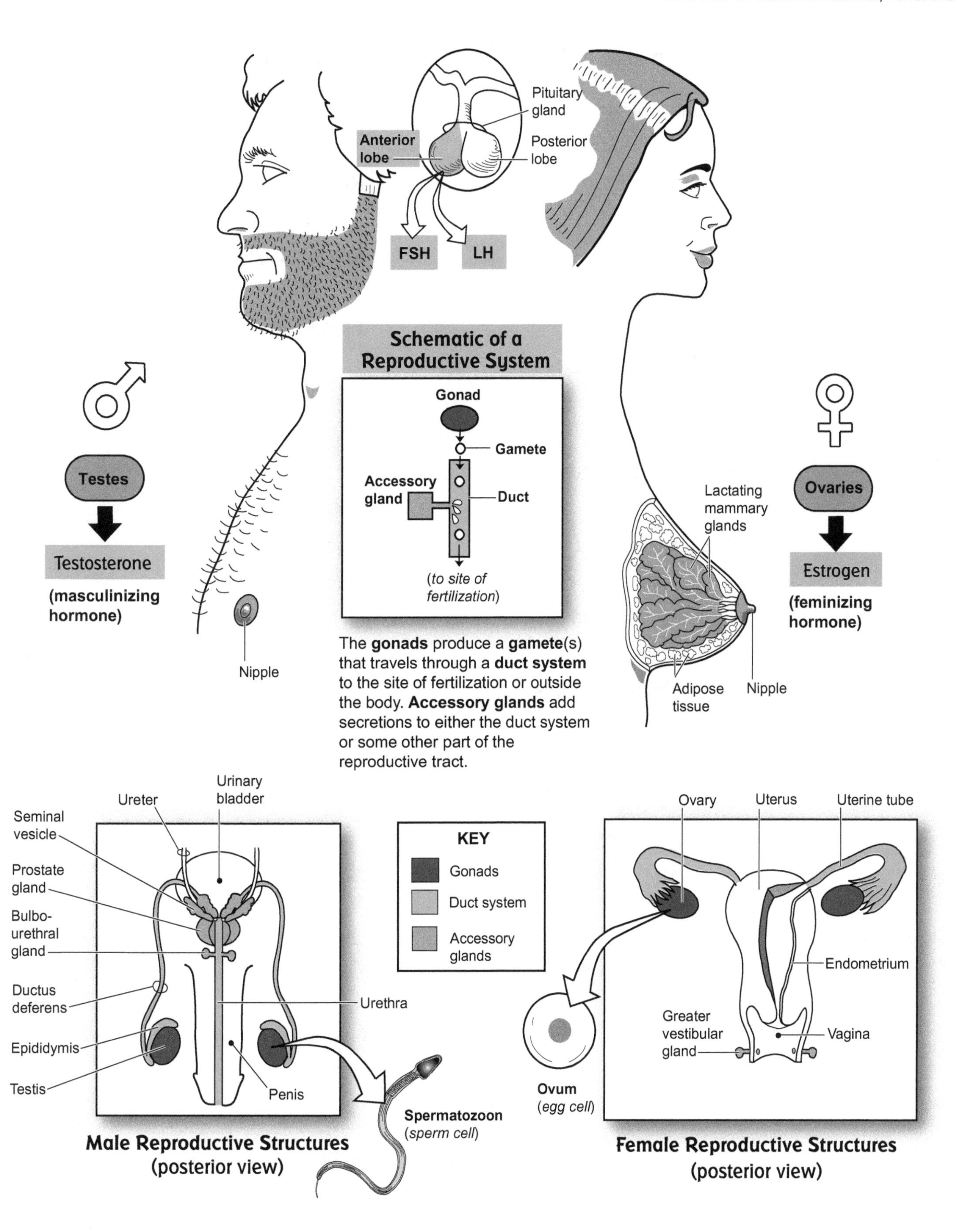

The **gonads** produce a **gamete**(s) that travels through a **duct system** to the site of fertilization or outside the body. **Accessory glands** add secretions to either the duct system or some other part of the reproductive tract.

Male Reproductive Structures (posterior view)

Female Reproductive Structures (posterior view)

Detailed Pathway

1. Lumen of seminiferous tubule
2. Rete testis
3. Efferent ductules
4. Epididymis

5. Ductus (*vas*) deferens

6. Ampulla of ductus deferens

7. Ejaculatory duct

8. Prostatic urethra

9. Membranous urethra

10. Penile urethra

11. External urethral orifice

Simplified Pathway

Seminiferous tubule

Rete Testis

Efferent ductules

Epididymis

Ductus deferens

Ejaculatory duct

Urethra

Study Tip

Use this mnemonic for the simplified pathway shown above:

"***S***ome ***R***eally ***E***lderly ***E***lephants ***D***on't ***E***ven ***U***rinate!"

Key to Illustration		
A. Spermatogonium	C. Secondary spermatocyte	E. Spermatozoon (*sperm cell*)
B. Primary spermatocyte	D. Spermatid	

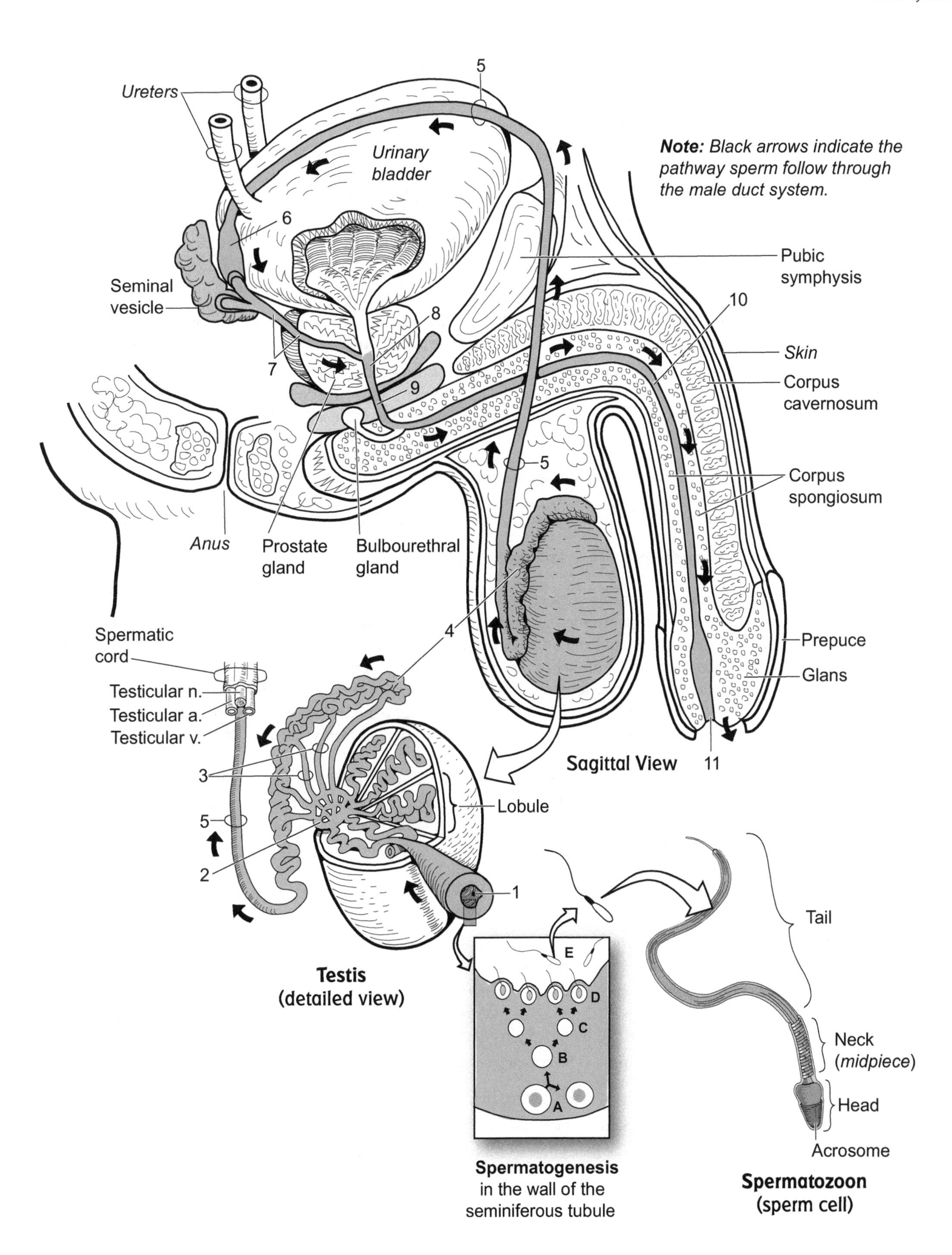

Testis (detailed view)

Spermatogenesis in the wall of the seminiferous tubule

Spermatozoon (sperm cell)

Description

Spermatogenesis is the process of developing sperm cells in the **testes** of the male. It begins for the first time in **puberty** and continues throughout adulthood. The process occurs within the walls of the **seminiferous tubules** in the testes and takes about 2 to 2.5 months to complete. It begins near the outer portion of the seminiferous tubules, proceeds toward the center, and finally ends when the sperm cells are released into the **lumen** of the seminiferous tubules. One of the hormones that regulates spermatogenesis is testosterone. It is secreted by **interstitial cells** (*cells of Leydig*), located between the seminiferous tubules.

Sustentacular cells (*Sertoli cells*) compose the walls of the seminiferous tubules. They physically support and nourish the developing sperm cells (DSCs), control the movement of the DSCs, remove wastes, produce the fluid that fills the lumen of the seminiferous tubules, and are involved in other functions. Each sustentacular cell is joined tightly to another by **tight junctions** like rivets that anchor cell membranes together (see p. 54). This tight seal helps maintain the **blood testis barrier**, which has two specific functions:

1. It protects developing sperm cells from potentially harmful substances by preventing proteins and other large molecules in the blood from coming in direct contact with developing sperm cells.
2. It prevents the immune system from being exposed to sperm cell antigens not found on any other body cells. If this were to occur, the immune system would view sperm cells mistakenly as foreign and respond by making antisperm antibodies in an effort to destroy the sperm cells.

Stages

This flowchart shows the **stages in spermatogenesis:**

Spermatogonium (SG) → primary spermatocyte (PS) → secondary spermatocyte (SS) →
early spermatids (EST) → late spermatids (LST) → sperm cells (S)

The **spermatogonium** is a **stem cell** found in the outer wall of the seminiferous tubules within the testes. It undergoes cell division to produce two new copies of itself. One of these new cells remains in the outer wall as a new spermatogonium, and the other differentiates into a slightly larger cell called a **primary spermatocyte.** This cell contains the total number or **diploid** (**2n**) number of chromosomes found in all human body cells: 46.

The primary spermatocyte enters a special process called **meiosis**, in which a cell undergoes two cell divisions to produce four final cells, *each containing half the number of chromosomes as the original cell.* This half-set of chromosomes is called the **haploid** (**n**) **number**. The only purpose of meiosis is to produce **gametes** (sperm, ova).

Because the total chromosome number is reduced by half, meiosis is called "reduction division." The primary spermatocyte divides to produce two **secondary spermatocytes**. Each of these cells contains a set of 23 replicated chromosomes. Each secondary spermatocyte then divides to produce two new **spermatids**. In total, four spermatids are produced, each containing the haploid (n) number, or 23 chromosomes. Last, the spermatids develop into spermatozoa (sperm cells).

Spermiogenesis is the last part of the spermatogenesis process, in which a spermatid is transformed into a sperm cell. Highlights of this metamorphosis are:

1. **Acrosome formation:** A sac covers the head of the sperm cell and contains the digestive enzymes used to penetrate the outer portion of the ova in the fertilization process.
2. **Mitochondrial reproduction:** The mitochondria replicate themselves and cluster to form a tightly coiled spiral around the neck of the sperm cell. They will be used to supply large amounts of ATP to propel the sperm cell by moving its flagellum.
3. **Flagellum formation:** The centrioles form the microtubules that make up the flagellum.

The **spermatozoon** or **sperm cell** has three main parts:

1. **Head:** has a nucleus that contains the haploid (n) number of chromosomes (23).
2. **Midpiece:** contains many mitochondria that produce large amounts of ATP to fuel the whipping of the tail.
3. **Tail** (*flagellum*): contains microtubules and is used to propel the sperm cell.

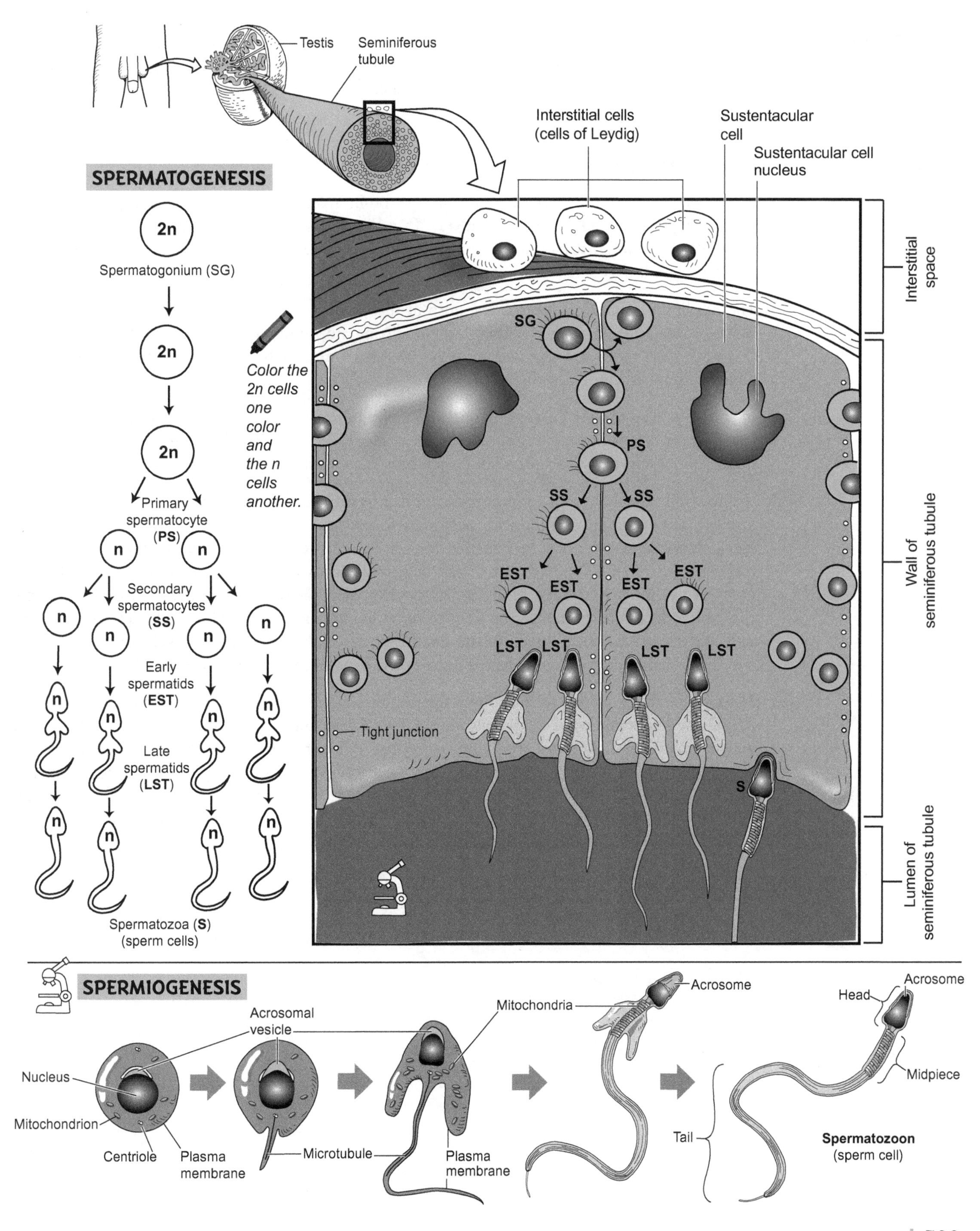
Testis
Seminiferous tubule
SPERMATOGENESIS
2n
Spermatogonium (SG)
2n
2n
Primary spermatocyte (PS)
n
n
Secondary spermatocytes (SS)
n
n
n
n
Early spermatids (EST)
Late spermatids (LST)
Spermatozoa (S) (sperm cells)
Color the 2n cells one color and the n cells another.
Interstitial cells (cells of Leydig)
Sustentacular cell
Sustentacular cell nucleus
Interstitial space
Wall of seminiferous tubule
Lumen of seminiferous tubule
SG
PS
SS
SS
EST
EST
EST
EST
LST
LST
LST
LST
S
Tight junction
SPERMIOGENESIS
Acrosomal vesicle
Nucleus
Mitochondrion
Centriole
Plasma membrane
Microtubule
Mitochondria
Plasma membrane
Acrosome
Tail
Head
Acrosome
Midpiece
Spermatozoon (sperm cell)

Description

Spermatogenesis, the development of sperm cells, occurs within the seminiferous tubules in the male testes. This process begins at puberty and continues throughout adult life. The group of hormones that regulates this process is the same basic set that regulates the ovarian cycle in the female (see p. 318).

Stages

A step-by-step description of this process is given below:

① It all begins when the hypothalamus in the brain releases into the blood a **peptide** called **gonadotropin-releasing hormone** (**GnRH**). In adult males, GnRH is released in regular bursts every 60–90 minutes. It travels through the blood to target cells within the anterior lobe of the pituitary gland, which has receptors for GnRH.

② The binding of GnRH induces a response in these cells to make two **gonadotropin** hormones: **follicle-stimulating hormone** (**FSH**) and **luteinizing hormone** (**LH**). Like any hormone, these chemical messengers are released into the bloodstream and travel to their respective targets, which contain receptors for these hormones.

③ In the male, **FSH** targets the **sustentacular cells** within the walls of the seminiferous tubules in the testes. These cells support and nourish the spermatogenetic cells. The binding of FSH also induces these cells to secrete two proteins: **androgen-binding protein** (**ABP**) and **inhibin.**

④ At the same time, **LH** binds to receptors on the **interstitial cells**, located between the seminiferous tubules. After binding, LH induces these cells to produce the hormone **testosterone**, which has many different targets. It is responsible for producing the secondary sex characteristics in the male (deepening of voice, enlargement of the genitals, growth of body hair, among other things). In this case, the testosterone binds to the protein **ABP** to form a complex. Once formed, this complex induces the spermatogenic cells to develop into sperm cells. The newly developed sperm cells are released into the lumen of the seminiferous tubule, where they travel through the male duct system until they are released from the body through the external orifice of the penis.

⑤ Negative feedback is used to shut down the pituitary gland. As the levels of **testosterone** rise, they inhibit the hypothalamus and the pituitary gland. The result is that production of GnRH ceases, which, in turn, leads to no production of either FSH or LH. This inhibition is enhanced by production of the hormone **inhibin** by the sustentacular cells. It has the same effect on the hypothalamus and pituitary as did testosterone. Once the testosterone and inhibin levels fall, inhibition is lost, and the cycle begins again. In the adult male, this cycle results in a steady level of testosterone production that fluctuates within a normal range.

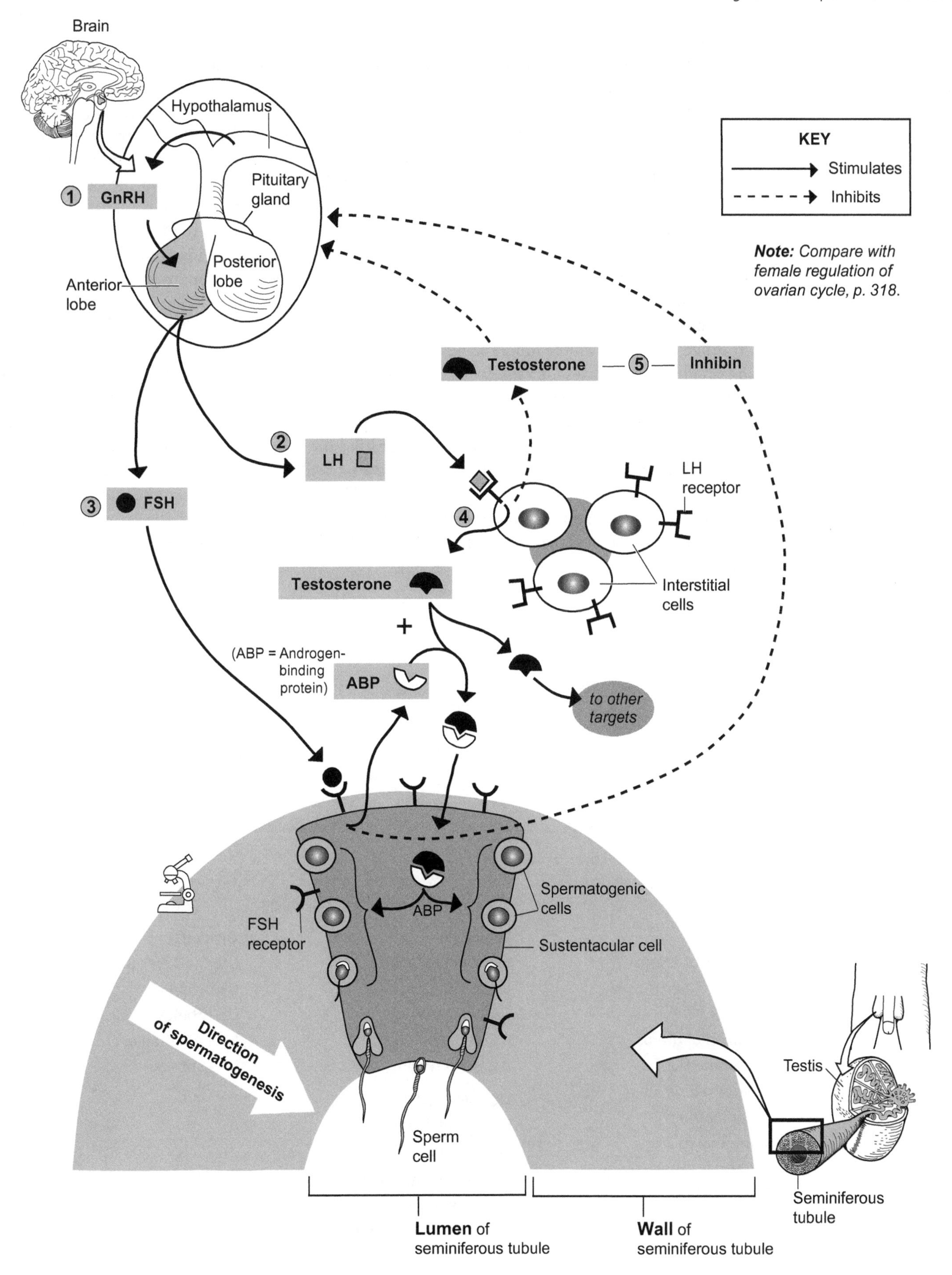
Brain
Hypothalamus
① GnRH
Pituitary gland
Anterior lobe
Posterior lobe
KEY
Stimulates
Inhibits
Note: Compare with female regulation of ovarian cycle, p. 318.
Testosterone — ⑤ — Inhibin
② LH
③ FSH
LH receptor
④
Testosterone
Interstitial cells
+
(ABP = Androgen-binding protein)
ABP
to other targets
Spermatogenic cells
ABP
FSH receptor
Sustentacular cell
Direction of spermatogenesis
Testis
Sperm cell
Seminiferous tubule
Lumen of seminiferous tubule
Wall of seminiferous tubule

Description

The **uterus** (*womb*) is a hollow, muscular organ divided into three regional areas: (1) **fundus,** (2) **body,** and (3) **cervix.** The **fundus** is the most superior portion of the uterus, the main portion is called the **body,** and the narrowed, neck-like portion that extends into the vagina is called the **cervix.** The wall of the uterus is made of three layers, from innermost to outermost: the endometrium, myometrium, and perimetrium. The **endometrium,** or mucosal lining, is made of simple columnar epithelium and an underlying vascular connective tissue. This entire layer thickens during a normal menstrual cycle in preparation for the implantation of an embryo. If no implantation occurs, this layer is sloughed off at the end of the menstrual cycle. The thick middle layer, the **myometrium,** is made of multiple layers of smooth muscle. The myometrium is hormonally stimulated to contract during childbirth to help move the baby out of the uterus. The outermost layer, the fibrous connective **perimetrium,** is the same as the **visceral peritoneum.**

The **vagina** is a thick, muscular tube that extends from the cervix to outside the body. It is lined with stratified squamous epithelium. The penis enters this passageway during sexual intercourse.

The small, lumpy **ovaries** are loosely held in place within the abdomen by various connective tissue ligaments. The **ovarian ligament** anchors the medial side of each ovary to the lateral side of the uterus, the **suspensory ligament** anchors the ovaries to the pelvic wall, and the wide, flat, **broad ligament** extends like a tarp over the uterus and ovaries and cradles the vagina, uterus, and uterine tubes.

The **uterine tubes** (consisting of the *fallopian tubes* and *oviducts*) serve to transport a female gamete (*egg cell; ova*) from the ovary to the uterus. The proximal end (near the uterus) is a long, narrow tube called the **isthmus.** The distal end widens and curves around the ovary to form the **ampulla.** The ampulla becomes the funnel-shaped **infundibulum,** which has fingerlike extensions called **fimbriae.**

Key to Illustration

Uterus
1. Fundus of uterus
2. Body of uterus
3. Cervix
4. Lumen of uterus
5. Endometrium
6. Myometrium
7. Perimetrium
8. Internal os
9. Cervical canal
10. External os

Vagina (V)

Blood Vessels (B)
B1. Ovarian artery
B2. Ovarian vein
B3. Uterine artery
B4. Uterine vein

Ligaments (L)
L1. Round ligament
L2. Suspensory ligament
L3. Ovarian ligament
L4. Broad ligament
L5. Uterosacral ligament

Ovary (O)

Uterine Tubes (U)
U1. Ampulla
U2. Isthmus
U3. Infundibulum
U4. Fimbriae

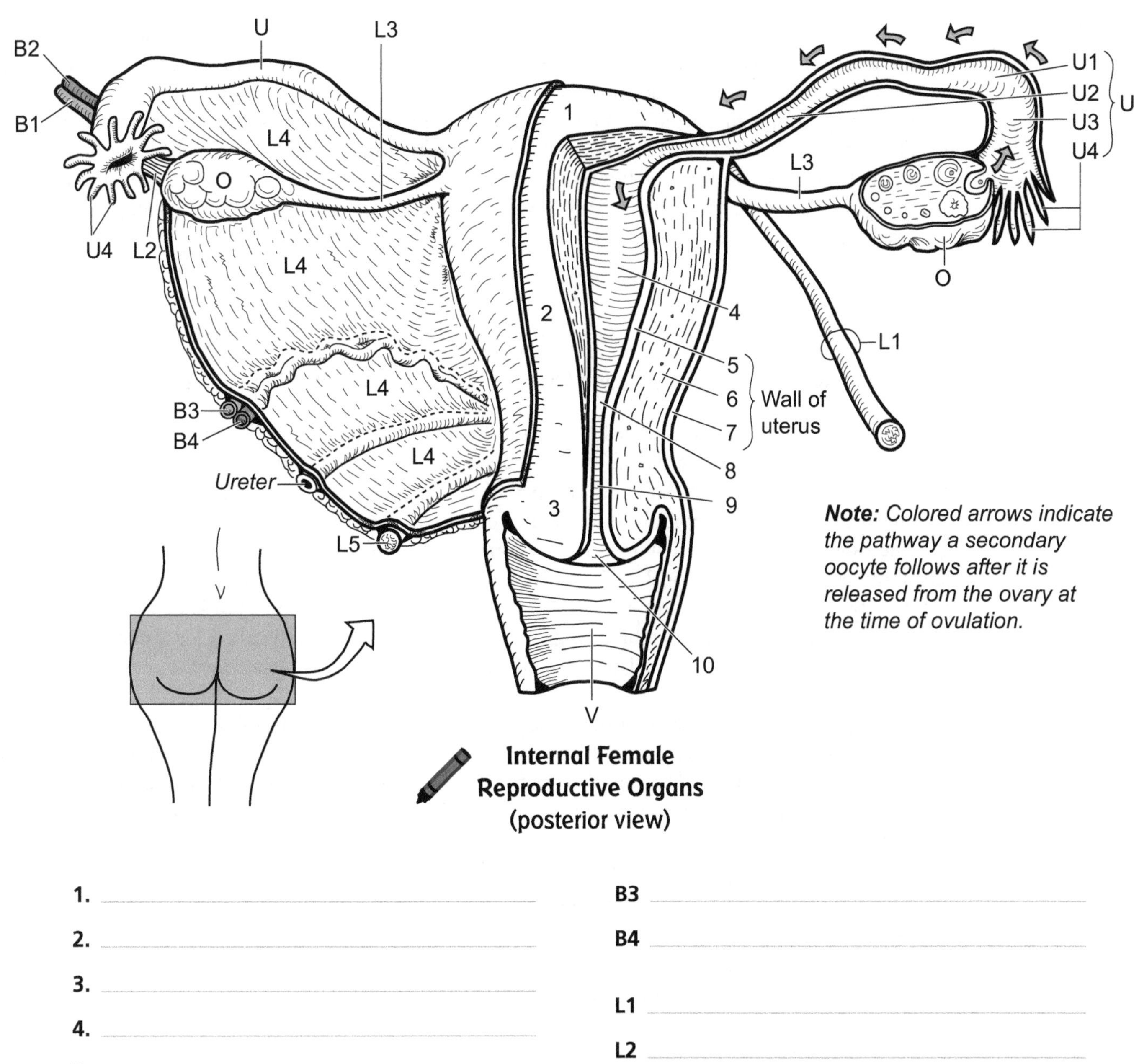

Note: Colored arrows indicate the pathway a secondary oocyte follows after it is released from the ovary at the time of ovulation.

Internal Female Reproductive Organs
(posterior view)

1. ______
2. ______
3. ______
4. ______
5. ______
6. ______
7. ______
8. ______
9. ______
10. ______

V ______

B1 ______
B2 ______
B3 ______
B4 ______

L1 ______
L2 ______
L3 ______
L4 ______
L5 ______

O ______

U1 ______
U2 ______
U3 ______
U4 ______

Description

Oogenesis is the process of developing **ova** (*egg cells*) within the ovaries of the female. Unlike sperm production in the male, this process begins during fetal development instead of at puberty. It involves two processes occurring together: **meiosis** and **follicle maturation.**

Imagine putting a marble in a balloon and then filling that balloon with water. In this analogy, the marble is the cell that has to go through meiosis to produce an ovum, and the water balloon is like the follicle that has to grow and develop by filling itself with fluid. It is important to remember that both of these processes occur together. In other words, meiosis occurs inside the maturing follicles. Let's consider each process separately.

Meiosis

This flowchart shows the stages of **meiosis** to produce an **ovum** within a follicle:

Oogonium (OG) → primary oocyte (PO) → secondary oocyte (SO) → zygote (if fertilized)

Meiosis is a type of cell division used solely to produce **gametes** (*sperm, ova*). It is called "reduction division" because during this process a cell undergoes two cell divisions to produce four final cells that each contain half the number of chromosomes as the original cell. This half-set of chromosomes, called the **haploid (n) number**, is equal to 23 chromosomes in humans.

The **oogonium** (OG) is a stem cell that contains the full set of chromosomes called the **diploid (2n) number**, or 46 ($2 \times n$ or 2×23). It is found in the outer region of the ovaries and undergoes cell division during fetal development to produce millions of cloned copies of itself. Most of these are degenerated, but some are stimulated to grow into a slightly larger cell called a **primary oocyte** (**PO**). Though they begin their first cell division of meiosis (meiosis I) in fetal development, *they do not complete it.*

At birth, about 400,000 to 4,000,000 oogonia (sing., *oogonium*) and POs remain in both ovaries. By puberty, about 40,000 remain, and the meiosis process is now strictly regulated by hormones. At this time, one PO is triggered to complete its first cell division of meiosis (meiosis I) every month. This cell division results in the formation of two unequally sized cells—a large, viable **secondary oocyte** (**SO**) and a small, nonfunctional cell called a **polar body.** The **SO** contains 23 replicated chromosomes. It will undergo its second cell division (meiosis II) only if it is fertilized by a sperm cell. If this occurs, the result is two new cells: a fertilized egg called a **zygote** and a small, nonfunctional **polar body.**

Follicle Maturation

This flowchart shows the stages in **follicle maturation:**

Primordial follicle → primary follicle → secondary follicle → tertiary follicle

Follicle maturation is like a water balloon filling with water until it bursts. The primitive **primordial follicles** are located in the outer region of the ovaries. They consist of a primary oocyte surrounded by a single layer of follicular cells. During the fetal stage, the primordial follicles are transformed into **primary follicles** by gradually thickening the follicle wall with multiple layers of cells around the primary oocyte.

As maturation continues, cells in the wall of the follicle, called granulosa cells, begin to secrete **follicular fluid**, which fills a potential space called the **antrum.** This marks the formation of the **secondary follicles** during adulthood. Like the expansion of a water balloon filling with fluid, the secondary follicle increases in size to become a **tertiary follicle** (*Graafian follicle; mature follicle; vesicular follicle*). At the same time, the primary oocyte has undergone its first meiotic division and produced a secondary oocyte and first polar body.

The tertiary follicle soon ruptures because of a hormonal surge and releases the **secondary oocyte** from the ovary in an event called **ovulation.** Last, the damaged follicle forms a temporary gland called a **corpus luteum** that produces the hormones progesterone and estrogen.

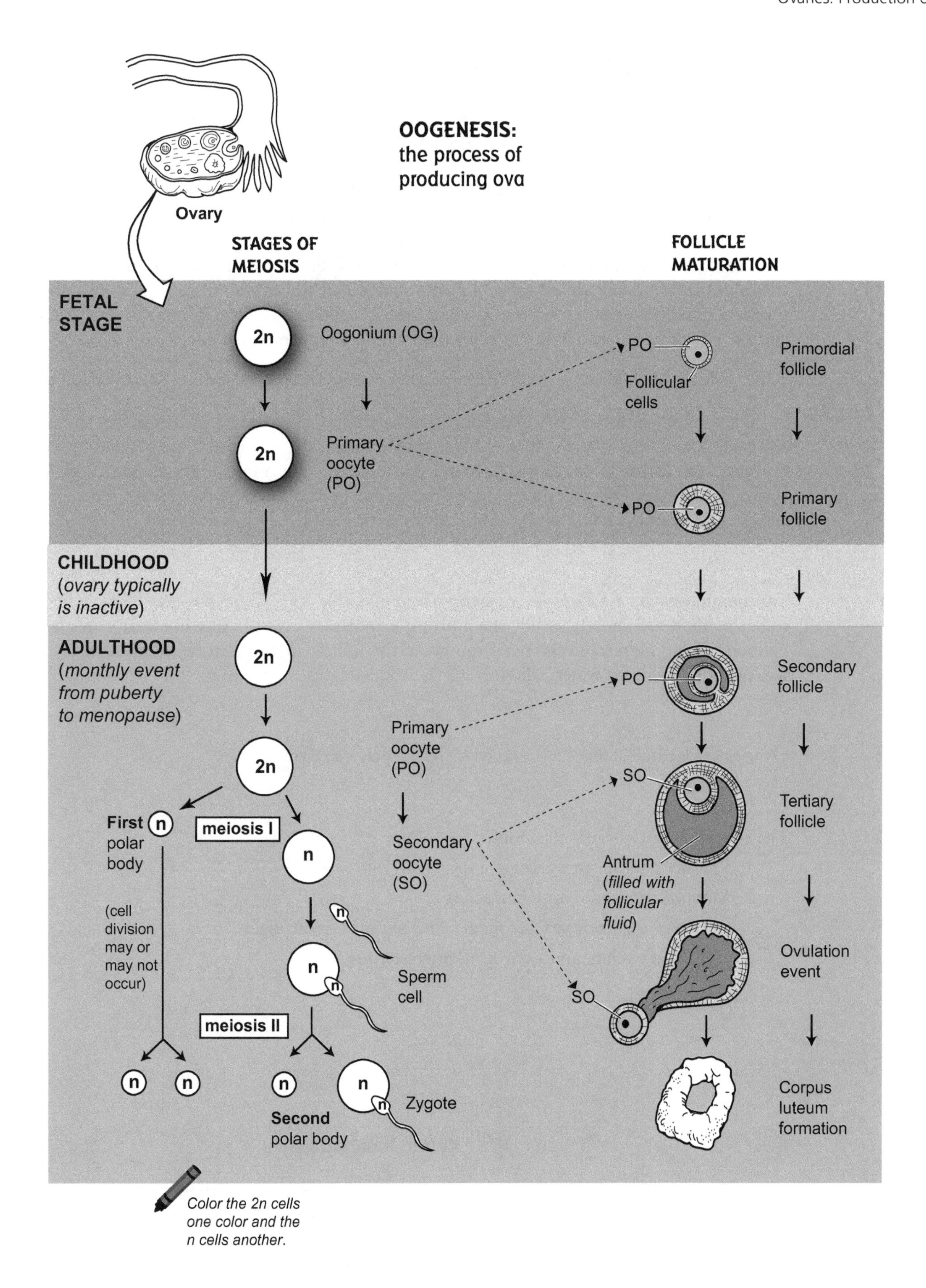
OOGENESIS:
the process of producing ova
Ovary
STAGES OF MEIOSIS
FOLLICLE MATURATION
FETAL STAGE
2n
Oogonium (OG)
2n
Primary oocyte (PO)
PO
Follicular cells
Primordial follicle
PO
Primary follicle
CHILDHOOD
(ovary typically is inactive)
ADULTHOOD
(monthly event from puberty to menopause)
2n
2n
Primary oocyte (PO)
PO
Secondary follicle
First polar body
n
meiosis I
n
Secondary oocyte (SO)
SO
Tertiary follicle
Antrum (filled with follicular fluid)
(cell division may or may not occur)
n
n
Sperm cell
SO
Ovulation event
meiosis II
n
n
n
n
n
Zygote
Second polar body
Corpus luteum formation
Color the 2n cells one color and the n cells another.

Description

The female has two lumpy, oval-shaped structures called **ovaries**, each measuring about 5 cm. in length. Blood is brought to the ovary by an **ovarian artery** and is drained by an **ovarian vein.** The surface of the ovary is covered by a simple cuboidal epithelial layer called the **germinal epithelium.** The two regional areas are the outer **cortex** and the inner **medulla.** Production of **gametes** (*sex cells*) occurs in the **cortex.**

Ideally, the ovaries in a sexually mature female alternate to produce one ovum each month. The specific process for developing an **ovum** (*egg cell*) is called **oogenesis.** This is a part of the ovarian cycle, which refers to all the processes that occur in the ovary during this monthly event.

The ova are produced in special chambers called **follicles.** The process begins when a hormone stimulates an immature follicle called a **primordial follicle** to begin to mature. The wall of the follicle thickens, and cells within it begin to produce a fluid called **follicular fluid** that fills a space called the **antrum.** As the follicle expands, it goes through the following progression:

Primordial follicle → primary follicle → secondary follicle → tertiary follicle

Once a **tertiary follicle** has been formed, a hormonal surge causes it to rupture, and an ovum is released in an event called **ovulation.** The specific name for this ovum is a **secondary oocyte.** The broken tissue within the tertiary follicle thickens and becomes a temporary endocrine gland called a **corpus luteum** that produces a mixture of **estrogens** and **progestins.** Gradually the corpus luteum shrinks and degrades into a small mass of scar tissue called the **corpus albicans.**

Analogy

The **maturation of a follicle** in the ovary is comparable to a **water balloon filling with water.** As follicular fluid accumulates inside the antrum, it increases pressure just like water inside the water balloon. When the pressure becomes too great, the follicle ruptures at the **moment of ovulation** just like the **bursting of the water balloon.**

Location

Ovaries are located along the lateral edges of the pelvic cavity.

Function

The ovaries have two basic functions:

1. Produce **gametes** (sex cells)
2. Manufacture and release **hormones**
   - Follicles (primary, secondary, and tertiary) → **estrogen**
   - Corpus luteum → **estrogen, progesterone**

Key to Illustration

1. Primary follicle
2. Secondary follicle
3. Tertiary follicle
4. Ovulation
5. Early corpus luteum (*still forming*)
6. Corpus luteum
7. Corpus albicans

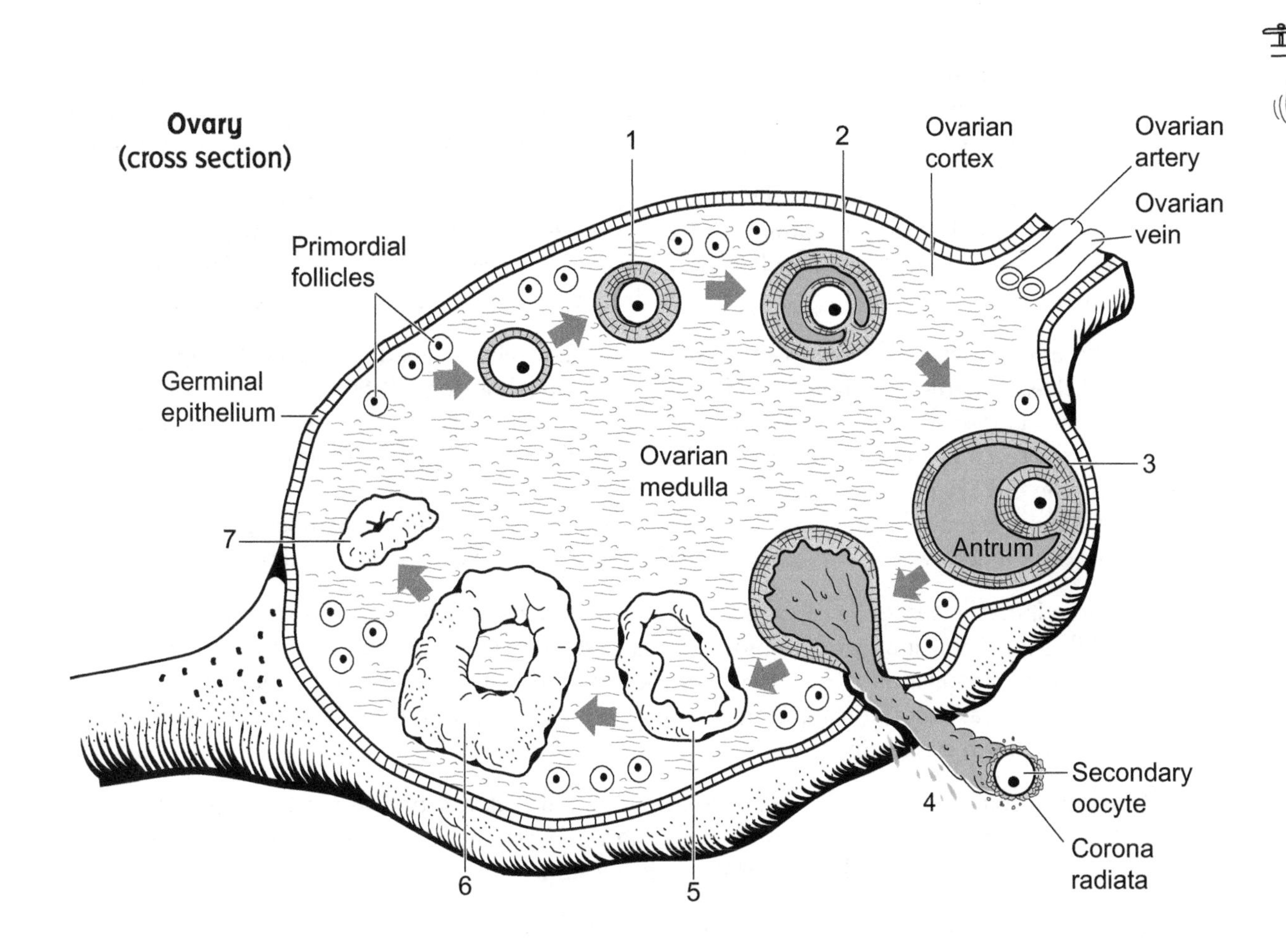

The maturation of a follicle is like...

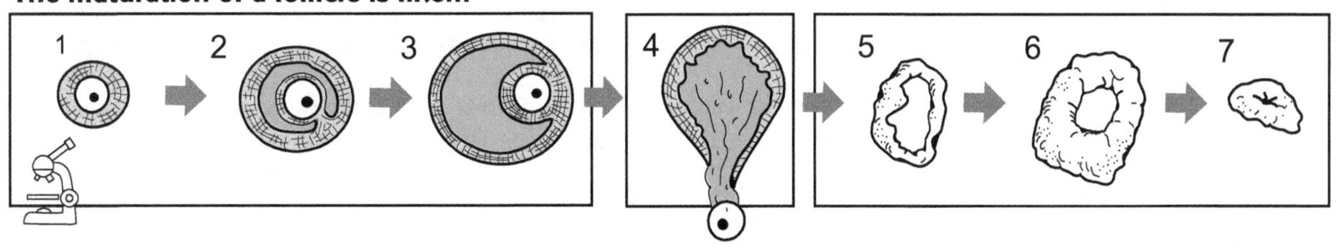

...a water balloon filling up with water and then bursting.

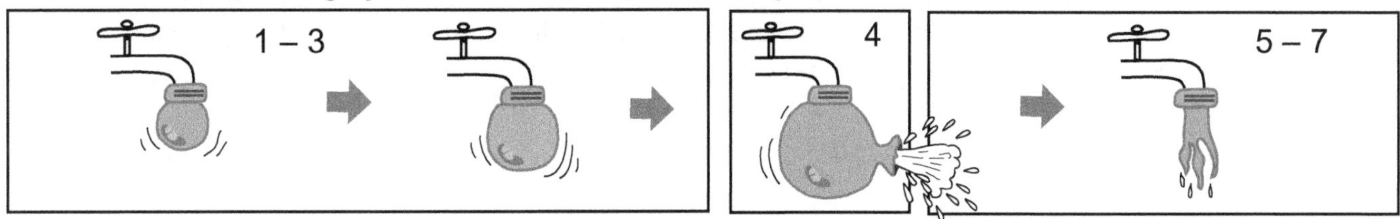

Write the correct term in each box:

1	2	3	4	5	6	7

Description

The **ovarian cycle** refers to the regular, monthly events that occur within the ovary of a sexually active female. The group of hormones that regulates this process is the same basic set that regulates **spermatogenesis** in the male (see p. 310). A step-by-step description of this process is given below:

① It all begins when the hypothalamus in the brain releases into the blood a **peptide** called **gonadotropin-releasing hormone** (**GnRH**). It travels through the blood to target cells within the anterior lobe of the pituitary gland because they have receptors for GnRH. Unlike in the male, GnRH is not released in regular pulses. Instead, the amount released fluctuates in the female.

② The binding of GnRH induces a response in these cells to make two **gonadotropin** hormones: **follicle-stimulating hormone** (**FSH**) and **luteinizing hormone** (**LH**). Like any hormone, these chemical messengers are released into the bloodstream and travel to their respective targets, which contain receptors for these hormones. In the female, FSH targets the follicles within the ovary. The binding of FSH induces these follicles to mature and increase in size as a result of the production of a fluid called follicular fluid. As the fluid accumulates within a chamber called the antrum, it increases fluid pressure within the follicle. Meanwhile, LH causes the follicles to produce **estrogen.**

③ These initial lower levels of estrogen inhibit the anterior pituitary from *releasing* any hormones while *stimulating* it to produce mostly LH and some FSH within the cells of the anterior lobe. It is similar to stepping on the gas pedal of a car placed in neutral. Though you may be revving the engine, the car is not moving. Similarly, though the cells may be actively producing gonadotropins, they are not releasing any yet.

④ As the **tertiary** (*Graafian, vesicular*) **follicle** is established, it produces more estrogen. These higher levels have the opposite effect of the initial lower levels: they *stimulate* the anterior lobe to release its accumulated LH (and some FSH) in one big surge.

⑤ This one- to two-day spike in LH levels in the blood triggers many events: First, it stimulates formation of the secondary oocyte within the tertiary follicle.

⑥ **LH** triggers rupturing of the tertiary follicle and release of the secondary oocyte in an event called **ovulation.** Estrogen levels fall slightly after ovulation because of damage to the tertiary follicle.

⑦ The LH surge also changes the damaged follicle into a **corpus luteum,** which functions as a temporary endocrine gland, producing **estrogen, progesterone,** and **inhibin.**

⑧ Working together, when these three hormones reach a critical level, they serve to inhibit the hypothalamus and pituitary gland. This results in no release of GnRH, which translates into no release of FSH or LH. As a result, the corpus luteum shrinks to become a small, inactive mass of scar tissue, called the **corpus albicans.** With the loss of the corpus luteum, the estrogen, progesterone, and inhibin levels fall. As a result, inhibition is lost, and the hypothalamus again begins to release GnRH to start another ovarian cycle.

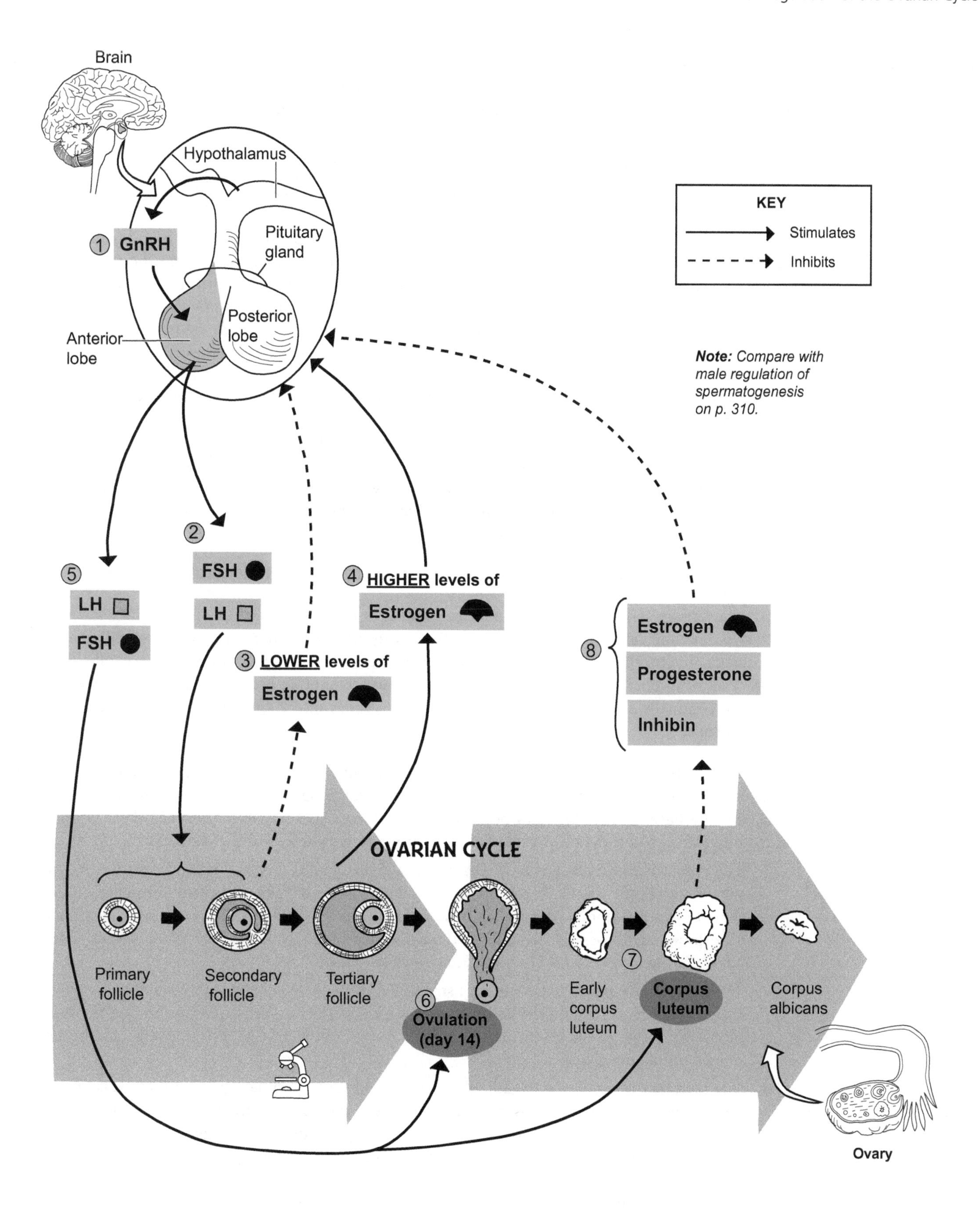
Brain
Hypothalamus
① GnRH
Pituitary gland
Posterior lobe
Anterior lobe
KEY
Stimulates
Inhibits
Note: Compare with male regulation of spermatogenesis on p. 310.
②
FSH
LH
⑤
LH
FSH
④ HIGHER levels of
Estrogen
③ LOWER levels of
Estrogen
⑧
Estrogen
Progesterone
Inhibin
OVARIAN CYCLE
Primary follicle
Secondary follicle
Tertiary follicle
⑥ Ovulation (day 14)
Early corpus luteum
⑦ Corpus luteum
Corpus albicans
Ovary

Description

In a sexually mature female, a predictable series of changes occurs every month within the ovaries and the uterus. These processes are referred to separately as the **ovarian cycle** and the **uterine cycle** (or menstrual cycle) and are regulated by specific hormones. These processes are linked by a series of cascading events. Here is an overview:

Hypothalamus → GnRH → anterior lobe of pituitary gland → FSH, LH → ovary → estrogen, progesterone → uterus

It all begins when the hypothalamus in the brain releases a **peptide** into the blood, called **gonadotropin-releasing hormone** (**GnRH**). It travels through the blood to target cells within the anterior lobe of the pituitary gland because they have receptors for GnRH. The binding of GnRH to its receptor induces a response in these cells to make the hormones **follicle-stimulating hormone** (**FSH**) and **leuteinizing hormone** (**LH**). Like any hormone, these chemical messengers are released into the bloodstream and travel to their respective targets. Target sites always contain receptors for these hormones.

Ovarian Cycle

The **ovarian cycle** refers to all the changes that occur within the ovary. Ideally, the ovaries in a sexually mature female alternate to produce one ovum each month. The process for developing an **ovum** (*egg cell*) is called **oogenesis.** This is a part of the ovarian cycle, which refers to all the processes within the ovary during this monthly event. The ova are produced in special chambers called **follicles.** The process begins when **FSH** stimulates a **primary follicle** to begin to mature. The wall of the follicle thickens, and cells within it begin to produce a fluid called **follicular fluid** that fills a space called the **antrum.** As the follicle expands, it goes through the following progression:

Primary follicle → secondary follicle → tertiary follicle

After a **tertiary** (*Graafian, vesicular, mature*) **follicle** is established, an **LH surge** occurs, lasting about one to two days and causing the follicle to rupture and release an ovum in an event called called **ovulation.** The specific name for this ovum is a **secondary oocyte.** The broken tissue within the tertiary follicle thickens and becomes a temporary endocrine gland called a **corpus luteum** that produces a mixture of **estrogens** and **progestins.** Gradually the corpus luteum shrinks and degrades into a small mass of scar tissue called the **corpus albicans.**

Uterine Cycle

The **uterine cycle** refers to changes within the **endometrium**, the innermost lining of the uterus. The endometrium is subdivided into two layers—the **functional zone**, which lines the lumen of the uterine cavity, and the **basal zone**, which lies beneath it. The entire cycle averages about 28 days. This cycle is divided into three phases: the **menstrual phase**, the **proliferative phase**, and the **secretory phase.**

1. The **menstrual phase** (days 1–5). During this phase, blood flow to the endometrium is reduced, which starves the tissues in the functional zone of their oxygen and other nutrients. As a result, these tissues deteriorate. When they break away from the uterine lining, it causes some blood loss. The sloughing off of these tissues and the associated blood, called **menstruation**, lasts one to seven days. During menstruation the basal zone remains intact.
2. The **proliferative phase** (days 6–13). Estrogens induce all the changes throughout this phase. The major event is to reestablish the functional zone lost during menstruation. Epithelial tissues and blood vessels grow back into the functional zone. **Endometrial glands** appear, which produce a mucus containing glycogen.
3. The **secretory phase** (days 15–28). This last and longest phase covers about 14 days. The tissues within the functional zone thicken and increase in vascularization, while endometrial glands enlarge and produce more mucus. The combination of estrogen and progesterone working together are responsible for these changes. The purpose of this thickening of the endometrium is to prepare for possible implantation of a human embryo into the functional layer of the endometrium. The glycogen in the mucus from the endometrial glands will serve to nourish the developing embryo. If no implantation occurs, both the ovarian and uterine cycles begin once again with release of GnRH from the pituitary gland.

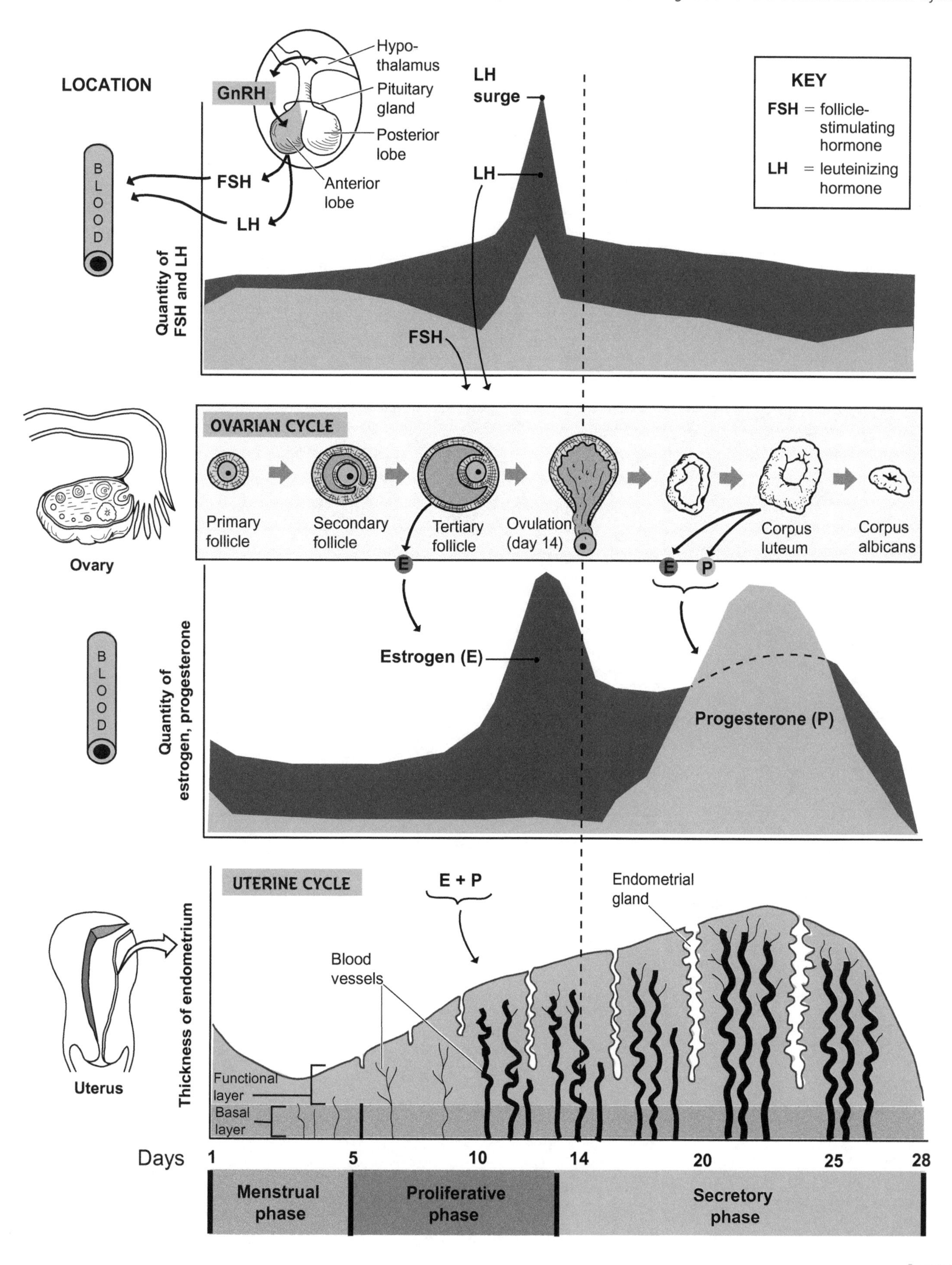
LOCATION
GnRH
Hypo-
thalamus
Pituitary
gland
Posterior
lobe
Anterior
lobe
FSH
LH
B L O O D
LH
surge
LH
FSH
Quantity of
FSH and LH
KEY
FSH = follicle-
stimulating
hormone
LH = leuteinizing
hormone
OVARIAN CYCLE
Ovary
Primary
follicle
Secondary
follicle
Tertiary
follicle
Ovulation
(day 14)
Corpus
luteum
Corpus
albicans
E
E
P
B L O O D
Quantity of
estrogen, progesterone
Estrogen (E)
Progesterone (P)
UTERINE CYCLE
E + P
Endometrial
gland
Blood
vessels
Uterus
Thickness of endometrium
Functional
layer
Basal
layer
Days
1
5
10
14
20
25
28
Menstrual
phase
Proliferative
phase
Secretory
phase

Notes

Glossary of Prefixes and Suffixes

Element	Definition and Example
a-	absent, deficient, or without: atrophy
ab-	off, away from: abduct
abdomin-	abdomen
-able	capable of: viable
ac-	toward, to: actin
acou-	hear: acoustic
ad-	denoting to, toward: adduct
af-	movement toward a central point: afferent artery
alba-	pale or white: linea alba
-alg	pain: neuralgia
ambi-	both: ambidextrous
angi-	pertaining to vessel: angiogram
ante-	before: antebrachium
anti-	against: anticoagulant
aqua-	water: aqueous
archi-	to be first: archeteron
arthri-	joint: arthritis
-asis	condition or state of: homeostasis
aud-	pertaining to ear: auditory
auto-	self: autolysis
bi-	two: biceps
bio-	life: biology
blast-	generative or germ bud: osteoblast
brachi-	arm: brachialis
brachy-	short: brachydont
brady-	slow: bradycardia
bucc-	cheek: buccal cavity
cac-	bad, ill: cachexia
calci-	stone: calcification
capit-	head: capitate
carcin-	cancer: carcinogenic
cardi-	heart: cardiac
cata-	lower, under, against: catabolism
caud-	tail: cauda equina
cephal-	head: cephalis
cerebro-	brain: cerebrospinal fluid
chol-	bile: cholic
chondr-	cartilage: chondrocyte
chrom-	color: chromosome
-cid(e)	destroy: germicide
circum-	around: circumduct
-cis	cut, kill: excision
co-	together: copulation
coel-	hollow cavity: coelom
-coel	swelling, and enlarged space or cavity: blastocoel
con-	with, together: congenital
contra-	against, opposite: contraception
corn-	denoting hardness: cornified
corp-	body: corpus
crypt-	hidden: cryptorchism
cyan-	blue color: cyanosis
cysto-	sac or bladder: cystoscope
cyto-	cell: cytology
de-	down, from: descent
derm-	skin: dermatology
di-	two: diarthrotic
dipl-	double: diploid
dis-	apart, away from: disarticulate
duct-	lead, conduct: ductus deferens
dur-	hard: dura mater
-dynia	pain: gastrodynia
dys-	bad, difficult, painful: dysentery
e-	out, from: eccrine
ecto-	outside, outer, external: ectoderm
-ectomy	surgical removal: tonsillectomy
ede-	swelling: edema
-emia	pertaining to a condition of the blood: lipemia
end-	within: endoderm
entero-	intestine: enteritis
epi-	upon, in addition: epidermis
erythro-	red: erythrocyte
ex-	out of: excise
exo-	outside: exocrine
extra-	outside of, beyond, in addition: extracellular
fasci-	band: fascia
febr-	fever: febrile
-ferent	bear, carry: efferent arteriole
fiss-	split: fissure
for-	opening: foramen
-form	shape: fusiform
gastro-	stomach: gastrointestinal
-gen	an agent that produces or originates: pathogen
-genic	produced from, producing: carcinogenic
gloss-	tongue: glossopharyngeal
glyco-	sugar: glycogen
-gram	a record, recording: myogram
gran-	grain, particle: agranulocyte
-graph	instrument for recording: electrocardiograph
grav-	heavy: gravid
gyn-	female sex: gynocology
haplo-	simple or single: haploid
hema(o)-	blood: hematology
hemi-	half: hemisphere
hepat-	liver: hepatic portal
hetero-	other, different: heterosexual
histo-	tissue: histology
holo-	whole, entire: holocrine
homo-	same, alike: homologous
hydro-	water: hydrophilic
hyper-	beyond, above, excessive: hypertension
hypo-	under, below: hypoglycemia
-ia	state or condition: hypoglycemia
-iatrics	medical specialties: pediatrics
idio-	self, separate, distinct: idiopathic
ilio-	ilium: iliosacral
infra-	beneath: infraspinatus

Element	Definition and Example
inter-	among, between: intercellular
intra-	inside, within: intracellular
-ion	process: acromion
iso-	equal, like: isotonic
-ism	condition or state: rheumatism
-itis	inflammation: meningitis
labi-	lip: labium major
lacri-	tears: lacrimal
later-	side: lateral
leuc-	white: leucocyte
lip-	fat: lipid
-logy	science of: cytology
-lysis	solution, dissolve: hemolysis
macro-	large, great: macrophage
mal-	bad, abnormal, disorder: malignant
medi-	middle: medial
mega-	great, large: megakaryocyte
meso-	middle or moderate: mesoderm
meta-	after, beyond: metatarsal
micro-	small: microscope
mito-	thread: mitosis
mono-	alone, one, single: monocyte
mons-	mountain: mons pubis
morph-	form, shape: morphology
multi-	many, much: multicellular
myo-	muscle: myofibril
narc-	numbness, stupor: narcotic
necro-	corpse, dead: necrosis
neo-	new, young: neonatal
nephro-	kidney: nephron
neuro-	nerve: neurolemma
noto-	back: notochord
ob-	against, toward, in front of: obturator
oc-	against: occlusion
-oid	resembling, likeness: sigmoid
oligo-	few, small: oligodendrocyte
-oma	tumor: lymphoma
oo-	egg: oocyte
or-	mouth: oral
orchi-	testicles: cryptorchidism
-ory	pertaining to: sensory
-ose	full of: adipose
osteo-	bone: osteocyte
oto	ear: otolith
ovo-	egg: ovum

Element	Definition and Example
par-	give birth to, bear: parturition
para-	near, beyond, beside: paranasal
path-	disease, that which undergoes sickness: pathology
-pathy	abnormality, disease: neuropathy
ped-	children: pediatrician
pen-	need, lack: penicillin
-penia	deficiency: thrombocytopenia
per-	through: percutaneous
peri-	near, around: pericardium
phag-	to eat: phagocyte
-phil	have an affinity for: neutrophil
phlebo-	vein: phlebitis
-phobe	abnormal fear, dread: hydrophobe
-plasty	reconstruction of: rhinoplasty
platy-	flat, side: platysma
-plegia	stroke, paralysis: paraplegia
-pnea	to breathe: apnea
pneumato-	breathing: pneumonia
pod-	foot: podiatry
-poiesis	formation of: hematopoiesis
poly-	many, much: polyploid
post-	after, behind: postnatal
pre-	before in time or place: prenatal
prim-	first: primitive
pro-	before in time or place: prosect
proct-	anus: proctology
pseudo-	false: pseudostratified
psycho-	mental: psychology
pyo-	pus: pyoculture
quad-	fourfold: quadriceps femoris
re-	back, again: repolarization
rect-	straight: rectus abdominis
reno-	kidney: renal
rete-	network: retina
retro-	backward: retroperitoneal
rhin-	nose: rhinitis
-rrhage	excessive flow: hemorrhage
-rrhea	flow, or discharge: diarrhea
sanguin-	blood: sanguiferous
sarc-	flesh: sarcoplasm
-scope	instrument for examination of a part: stethoscope
-sect	cut: dissect
semi-	half: semilunar
serrate-	saw-edged: serratus anterior

Element	Definition and Example
-sis	state or condition: dialysis
steno-	narrow: stenohaline
-stomy	surgical opening: tracheostomy
sub-	under, beneath, below: subcutaneous
super-	above, beyond, upper: superficial
supra-	above, over: suprarenal
syn (sym)	together, joined, with: synapse
tachy-	swift, rapid: tachometer
tele-	far: telencephalon
tens-	stretch: tensor fasciae latae
tetra-	four: tetrad
therm-	heat: thermogram
thorac-	chest: thoracic cavity
thrombo-	lump, clot: thrombocyte
-tomy	cut: appendectomy
tox-	poison: toxic
tract-	draw, drag: traction
trans-	across, over: transfuse
tri-	three: trigone
trich-	hair: trichology
-trophy	a state relating to nutrition: hypertrophy
-tropic	turning toward, changing: gonadotropic
ultra-	beyond, excess: ultrasonic
uni-	one: unicellular
-uria	urine: polyuria
uro-	urine, urinary organs or tract: uroscope
vas-	vessel: vasoconstriction
vermi-	worm: vermiform
viscer-	organ: visceral
vit-	life: vitamin
zoo-	animal: zoology
zygo-	union, join: zygote

Glossary

A

abdomen (AB-doh-men) The region between the diaphragm and the pelvis.

absorption (ab-ZORP-shun) The passage of gases, liquids, or solutes through a membrane.

acetylcholine (ACh) (ass-ee-til-KOH-leen) A neurotransmitter secreted by various neurons into synapses; may have an excitatory effect (neuromuscular junction of skeletal muscles) or an inhibitory effect (neuromuscular junction of cardiac muscle tissue).

acetylcholinesterase (AChase) (ass-ee-til-koe-lin-ESS-ter-ase) An enzyme that rapidly inactivates the acetylcholine bound to postsynaptic receptors.

acidic (ah-SID-ik) Describes a solution in which the pH is less than 7; having a relatively high concentration of hydrogen ions.

acrosome (AK-roh-sohm) A cap-like structure on the head of a sperm cell; produces enzymes for penetrating eggs.

actin (AK-tin) A thin protein filament found in skeletal muscle cells; protein component of microfilaments.

action potential an electrical signal that advances along the plasma membrane of an excitable cell, such as a neuron or a muscle cell; used to communicate over long or short distances.

active transport A carrier-mediated process in which cellular energy is used to move molecules against their concentration gradient through a plasma membrane.

adenosine diphosphate (ah-DEN-oh-seen dye-FAHS-fate) An energy molecule composed of an adenine base, a ribose sugar, and two phosphate groups.

adenosine triphosphatase (ATPase) (ah-DEN-oh-seen try-FAHS-fate-ays) (AY-tee-pee-ase) An enzyme that catalyzes the hydrolysis of ATP to form ADP, a phosphate group, and a net release of free energy.

adenosine triphosphate (ATP) (ah-DEN-oh-seen try-FAHS-fate) The energy currency of a cell; composed of an adenine base, a ribose sugar, and three phosphate groups.

adenylate cyclase (ah-DEN-il-ayt SYE-klayz) An enzyme that catalyzes the conversion of ATP to cyclic AMP (cAMP).

adipocyte (AD-i-poh-syte) A fat cell.

adrenergic neuron (ad-ren-ER-jik NOO-ron) A neuron that secretes norepinephrine (*noradrenaline*) or epinephrine (*adrenaline*) as its neurotransmitter.

aerobic Any process that requires oxygen.

afferent (AF-fer-ent) Toward; opposite of efferent.

aldosterone (al-DAH-stair-ohn) A mineralocorticoid; a hormone produced by the adrenal cortex that acts on the kidneys to increase sodium levels in the blood and increase excretion of potassium in urine.

alkaline (AL-kah-lin) Refers to a base, a pH greater than 7; having a relatively low concentration of hydrogen ions.

alveolus/alveoli (al-VEE-oh-luss/al-VEE-oh-lye) Delicate air sacs in the lungs where gas exchange occurs.

amino acid (ah-MEE-no ASS-id) The structural unit of a protein.

amnion (AM-nee-on) One of the extraembryonic membranes; develops around the embryo/fetus, forming the amniotic cavity.

amniotic fluid (am-nee-OT-ik fluid) The fluid that surrounds and cushions the developing embryo and fetus.

amniotic sac (am-nee-OT-ik sak) Fluid-filled chamber in which the embryo floats during development.

amylase (AM-eh-layz) An enzyme that breaks down polysaccharides; produced by salivary glands and pancreas.

anabolism (ah-NAB-oh-liz-em) Synthesis reactions that require energy to join small molecules to form more complex molecules; for example, amino acids bonding to make proteins.

anaerobic (an-air-OH-bik) Any process that does not require oxygen.

anaphase (AN-nah-fayz) Stage of mitosis when the chromatid pairs separate and the daughter chromosomes move toward the opposite ends of the cell.

antagonist (an-TA-gun-ist) Something that has an opposing action on something else, such as a muscle, hormone, or drug.

anterior (an-TEER-ee-or) Toward the front or ventral; opposite of posterior.

antibody Proteins produced by plasma cells in response to specific antigens; antibodies combine with antigens to render a pathogen harmless; also known as immunoglobulin (Ig).

antigen (AN-tih-jen) Means "*anti*body *gen*erating substances"; macromolecules that induce the immune system to make antibodies.

antrum (AN-trum) Central chamber.

anus (AY-nus) External opening at the end of the rectum.

aorta (ay-OR-tah) Largest artery in the body; carries oxygenated blood from left ventricle and into the systemic circuit.

apex (AY-peks) Pointed tip of a structure.

apical (AY-pik-al) Located near or relating to the apex or pointed structure; opposite of basal.

apnea (APP-nee-ah) Temporary cessation of breathing at the end of a normal expiration.

aponeurosis (ap-oh-nyoo-ROH-sis) Broad, flat collagenous sheets that may serve as anchor points for skeletal muscle.

appendix (ah-PEND-diks) Small organ connected to the cecum of the large intestine.

arachnoid (ah-RAK-noyd) Middle meninges that surround the CSF and protect the brain and spinal cord.

arbor vitae (AR-bor VYE-tay) Central area of white matter in the cerebellum.

arteries (AR-ter-eez) Blood vessels that carry blood away from the heart.

arteriole (ar-TEER-ee-ohl) Microscopic blood vessel that connects small arteries to capillaries.

atom (AT-om) The smallest particle of a chemical element that displays the properties of that element.

atrioventricular (AV) (ay-tree-oh-ven-TRIK-yoo-lar) **bundle** Group of specialized cardiac muscle cells that extends from the AV node to the Purkinje fibers; coordinates contraction of the heart muscle.

atrioventricular (ay-tree-oh-ven-TRIK-yoo-lar) **(AV) node** Secondary pacemaker of the heart; a small mass of specialized cardiac muscle cells that plays a role in the electrical conduction system within the heart.

atrioventricular (AV) (ay-tree-oh-ven-TRIK-yoo-lar) **valves** Pair of heart valves located between the atria and the ventricles; tricuspid and bicuspid; prevent backflow of blood into atria during ventricular contraction.

atrium (AY-tree-um) An upper chamber of the heart; receives blood from the pulmonary or systemic circuit.

auditory (AW-di-toh-ree) Pertaining to the sense of hearing.

auricle (AW-ri-kul) Curved, flexible upper portion of the ear; also, expandable flap-like structure of an atrium in the heart.

autonomic (aw-toh-NAHM-ik) **nervous system (ANS)** The part of the nervous system that governs itself without our conscious knowledge; divided into sympathetic and parasympathetic divisions; controls organs through reflex pathways that link to smooth muscle, cardiac muscle, and glands.

axon hillock (AK-son HILL-ok) Funnel-shaped portion of neural cell body from which the axon extends.

axons (AK-sonz) Long, single processes of neurons that conduct nerve impulses toward the synaptic knobs.

B

basal (BAY-sal) Located at or near the base of a structure; opposite of apical.

basement membrane (BAYSE-ment MEM-brayne) A layer of protein fibers that connects the epithelium to the underlying connective tissue.

basic *See alkaline.*

basophil (BAY-so-fil) White blood cell; releases histamine into damaged tissue.

bicuspid valve (bye-KUSS-pid valve) The left atrioventricular (AV) valve in the heart; also known as the mitral valve.

bile A secretion of the liver stored in the gallbladder and released into the duodenum; physically breaks down fat into smaller droplets.

bilirubin (BILL-y-roo-bin) Orange or yellow pigment found in bile and derived from hemoglobin catabolism.

bipolar neuron (bye-POH-lar NOO-ron) Nerve cell with two distinct processes; one dendrite, one axon.

blastocyst (BLASS-toh-sist) Early stage in embryonic development; a hollow ball of cells consisting of an inner cell mass and an outer cell mass.

blastomere (BLASS-toh-meer) The first cleavage division, which produces a pre-embryo consisting of two identical cells.

bolus (BOH-luss) A ball or mass of chewed food that passes from the mouth to the stomach.

Bowman's capsule *See glomerular capsule.*

Boyle's law A physics law stating that the volume of a gas is inversely proportional to its pressure.

brachial (BRAY-kee-al) Pertaining to the upper limb between shoulder and elbow.

brainstem (brayn-stem) Part of the brain that contains important processing centers; consists of the medulla oblongata, pons, and midbrain.

bronchiole (BRONG-kee-ohl) A small, tubelike branch of a bronchus; lacks cartilaginous supports, but wall contains smooth muscle.

bronchus (BRONG-kuss) A branch of the bronchial tree between the trachea and the bronchioles.

buccal (BUK-al) Pertaining to the cheek.

bulbourethral glands (BUL-boh-yoo-REE-thral glands) Small mucus glands located at base of the penis; secretions lubricate the urethra.

bundle branches Two bands of specialized cardiac muscle cells that transmit impulses from the AV bundle to the Purkinje fibers.

bundle of HIS *See atrioventricular (AV) bundle.*

C

calcification (kal-sih-fih-KAY-shun) Process of hardening a tissue with deposits of calcium salts.

calcitonin (CT) (kal-sih-TOE-nin) A hormone produced by the thyroid gland that serves to decrease calcium levels in the blood.

canaliculus/canaliculi (kan-ah-LIK-yoo-luss/kan-ah-LIK-yoo-lye) Microscopic channels between cells, found in compact bone and liver; in compact bone, canaliculi allow diffusion of nutrients and wastes; in liver, bile canaliculi transport bile to bile ducts.

capillary (KAP-i-lair-ee) Smallest, simplest blood vessel; microscopic; connects arterioles and venules.

carbohydrate (kar-boh-HYE-drayt) Organic compound containing carbon, oxygen, and hydrogen; sugars, starches, and cellulose.

cardiac (KAR-dee-ak) **cycle** A complete heartbeat or pumping cycle consisting of systole and diastole.

cardiac (KAR-dee-ak) **output** The volume of blood pumped by one heart ventricle each minute.

carotid artery (kah-ROT-id AR-ter-ee) The large artery of the neck that provides a major blood supply to the brain.

catabolism (kah-TAB-oh-liz-em) Chemical reactions that break down complex organic compounds into simpler ones with a net release of energy; for example, proteins broken down into amino acids.

catecholamines (kat-eh-KOLE-ah-meenz) A class of neuro-transmitters synthesized from the amino acid tyrosine; include norepinephrine, epinephrine, and dopamine; may play a role in sleep, motor function, regulating mood, and pleasure recognition.

caudal (KAW-dal) The tail.

CCK *See cholecystokinin.*

cecum (SEE-kum) The pouch located at the beginning of the large intestine.

cell (sell) The basic unit of life.

cellular respiration Metabolic process by which carbohydrates, fatty acids, and amino acids are broken down to produce ATP.

centriole (SEN-tree-ohl) Tiny, cylindrical organelle of a cell; involved with spindle formation during mitosis.

centromere (SEN-troh-meer) The region where two chromatids are connected together during the early stages of cell division.

centrosome (SEN-troh-sohm) The region of cytoplasm that coordinates the activities of centrioles.

cerebellum (sair-eh-BELL-um) Second largest part of the brain; coordinates and refines learned movement patterns.

cerebrospinal (SAIR-eh-broh-SPY-nal) **fluid (CSF)** Liquid found in the central nervous system by filtering and processing blood plasma; circulates in the ventricles, central canal, and subarachnoid space of the brain and spinal cord.

cerebrum (SAIR-eh-brum) Largest region of the brain; origin of conscious thoughts and all intellectual functions; controls sensory and motor integration.

cervical (SER-vih-kal) Pertaining to the neck.

cervix (SER-viks) Neck-like structure at the inferior portion of the uterus; projects into vagina.

chief cells (CHEEF sells) Cells in the gastric mucosa that primarily secrete pepsinogen, which later is converted into the protein-digesting enzyme called pepsin.

cholecystokinin (CCK) (koh-lee-sis-toh-KYE-nin) A hormone produced by the duodenal mucosa that stimulates contraction of the gallbladder and secretion of pancreatic juice rich in digestive enzymes.

cholesterol (ko-LES-ter-all) Steroid compound produced by animals; found in their plasma membranes.

cholinergic neuron (koh-leh-NER-jik NOO-ron) Neuron that secretes acetylcholine as its neurotransmitter.

chondrocyte (KON-droh-syte) A cartilage cell.

chorion (KOH-ree-on) A membrane consisting of the mesoderm and trophoblast; develops into a membrane of the placenta.

choroid (KOH-royd) Middle vascular layer of the eye.

chromatid (KROH-mah-tid) Either of two daughter strands of chromosomes joined by a single centromere.

chromatin (KROH-mah-tin) Chromosomal material that is loosely coiled, forming a tangle of fine filaments with a grainy appearance.

chromosomes (KROH-moh-sohms) Tightly compacted structures that contain coiled DNA wrapped around histone proteins; normal human body cells contain 46 chromosomes; term means "colored body."

chyme (kyme) A soupy, viscous mixture of ingested substances and gastric juices leaving the stomach.

cilia (SIL-ee-ah) Long folds of the plasma membrane of an epithelial cell that contains microtubules; move back and forth to propel fluids or secretions across cell surfaces.

ciliary body (SIL-ee-air-ee body) A muscular structure that surrounds the perimeter of the lens of the eye and attaches to it through the suspensory ligaments.

citric acid cycle (SIT-rik ASS-id SYE-kul) An aerobic chemical cycle that begins with the formation of citric acid and results in the formation of oxaloacetic acid; in the process, ATP is produced and carbon dioxide is released; occurs in the mitochondrion.

clitoris (KLIT-oh-ris) Small female organ composed of erectile tissue located behind the junction of the labia majora.

collagen (KAHL-ah-jen) The most common type of protein fiber found in connective tissues; serves to strengthen tissues.

colon (KOH-lon) The large intestine.

common bile duct (KOM-mon byle dukt) Formed by the union of the cystic duct and the hepatic duct; carries bile from the liver and gallbladder to the duodenum.

compound (KOM-pound) A substance formed by the union of two or more elements.

connective tissue (koh-NEK-tiv TISH-yoo) One of the four major tissue types; serves to give structural support to other tissues and organs in the body; most contains cells, protein fibers, and ground substance.

convolution *See gyrus.*

corona radiata (koh-ROHN-ah ray-dee-AY-tah) Follicular cells that surround the oocyte.

corpus/corpora (KOHR-pus/KOHR-pohr-ah) Body.

corpus albicans (KOHR-pus AL-bi-kans) Pale scar tissue in the ovaries that replaces the nonfunctional corpus luteum.

corpus callosum (KOHR-pus kah-LOH-sum) Area of the brain that links the right and left cerebral hemispheres.

corpus cavernosum (KOHR-pus kav-er-NO-sum) Two columns of erectile tissue that extend along the length of the penis.

corpus luteum (KOHR-pus LOO-tee-um) An ovarian structure transformed from a ruptured follicle; secretes estrogen and progesterone.

corpus spongiosum (KOHR-pus spun-jee-OH-sum) Erectile body that surrounds the urethra.

cortex (KOHR-teks) Outer part of an organ; adrenal cortex.

cortisol (KOHR-tih-sawl) A glucocorticoid; a hormone produced by the adrenal cortex that helps regulate responses to stress; also known as hydrocortisone.

Cowper's glands *See bulbourethral glands.*

cyclic adenosine monophosphate (SIK-lik ah-DEN-oh-seen mon-oh-FAHS-fate) **(cAMP)** A second messenger formed from ATP; composed of an adenine base, a ribose sugar, and one phosphate group; used for signal transduction.

cystic duct (SIS-tik dukt) A tube that leads from the gallbladder toward the liver; unites with the common hepatic duct to form the common bile duct.

cytokinesis (SYE-toe-kih-nee-siss) Physical division of the cytoplasm during cell division.

cytoplasm (SYE-toh-plaz-em) Gel-like material between the nucleus and the plasma membrane; includes cytosol and organelles.

cytoskeleton (sye-toh-SKEL-eh-ton) An integrated network of microtubules and microfilaments in the cytoplasm that gives shape and provides support to the cell and its organelles.

cytosol (SYE-toh-sawl) Fluid portion of the cytoplasm.

D

decidua basalis (dih-SID-yoo-ah bah-SAY-lis) Area of the endometrium that develops into the maternal part of the placenta.

deep Away from the surface; opposite of superficial.

dendrite (DEN-dryte) A long cellular extension of a neuron that responds directly to stimuli.

diaphragm (DYE-ah-fram) Dome-shaped muscle that separates the thoracic and abdominal cavities; a major muscle involved with respiration.

diaphysis (dye-AF-i-sis) Tubular shaft of a long bone.

diastole (dye-ASS-toh-lee) Relaxation of both atria and both ventricles in the heart; opposite of systole.

diffusion (dih-FYOO-shun) The net movement of substances from an area of high concentration to an area of low concentration.

disaccharide (dye-SAK-ah-ryde) A double-unit sugar formed by bonding a pair of monosaccharides; *ex:* lactose, sucrose.

distal (DIS-tal) Refers to the region or reference away from an attached base; opposite of proximal.

distal convoluted tubule (DCT) (DIS-tal KON-voh-loo-ted TOOB-yool) The part of the nephron distal to the ascending limb of the nephron loop; site for active secretion and selective reabsorption of ions and other substances.

dorsal (DOR-sal) Pertaining to the back; posterior; opposite of ventral.

ductus deferens (DUK-tus DEF-er-ens) A smooth muscular tube that propels sperm from the epididymis to the ejaculatory duct; also known as vas deferens.

duodenum (doo-oh-DEE-num) First and shortest segment of the small intestine—about 10 inches long; receives chyme from the stomach and digestive secretions from the pancreas.

dura mater (DOO-rah MAH-ter) Outermost layer of the meninges.

E

efferent (EF-fer-ent) Away from; opposite of afferent.

ejaculatory duct (ee-JAK-yoo-lah-toh-ree dukt) Short passageway that allows sperm to enter the urethra.

electrocardiogram (ECG or EKG) (ee-lek-troh-KAR-dee-oh-gram) A graphic record of the heart's electrical activity or conduction of impulses; used to evaluate the heart's action potential.

electroencephalogram (EEG) (ee-lek-troh-en-SEF-ah-loe-gram) A graphic record of brain electrical potentials; used to evaluate nerve tissue function and diagnose specific disorders (such as epilepsy).

electrolytes (ee-LEK-troh-lytes) Inorganic substances that break up in solution to form ions and conduct electricity; include acids, bases, and salts.

electrons (ee-LEK-trons) Negatively charged subatomic particles in atoms; located in energy shells outside the nucleus of the atom.

electron transport system (ETS) (ee-LEK-tron TRANS-port SIS-tem) Transfer of electrons along a series of membrane-bound electron carrier molecules in the mitochondria; energy from this process is used to synthesize ATP; aerobic process that produces water.

embolus (EM-boe-luss) A clot that dislodges and circulates through the bloodstream.

embryo (EM-bree-oh) A stage of human development beginning at fertilization and ending at the start of the eighth developmental week.

emulsification (ee-mul-sih-fih-KAY-shun) The process by which bile breaks up fats.

endocardium (en-doh-KAR-dee-um) Simple squamous epithelial inner layer of the heart.

endocytosis (en-doh-sye-TOH-sis) An active transport mechanism that allows extracellular substances to enter the cell by phagocytosis, pinocytosis, or receptor-mediated endocytosis.

endometrium (en-doh-MEE-tree-um) Innermost, glandular layer of the uterine wall.

endomysium (en-doh-MISH-ee-um) Inner layer of connective tissue that surrounds each skeletal muscle fiber.

endoneurium (en-doh-NOO-ree-um) A layer of connective tissue that surrounds individual axons.

endoplasmic reticulum (en-doh-PLAS-mik re-TIK-yoo-lum) The network of intracellular membranes that synthesizes and manufactures membrane-bound proteins.

endosteum (en-DOS-tee-um) A membrane that lines the marrow cavity inside a bone.

enzyme (EN-zyme) A biological catalyst.

eosinophil (ee-oh-SIN-oh-fil) A phagocytic white blood cell; numbers increase during allergic reaction.

epicardium (ep-i-KAR-dee-um) Outer covering of the heart; also called the visceral pericardium.

epididymis (ep-i-DID-i-miss) A long, coiled and twisted tubule that lies along the posterior border of the testis; stores sperm and facilitates their maturation.

epiglottis (ep-i-GLOT-iss) Flap of elastic cartilage that folds back over the larynx during swallowing.

epimysium (ep-i-MISH-ee-um) Connective tissue layer that surrounds the entire skeletal muscle.

epineurium (ep-i-NOO-ree-um) Outermost fibrous connective tissue sheath that surrounds a peripheral nerve.

epithelial tissue (ep-i-THEE-lee-al TISH-yoo) One of the four major tissue types; serves to cover exposed body surfaces and lines internal cavities and passageways.

erythrocyte (e-RITH-roh-syte) Red blood cell.

esophagus (eh-SOF-ah-gus) Hollow, muscular tube that transports food and liquid from the pharynx to the stomach.

estrogens (ES-troh-jens) Steroid hormones produced by the ovaries; dominant sex hormones in females.

exocytosis (eks-oh-sye-TOH-sis) An active transport mechanism that involves movement of vesicle-bound substances out of the cell; vesicles fuse with the plasma membrane and contents are released outside the cell.

extracelluar fluid (ECF) Liquid located outside body cells, such as plasma, lymph, and interstitial fluid.

F

facilitated diffusion A passive transport process wherein a substance binds to a carrier molecule in the plasma membrane and is moved from an area of higher concentration to an area of lower concentration.

fallopian tubes *See uterine tubes.*

fascia (FAY-sha) A connective tissue sheath consisting of fibrous tissue and fat; unites skin to underlying tissue.

fascicle (FASS-i-kul) A single bundle.

feces (FEE-seez) The waste material eliminated from the large intestine; residue of digestion; also known as *stool.*

fetus (FEE-tus) The name given to the unborn young from the eighth week of pregnancy to birth.

fibroblasts (FYE-broh-blasts) The most abundant fixed cells in connective tissue proper.

fight-or-flight response A group of reactions that ready the body to engage in maximum muscular exertion needed to deal with a perceived threat.

filtrate Fluid produced by the process of filtration, such as that made by the glomeruli in the kidneys.

filtration A process that uses hydrostatic pressure to move water and solutes through a membrane.

fimbria/fimbriae (FIM-bree-ah/FIM-bree-ee) Fingerlike projections, such as those found on the ends of the uterine tubes nearer the ovaries.

first messenger Typically, a nonsteroid hormone that binds to its plasma membrane receptor, thereby inducing a response in the target cell.

fissure (FISH-ur) Deep groove.

follicle (FOL-lih-kul)**-stimulating hormone (FSH)** A hormone that stimulates structures within the ovaries and primary follicles to grow toward maturity; stimulates follicle cells to synthesize and secrete estrogen in the female and stimulates development of the seminiferous tubules in the male.

frenulum (FREN-yoo-lum) Small bridle; thin, flat fold of mucous membrane between two structures, such as the lingual frenulum that anchors the body of the tongue to the floor of the oral cavity.

frontal (FRUN-tal) Toward the anterior.

frontal plane (FRUN-tal playne) Runs parallel to the long axis of the body; divides the body into anterior and posterior sections.

G

gallbladder Pear-shaped organ attached to the liver that stores and concentrates bile.

gametes (GAM-eets) Mature male or female reproductive cells—sperm or secondary oocyte.

ganglion (GANG-lyun) A collection of neuronal cell bodies outside the CNS.

gastric (GAS-trik) Pertaining to the stomach.

gastric juice (GAS-trik joos) A mixture of secretions that contain water, mucus, hydrochloric acid, and the enzyme pepsin; secreted by exocrine gastric glands.

gastrin (GAS-trin) The hormone that stimulates the secretion of both parietal and chief cells in the stomach.

G cells (jee sells) Cells found in the stomach; secrete the hormone gastrin.

genitalia (jen-i-TAYL-yah) The reproductive organs.

glomerular (gloh-MARE-yoo-lar) **capsule** Cup-shaped mouth of the nephron; surrounds the glomerulus; receives the glomerular filtrate; also known as Bowman's capsule.

glomerulus/glomeruli (gloh-MARE-yoo-lus/gloh-MARE-yoo-lye) Cluster of capillaries tucked into the glomerular capsule; forms the glomerular filtrate.

glucagon (GLOO-kah-gon) A hormone produced by alpha cells in the pancreas; increases blood glucose levels.

glucocorticoids (gloo-koe-KORE-tih-koyds) A group of hormones produced by the adrenal cortex that help regulate normal glucose levels and increase resistance to stress; *ex:* cortisol.

gluconeogenesis (gloo-koh-nee-oh-JEN-eh-sis) The synthesis of glucose from lipid or protein.

glucose (GLOO-kohs) A monosaccharide; principal source of energy for the cells.

glycogen (GLYE-koh-jen) A polysaccharide consisting of a long chain of glucose units; found mainly in liver and skeletal muscle cells.

glycolysis (glye-KOL-ih-sis) An anaerobic process that breaks down a glucose molecule into two pyruvic acid molecules; ATP is produced; occurs in the cytoplasm.

Golgi complex (GOL-jee KOM-pleks) Membranous cellular organelle that stores, alters, and packages secretory products such as proteins; gives rise to lysosomes and secretory vesicles; also known as Golgi apparatus or Golgi body.

G protein Membrane protein that reacts with guanosine triphosphate (GTP); activates the membrane-bound enzyme adenylate cyclase.

Graafian follicle *See tertiary follicle.*

graded potential (GRAY-ded poh-TEN-shul) A small, local change in the membrane potential; used to communicate *only* over short distances.

gray matter (gray MAT-er) Regions in the brain and spinal cord containing nerve cell bodies, glial cells, and unmyelinated axons.

growth hormone Hormone produced by the anterior pituitary; promotes bodily growth indirectly by stimulating the liver to produce certain growth factors; also known as somatotropin.

gyrus/gyri (JYE-russ/JYE-rye) Elevated ridge.

H

hard palate (hard PAL-let) Bony roof of the mouth.

haustrum/haustra (HAWS-trum/HAWS-trah) Series of pouches in the wall of the colon; permits distention and elongation.

hemoglobin (hee-mo-GLOH-bin) A protein containing iron in red blood cells; functions to transport oxygen and some carbon dioxide in the blood.

hepatic duct (heh-PAT-ik dukt) Collects bile from all of the bile ducts of the liver lobes, which unite to form common bile duct.

hepatic portal vein (heh-PAT-ik POR-tal vane) Delivers blood to the liver.

hepatic triads (heh-PAT-ik TRYE-ads) Portal areas located at each of the six corners of the hepatic lobule; each triad consists of the following three structures: (1) branch of hepatic portal vein, (2) branch of hepatic artery proper, and (3) branch of the bile duct.

hepatocytes (heh-PAT-oh-sytes) Liver cells.

hepatopancreatic sphincter (heh-PAT-oh-pan-kree-AT-ik SFINGK-ter) Muscular ring that surrounds the lumen of the common bile duct and the duodenal ampulla; seals off the passageway and prevents bile from entering the small intestine; also known as sphincter of Oddi.

hilus (HYE-luss) A depression on an organ that serves as an entrance site for blood vessels, lymphatic vessels, and/or nerves; *ex.:* renal hilus.

homeostasis (hoh-mee-oh-STAY-sis) The process of maintaining the relatively stable state in the body's internal environment despite regular changes in the external environment.

hormone (HOR-mohn) Chemical messenger; secreted by an endocrine gland and transported through the blood.

hydroxyapatite (hye-DROK-see-ap-ah-tyte) The chief structural component of bone; formed by the interaction of calcium phosphate and calcium hydroxide.

hyperpolarization (hi-per-pohl-er-iz-A-shun) An increase in the membrane potential causing the measured voltage to become more negative.

hypertonic (hye-per-TON-ik) **solution** Solution that contains a higher concentration of nonpenetrating solutes than the adjacent cell; causes cell to shrink.

hypodermis (hye-poh-DER-mis) Layer of loose connective tissue that separates the skin from underlying tissue and organs; also known as subcutaneous layer.

hypothalamus (hye-poh-THAL-ah-mus) Part of the diencephalon in the brain; contains control centers regulating autonomic functions, emotions, and hormone production.

hypotonic (hye-poh-TON-ik) **solution** Solution that contains a lower concentration of nonpenetrating solutes than the adjacent cell; causes cell to swell.

I

ileum (IL-ee-um) Final 12-foot segment of the small intestine.

inferior (in-FEER-ee-or) Below; opposite of superior.

inferior vena cava (inferior VEE-nah KAY-vah) One of the great veins; carries blood to the right atrium from the trunk and lower extremities.

inguinal (ING-gwih-nal) Pertaining to the groin.

inhibin (in-HIB-in) A hormone secreted by the testes and ovaries that inhibits the release of follicle-stimulating hormone (FSH) from the anterior pituitary.

inner ear Region of the ear that contains the receptors of equilibrium and hearing.

inorganic (in-or-GAN-ik) Pertaining to compounds that lack hydrocarbons (carbon atoms bonded to hydrogen atoms); *ex.:* water, salts.

insulin A hormone produced by the pancreas; promotes the movement of glucose, amino acids, and fatty acids out of the blood and into body cells.

intercalated disks (in-TER-kah-lay-ted disks) Specialized regions that form connections between cardiac muscle cells.

interdigitate (in-ter-DIJ-ih-tayt) To interlock.

intermediate filaments (in-ter-MEE-dee-it FIL-ah-ments) Part of the cell's cytoskeleton; provide strength, stabilize the position of the organelles, and transport material within cells.

interneuron (in-ter-NOO-ron) Short neurons in the central nervous system located between sensory and motor neurons; association neuron.

interphase (IN-ter-fayz) Longest stage in the cell life cycle; when the cell prepares for division; DNA is replicated in this stage.

interstitial (in-ter-STISH-al) **fluid** Solution that fills the spaces between cells.

intervertebral disk (in-ter-VER-tee-bral disk) Pad of fibrocartilage that separates and cushions the vertebrae.

intra-alveolar (in-trah-al-VEE-oh-lar) **pressure** Air pressure inside the alveoli of the lungs.

intracellular fluid (ICF) Liquid located within the cells.

ion (EYE-on) A charged particle; *ex.:* Na^+.

isotonic (eye-soh-TON-ik) **solution** Solution that contains the same concentration of nonpenetrating solutes as the adjacent cell; causes cells to neither shrink nor swell.

J

jejunum (jeh-JOO-num) 8-foot middle segment of the small intestine between the duodenum and the ileum.

K

kidney (KID-nee) Major organ of the urinary system; filters waste products from the blood.

Krebs cycle *See citric acid cycle.*

L

labia (LAY-bee-ah) "Lips"; skin folds on the sides of the opening to the vagina.

lacrimal gland (LAK-rih-mal gland) Almond-shaped tear gland.

lacteal (LAK-tee-al) Lymphatic capillary within a villus of the small intestine.

lactic acid A waste product of glucose metabolism as a result of insufficient oxygen supply; produced in muscles after an intense workout.

lacuna (lah-KOO-nah) A small space where bone or cartilage cells are found.

lamella/lamellae (lah-MEL-ah/lah-MEL-ee) Layer of calcified bone matrix.

lamina propria (LAY-min-ah PRO-pree-ah) The loose connective tissue component of a mucous membrane.

laryngopharynx (lair-ring-go-FAIR-inks) Portion of the pharynx lying between the hyoid bone and the entrance to the esophagus.

larynx (LAIR-inks) Cartilaginous structure that surrounds and protects the glottis; voice box.

lateral (LAT-er-al) Away from the midline of the body; opposite of medial.

leukocyte (LOO-koh-syte) White blood cell.

ligament (LIG-ah-ment) A type of connective tissue that connects bone to bone.

lipase (LYE-payse) A pancreatic enzyme that breaks down lipids.

lipids (LIP-idz) Organic molecules containing carbon, hydrogen, and oxygen, such as in fats and oils.

lobule (LOB-yool) A small lobe.

loop of Henle *See nephron loop.*

lumen (LOO-men) Hollow area within a tube.

lungs The pair of respiratory organs that exchanges oxygen and carbon dioxide in the blood.

luteinizing (loo-tee-in-EYE-zing) **hormone (LH)** A hormone produced by the anterior pituitary; in females, acts on ovaries to stimulate ovulation and progesterone secretion from corpus luteum; in males, acts on testes to stimulate the synthesis and secretion of testosterone.

lymph (limf) Fluid transported by the lymphatic system; similar to plasma but contains lower concentration of proteins.

lymph node (limf node) Small oval organ that filters and purifies lymph before it reaches the venous system.

lymphocyte (LIM-foh-syte) Primary cell of the lymphatic system; a type of white blood cell.

lysosome (LYE-so-sohm) Cell organelle that contains digestive enzymes.

M

macromolecules (mak-roh-MOL-eh-kyools) Large, complex molecules made from simpler molecules; *ex.:* proteins, nucleic acids.

matrix (MAY-triks) Extracellular substance of a connective tissue.

meatus (mee-AY-tus) Tubelike opening or channel.

medial (MEE-dee-al) Toward the midline longitudinal axis of the body; opposite of lateral.

median plane (MEE-dee-an plane) A section that passes along the midline and divides the body into right and left halves.

mediastinum (mee-dee-as-STYE-num) Region of the thoracic cavity between the lungs.

medulla (meh-DUL-ah) Inner portion of an organ; *ex.:* adrenal medulla.

medulla oblongata (meh-DUL-ah ob-long-GAH-tah) Most inferior portion of the brainstem; site of vital control centers for respiratory and cardiovascular systems.

medullary cavity (MED-oo-lair-ee cavity) Hollow space inside the shaft of the bone; marrow cavity.

meiosis (my-OH-sis) Type of cell division that results in the formation of sperm or ova that contain half the number of chromosomes as the parent cell; occurs in the testes and ovaries.

melanin (MEL-ah-nin) Brown skin pigment that absorbs ultraviolet radiation.

melanocytes (MEL-ah-noh-sytes) A fixed cell that synthesizes and stores melanin.

melatonin (mel-ah-TOE-nin) A hormone produced by the pineal gland; involved in regulating the body's sleep-wake cycles.

membrane potential (MEM-brayn po-TEN-shul) A separation of positive and negative charges across a plasma membrane that results in an electrical potential or a voltage.

meninx/meninges (ME-ninks/meh-NIN-jeez) One of the three membranes surrounding the spinal cord or brain.

menstrual phase (MEN-stroo-al fayz) The first phase of the uterine cycle, characterized by menstrual bleeding as a result of shedding the uterine lining called the endometrium.

metabolism (meh-TAB-oh-liz-em) Sum of all chemical activities in the body.

metaphase (MET-ah-fayz) Stage of mitosis evident when chromatids line up at the middle of the cell.

microfilaments (my-KROH-FIL-ah-ments) Slender protein strands that make up the framework of the cytoskeleton; also called actin.

microtubules (my-kroh-TOOB-yools) Hollow protein tubes found in all body cells; framework of the cytoskeleton.

microvillus/microvilli (my-kroh-VIL-luss/my-kroh-VIL-eye) Small, finger-shaped projections of the plasma membrane of an epithelial cell that contain microtubules; function to increase surface area for better absorption.

midbrain Portion of the brainstem; provides pathways between brainstem and cerebrum.

mineraloccorticoids (min-er-al-oh-KOR-tih-koyds) A group of hormones produced by the adrenal cortex that regulate mineral homeostasis, such as sodium and potassium levels in the blood; *ex.:* aldosterone.

mitochondrion/mitochondria (my-toh-KON-dree-ohn/my-toh-KON-dree-ah) Cellular organelle that is the site of aerobic cellular respiration; produces ATP for the cell.

mitosis (my-TOH-sis) Division of the cell nucleus that results in the formation of two new daughter cells that contain the same type and number of chromosomes as the parent cell; divided into four stages: prophase, metaphase, anaphase, and telophase.

molecules (MOL-eh-kyools) Compound consisting of two or more atoms held together by chemical bonds.

monocyte (MON-oh-syte) A phagocytic white blood cell.

monosaccharide (mon-oh-SAK-ah-ryde) A simple, single-unit sugar such as glucose or fructose.

monosynaptic reflex (mon-oh-sin-AP-tic RE-flex) A reflex pathway that does not contain any interneurons between the sensory and motor neurons.

morula (MOR-yoo-lah) A solid ball of cells formed from mitotic divisions of the blastomeres.

motility (moh-TIL-i-tee) Contraction of smooth muscle in the wall of the digestive tract for the purpose of mixing and propelling digested substances; includes the processes of peristalsis and segmentation.

mucosa (myoo-KOH-sah) Mucous membrane.

mucous (adj.) (MYOO-kuss) Describes lubricating secretions along the digestive, respiratory, urinary, and reproductive tracts.

mucus (n.) (MYOO-kuss) A thick, gel-like substance secreted by glands in the digestive, respiratory, urinary, and reproductive tracts.

multipolar neuron (mul-tee-PO-lar NOO-ron) A neuron that has several dendrites and a single axon.

myelin (MY-eh-lin) A membranous insulation consisting of many layers of glial cell membrane around the axon of a nerve cell; improves the speed of impulse conduction.

myelination (my-eh-lih-NAY-shun) The process of forming a myelin sheath.

myocardium (my-oh-KAR-dee-um) The cardiac muscle in the wall of the heart.

myofibril (my-oh-FYE-brill) Slender fibers found in skeletal muscle and extending the length of the cell; composed of thick and thin filaments.

myometrium (my-oh-MEE-tree-um) The middle muscular layer of the uterine wall.

myosin (MY-oh-syn) Thick protein filament found in skeletal muscle cells.

N

nasal (NAY-zal) Pertaining to the nose.

nasopharynx (nay-zoh-FAIR-inks) The superior portion of the larynx.

nephron (NEF-ron) The basic functional unit of the kidney.

nephron loop (NEF-ron loop) A subdivision of the nephron; located between the proximal convoluted tubule and the distal convoluted tubule; also known as the loop of Henle.

nerves Long, cable-like structures that extend throughout the body; bundles of peripheral neurons held together by several layers of connective tissues.

neurilemma (noo-rih-LEM-mah) The cytoplasmic covering around the axon provided by the Schwann cells; found only in the peripheral nervous system.

neuroglia (noo-ROG-lee-ah) A class of cell types in neural tissue that provides nutrients and a supporting framework to neural tissue; does not transmit impulses.

neurolemmocyte (noo-ROH-lem-moh-syte) A neuroglial cell that wraps itself around peripheral axons to form a protective covering.

neurons (NOO-rons) Conducting cells of the nervous system; each consists of a cell body, one or more dendrites, and a single axon.

neurotransmitters (noo-roh-trans-MIH-terz) Chemical messengers released by neurons for the purpose of stimulating or inhibiting target cells.

neutron (NOO-tron) Subatomic particle in the nucleus of an atom; no electrical charge.

neutrophil (NOO-troh-fil) Phagocytic white blood cell; produced in bone marrow.

Nissl bodies (NISS-ul bodies) Ribosomal clusters found in neuron cell bodies.

node of Ranvier (node of rahn-vee-AY) Gaps of exposed axon not covered by a myelin sheath; appear at regular intervals in some myelinated axons.

norepinephrine (nor-ep-ih-NEF-rin) (**NE**) A neurotransmitter and a hormone; involved in regulating autonomic nervous system activity; also known as noradrenaline.

nuclease (NOO-klee-ayz) An enzyme that catalyzes the hydrolysis of nucleic acids.

nucleic acid (noo-KLAY-ik ASS-id) Organic molecule found in the nucleus; made up of functional units called nucleotides; *ex.:* RNA, DNA.

nucleus Prominent organelle of the cell; contains DNA molecules.

O

oblique (oh-BLEEK) Diagonal.

olfaction (ohl-FAK-shun) The sense of smell.

oocyte (OH-oh-syte) Immature stage of the female gamete.

oogenesis (oh-oh-JEN-eh-sis) Production of female gametes or ova; occurs in the ovaries.

optic chiasma (OP-tik kye-AS-mah) Area near the diencephalon where the optic nerves cross over each other to eventually connect to opposite sides of the brain.

optic nerve (OP-tik nerve) Carries visual information from the retina of the eyes to the brain.

oral (OR-al) Pertaining to the mouth.

oral cavity (OR-al KAV-i-tee) Space within the mouth; includes lips, cheeks, tongue and its muscles, and hard and soft palates.

orbital (OR-bi-tal) Pertaining to the eye.

organ (OR-gan) A group of tissues that performs a specific function.

organelle (or-gah-NELL) Small internal structures in cells; divided into membranous and nonmembranous types; *ex.:* ribosome.

organic (or-GAN-ik) Pertaining to compounds that contain hydrocarbons (carbon atoms bonded to hydrogen atoms); *ex.:* carbohydrates, proteins.

organism (OR-gan-ism) An individual living thing.

organ system (OR-gan SISS-tem) Varying number and kinds of organs arranged in a way to perform complex functions; most complex organizational unit of the body; *ex.:* respiratory system.

oropharynx (oh-roh-FAIR-inks) Portion of the pharynx that extends between the soft palate and the base of the tongue at the level of the hyoid bone.

osmosis (os-MOH-sis) Movement of water across a semipermeable membrane toward the solution containing the higher solute concentration.

osteoblasts (OS-tee-oh-blasts) Cells that lay down the specialized matrix of bone; responsible for the production of new bone.

osteoclasts (OS-tee-oh-clasts) Cells that decompose the matrix of bone.

osteocytes (OS-tee-oh-sytes) Mature bone cells.

ovary (OH-vah-ree) Female gonad; produces ova.

ovulation (ov-yoo-LAY-shun) Release of an egg cell called a secondary oocyte from a tertiary follicle in the ovary.

ovum/ova (OH-vum/OH-vah) Female sex cell; egg cell.

oxidation Loss of one or more electrons from an atom or molecule.

P

palpate (PAL-payt) To examine by feeling part of the body.

pancreas (PAN-kree-ass) Slender, elongated organ that lies in the abdominopelvic cavity; has the dual function of manufacturing digestive enzymes and making hormones such as insulin.

pancreatic juice A mixture of secretions that contains mostly water, salts, sodium bicarbonate, and various digestive enzymes; secreted by the exocrine acinar cells of the pancreas.

papilla (pah-PILL-ah) Nipple-shaped mounds.

parathyroid hormone (PTH) Hormone secreted by the parathyroid glands that is the primary regulator of calcium homeostasis; increases calcium levels in the blood.

parietal (pah-RYE-i-tal) Relating to or forming the wall of an organ or cavity.

parietal cells (pah-RYE-i-tal sells) Cells located in the stomach; secrete hydrochloric acid.

pectoral (PEK-toh-ral) Pertaining to the chest; between sternal and axillary regions.

pepsin (pep-SYN) Enzyme made by chief cells in the stomach; functions to digest proteins.

pepsinogen (pep-SYN-oh-jen) An inactive enzyme secreted by chief cells in the stomach; converted to the active enzyme pepsin during gastric digestion.

perimysium (pair-i-MISH-ee-um) A connective tissue layer that divides the skeletal muscle into a series of compartments or bundles called fascicles.

perineurium (pair-i-NOO-ree-um) Connective tissue sheath that surrounds a bundle of nerve fibers within a nerve and holds them together.

peristalsis (pair-ih-STAWL-siss) Wavelike ripple of muscular contractions along the wall layer of a hollow organ, such as the small intestine.

peroxisomes (per-AHK-si-sohms) Cell organelles that absorb and neutralize toxins.

pH Means "potential Hydrogen scale"; a measure of the concentration of hydrogen ions in a solution; scale ranges in value from 0–14 where 7 is the neutral point; values below 7 are acidic and those above 7 are alkaline.

phagocytosis (fag-oh-sye-TOH-siss) Type of endocytosis; process in which microorganisms or other large particles are engulfed by the plasma membrane and enter the cell.

pharynx (FAIR-inks) Passageway that connects the nose, mouth, and throat; commonly called the throat.

physiology (fiz-ee-AHL-oh-jee) Science that examines the function of the living organism and its parts.

pia mater (PEE-ah MAH-ter) The highly vascular, innermost layer of the meninges.

pinocytosis (pin-oh-sye-TOH-sis) Type of endocytosis; process in which a plasma membrane invaginates to capture some extracellular fluid and brings it into the cytoplasm in a membranous vesicle.

pituitary (pih-TOO-i-tair-ee) **gland** Small endocrine gland located near the base of the brain that releases many different hormones; consists of two separate glands: anterior pituitary and posterior pituitary; referred to as the "master gland" because many of its hormones control other endocrine glands in the body.

placenta (plah-SEN-tah) The specialized organ within the uterus that supports embryonic and fetal development; also called the afterbirth.

plasma Fluid portion of the blood.

plica/plicae (PLYE-kah/PLYE-kee) Transverse folds of the intestinal lining.

polysaccharide (pol-ee-SAK-ah-ryde) An organic macromolucule; a complex sugar formed by bonding many monosaccharides in a long chain, such as starch or cellulose.

polysynaptic reflex (pahl-y-sin-AP-tic RE-flex) A reflex pathway that contains interneurons between the sensory neurons and motor neurons.

pons (ponz) Part of the brainstem; connects the brainstem to the cerebellum; contains relay centers and is involved with somatic and visceral motor control.

positive feedback A physiological mechanism that tends to amplify or reinforce the change in internal environment.

posterior (pos-TEER-ee-or) Back; behind; opposite of anterior.

postganglionic neuron (post-gang-glee-ON-ik NOO-ron) The second neuron in the motor output portion of an autonomic reflex pathway; connects a ganglion to smooth muscle or cardiac muscle, or a gland.

preganglionic neuron (pree-gang-glee-ON-ik NOO-ron) The first neuron in the motor output portion of an autonomic reflex pathway; connects the central nervous system to a ganglion.

primary active transport Refers to the transport of a specific substance across the plasma membrane, against its concentration gradient; typically involves a membrane protein, referred to as a "pump," that uses ATP.

progesterone (pro-JES-ter-ohn) Female hormone produced by the ovaries; prepares the endometrium for possible implantation of the embryo and stimulates milk secretion in mammary glands.

prolactin (pro-LAHK-tin) **(PRL)** Hormone secreted by the anterior pituitary gland; stimulates milk secretion in mammary glands.

proliferative phase (PROH-lif-eh-rah-tiv fayz) Second phase of the uterine cycle; involves reestablishing a portion of the endometrium.

prophase (PRO-fayz) Initial stage of mitosis; evident when the chromosomes become visible.

prostate gland (PROSS-tayt) Small, muscular, rounded organ; produces a weak acidic secretion that contributes about 30% of the volume of semen.

protease (PRO-tee-ayz) Enzyme that catalyzes the breakdown of proteins into intermediate compounds.

protein (PRO-teen) Organic macromolecule with a long, complex polymer structure and a variety of functions; structural units are amino acids.

proton (PRO-ton) Positively charged subatomic particle; located in the nucleus of an atom.

proximal (PROK-sih-mal) Refers to the region or reference toward an attached base; opposite of distal.

proximal convoluted tubule (PROK-sih-mal kon-voh-LOO-ted TOOB-yool) Part of the nephron in the kidney; located between the glomerular capsule and the nephron loop; primary site of reabsorption of nutrients from the filtrate.

pubic (PYOO-bik) Relating to the region of the pubis.

pudendum (pyoo-DEN-dum) Region enclosing the female external genitalia; usually called the vulva.

Purkinje (pur-KIN-jee) **fibers** Specialized cardiac muscle cells that extend out to the lateral walls of the ventricles and papillary muscles; conduct impulses from the AV bundle, stimulating heart contraction.

pyloric sphincter (pye-LOR-ik SFINGK-ter) A muscular structure that regulates the release of chyme from the stomach to the duodenum.

R

reabsorption Second step in urine formation; movement of substances out of the renal tubules and into the blood.

rectum (REK-tum) Last segment of the large intestine; end of the digestive tract.

rectus (REK-tus) Straight.

reduction Gain of one or more electrons by an atom or molecule.

reflex A fast, automatic response to a stimulus.

renal (REE-nal) Pertaining to the kidney.

renin (REH-nin) An enzyme produced by the nephrons in the kidneys; converts angiotensinogen to angiotensin I in the plasma.

resting potential *See membrane potential.*

rete testis (REE-tee TES-tis) A maze of passageways formed from the seminiferous tubules and located within the mediastinum of the testis.

ribosomes (RYE-boh-sohms) Cellular organelles that are the sites of protein synthesis; composed of two subunits.

round window (round WIN-doh) A thin, membranous partition that separates the cochlear chambers from the air spaces of the middle ear.

rugae (ROO-gee) Longitudinal folds that line the inner wall of the stomach.

S

saggital (SAJ-i-tal) Runs along with the long axis of the body.

sagittal plane (SAJ-i-tal plane) Extends from anterior to posterior and divides the body into right and left sections.

sarcolemma (sar-koh-LEM-ah) Plasma membrane of a skeletal muscle cell.

sarcomere (SAR-koh-meer) Contractile unit of a muscle cell.

Schwann cell *See neurolemmocyte.*

scrotum (SKRO-tum) Pouch-like sac, divided into two chambers; contains the testes.

sebaceous gland (seh-BAY-shus gland) Holocrine gland that secretes waxy, oily sebum into the hair follicles.

secondary active transport Typically involves cotransport of substances across the plasma membrane; does not use ATP; relies on the concentration gradients established by primary active transport.

secondary oocyte (SEK-on-dair-ee OH-oh-syte) Ovum released at ovulation.

second messenger An intracellular molecule, such as cyclic AMP, produced in response to a first messenger (*ex.:* hormone) binding to its plasma membrane receptor; triggers a chemical chain reaction within the target cell, causing cellular changes.

secretin (seh-KREE-tin) A digestive hormone produced by duodenal mucosa; stimulates secretion of pancreatic juice rich in bicarbonate ions.

secretion Production and release of a chemical substance from a cell; *ex.:* neurotransmitters from a neuron.

secretory phase (SEEK-reh-toh-ree fayz) The last and longest phase of the uterine cycle; characterized by thickening of the endometrium in preparation for possible implantation of a human embryo.

segmentation Mixing movement; occurs when digestive reflexes cause a forward-and-backward movement within a single region of the digestive tract.

semen (SEE-men) Fluid released from the penis during ejaculation; contains sperm and seminal fluid produced by various glands.

semilunar valve (sem-i-LOO-nar valv) Pair of heart valves located at the exit point of each ventricle; each valve has three pouch-like flaps; consists of the pulmonary and aortic valves.

seminiferous tubules (seh-mih-NIF-er-us TOOB-yools) A tightly coiled structure located in each lobule of the testis; site where spermatozoa develops.

sensory transduction (SEN-sohr-y trans-DUCK-shun) The conversion of energy from a stimulus into an action potential in a sensory neuron.

septum/septae (SEP-tum/SEP-tee) A terminal partition that divides an organ.

serosa (seh-ROH-sah) A serous membrane that covers most of the digestive tract.

simple diffusion *See diffusion.*

sinoatrial (sye-NOH-AY-tree-al) **(SA) node** Primary pacemaker of the heart; specialized mass of cardiac muscle cells located in the wall of the right atrium that plays a role in electrical conduction system within the heart.

sinus (SYE-nus) A cavity or space in a tissue; a large dilated vein.

sinusoid (SYE-nu-soyd) Large, permeable blood capillaries found in organs such as the liver, spleen, and bone marrow; specialized to allow large proteins to smoothly move in or out of the blood.

sodium-potassium pump Protein pump that uses active transport; located in the plasma membrane of most cells; exchanges three sodium ions moved out of the cell for two potassium ions brought into the cell; also known as the sodium-potassium ATPase.

soft palate (soft PAL-let) The portion of the roof of the mouth that lies posterior to the hard palate.

solute Dissolved substance(s) in a solution; *ex:* salt.

solution (suh-LOO-shun) Mixture in which a solute is dissolved in a solvent; *ex.:* salt water.

solvent Liquid portion of a solution; dissolves the solute; *ex:* water.

soma (SO-mah) Cell body.

sperm *See spermatozoon.*

spermatic cord (sper-MAT-ik cord) A layer of fascia, tough connective tissue, and muscle surrounding the ductus deferens and the blood vessels and nerves of the testes.

spermatogenesis (sper-mah-toh-JEN-eh-sis) Production of sperm cells; occurs in the testes.

spermatozoon/spermatozoa (sper-mah-tah-ZOH-ohn/sper-mah-tah-ZOH-ah) Sperm cell; male gamete or sex cell.

spindle fiber (SPIN-dul FYE-ber) Network of tubules in the cell that extends between the centriole pairs.

stem cells Nonspecialized cells that have the ability to divide continually to form new types of more specialized cells.

steroids A large class of lipids characterized by a carbon ring chemical structure; act as components for cells and regulate body function; *ex.:* cholesterol.

stimulus Any agent that induces a response; causes a change in an excitable tissue.

striation (STRYE-ay-shun) A striped or banded appearance such as that found in skeletal muscle cells and cardiac muscle cells.

submucosa (sub-myoo-KOH-sah) A layer of loose connective tissue in the wall of the digestive tract; large blood vessels, lymphatic vessels, and nerve fibers are found here.

sulcus (SUL-kus) A shallow depression.

superficial (soo-per-FISH-al) On the surface; opposite of deep.

superior (soo-PEER-ee-or) Above; opposite of inferior.

surface tension The force of attraction between water molecules as a result of hydrogen bonding; creates a thin surface film on water.

surfactant (sur-FAK-tant) Chemical mixture that contains lipids and proteins and coats the inner surface of each alveolus in the lungs; reduces surface tension.

suture (SOO-chur) The boundary between the skull bones; immovable joint.

synapse (SIN-aps) Specialized junction where a neuron communicates with another cell such as another neuron, a muscle cell, or a glandular cell.

synaptic cleft (sin-AP-tic kleft) Within the synapse, the extracellular space between the membrane of the neuron and the membrane of the other cell with which it is communicating.

synaptic knob (sin-AP-tic nahb) The bulbous structure at the end of a neuron's axon where neurotransmitters are secreted; also known as axon terminal, synaptic terminal, and synaptic end bulb.

systole (SISS-toh-lee) Contraction of both atria and both ventricles in the heart; opposite of diastole.

T

target cell A general term for any cell affected by a hormone because it contains a receptor for that hormone.

telophase (TEL-oh-fayz) Stage of mitosis; nuclear membrane forms, nuclei enlarge, and chromosomes gradually uncoil and disappear.

tendon (TEN-don) A cable-like, fibrous connective tissue that connects skeletal muscle to bone.

tertiary follicle (TER-shee-air-ee FOL-lih-kul) Mature ovum in ovary; also known as Graafian follicle or vesicular follicle.

testis/testes (TES-tiss/TES-teez) The male gonad that produces sperm.

testosterone Principal male sex hormone; produced by interstitial cells in the testes; responsible for growth and maintenance of male sexual characteristics and for sperm production.

thalamus (THAL-ah-mus) Located in diencephalon of the brain; final relay point for ascending sensory information that will be sent to the cerebrum.

threshold stimulus Minimum stimulus needed to produce a response in an excitable cell.

thrombocyte (THROM-boh-syte) A blood cell that has a role in clotting: also known as a platelet.

thrombus (THROM-bus) Stationary blood clot.

thyroid gland (THY-royd gland) Endocrine gland located near the trachea in the neck; produces the hormones T_3, T_4 and calcitonin.

thyroid-stimulating hormone (TSH) Hormone secreted by anterior pituitary gland; stimulates thyroid gland to produce the hormones T_3 and T_4.

thyroxin (T_4) (thy-ROK-syn) Hormone produced by the thyroid gland; influences metabolic rate, growth, and development.

tissue (TISH-yoo) Group of similar cells that perform a common function.

tonsil (TAHN-sil) A large nodule containing masses of lymphoid tissue; located in the walls of the pharynx.

trachea (TRAY-kee-ah) Long tube between the larynx and the primary bronchi that serves as air passageway; windpipe.

transverse plane (TRANS-vers plane) A division that lies at right angles to the long axis of the body; divides the body into superior and inferior sections.

tricuspid valve (try-KUS-pid valv) The right atrioventricular (AV) valve in the heart; located between the right atrium and right ventricle.

triiodothyronine (T_3) (try-eye-oh-doh-THY-roe-neen) Hormone produced by the thyroid gland; regulates metabolic rate, growth, and development.

U

umbilical cord (um-BIL-i-kul cord) The vascular structure that connects the fetus to the placenta.

unipolar neuron (YOO-nee-POH-lar NOO-ron) A neuron in which the dendrite and axonal processes are continuous, and the cell body lies off to the side.

ureters (YOOR-eh-ters) Pair of long, slender, muscular tubes that transports urine from the kidneys to the urinary bladder.

urethra (yoo-REE-thrah) A passageway that transports urine from the neck of the urinary bladder to the exterior; in males, also acts as a passageway for spermatozoa to the exterior.

uterine tubes (YOO-ter-in tubes) Hollow muscular tubes that transport a secondary oocyte to the uterus; also known as oviducts or fallopian tubes.

uterus (YOO-ter-us) Hollow muscular organ that provides mechanical protection, nutritional support, and waste removal for a developing embryo.

V

vagina (vah-JYE-nah) Muscular passageway in the female that connects the uterus with the exterior genitalia.

vascular (vas-KYOO-ler) Pertaining to the blood vessels.

vas deferens *See ductus deferens.*

veins (vanes) Vessels that collect blood from tissues and organs and return it to the heart.

ventral (VEN-tral) Anterior or belly side in humans; opposite of dorsal.

ventricle (VEN-tri-kul) Cavity or chamber.

venule (VEN-yool) Microscopic blood vessel that connects capillaries to small veins.

villus (VIL-us) Fingerlike extension of mucous membrane of small intestine.

W

white matter (whyte MAT-ter) Nerve fibers covered with the myelin sheath.

Z

zygote (ZYE-goht) Fertilized egg cell.

Index